Atomic and Molecular Processes in Fusion Edge Plasmas

Atomic and Molecular Processes in Fusion Edge Plasmas

Edited by
R. K. Janev

PLENUM PRESS • NEW YORK AND LONDON

Library of Congress Cataloging-in-Publication Data

Janev, R. K. (Ratko K.), 1939-
  Atomic and molecular processes in fusion edge plasmas / edited by
R.K. Janev.
     p.   cm.
  Includes bibliographical references and index.
  ISBN 0-306-45043-7
  1. Plasma confinement. 2. Plasma stability. 3. Plasma
instabilities.   I. Title.
QC718.5.C65J35 1995
621.4'84--dc20   95-25127
                                                                CIP

ISBN 0-306-45043-7

© 1995 Plenum Press, New York
A Division of Plenum Publishing Corporation
233 Spring Street, New York, N. Y. 10013

10 9 8 7 6 5 4 3 2 1

All rights reserved

No part of this book may be reproduced, stored in a retrieval system, or transmitted in any form or by any means, electronic, mechanical, photocopying, microfilming, recording, or otherwise, without written permission from the Publisher

Printed in the United States of America

Contributors

P. B. Armentrout, Department of Chemistry, University of Utah, Salt Lake City, Utah 84112

J. Botero, International Atomic Energy Agency, A-1400 Vienna, Austria

F. Brouillard, Laboratoire de Physique Atomique et Moléculaire, Université Catholique de Louvain, Louvain-la-Neuve, Belgium

M. Capitelli, Departimento di Chimica, Università di Bari and Centro di Studio per la Chimica dei Plasmi del CNR, Bari, Italy

R. Celiberto, Departimento di Chimica, Università di Bari and Centro di Studio per la Chimica dei Plasmi del CNR, Bari, Italy

P. Defrance, Department of Physics, Université Catholique de Louvain, B-1348 Louvain-la-Neuve, Belgium

M. Duponchelle, Department of Physics, Université Catholique de Louvain, B-1348 Louvain-la-Neuve, Belgium

W. Fritsch, Bereich Theoretische Physik, Hahn-Meitner-Institut, D-14109 Berlin, Germany

Yukap Hahn, Department of Physics, University of Connecticut, Storrs, Connecticut 06269 and CTR, Storrs, Connecticut 06268

R. K. Janev, International Atomic Energy Agency, A-1400 Vienna, Austria

Isik Kanik, Jet Propulsion Laboratory, California Institute of Technology, Pasadena, California 91109

Yong-Ki Kim, National Institute of Standards and Technology, Gaithersburg, Maryland 20899

F. Linder, Department of Physics, University of Kaiserslautern, D-67653 Kaiserslautern, Germany

T. D. Märk, Institut für Ionenphysik, Leopold Franzens Universität, A-6020 Innsbruck, Austria

J. B. A. Mitchell, Department of Physics, University of Western Ontario, London, Ontario, Canada N6A 3K7

D. L. Moores, Department of Physics and Astronomy, University College London, London WC1E6BT, United Kingdom

S. Yu. Ovchinnikov, Department of Physics and Astronomy, University of Tennessee, Knoxville, Tennessee 37996-1501; *permanent address:* Ioffe Physical Technical Institute, St. Petersburg, Russia

S. V. Passovets, Department of Physics and Astronomy, University of Tennessee, Knoxville, Tennessee, 37996-1501; *permanent address:* Ioffe Physical Technical Institute, St. Petersburg, Russia

Michael S. Pindzola, Department of Physics, Auburn University, Auburn, Alabama 36849

Anil K. Pradhan, Department of Astronomy, Ohio State University, Columbus, Ohio 43210

D. R. Schultz, Physics Division, Oak Ridge National Laboratory, Oak Ridge, Tennessee 37831-6373

Hiroyuki Tawara, National Institute for Fusion Science, Nagoya 464-01, Japan

Swaraj S. Tayal, Department of Physics and Center for Theoretical Studies of Physical Systems, Clark Atlanta University, Atlanta, Georgia 30314

Sandor Trajmar, Jet Propulsion Laboratory, California Institute of Technology, Pasadena, California 91109

X. Urbain, Laboratoire de Physique Atomique et Moléculaire, Université Catholique de Louvain, Louvain-la-Neuve, Belgium

W. L. Wiese, National Institute of Standards and Technology, Gaithersburg, Maryland 20899

H.P. Winter, Institut für Allgemeine Physik, Technische Universität Wien, A-1040 Vienna, Austria

Preface

The recent progress of thermonuclear fusion research based on the magnetic confinement of high-temperature plasmas, as well as the design of the first experimental fusion tokamak reactors, has demonstrated that the physical conditions at the plasma periphery play a decisive role in achieving, maintaining, and controlling the thermonuclear burn. Because it is an interface between the hot burning deuterium–tritium plasma and the cold material walls of the reactor vessel, the boundary (or the edge) plasma has to fulfill many functions related to the protection of the reactor walls from the intense particle and power fluxes generated in the reactor burning zone, protection of the central plasma from contamination by nonhydrogenic wall impurities (which dilute the thermonuclear fuel and degrade the burn conditions), exhaust of the thermal plasma power and the reactor ash (thermalized helium), etc. These functions of the boundary plasma can be accomplished by suitable modification of the configuration of the confining magnetic field in the edge region and by an appropriate use of the radiative and collisional properties of atomic, ionic, and molecular species present (or deliberately introduced) in the plasma edge region.

In order to understand the role of atomic and molecular radiative and collision processes in the plasma edge region, their effects on plasma properties and dynamics, and particularly their use in the control of physical conditions in this plasma region, it is essential to have an in-depth knowledge of the physics of these processes and detailed information on their quantitative characteristics (transition rates, cross sections, reaction rate coefficients, etc.). Because of the low temperatures prevailing in the plasma edge region, the edge plasma composition contains significant fractions of neutral hydrogen and low-charged atomic and molecular impurities produced in plasma–wall interactions. The range of collision processes involving plasma electrons and ions, hydrogenic neutrals, and wall impurities is extremely large. The most important of these processes, from the standpoint of their effects on plasma edge properties and behavior (e.g., ionization balance, plasma energy and momentum losses, plasma transport, etc.), are the electron impact excitation and ionization of plasma edge atoms, ions, and molecules, electron–ion recombination, elastic and momentum transfer collisions of plasma electrons and ions with the plasma edge neutrals, dissociative processes in electron–molecule (molecular ion) collisions, electron capture in ion–atom (molecule) collisions, and energy transfer and reactive ion–molecule collisions.

This book contains an account of the most important atomic and molecular processes taking place in fusion edge plasmas and of the spectroscopic characteristics of edge plasma constituents. Each chapter describes the basic physics of a specific type or class of processes, gives an overview of the current research on the subject, and presents a significant amount of quantitative information on the characteristics of the processes considered, including the results of most recent research. The primary purpose of this book is to provide the fusion plasma researchers with a source of critically assessed information on atomic and molecular processes in edge plasmas. This information is indispensable in the interpretation of experimental observations and in the modeling and diagnostics of fusion edge plasmas. It is hoped, however, that the information presented can also be useful to the researchers in atomic collision physics, providing them with an overview of spectroscopic and atomic collision data needs in controlled fusion research.

It is a great pleasure to express my deep gratitude to all contributors to this volume for their dedicated effort and friendly cooperation in mastering the content of individual chapters and ensuring the comprehensive character of the book.

R. K. Janev

Vienna, Austria

Contents

Chapter 1

Basic Properties of Fusion Edge Plasmas and Role of Atomic and Molecular Processes

R. K. Janev

1. Introduction 1
2. Basic Plasma Edge Configurations 3
3. Parameters and Composition of Edge Plasmas 6
4. Atomic and Molecular Processes in the Plasma Edge 9
5. Role of Atomic and Molecular Processes in Plasma Edge Physics 10
6. Conclusion 12
   References 13

Chapter 2

Spectroscopic Processes and Data for Fusion Edge Plasmas

W. L. Wiese

1. Introduction 15
2. Status of Research on Atomic Structure Data 17
3. Description of Principal Methods to Determine Transition Probabilities . . 18
   3.1. Theoretical Methods 18

3.2. Experimental Methods 19
4. Availability of Spectroscopic Data 21
5. Numerical Spectroscopic Data Bases 22
   5.1. Wavelength Tables 22
   5.2. Energy Level Tables 23
   5.3. Transition Probability Tables 24
   5.4. Comprehensive Recent Determinations of Wavelength and Transition Probability Data of Interest for Fusion Edge Plasmas 25
   5.5. Spectroscopic Data for Molecules 25
6. Summary 26
   References 28

Chapter 3

Elastic and Excitation Electron Collisions with Atoms

Sandor Trajmar and Isik Kanik

1. Introduction 31
2. Definition of Cross Sections 32
3. Experimental Methods 33
   3.1. Differential Cross Sections 34
   3.2. Integral Cross Sections 36
   3.3. Total Scattering Cross Sections 37
4. Review of Cross Section Data 37
   4.1. Primary Species: H, He 37
   4.2. Common Impurities: C, O 52
   4.3. Metallic Impurities: Be, Al, Ti, Cr, Fe, Ni, Cu, Ga, Mo, Ta, W, V, and Zr 53
   4.4. Diagnostic Species: Li, Ne, Ar, Kr, and Xe 54
   References 55

Chapter 4

Electron Impact Ionization of Plasma Edge Atoms

T. D. Märk

1. Introduction 59
2. Electron Impact Ionization: Mechanisms and Definitions 60
3. Total Electron Impact Ionization Cross Sections of Atoms and Molecules 63
   3.1. Experimental Methods and Techniques 63
   3.2. Theoretical Considerations 65

| 3.3. Consistency Checks 68
| 3.4. Recommended Total Ionization Cross Sections 69
| 4. Partial Electron Impact Ionization Cross Sections 74
| 4.1. Experimental Methods and Techniques 74
| 4.2. Theoretical Considerations 77
| 4.3. State-Selected Partial Ionization Cross Sections 79
| 4.4. Recommended Partial Cross Sections 81
| References 86

Chapter 5

Electron–Ion Recombination Processes in Plasmas

Yukap Hahn

| 1. Introduction 91
| 2. Theory of Electron–Ion Recombination in Plasmas 94
| 3. Radiative Recombination and Scaled Rates 97
| 4. Dielectronic Recombination for the Ground States 101
| 5. Plasma Field Effects and Rate Equations 110
| 6. Summary and Conclusions 114
| References 116

Chapter 6

Excitation of Atomic Ions by Electron Impact

Swaraj S. Tayal, Anil K. Pradhan, and Michael S. Pindzola

| 1. Introduction 119
| 2. Carbon and Oxygen Ions 121
| 2.1. General Considerations 121
| 2.2. O^+ Cross Sections 129
| 2.3. O^{2+} Cross Sections 131
| 2.4. C^+ and O^{3+} Cross Sections 132
| 2.5. C^{2+} and O^{4+} Cross Sections 132
| 2.6. C^{3+} and O^{5+} Cross Sections 133
| 2.7. C^{4+} and O^{6+} Cross Sections 134
| 2.8. C^{5+} and O^{7+} Cross Sections 135
| 3. Iron Ions 135
| 3.1. General Considerations 135
| 3.2. Fe^+ Cross Sections 136

3.3. Fe^{2+} Cross Sections	139
3.4. Fe^{3+} Cross Sections	140
3.5. Fe^{5+} Cross Sections	140
3.6. Fe^{6+} Cross Sections	140
3.7. Fe^{7+} Cross Sections	141
3.8. Further Considerations	141
4. Rare-Gas Ions	141
4.1. General Considerations	141
4.2. Ar^{7+} Cross Sections	142
4.3. Ar^{6+} Cross Sections	144
4.4. Kr^{6+} Cross Sections	145
5. Conclusions	145
References	147

Chapter 7

Ionization of Atomic Ions by Electron Impact

P. Defrance, M. Duponchelle, and D. L. Moores

1. Introduction: Types of Ionization Processes	153
2. Electron Impact Ionization: Theoretical Methods	156
2.1. Direct, Single Ionization	156
2.2. Indirect Ionization	160
2.3. Multiple Ionization	164
3. Experimental Methods	166
3.1. Crossed Electron–Ion Beam Experiments	166
3.2. Electron Beam Ion Source and Trap	169
4. Cross Sections	170
4.1. Hydrogen Isoelectronic Sequence	170
4.2. Helium Isoelectronic Sequence	171
4.3. Lithium Isoelectronic Sequence	172
4.4. Be Sequence: B^+, C^{2+}, O^{4+}, and Ne^{6+}	172
4.5. B Sequence: C^+ and O^{3+}	172
4.6. O^+ and O^{2+}	172
4.7. Rare Gases	172
4.8. Metallic Ions	177
5. Parametric Representation of the Cross Sections	182
6. Conclusions	191
References	191

Chapter 8
The Dependence of Electron Impact Excitation and Ionization Cross Sections of H_2 and D_2 Molecules on Vibrational Quantum Number
M. Capitelli and R. Celiberto

1. Introduction 195
2. Resonant Vibrational Excitation 197
3. Dissociative Attachment Cross Section 199
4. Electronic Excitation 200
5. Dissociation Processes 211
   5.1. Allowed Transitions 211
   5.2. Spin-Forbidden Transitions 213
6. Ionization 216
7. Electronic Excitation from Electronically Excited States 220
8. Conclusion 222
   References 223

Chapter 9
Electron–Molecular Ion Collisions
J. B. A. Mitchell

1. Introduction 225
2. H_2^+ 226
   2.1. Dissociative Recombination 230
   2.2. Dissociative Excitation 238
   2.3. Dissociative Ionization 240
   2.4. Ion-Pair Formation 241
3. H_3^+ 243
   3.1. Dissociative Recombination 245
   3.2. Dissociative Excitation 251
   3.3. Ion-Pair Formation 251
4. O_2^+ 252
   4.1. Dissociative Recombination 252
   4.2. Dissociative Excitation 256
5. CO^+ 256
   5.1. Dissociative Recombination 256
   5.2. Dissociative Excitation 258
6. CO_2^+ 259

7. Summary 259
   References 260

Chapter 10
Energy and Angular Distributions of Secondary Electrons Produced by Electron Impact Ionization
Yong-Ki Kim

1. Introduction 263
2. Qualitative Considerations 265
3. Analytical Model for Energy Distributions of Secondary Electrons 267
4. Analytical Model for Angular Distributions of Secondary Electrons 269
5. Comparisons with Experiment 271
6. Concluding Remarks 275
   References 276

Chapter 11
Elastic and Related Cross Sections for Low-Energy Collisions among Hydrogen and Helium Ions, Neutrals, and Isotopes
D. R. Schultz, S. Yu. Ovchinnikov, and S. V. Passovets

1. Introduction 279
2. The Elastic and Related Cross Sections 281
   2.1. Theoretical Approaches 282
   2.2. Related Cross Sections 284
3. The Semiclassical Method 286
   3.1. The Massey–Mohr Approximation 286
   3.2. Practical Computational Schemes 289
   3.3. Asymptotic Behavior 292
4. Specific Cross Sections 293
   4.1. $H^+ + H$ 293
   4.2. $D^+ + D$ 295
   4.3. $H^+ + D$, $H^+ + T$, $D^+ + H$, $D^+ + T$, $T^+ + H$, $T^+ + D$, and $T^+ + T$ 298
   4.4. $H + H$ 299
   4.5. $H + He$ and $D + He$ 301
   4.6. H^+, H_3^+, H, H^-, and $H_2 + H_2$ 303
   4.7. $H^+ + He$, $H^+ + H_2$, $He^+ + He$, and $He^{2+} + He$ 306
5. Conclusions 306
   References 306

Chapter 12

Rearrangement Processes Involving Hydrogen and Helium Atoms and Ions

F. Brouillard and X. Urbain

1. Introduction 309
2. Experimental Methods 310
   2.1. The Interaction of Two Beams 310
   2.2. Control of the Reactants 312
   2.3. Specification of the Process 313
   2.4. Detection of Reaction Products 314
3. Cross Sections 314
   3.1. Collisions of H—Production of H^- 314
   3.2. Collisions of H—Production of H^+ 316
   3.3. Collisions of H^+—Production of H 318
   3.4. Collisions of H^+—Production of H^- 319
   3.5. Collisions of He^{2+} 319
   3.6. Collisions of He^+ 322
   3.7. Collisions of He 323
   3.8. Collisions of H^- 326
   3.9. Collisions of H_2—Ionization and Dissociation 328
   3.10. Collisions of H_2^+—Ionization, Dissociation, and Charge Exchange 330
   3.11. Associative and Penning Ionization 332
References 334

Chapter 13

Electron Capture Processes in Slow Collisions of Plasma Impurity Ions with H, H_2, and He

R. K. Janev, H.P. Winter, and W. Fritsch

1. Introduction 341
2. Methods in Studies of Low-Energy Electron Capture Processes 344
   2.1. Experimental Methods 344
   2.2. Theoretical Methods 351
3. Total Electron Capture 356
   3.1. Collisions with Atomic Hydrogen 357
   3.2. Collisions with Helium Atoms 367
   3.3. Collisions with Molecular Hydrogen 376
4. State-Selective Electron Capture 382
   4.1. General Considerations: n-Distributions and l-Distributions 382
   4.2. Collisions with Hydrogen Atoms 384

4.3. Collisions with He Atoms 387
4.4. Collisions with H_2 Molecules 389
5. Conclusion 390
References 391

Chapter 14
Reactive Ion–Molecule Collisions Involving Hydrogen and Helium
F. Linder, R. K. Janev, and J. Botero

1. Introduction 397
2. Experimental Methods 399
3. Total Cross Sections for Particle Rearrangement Collisions 403
   3.1. Electron Transfer Reactions 404
   3.2. Particle Interchange Reactions 409
4. State-Selective Cross Section Measurements 423
   4.1. Reactions in Hydrogen Ion–Molecule Systems 424
   4.2. Reactions in Hydrogen-Helium Ion-Molecule Systems 426
   4.3. Energy and Angular Distribution of Reaction Products 428
5. Summary and Conclusions 429
References 430

Chapter 15
Particle Interchange Reactions Involving Plasma Impurity Ions and H_2, D_2, and HD
P. B. Armentrout and J. Botero

1. Introduction 433
2. Experimental Description 434
   2.1. General Considerations 434
   2.2. The Octopole Ion-Beam Guide 435
   2.3. Kinetic Energy Scale and Doppler Broadening 436
   2.4. Ion Sources 436
3. Theoretical Considerations 437
4. Results 440
   4.1. Carbon, Oxygen, and Silicon 440
   4.2. Metals 445
5. Analytic Representation 453
6. Discussion 457
References 459

Chapter 16
Electron Collision Processes Involving Hydrocarbons
Hiroyuki Tawara

1. Introduction 461
   1.1. Physical Sputtering 462
   1.2. Chemical Sputtering 464
2. Important Collision Processes Involving Hydrocarbon Molecules 466
3. Experimental Techniques and Their Features 468
4. Present Status of Electron Collision Data for
   Hydrocarbon Molecules 470
   4.1. Dissociation and Ionization or Ion and Neutral Particle Production 472
   4.2. Energy Distributions of Product Ion and Neutral Species 481
   4.3. Dissociative Recombination and Dissociation/Ionization of
        Hydrocarbon Molecular Ions 484
   4.4. Photon Emission 485
5. Summary, Further Data Needs, and Recommended Work 493
   References 495

Index **497**

Chapter 1

Basic Properties of Fusion Edge Plasmas and Role of Atomic and Molecular Processes

R. K. Janev

1. INTRODUCTION

The self-sustained thermonuclear burn of a deuterium–tritium (D–T) plasma in a fusion reactor relies on maintaining certain stringent conditions imposed on the plasma parameters (temperature and density), plasma energy confinement time, and power generating and loss processes. The plasma burn condition requires that the power density of fusion-born alpha particles exceeds the sum of the densities of all radiative and thermal power losses. While the thermal power losses are determined by collective plasma transport phenomena, the radiation losses (such as bremsstrahlung and line radiation) are determined by collisional and radiative atomic processes. Nonhydrogenic species (impurities) present in the plasma may substantially increase the radiation losses and, above certain critical amounts (e.g., 1% of plasma density for Fe and 0.1% of plasma density for W), may extinguish the burn. The fusion-born alpha particles carry about one-fifth of the total fusion power generated by D–T thermonuclear reactions, and only a small part of it is used to sustain the thermonuclear burn or is lost by bremsstrahlung radiation. The remaining power, as well as the alpha particles themselves, have to be removed

---

R. K. JANEV • International Atomic Energy Agency, A-1400 Vienna, Austria.

Atomic and Molecular Processes in Fusion Edge Plasmas, edited by R. K. Janev. Plenum Press, New York, 1995.

from the burning reactor zone in order to avoid plasma overheating and poisoning. The accumulation of alpha particles in the reacting zone (He ash) dilutes the thermonuclear fuel, degrades the burning conditions, and, above certain levels, may extinguish the plasma burn. Under steady-state conditions, the alpha particles have to be removed from the reactor at a rate equal to that of their production.

In a fusion D–T reactor based on the tokamak concept, the generation of impurities, undesired in the central plasma zone, takes place at the plasma boundary owing to interactions of the plasma particles with the walls of the containment vessel. The helium ash and thermal plasma power can only be exhausted from the fusion reactor through the plasma edge region. Control of the impurity ingress into the plasma and of the thermal power and alpha particle exhaust can be achieved by an appropriate control of the physical conditions in the plasma boundary region. As we have seen, this control is essential for maintaining the plasma burning conditions in the central plasma region.

The physical conditions at the plasma boundary of a tokamak fusion reactor are determined by the radial power and particle fluxes coming from the central plasma region, by the plasma–wall interaction processes, and by the radiative and collision processes among the atomic and molecular species present in this region. Control of these conditions, and eventually their tailoring to minimize the impurity ingress and optimize the power and particle exhaust, is possible only on the basis of a detailed knowledge of all plasma, surface, and atomic processes occurring in the plasma edge region. From a broader point of view, the plasma edge region can be defined as the region at the plasma periphery in which the effects of plasma coupling with the reactor material boundaries and released atomic species are strong. In a tokamak magnetic configuration, this is the region immediately inboard the last closed magnetic surface (see Section 2) and the entire region outside this surface.

In this chapter we will describe the basic properties of the plasma in the edge region, the most important physical processes taking place in this region, and, in particular, the role of atomic and molecular processes in achieving the desired properties of the edge plasma. There exist several comprehensive reviews describing the plasma edge physics[1-3] and the role of atomic and molecular processes in edge plasmas.[4-7] Therefore, we will confine our discussions to those aspects that are relevant mainly for the impurity control and particle and power exhaust problems. We start first with a description of the most common plasma edge configurations in tokamaks, designed to control the plasma edge conditions.[†]

[†] In our discussions, the term "hydrogen" will be used to denote the D and T isotopes, and the term "proton" represents the ions of these isotopes.

2. BASIC PLASMA EDGE CONFIGURATIONS

The plasma confinement in toroidal magnetic configurations (such as tokamak, stellarator, etc.) results from a number of constraints imposed by the magnetic field on the plasma particle motion. The conservation of magnetic flux within any plasma cross section perpendicular to the toroidal direction and the Lorentz force acting on charged plasma particles restrict the plasma particle motion on self-closed magnetic flux surfaces. The confinement is, however, not perfect. Multiple long-range Coulomb collisions of plasma particles and a large variety of magnetohydrodynamic instabilities induce plasma particle and energy transport across the magnetic flux surfaces, thus establishing contact of the plasma with the walls of the containment vessel. The impurities generated in plasma–wall interactions can radially diffuse into the plasma and contaminate it.

In order to minimize the direct plasma–wall contact and reduce the ingress of impurities into the plasma, the magnetic field in the boundary region has to be appropriately modified. In tokamak configurations, this is most often achieved either by introducing a mechanical plasma limiter on a finite segment of the vessel wall or by superimposing an additional magnetic field at the plasma periphery that diverts the field lines of the peripheral magnetic surfaces to close outside the main torus (magnetic limiter).

The mechanical plasma limiter may take the form of either a poloidal ring or a toroidal "belt". A schematic view of a toroidal belt limiter is shown in Fig. 1, representing the poloidal cross section of a D-shaped torus. The magnetic flux surface touching the limiter tip is the last (outermost) closed flux surface (LCFS). The magnetic surfaces outside the LCFS are "open" and intercept the boundary of the limiter (i.e., strike a material "target").

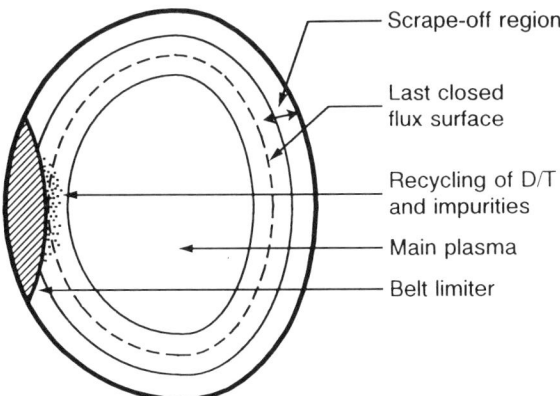

Figure 1. Schematic view of a belt limiter configuration.

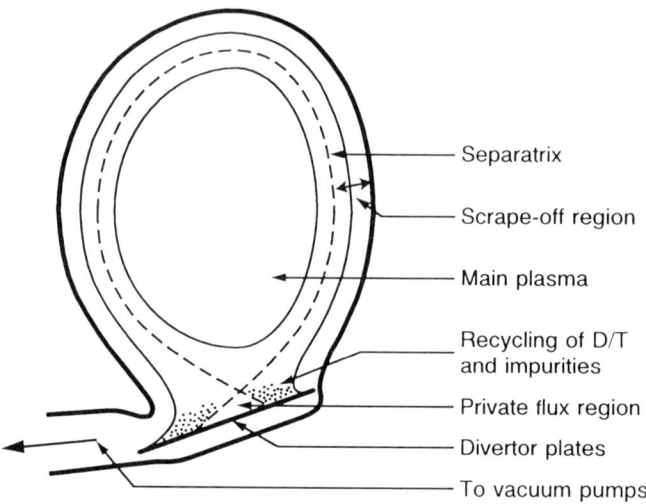

Figure 2. Schematic view of a single-null poloidal divertor configuration.

The most common magnetic limiter configuration in tokamaks is the poloidal divertor. This configuration can be obtained by adding a toroidal coil outside the torus, carrying a current in the same direction as the plasma current. The superposition of the poloidal magnetic fields of the divertor coil and plasma current creates a poloidal magnetic null (X-point). The magnetic flux surface passing through the X-point is the LCFS for this configuration (separatrix). The open (diverted) magnetic flux surfaces can be conveyed some distance away from the X-point (e.g., in a separate divertor chamber), where they can intercept certain material boundaries (divertor plates). Sometimes two divertor coils are used in order to create a double-null divertor. A schematic view of a single-null poloidal divertor configuration is shown in Fig. 2. The plasma region outside the separatrix is called the scrape-off layer, and the region beneath the X-point is the private flux region of the divertor.

The material targets on which the open magnetic field lines end have temperatures much lower than the plasma temperature at the last closed magnetic surface. The large plasma temperature gradients in the region between the LCFS and the target induce a strong conductive plasma flow in the direction parallel to the open magnetic field lines. This intense longitudinal plasma flow reduces strongly the radial plasma fluxes diffusing toward the vessel walls in the scrape-off region and, thereby, the effects of their interactions with the wall materials. The reduced amounts of impurity atoms released from the walls, after being ionized in the scrape-off region, become part of the conductive parallel plasma flow in this region.

The scrape-off region thus shields the main plasma (inside the LCFS) from wall impurity influxes and shields the vessel walls from the radial plasma fluxes. Due to the rapid decrease of all plasma parameters (temperature, density) in the radial direction beyond the separatrix, the radial extension of the most intense parallel plasma flow is limited to a narrow scrape-off layer (SOL) of the order of a few centimeters.

From the point of view of impurity shielding, the poloidal divertor configuration has distinct advantages over the limiter configuration. In the limiter case, impurities released near the limiter tip can readily penetrate into the main plasma and be ionized inboard the LCFS. Their transport there is governed by a variety of transport mechanisms,[8,9] the interplay of which may lead to impurity accumulation in the main plasma region near the toroidal magnetic axis. In the case of a poloidal divertor, the target plates intercepting the intense SOL plasma power and particle fluxes can be placed sufficiently far from the X-point so that the penetration of released impurities into the main plasma becomes much less probable. Moreover, because the plasma in the divertor chamber is far from the last closed magnetic surface, its parameters can be controlled by external means with the purpose of minimizing both the wall impurity generation and impurity penetration into the main plasma. The increase of neutral hydrogen density in this region (e.g., by introducing cold molecular hydrogen gas externally) is the simplest way to decrease the divertor plasma temperature, and, thereby, the sputtering of divertor target materials, and to increase the residence time of impurities in the divertor. The buildup of high neutral hydrogen densities in the divertor region results also from the hydrogen recycling in this region. Plasma ions interacting with the divertor plate can return back into the divertor plasma as neutrals (due to backscattering, desorption, and other particle-surface processes). In the plasma region, these neutrals are re-ionized and flow again toward the divertor plates. Under steady-state conditions (when divertor plate material is saturated with hydrogen), the ratio of the returning neutral particle flux to the plasma ion flux (i.e., the recycling coefficient) is unity (or very close to it). The recycling process evidently increases the neutral hydrogen density in the divertor chamber (particularly in front of the plates), but it also amplifies the plasma particle flux on the plates. Since during the recycling process the plasma energy is continuously expended on re-ionization of the cold neutrals coming from the plates (as well as on excitation, molecular dissociation, and other inelastic processes), the plasma temperature near divertor plates may become very low (of the order of several electron volts, under high recycling conditions). At such temperatures, the majority of plasma ions have energies below the threshold of many surface erosion processes (e.g., that for physical sputtering).

The high recycling regime in front of divertor plates can be achieved because of the intense conductive plasma ion motion along the open magnetic field lines. In the limiter configuration, a significant portion of neutralized plasma ions in the SOL region near the limiter tip enter the main plasma, and their return to the limiter surface can be achieved only by cross-field transport. Most intense recycling in the limiter configuration occurs in the outermost part of the SOL, where the plasma ion fluxes are weaker. Therefore, the potential of the limiter configuration to sustain intense localized hydrogen recycling is much lower than that of the divertor.

With respect to the power and particle exhaust, the divertor configuration also has significant advantages over the limiter configuration. As we have seen, the hydrogen recycling process is a powerful mechanism of plasma energy exhaust via the inelastic collisions of plasma particles (mainly electrons) with the cold neutrals. The efficiency of this mechanism is directly related to the intensity of recycling and is, therefore, higher in the divertor configuration than in the limiter configuration. Recycling losses alone are, however, insufficient to exhaust the high thermal power flowing into the scrape-off region of a fusion reactor. Other mechanisms of plasma power exhaust will be discussed in Section 5.

Both the limiter and the divertor configuration can also be used for the exhaust of helium from a reactor, as well as of the impurities which tend to accumulate in the near-target region. Alpha particles entering the scrape-off region together with the radial plasma fluxes are subjected to the dynamics of particle transport in that region and reach the intercepting targets. After neutralization on the targets, helium diffuses in the local edge plasma with a mean free path much larger than that for hydrogen atoms (due to its small electron-impact ionization and proton-impact charge exchange cross sections). In order to prevent its diffusion back into the main plasma, helium has to be pumped out from the divertor region (together with other impurities and thermalized hydrogen). This method is also used in the limiter configuration (pumped limiters).

More detailed information on power and particle exhaust issues in the tokamak fusion reactor can be found elsewhere.[5,10,11] In the remainder of this chapter we will confine our discussions to the edge plasma properties in a poloidal divertor configuration.

3. PARAMETERS AND COMPOSITION OF EDGE PLASMAS

In a poloidal divertor configuration, the plasma edge region is defined by the region outside the separatrix together with a layer immediately inboard the separatrix to which the wall-released neutrals can penetrate. Due to the fast increase of plasma temperature toward the magnetic axis (plasma center), neutral wall impurities traversing the separatrix are swiftly ionized and generate intense radiation in a layer inboard the separatrix (radiating layer). The plasma edge region outside the separatrix consists of the scrape-off region (extending up to the vessel walls), the divertor channel, and the divertor chamber (in which the divertor plates are located). The parameters (temperature, density) and chemical composition of the plasma in these three regions may differ substantially.

Basic Properties of Fusion Edge Plasmas

In a fusion reactor, the central plasma temperature and density are of the order of $T(0) \approx 10\text{--}20$ keV and $n(0) \approx 10^{14}$ cm^{-3}, respectively, and they decrease in the radial direction (roughly parabolically) to attain at the separatrix typical values of the order of $T(r_s) \approx 200\text{--}300$ eV and $n(r_s) \approx 10^{13}$ cm^{-3}. Because of the powerful convection in the scrape-off layer (having a radial extension of ~1 cm) in the direction parallel to the magnetic field, the scrape-off plasma temperature and density decay (roughly) exponentially in the radial direction, attaining values $T(r_w) \sim 5\text{--}10$ eV and $n(r_w) \approx 10^{12}$ cm^{-3} near the vessel walls.

The plasma parameters in the divertor region (divertor channel and divertor chamber) are determined by the power and particle fluxes entering the SOL, by the temperature gradients along the SOL, by the amount of power radiated by the impurities in this region, and by the recycling losses. Under a high recycling regime, the plasma temperature in front of the divertor plates can be reduced to about 10–30 eV if the power entering the scrape-off region is of the order of 100 MW. The same recycling process, however, may increase the plasma density in front of the divertor plates by an order of magnitude with respect to its value at the separatrix. The neutral hydrogen density in this region is also high and comparable to plasma density. The plasma temperature in the divertor region can further be decreased if the divertor plasma is "seeded" with strongly radiating and recycling impurities (such as Ne and Ar) in a controlled fashion. Another way of decreasing the divertor plasma temperature is to extract the plasma energy not from its electron components (as in the recycling and impurity radiation cooling), but from its ion components by enhanced proton–hydrogen charge exchange collisions[12] (see Section 5).

Most of the neutral hydrogen released from the divertor plates in the recycling process is in the molecular form, and the rate of its dissociation depends on the divertor plasma temperature. In the regions where the plasma temperature is below a few electron volts, the lifetime of molecular hydrogen prior to its dissociation may be long. Molecular hydrogen may also be deliberately introduced into the divertor chamber for plasma cooling purposes. Wall impurities, generated by the plasma–surface interactions, are important constituents of the divertor edge plasma.

The most important impurity-generating processes are thermal and particle-impact-induced desorption, physical sputtering, and evaporation (due to excessive localized thermal loads). For carbon-based wall and divertor plate materials, chemical sputtering and radiation-enhanced sublimation are the most important impurity-generating mechanisms. A detailed account of the physics and efficiency of these processes, under various plasma–surface interaction conditions, is given elsewhere.[13]

The impurity composition of the edge plasma obviously depends on the materials used for the plasma-facing components (the first wall, its protective tiles, divertor plates, antennas for radio-frequency heating, etc.) in a particular device. In the currently operating large tokamak devices, such as JET, TFTR, JT-60U, Tore Supra, and DIII-D, the range of materials used is fairly broad and covers elements with both low and medium atomic numbers. The materials that are currently being considered for the plasma-facing components in the design of next-generation fusion devices [such as the International Thermonuclear Experimental Reactor (ITER)] include beryllium, boronized beryllium, carbon composites, or carbon-based material (C+B, C+Ti, SiC) for the divertor plates and first-wall protecting tiles, and stainless steel, vanadium, and Ni-based steels as the first-wall base materials.[11] If temperatures below ~10 eV can be achieved in the divertor region, high-Z materials such as W, Mo, and Ta are also considered as candidate divertor plate materials (on a Cu-based structure).[11] Therefore, the expected atomic impurities in the plasma edge of a fusion reactor include Be, B, C, Al, Si, Mg, Ti, Cr, Fe, Ni, Cu, Mo, Nb, Ta, and W; oxygen should be added to this list as an omnipresent common impurity. Under the low-temperature conditions prevailing in the divertor plasma, molecular impurities can also exist in this region, mostly in the form of oxides and hydrides. If carbon is used as plasma-facing material, the most abundant plasma edge molecular impurities would be CO, CO_2, CH_4, C_2H_2, and other hydrocarbons C_kH_n. These molecules are usually formed on the surface and released by thermal or particle-impact-induced desorption. Water molecules are also present in the plasma edge.

As we have mentioned in the previous section, impurities can be deliberately introduced in the plasma edge region for radiative plasma cooling purposes. These impurities should have an atomic number in the range $Z = 10$–20 and should have the capability to recycle between the plasma and the walls. Neon and argon are considered as the most suitable candidates. Other impurities (such as Li) can also be introduced (as beams) in the divertor region for plasma diagnostic purposes.

Because of the low scrape-off and divertor plasma temperatures (1–200 eV), the atomic impurities in the plasma edge exist in relatively low charge states q. For the low-Z impurities ($Z \leq 10$), q may be in the range 1–4, while for the high-Z impurities, q may be as high as ~8–10. In these charge states, impurities have a large radiating potential and can provide a strong plasma edge cooling. Molecular impurities at the edge are usually in the $q = 1$ charge state.

The relative abundance of impurities in the plasma edge may vary depending on the part of the edge region (scrape-off, divertor chamber), neutral gas content, construction details (radial dimensions of the divertor throat, inclined plates, etc.), and desired level of impurity radiation cooling of the plasma edge, but it is ultimately constrained by the tolerable levels of impurities penetrating the main plasma region. For the low-Z impurities, these levels are of the order of several percent of plasma density; for medium-Z impurities, the tolerable concentrations are below 1%; and for high-Z impurities, these levels are below 0.1%. The molecular impurity concentrations in the divertor region are about 0.5% of plasma density, except for the CH_4 and C_2H_2 impurities, which may reach levels of about 1%.

4. ATOMIC AND MOLECULAR PROCESSES IN THE PLASMA EDGE

The variety of atomic and molecular species present in the plasma edge, coupled with the relatively broad range of charge states of impurity ions, makes the collision physics of the edge plasma very complex. Because of the high neutral densities in the divertor region, the characteristic collision times may be shorter than or comparable to the radiative lifetimes of excited species, and the collision processes of these species add to the complexity of plasma edge collision physics.

The importance of specific collision processes for the global plasma edge phenomena (such as impurity and neutral particle transport, radiation losses, etc.) can be judged by the relative magnitude of their rates and by the ratio of their characteristic collision times and the residence times of collision partners in the plasma edge region. Based on the fact that the plasma electrons and ions, and also the neutral hydrogen atoms and molecules in the divertor chamber, are by far the most abundant plasma edge species, all collision processes among them are of primary importance for the edge plasma properties and dynamics. Helium in the plasma edge can have concentrations on the level of 10% of plasma density, and its collision processes with plasma electrons, ions, and hydrogen neutrals are of great importance for its transport and removal from the edge. The importance of collision processes of plasma edge impurities with the primary plasma edge constituents (plasma electrons and ions, neutral hydrogen and helium) increases with the increase of their relative concentration and residence time. The residence time of impurity ions in the SOL region is determined by its length and the velocity of the plasma flow. For a tokamak with major radius of 4–5 m and flow velocities somewhat below the sonic speed, the ion residence time is about 10^{-3}–10^{-4} s. Outside the scrape-off layer, the residence time of impurities is determined by their recycling and may attain much higher values. On the other hand, for plasma temperatures of 20–50 eV, the electron–ion excitation and ionization processes have rate coefficients of the order of 10^{-8} cm^3/s, which for a plasma density of about 10^{13} cm^{-3} gives a characteristic time for the process of about 10^{-5} s. Therefore, the electron–ion inelastic processes are important in the plasma edge. The frequency of heavy-particle collisions is, however, about one to two orders of magnitude smaller than that of electron collisions, and, therefore, the heavy-particle collisions in the edge plasmas are, generally speaking, less important than those involving electron impact. Exceptions to this rule are the resonant and quasi-resonant heavy-particle collision processes that conserve (or nearly conserve) the internal electronic energy of the colliding system or are strongly exothermic. The rate coefficients of these processes may be of the order of 10^{-6}–10^{-7} cm^3/s.

The most important electron impact processes in the plasma edge include excitation and ionization of hydrogen and impurity atoms and ions from their ground and first several excited states, radiative and dielectronic recombination on impurity ions, direct and dissociative excitation and ionization of hydrogen and impurity molecules and their ions, and dissociative recombination on molecular ions. It should be emphasized that the above electron collision processes with vibrationally excited molecular species usually have much larger cross sections than in the case when these species are in their ground vibrational state. For many plasma research applications, the important information associated with the electron–atom (ion) or electron–molecule (molecular ion) processes is not only the total or partial cross sections (or the corresponding rate coefficients), but also their differential characteristics (e.g., angular and energy distributions of reaction products).

The most important heavy-particle collision processes in the plasma edge are those which have resonant or quasi-resonant character. The electron capture in proton–hydrogen atom or proton–oxygen atom collisions, momentum transfer in these and other ion–atom systems, and proton-impact-induced transitions between fine-structure components of atoms and their ions are typical examples of such processes. Energy transfer (ro-vibrational excitation) and particle interchange reactions in ion–molecule collisions may also proceed with high probabilities. Charge exchange collisions involving excited neutrals and/or multiply charged ions also have a quasi-resonant character. The cross sections of these processes at low energies may be sufficiently large to compensate for the low heavy-particle collision frequency, resulting in rates comparable to or higher than those of electron impact processes. Moreover, most of the inelastic electron collision processes have a certain threshold, whereas for the resonant and highly exothermic heavy-particle reactions such energy constraint does not exist.

5. ROLE OF ATOMIC AND MOLECULAR PROCESSES IN PLASMA EDGE PHYSICS

Since in atomic and molecular processes energy and momentum are transferred between the colliding particles, and since in many of them also the charge state of colliding particles is changed, it is obvious that these processes should play an important role in the plasma energy, momentum, and particle transport in the plasma edge. For the same reasons, they should play a similar role in the neutral particle transport in the divertor region. Indeed, the plasma transport in the edge is described by a set of coupled fluid equations expressing the conservation of plasma particles, momentum, and energy.[14] The effects of atomic and molecular collisions on these plasma quantities appear as source/sink terms in the fluid equations. The impurity transport in the plasma edge, dominated by the conduction flow along the magnetic field lines, is also described by a set of fluid equations[15] (for each charge state of each impurity), and the collision processes of impurity ions with plasma electrons and divertor neutrals also affect their transport. Finally, the transport of neutral particles in the divertor region has to be described by the kinetic Boltzmann equation, which is usually solved by Monte Carlo numerical techniques.[16] Because

Basic Properties of Fusion Edge Plasmas

no electromagnetic forces are involved in the neutral particle transport, atomic and molecular collisions have a dominant influence on this transport. Because the collisions involve both charged and neutral particles, it is obvious that the plasma transport, impurity transports, and neutral particle transport are strongly mutually coupled. While the fluid equations for the plasma particles determine the distributions of plasma temperature and density, the impurity and neutral particle transport equations determine the distribution of source/sink terms in the plasma transport equations. Because of the formidable numerical complexity, self-consistent descriptions of plasma, impurity, and neutral particle transport have emerged only very recently.[17]

The role of atomic and molecular processes in resolving the problem of thermal plasma power exhaust in a divertor configuration can be appreciated if one keeps in mind that the overwhelming part of the plasma power and particle fluxes traversing the separatrix is directed by the scrape-off layer to a relatively small area on the divertor plates. For plasma powers of the order of 100–200 MW entering the SOL, the thermal load delivered on these areas can be 30–50 MW/m^2. The available materials for divertor plates can sustain only loads about 10 times smaller than this without significant distortion of their thermomechanical properties and physical integrity. The reduction of thermal power to the tolerable levels of ~1–3 MW/m^2 should be accomplished in the scrape-off and divertor plasma edge regions. There have been many proposals regarding means of exhausting the thermal power (for critical reviews, see Refs. 10 and 11), most of them based on the inelastic energy transfer collisions of plasma electrons and ions with other constituents of the edge plasma.

The kinetic energy of plasma electrons can be decreased in excitation, ionization, and dissociative collisions with the plasma neutrals and in excitation, ionization, and recombination collision processes with ionized plasma impurities. The electron energy losses in excitation and recombination processes are transformed into radiation, which readily escapes from the plasma. The intense energy exchange between plasma electrons and ions (due to the long-range Coulomb collisions) reduces the plasma temperature (radiative cooling). Significant density levels of plasma edge neutrals and impurities are required for this cooling mechanism to be efficient. The intense hydrogen and impurity recycling in front of the divertor plates is extremely beneficial for the enhancement of electron energy losses. However, as mentioned earlier, the upstream extension of the high-recycling zone is rather small (determined by the ionization length of the neutrals), and much of the incoming power consumed by ionization of recycling neutrals is redeposited onto the target plates. Increasing the thickness of the recycling zone would require higher recycling intensities, with a consequent increase in the flux amplification factor. A high-recycling regime, therefore, cannot reduce the power load on divertor plates to levels consistent with the thermomechanical capabilities of the available materials.[12] In order to achieve further reduction of the power load, energy and momentum should be extracted from the plasma ions. This can be accomplished by charge exchange and momentum transfer collisions of plasma ions with neutral hydrogen atoms and molecules, not in the small recycling zone, but rather along the entire divertor channel. In these collisions, the fast plasma ions are converted into fast neutrals which do not follow the magnetic field lines and dissipate their energy and momentum on the side walls of the divertor. Higher neutral densities and longer divertor channels are required for this concept to be effective.[12,18] The energy and momentum of fast-charge-exchange neutrals can also be dissipated on a circulating neutral hydrogen gas in the divertor chamber, so that the power deposition on the divertor walls is again reduced. Efficient penetration of cold neutrals in the plasma flow channel requires densities of the neutrals in the divertor region be appreciably higher than the plasma density. Because of the large area on which the plasma energy and momentum are dispersed, the power loads on the divertor side walls and divertor plates are drastically reduced and can be decreased below 1 MW/m^2. In all power exhaust scenarios, intense plasma edge radiative cooling (by the hydrogen neutrals or impurities) is considered unavoidable.

The role of plasma edge atomic collision processes in the helium exhaust problem is also significant. Helium ions entering the divertor region are neutralized on the divertor plates and return back into the divertor plasma as atoms. Helium recycles between the walls and the plasma in much the same way as the hydrogen, but its ionization length is larger than that for hydrogen, and its recycling coefficient is considerably smaller. Nevertheless, a helium "enrichment" effect occurs, which is beneficial for its pumping out of the divertor chamber, but which also increases the probability of helium returning to the main plasma. The helium transport in the divertor region is strongly affected by the charge exchange and momentum transfer collisions of helium ions with the neutral atomic and molecular hydrogen, and also by the helium particle collisions with the divertor chamber walls. The charge exchange collisions increase the residence time of helium in the divertor, whereas the collisions with the chamber walls and momentum transfer collision have the opposite effect. Coupled with an appropriate geometry of the divertor plates and walls, these processes direct the helium transport toward the pumping ducts and, thus, relax the requirements on the pumping speed.

6. CONCLUSION

The successful design and operation of a fusion reactor relies on resolving the critical issues of impurity control, plasma thermal power exhaust, and removal of the helium ash from the reactor. The majority of current reactor design concepts seek a solution of these issues in the control of physical conditions of the plasma edge region. Atomic and molecular collision processes taking place in this region, together with the plasma–wall interaction processes, have a strong impact on the edge plasma properties and dynamics. A detailed knowledge of the characteristics of these processes (cross sections, reaction rate coefficients) is required in order to achieve an understanding of the edge plasma behavior and provide a predictive modeling tool for fusion reactor design studies.

The existing quantitative information on the atomic and molecular processes in the plasma edge, although impressive in its extent, is still insufficient to provide a basis for an accurate description of all radiative and collisional phenomena that influence the edge plasma behavior.[19,20] Establishment of a complete data base for the radiative and collisional atomic and molecular processes in the plasma edge is a significant challenge for the atomic physics research community.

REFERENCES

1. P. C. Stangeby and G. M. McCracken, *Nucl. Fusion* **30**, 1225 (1990).
2. D. E. Post and K. Lackner, in *Physics of Plasma–Wall Interactions in Controlled Fusion* (D. E. Post and R. Behrisch, eds.), Plenum, New York (1984), p. 627.
3. M. F. A. Harrison, E. S. Hotson, J. G. Morgan, and G. P. Maddison, Plasma Edge Physics for NET/INTOR, Report CLM-P761, United Kingdom Atomic Energy Authority, Culham Laboratory, Abingdon, U.K., 1985.
4. M. F. A. Harrison, in *Atomic Processes in Electron–Ion and Ion–Ion Collisions* (F. Brouillard, ed.), Plenum, New York (1986), p. 421.
5. M. F. A. Harrison, in *Atomic and Plasma-Material Interaction Processes in Controlled Thermonuclear Fusion* (R. K. Janev and H. W. Drawin, eds.), Elsevier, Amsterdam (1993), p. 285.
6. R. K. Janev, M. F. A. Harrison, and H. W. Drawin, *Nucl. Fusion* **29**, 109 (1989).
7. R. K. Janev, *Comments At. Mol. Phys.* **26**, 83 (1991).
8. K. Miyamoto, *Plasma Physics for Nuclear Fusion*, MIT Press, Cambridge, Massachusetts (1976).
9. R. A. Hulse, *Nucl. Technol./Fusion* **3**, 259 (1983).
10. International Tokamak Reactor, Phase IIA, Part 3, International Atomic Energy Agency, Vienna, 1988.
11. International Thermonuclear Experimental Reactor Concept Definition, Vol. 2, IAEA Documentation Series, No. 3, International Atomic Energy Agency, Vienna, 1989.
12. M. L. Watkins and P. H. Rebut, in *Controlled Fusion and Plasma Heating* (Proceedings of the 19th European Conference on Plasma Physics and Controlled Fusion, Innsbruck, 1992), Vol. 16C, Part II, European Physical Society, Geneva, (1992), p. 731.
13. J. Roth, in *Physics of Plasma–Wall Interactions in Controlled Fusion* (D. Post and R. Behrisch, eds.), Plenum, New York (1984), pp. 351 and 389.
14. S. I. Braginskii, in *Reviews of Plasma Physics*, Vol. 1 (M. A. Leontovich, ed.), Consultants Bureau, New York (1965), p. 205.
15. E. L. Vold, *Contrib. Plasma Phys.* **32**, 404 (1992).
16. D. Reiter, in *Atomic and Plasma-Material Interaction Processes in Controlled Thermonuclear Fusion* (R. K. Janev and H. W. Drawin, eds.), Elsevier, Amsterdam (1993).
17. D. Reiter, *J. Nucl. Mater.* **196–198**, 80 (1992).
18. K. Borrass and G. Janeschitz, Nucl. Fusion **34**, 1203 (1994).
19. H. Tawara and R. A. Phaneuf, *Comments At. Mol. Phys.* **21**, 177 (1988).
20. R. A. Phaneuf and R. K. Janev, in *Atomic and Plasma-Material Interaction Processes in Controlled Thermonuclear Fusion* (R. K. Janev and H. W. Drawin, eds.), Elsevier, Amsterdam (1993).

Chapter 2

Spectroscopic Processes and Data for Fusion Edge Plasmas

W. L. Wiese

1. INTRODUCTION

In this review of atomic structure data, all atomic species relevant to current and future fusion research facilities as compiled by Janev et al.[1,2] are included. His lists contain the following chemical elements, in order of increasing atomic number: H, D, He, Li, Be, B, C, O, Mg, Al, Si, Ti, V, Cr, Fe, Ni, Cu, Ga, Kr, Nb, Mo, Xe, Ta, and W. Data for the spectra of neutral atoms and ions up to about the 10th stage of ionization will be covered.

The emission of spectral radiation is of importance mainly for plasma diagnostics and plasma modeling. These two principal data needs put rather different demands on the data base. For diagnostic studies, the typical need is for very accurate data, but the needs are often limited to just a few or even a single spectral line or spectral feature. On the other hand, for modeling purposes usually a large body of atomic structure data is needed. Such data are normally not confined to one chemical element but may include numerous elements and stages of ionization, depending on the task at hand. For modeling work, atomic structure data of moderate accuracy for individual spectral features will often suffice.

W. L. WIESE • National Institute of Standards and Technology, Gaithersburg, Maryland 20899.

Atomic and Molecular Processes in Fusion Edge Plasmas, edited by R. K. Janev. Plenum Press, New York, 1995.

The basic atomic structure data are the transition energies, or wavelengths, of spectral lines with their spectroscopic classifications, the energy levels of atoms and ions, including their ionization energies, and the line strengths or atomic transition probabilities.

For emission studies of spectral transitions from a higher atomic level k to a lower level i under optically thin conditions, a basic relation of plasma spectroscopy,

$$\varepsilon_{ki} = \frac{1}{4\pi} A_{ki} N_k h\nu_{ik} \tag{1}$$

is applied. In this equation, ε_{ki} is the emission coefficient, A_{ki} is the atomic transition probability, N_k is the number density of ions or atoms in the excited state emitting the line (usually expressed as the number per cubic centimeter), and ν_{ik} is the frequency of the line. Measured line intensities I_{ki} are related to ε_{ki} by

$$I_{ki} = \int_0^l \varepsilon_{ki}\, dl \tag{2}$$

where l is the length of the observed plasma. For the identification of the species and stage of ionization, knowledge of the transition frequencies ν_{ik} is essential. The low-density plasmas of tokamak devices are in the collisional-radiative regime,[2,3] where the limiting factor for the rate of line emission is the collisional excitation rate from the ground (and metastable) states, and where therefore the atomic transition probability for spontaneous emission is not a critical quantity. For simplified plasma models including only the ground state (1) and the first resonance level (2), ε_{21} even becomes independent of A_{21}.[3] However, for detailed models including higher levels, the pertinent atomic transition probabilities must be known.

Also, for higher density plasmas, the knowledge of transition probabilities becomes essential. This fact appears to be of increasing importance in the plasma edge region, where higher plasma densities are desired for better power dissipation and thus where a regime of partial local thermodynamic equilibrium (partial LTE) for the populations of excited atomic levels is approached. In such a regime the distribution of atoms or ions in the various excited states is given by Boltzmann populations N_k, which are[4]

$$N_k = N_a \frac{g_k}{U_a(T)} \exp\left(\frac{-E_k}{k_B T}\right) \tag{3}$$

In this relation, E_k is the excitation energy of an atomic or ionic level k, k_B is the Boltzmann constant, and T is the excitation temperature; $g_k = 2J_k + 1$ is the statistical weight of level k. N_a is the total number density of species a, and $U_a(T)$ is its partition function. For such equilibrium conditions one may, for example, derive the excitation temperature from a plot of measured line intensities versus excitation energies. The key requirement for the application of this Boltzmann plot technique[5] is the availability of reliable atomic transition probabilities A.

2. STATUS OF RESEARCH ON ATOMIC STRUCTURE DATA

For the determination of wavelengths, or transition energies, and atomic energy levels, laboratory measurements have provided highly accurate data that only on rare occasions have been matched by calculations. Thus, almost all wavelength and energy level data in critical data compilations are observed or are derived from observed quantities. The measurement of wavelengths of spectral lines is the first step in establishing the structure of an atomic or ionic species. A subsequent analysis based on the Ritz combination principle[6]—utilizing the fundamental relation that all transition frequencies are differences between atomic energy levels—will yield the atomic structure, that is, the positions of various energy levels. The critical factor for an accurate analysis is the requirement that observational data for resonance lines as well as for numerous other transitions representing different combinations of energy levels are included.

For the accurate determination of wavelengths of unknown lines or for the determination of line shifts, it is essential to have wavelength standards available that are based on very accurate, interferometrically determined wavelengths of line-rich spectra. An extensive atlas of such secondary wavelength standards with about 6000 lines has been recently produced for platinum I and II.[7] A very useful compilation of earlier wavelength reference data covering the range 1500–25,000 Å has been a table of about 5400 highly accurate wavelengths produced by V. Kaufman and B. Edlén.[8]

Compared to the accurate and extensive knowledge of wavelengths and energy levels, our knowledge of atomic transition probabilities, or the equivalent term oscillator strengths, is still in a much more rudimentary stage. These data are orders of magnitude less accurate than those for wavelengths; data of accuracy better than ±50% are not even available for many lines. It has been estimated that transition probability data with uncertainties smaller than ±50% exist only for a small fraction, roughly about 5%, of the lines for which accurate wavelengths as well as spectroscopic classifications and energy levels are known.

For atomic transition probabilities, the data sources are quite different for different elements. For light elements, calculations are the main data source. Sophisticated calculational approaches have now often achieved uncertainties less than ±10% for many prominent lines (or multiplets). For more complex elements, starting approximately with scandium (atomic number $Z = 21$), reliable theoretical data are still scarce. The only exceptions are atoms and ions of simple structure, that is, those with one or two valence electrons outside closed shells such as the alkalies or alkaline earths, for which atomic transition probabilities have been calculated with high accuracy. Very complex spectra of species of fusion interest, for example, Fe I and Fe II, Ni I, Mo I, and W I, have been accurately determined by experiment only.

Since the uncertainties in atomic transition probability data are so large, it is of interest to briefly review the principal techniques that have produced such data so that a better understanding of the difficulties in getting accurate data is obtained.

3. DESCRIPTION OF PRINCIPAL METHODS TO DETERMINE TRANSITION PROBABILITIES

3.1. Theoretical Methods

Large-scale productions of atomic transition probabilities have become possible with the development of sophisticated atomic structure codes based on advanced quantum-mechanical approximations. For these theoretical approaches, several codes designed for large computers have been perfected and utilized for comprehensive numerical computations.

Probably the largest production of advanced atomic transition probability data has been carried out as part of the Opacity Project (OP).[9–12] This calculational approach makes use of the **R**-matrix method[12] and is an extension of the close-coupling (CC) approximation, which has been widely applied to calculate electron–ion or electron–atom collision data. In order to include calculations of transition probabilities for discrete transitions, the CC approximation has been extended to the case of an electron with negative energy, that is, an electron captured by a target ion and undergoing bound–bound transitions in the ion field. Thus, the wave function for some state of an ion or atom with $n + 1$ electrons is constructed from that of a system consisting of a target ion with n electrons—which is described with sophisticated configuration-interaction wave functions—plus an additional electron in its field bound to the ion.

The Opacity Project, an international collaboration that was formed in the early 1980s under the leadership of M. Seaton, has involved about 20 participating theoreticians from research groups in the United Kingdom, the United States, France, Germany, and Venezuela. The OP, now completed, has produced a large number of oscillator strengths for the lightest 10 elements through all stages of ionization and has also produced these data for selected heavier elements such as sodium, magnesium, silicon, sulfur, argon, titanium, and iron. For the main purpose of the Opacity Project, the detailed modeling of stellar opacities, calculations of *multiplet* oscillator strengths are sufficient, so that fine-structure calculations leading to data for individual spectral lines have not been part of the OP calculations. Therefore, if plasma diagnosticians require data for individual spectral lines, these must be obtained either from other calculations addressing fine structure or from the application of Russell–Saunders (or *LS*) coupling intensity fractions for lines within multiplets, provided it is ascertained that the LS coupling approximately holds.

It should be noted that the OP team has published a tabulation that contains their transition probability data and includes also selected results on photoionization cross sections, etc.[13] Also, new critical tables are being prepared at the U.S. National Institute of Standards and Technology for the first 10 elements, and these are in large part based on OP results.[14]

Another sophisticated atomic structure code is the CIV 3 code developed by Hibbert.[15] (CIV 3 stands for Configuration Interaction Code Version 3.) This code is one of two codes that have been utilized extensively in the OP calculations to obtain accurate target ion representations. It has also been applied to a fairly large number of transitions of neutral carbon,[16] nitrogen,[17] and oxygen[18] as well as singly ionized nitrogen[19] and oxygen.[20] These calculations comprise up to several hundred lines per spectrum and include most of the prominent lines. The calculations are more detailed than the OP work insofar as not only multiplet but also individual line data are obtained. This has been achieved by including spin–orbit and other relativistic terms of the Breit–Pauli type, in addition to the nonrelativistic electrostatic interactions (which would yield pure LS coupling data for the individual spectral lines). Thus, the line strengths have been produced in intermediate coupling, and also intersystem line data have been calculated. Some significant departures from LS coupling have indeed been obtained, especially for p–d transitions in carbon and nitrogen.

A few other sophisticated structure codes have also found applications, but on a more limited basis. One is the multiconfiguration Hartree–Fock code with Breit–Pauli corrections.[21] This code has been mostly applied to specific problem cases in light elements and has produced excellent results as seen from comparisons with the other approaches and experimental data.[22] Another approach is the SUPERSTRUCTURE code,[23] which has found similar applications for light elements and has also been employed for the description of target state functions in the Opacity Project. Again, this approach has produced results of similar accuracy as the other above-mentioned theoretical methods when configuration interaction has been included on an extensive scale.[24] Finally, the superposition-of-configurations (SOC) approach by Weiss has produced excellent results for Be-like and C-like systems.[25]

3.2. Experimental Methods

The two principal approaches for determining transition probabilities are emission and lifetime experiments, methods which are quite different but, because of that, also very complementary. The mean atomic lifetime τ_k of a level k is related to the transition probabilities A_{ki} by

$$\tau_k^{-1} = \sum_i A_{ki} \tag{4}$$

A values may be directly derived from τ_k when the sum is reduced to one term—which is the special case of resonance lines. In the normal case, lifetimes must be utilized in conjunction with emission measurements (which provide the individual A_{ki}'s) to yield transition probabilities for individual lines. Emission measurements are indeed capable of producing large numbers of data on an arbitrary, relative scale, but considerable difficulties are encountered in determining accurate data on an absolute basis. Because of this problem, lifetime measurements are often utilized to supply the absolute scale. Thus, in recent years, combinations of lifetime and emission measurements have become the method of choice.[26]

Emission measurements have been performed with a variety of light sources. For light elements of interest to fusion research, such as carbon and oxygen, wall-stabilized arcs[5] have proven to be very reliable steady-state sources, and numerous measurements have been carried out with them.[27,28] The arc work is limited to first and second spectra, however; for higher stages of ionization, pulsed plasma sources have to be used, such as, for example, the gas liner pinch.[29]

The introduction of heavier elements into plasmas in a reproducible manner and under steady-state conditions has often presented problems. For metals, compounds with high vapor pressures at room temperatures have been used, such as metal carbonyls and metal chlorides. In addition, the cathodes of hollow-cathode discharges have been lined with thin metal foils so that small quantities of pertinent metal atoms are sputtered into the plasma.[30] Such hollow-cathode discharges are not fully stable, and therefore Fourier transform spectrometers (FTS) have been used for the spectral observations, since the interferograms allow simultaneous measurements on all spectral elements. Also, the low-density hollow-cathode discharges cannot be assumed to be in partial local thermodynamic equilibrium, so that it would be very difficult to relate populations from various excited atomic states. Measurements of transition probabilities with these sources have thus been limited to "branching ratio" measurements but have been combined with lifetime measurements of the relevant excited atomic states to yield high-quality absolute data. This method has been applied to the complex spectra of important metals such as iron,[31] nickel,[32] niobium,[33] molybdenum,[34] tantalum,[30] and tungsten.[30] Such branching ratio measurements, as the name implies, are measurements of all emission lines—the "branches"—from a common upper level. For the above-cited metals, they have been done on a large scale and have been combined with equally extensive lifetime measurements, resulting in comprehensive measurements of transition probabilities for these spectra for as many as several thousand lines.

Atomic lifetime measurements have advanced to a high level of accuracy with the advent of tunable lasers that allow selective excitations of various atomic levels at low-density conditions.[30-34] This new approach has eliminated cascading effects and minimized other systematic errors, such as radiation trapping and collisional quenching, which were troublesome in earlier nonselective excitation work. (Cascading means here that the atomic level whose spontaneous radiative decay is studied is simultaneously repopulated with electrons dropping down from higher excited levels, which yields lifetimes that are too long.)

The new laser-induced fluorescence (LIF) lifetime techniques are not entirely free of problems either.[35,36] One of these arises from the circumstance that laser radiation is normally linearly polarized so that the excited atoms or ions are initially aligned. During the lifetime of excited atomic states, these atoms become gradually disaligned, for example, owing to stray magnetic fields. This may produce a time-dependent distortion in the recording of the radiative decay by the detector. These effects may be circumvented by using compensating magnetic field coils and by observing at the so-called magic angle.[35,36] Numerous lifetime measurements with these new tunable-laser excitation techniques—where these critical factors are taken into account—have been performed for a number of atomic ions of fusion edge plasma interest.

4. AVAILABILITY OF SPECTROSCOPIC DATA

In summary, atomic spectroscopy data are fairly plentiful and of high accuracy for the light elements from hydrogen through neon, that is, for atomic numbers $Z = 1–10$, with the exception of the transition probabilities for some species. This statement applies to the neutral atoms as well as all stages of ionization. For the moderately heavy elements in the range $Z = 11–28$ (sodium through nickel), significant gaps in the data start to appear, especially for lower ions of the iron group elements. The transition probability data for these species are quite incomplete, and the existing data are often of low quality. For the elements with atomic numbers between 29 and 42, many gaps exist in the data, and especially transition probabilities for low stages of ionization are very incomplete and rather inaccurate. However, Mo I ($Z = 42$) is a notable exception. For the heavier elements with atomic numbers 43 and higher, the available spectroscopic data are largely concentrated in the spectra of neutral and singly ionized atoms, as well as highly charged ions. Some reliable transition probability data are available for heavy, highly stripped ions of simple atomic structure such as alkali-like ions of the sodium and potassium sequences and ions of the copper isoelectronic sequence. In addition, very accurate data are available for hydrogen- and helium-like ions.

Generally, data on atomic transition probabilities are several orders of magnitude less accurate than those on wavelengths and energy levels, except for the hydrogen- and helium-like ions. Also, atomic transition probabilities of accuracy better than ±50% exist for only a few percent of all classified lines with known wavelengths.

Extensive bibliographical files for all spectroscopic quantities are maintained at two data centers at the U.S. National Institute of Standards and Technology (NIST), formerly the National Bureau of Standards (NBS). These files cover the literature from very early to current papers. In the data center on Atomic Energy Levels, bibliographies on the lines and analyses of atomic and ionic spectra have been published in special NBS publications through the year 1983,[37] and in the data center on Atomic Transition Probabilities, bibliographies have been published through March 1980.[38] After these dates, the bibliographic material has been recorded electronically, and general data bases are now being prepared for electronic dissemination. Also, since 1977, the NIST data centers have provided most of the spectroscopic information for the *International Bulletin on Atomic and Molecular Data for Fusion*,[39] a publication of the International Atomic Energy Agency (IAEA), in which up-to-date bibliographical coverage is maintained for all spectra of fusion interest.

It should be noted that the NIST bibliographic files are annotated. They contain coded information that specifies, for example, which type of data have been obtained (wavelengths, isotope shifts, hyperfine structure data, lifetimes, etc.), whether an experimental or a theoretical approach has been used, and if allowed (electric-dipole) or forbidden lines have been investigated.

5. NUMERICAL SPECTROSCOPIC DATA BASES

Critically evaluated numerical spectroscopic data bases have been published for more than 50 years, primarily at NIST. Also, some other scientists and institutions have made significant contributions, and these will be mentioned below. Unfortunately, comprehensive critical data tables are rarely up-to-date because the evaluation process takes considerable effort and time. However, past experience has shown that even somewhat older tables are still highly useful, since many data improvements amount to relatively small changes.

The various spectroscopic data tables and databases will now be reviewed. Also, new high-quality, comprehensive original work that is expected to be among the principal sources of future data tables will be discussed.

5.1. Wavelength Tables

The major recent wavelength tables are given in Table I. From top to bottom, this table lists two very general compilations covering basically all chemical elements (but only the lower stages of ionization)[40–42] and then three tables with more limited coverage. The last compilation listed, by D. E. Kelleher (NIST),[45] to be published in 1995, is based on material from earlier NIST (or NBS) compilations and will be made available in electronic form as a diskette for PC users and possibly for computer network users.

Table I. Recent Comprehensive Wavelength Tables

Coverage	Author(s)	Year	No. of lines	Reference(s)
All elements, first five spectra	Reader et al.	1980, 1992	47,000	40, 41
98 elements	Zaidel' et al.	1970	38,000	42
H through Kr ($Z = 1$–36), ultraviolet lines only	Kelly	1987	66,000	43
22 elements, first and second spectra	Striganov and Sventitskii	1968	~30,000	44
H through Ni ($Z = 1$–28)	Kelleher	1995	~30,000	45

Table II contains a listing of wavelength tables for specific elements or parts of elements.[46–57] We include the tables that have resulted from an ongoing collaboration between NIST and the Japan Atomic Energy Research Institute (JAERI), usually with T. Shirai as principal author.[51–57] These tables do not cover neutral atoms and lower ions but start at ionization stages between V and X and are therefore of some interest for the plasma edge region.

5.2. Energy Level Tables

Critical compilations of broad coverage are listed in Table III. The most comprehensive tabulations are three early volumes by C. E. Moore[58] published in 1949, 1952, and 1958 and reprinted in 1971. These compilations cover most of the chemical elements, with the exception of the rare earths. For numerous light elements, these tables are now superseded by more complete and somewhat more

Table II. Wavelength Tables for Specific Elements That Are of Fusion Interest

Element	Stage of ionization	Author(s)	Year	Reference
H	I	Moore (Gallagher, ed.)	1993	46
B	I	Odintzova and Striganov	1979	47
C, N, O	All	Moore (Gallagher, ed.)	1993	46
Mg	I–XII	Kaufman and Martin	1991	48
Al	I–XII	Kaufman and Martin	1991	49
Si	I–IV	Moore	1965/1967	50
Ti	V–XXII	Mori et al.	1986	51
V	V–XXIII	Shirai et al.	1992	52
Cr	V–XXIV	Shirai et al.	1993	53
Fe	VIII–XXVI	Shirai et al.	1990	54
Ni	IX–XXVIII	Shirai et al.	1987	55
Cu	X–XXIX	Shirai et al.	1991	56
Mo	VI–XLII	Shirai et al.	1987	57

Table III. Comprehensive Compilations of Atomic Energy Levels That Include Elements of Interest to Magnetic Fusion Research

Coverage	Author(s)	Year	Reference
H, C, N, O	Moore (Gallagher, ed.)	1993	46
Most elements	Moore	1949–1958	58
K through Ni ($Z = 19$–28)	Sugar and Corliss	1985	59
H through Ni	Kelleher	1995	45

accurate compilations which usually have been carried out for one specific element, and these are listed in Table IV. However, for a fairly large number of heavy elements beyond atomic number 30, this general compilation still is the only one.

The last compilation listed in Table III, the one by Kelleher[45] published in 1995, will be available on diskette for PC users as well as for users of computer networks.

5.3. Transition Probability Tables

In Table V all critical compilations that provide broad coverage of atomic transition probabilities are listed. The second compilation listed, covering the elements carbon, nitrogen, and oxygen through all stages of ionization (Ref. 14), is in press (*Journal of Physical and Chemical Reference Data*, Monograph Series). A diskette version for PC users is also planned.

Table IV. Tables of Atomic Energy Levels for Specific Elements of Interest to Fusion Research

Element	Stage of ionization	Author(s)	Year	Reference
He	I	Martin	1973, 1987	60
B	I	Odintzova and Striganov	1979	47
Mg	I–XII	Martin and Zalubas	1980	61
Al	I–XIII	Martin and Zalubas	1979	62
Si	I–XIV	Martin and Zalubas	1983	63
Ti	V–XXII	Mori *et al.*	1986	51
V	VI–XXIII	Shirai *et al.*	1992	52
Cr	V–XXIV	Shirai *et al.*	1993	53
Fe	VIII–XXVI	Shirai *et al.*	1990	54
Ni	IX–XXVIII	Shirai *et al.*	1987	55
Cu	I–XXIX	Sugar and Musgrove	1990	64
Kr	I–XXXVI	Sugar and Musgrove	1991	65
Mo	I–XLII	Sugar and Musgrove	1988	66

Table V. Comprehensive Critical Compilation of Atomic Transition Probabilities

Coverage	Author(s)	Year	No. of lines	Reference
Most elements (selected stages of ionization)	Fuhr and Wiese	1990	8,300	67
C, N, O	Wiese et al.	1995	~13,000	14
H through Ne ($Z = 1–10$)	Wiese et al.	1966	4,000	68
Na through Ca ($Z = 11–20$)	Wiese et al.	1969	5,000	69
Sc through Mn ($Z = 21–25$)	Martin et al.	1988	8,800	70
Fe through Ni ($Z = 26–28$)	Fuhr et al.	1988	9,500	71
H through Ni ($Z = 1–28$)	Kelleher	1995	~30,000	45

5.4. Comprehensive Recent Determinations of Wavelength and Transition Probability Data of Interest for Fusion Edge Plasmas

Table VI contains a list of very recent large-scale determinations of spectroscopic data for elements of magnetic fusion interest that were not included in any of the preceding data tables. These recent references thus supplement the critical tabulations and will most likely constitute key parts of future tables of spectroscopic data. These references contain hundreds, and sometimes thousands, of new values.

5.5. Spectroscopic Data for Molecules

A comprehensive review of the status of molecular spectroscopic tables and databases was written by Parkinson[77] in 1992. This review is mainly addressed to the interests and needs of the astrophysical community, but, fortunately, significant overlap with molecules of interest in fusion edge plasmas exists. Another, even

Table VI. Recent Large Data Tables for Spectra of Relevance to Fusion Edge Plasmas

Element and spectrum	Type(s) of data[a]	Author(s)	Year	Reference
Ti II	A	Savanov et al.	1990	72
Fe I	λ, E, A	O'Brian et al.	1991	31
Fe I	λ, A	Nave et al.	1994	73
Fe I–IV	A	Sawey and Berrington	1992	74
Fe II	A	Nahar and Pradhan	1994	75
Fe III	λ, E, A	Ekberg	1993	76
Nb I	λ, A	Duquette et al.	1986	33
Mo I	λ, E, A	Whaling and Brault	1988	34
Ta I	λ, A	Den Hartog et al.	1987	30
W I	λ, A	Den Hartog et al.	1987	30

[a]Notation: λ, wavelengths; E, energy levels; A, transition probabilities.

more up-to-date review of molecular data bases has been prepared by Jorgensen,[78] again mostly with respect to astrophysical needs and interests.

6. SUMMARY

In Table VII the major literature references on critical spectroscopic data compilations for elements of relevance to fusion edge plasmas are summarized. The most recent compilations as well as the important new original pieces of work are cited with respect to wavelength, energy level, and transition probability data. The arrangement of the chemical elements is in the usual order of increasing atomic number Z.

The quality of the wavelength and energy level data in the cited tables is uniformly excellent, with uncertainties rarely exceeding 1 part in 10^4, and with typical data being accurate to about 1 part in 10^6. However, as noted earlier, the quality of the transition probability data is still low, and only the data for hydrogen,

Table VII. Summary Bibliography: Most Recent Comprehensive Sources of Spectroscopic Data for Elements of Fusion Edge Plasma Interest

Atomic number (Z)	Element	Reference(s)		
		Wavelengths	Energy levels	Transition probabilities
1	H, D	46	46	45, 68
2	He	41, 43	60	45, 68
3	Li	41, 43	45	45, 68
4	Be	41, 43	45, 58	45, 68
5	B	47	45, 47, 58	45, 68
6	C	46	46	14, 45
8	O	46	46	14, 45
12	Mg	48	61	45, 69
13	Al	49	62	45, 69
14	Si	50	63	45, 69
22	Ti	41, 43, 51	51, 59	45, 70
23	V	41, 43, 52	52, 59	45, 70
24	Cr	41, 43, 53	53, 59	45, 70
26	Fe	54, 73	54, 59, 76	31, 45, 71, 73–75
28	Ni	41, 43, 55	55, 59	45, 71
29	Cu	56	64	67
31	Ga	41, 43	58	—
36	Kr	41, 43	65	67
41	Nb	33, 41	58	33
42	Mo	34, 41, 57	66	34
54	Xe	41	58	67
73	Ta	30, 41	58	30
74	W	30, 41	58	30

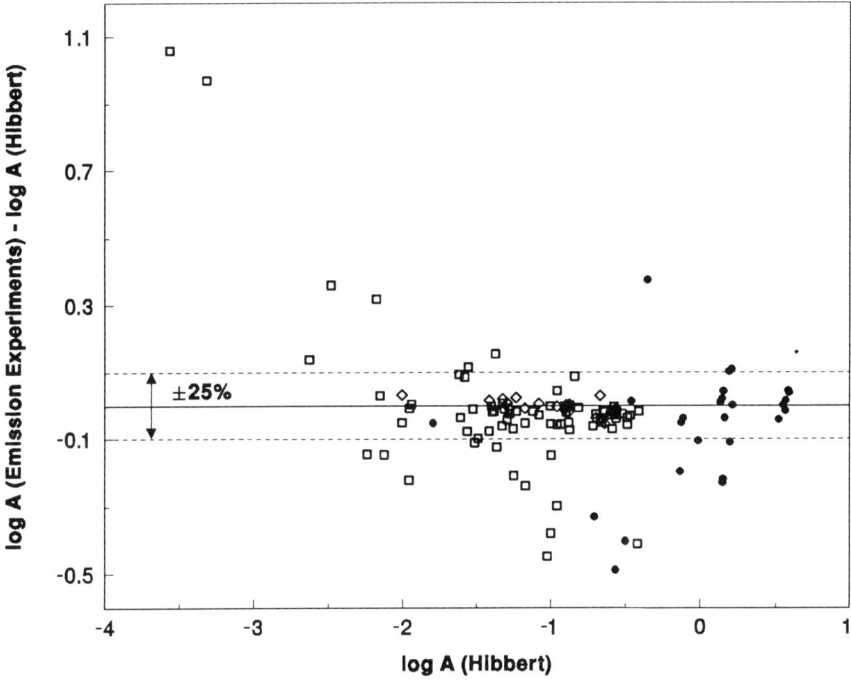

Figure 1. Comparison of atomic transition probability data for N I. Plotted are the *ratios* of the results of recent emission experiments to the results obtained from theory. The stabilized-arc emission experiments are those of Goldbach *et al.*[27] (●), Zhu *et al.*[28a] (◇), and Musielok *et al.*[28b] (□), and the theoretical data are from the configuration interaction calculations of Hibbert *et al.*[17]

helium, and hydrogen- and helium-like ions are essentially exact. For atoms and ions of simple atomic structure, such as the alkalies or the very light elements, the uncertainties for stronger lines are often smaller than ±10% and for some lines smaller than ±3%. However, for light elements with several equivalent electrons, for example, nitrogen and carbon, the uncertainties typically rise into the range of 10–25%. Figure 1 illustrates this situation for N I through a comparison of the best experimental and theoretical data. The majority of the differences between experiment and theory are in the above-cited range, but some larger discrepancies occur, especially for the weaker lines.

For medium-heavy and heavy elements, starting roughly at about $Z = 14$, the uncertainties in the transition probability data increase further and often reach ±50% and more. Thus, the situation for transition probabilities of metallic elements of fusion interest is not adequate. However, some exceptions exist for specific cases such as, for example, Fe I and Mo I, where comprehensive experimental measurements based on combined lifetime–emission work have provided accurate data.

REFERENCES

1. R. K. Janev, in: *Atomic and Plasma-Material Interaction Processes in Controlled Thermonuclear Fusion* (R. K. Janev and H. W. Drawin, eds.), Elsevier, Amsterdam (1993), p. 27.
2. R. K. Janev, in: *Review of Fundamental Processes and Applications of Atoms and Ions* (C. D. Lin, ed.), World Scientific, Singapore (1993), p. 1.
3. H. W. Drawin, in: *Atomic and Plasma-Material Interaction Processes in Controlled Thermonuclear Fusion* (R. K. Janev and H. W. Drawin, eds.), Elsevier, Amsterdam (1993), p. 45.
4. H. R. Griem, *Plasma Spectroscopy*, McGraw-Hill, New York (1964).
5. W. L. Wiese, in: *Methods of Experimental Physics*, Vol. 7B (W. Fite and B. Bederson, eds.), Academic Press, New York (1960), p. 307.
6. P. H. Heckmann and E. Traebert, *Introduction to the Spectroscopy of Atoms*, North-Holland, Amsterdam (1989).
7. J. E. Sansonetti, J. Reader, C. J. Sansonetti, and N. Acquista, *J. Res. Natl. Inst. Stand. Technol.* **97**, 1 (1992).
8. V. Kaufman and B. Edlén, *J. Phys. Chem. Ref. Data* **3**, 825 (1974).
9. M. J. Seaton, *J. Phys. B* **20**, 6363 (1987).
10. C. Mendoza, in *Atomic and Molecular Data for Space Astronomy* (P. L. Smith and W. L. Wiese, eds.), *Lecture Notes in Physics*, Vol. 407 Springer-Verlag, Berlin (1992), p. 85.
11. In a series of papers with the subtitle "Atomic Data for Opacity Calculations (ADOC)," partial results have been published by members of the OP team in *J. Phys. B*. The most recent paper is ADOC XXI, by A. Hibbert and M. P. Scott, *J. Phys. B* **27**, 1315 (1994).
12. K. A. Berrington, P. G. Burke, K. Butler, J. J. Seaton, P. J. Storey, K. T. Taylor, and Yu Yan, *J. Phys. B.* **20**, 6379 (1987).
13. Opacity Project Team, *The Opacity Project*, Vol. 1, Institute of Physics Publishers, Bristol, U.K. (1994).
14. W. L. Wiese, J. R. Fuhr, and T. M. Deters, *J. Phys. Chem. Ref. Data, Monograph Series* **7** (1995).
15. A. Hibbert, *Comput. Phys. Commun.* **9**, 141 (1975).
16. A. Hibbert, E. Biemont, M. Godefroid, and N. Vaeck, *Astron. Astrophys. Suppl. Ser.* **99**, 179 (1993).
17. A. Hibbert, E. Biemont, M. Godefroid, and N. Vaeck, *Astron. Astrophys. Suppl. Ser.* **88**, 505 (1991).
18. A. Hibbert, E. Biemont, M. Godefroid, and N. Vaeck, *J. Phys. B* **24**, 3943 (1991).
19. K. L. Bell, C. A. Ramsbottom, and A. Hibbert, *J. Phys. B* **25**, 1735 (1992).
20. K. L. Bell, A. Hibbert, R. P. Stafford, and B. M. McLaughlin, *Phys. Ser.* **50**, 343 (1994).
21. C. F. Fischer, *The Hartree–Fock Method for Atoms*, John Wiley & Sons, New York (1977).
22. C. Froese Fischer and H. P. Saha, *Phys. Scr.* **32**, 181 (1985).
23. W. Eissner, M. Jones, and H. Nussbaumer, *Comput. Phys. Commun.* **8**, 270 (1974).
24. E. Biemont and C. J. Zeippen, *Astron. Astrophys.* **265**, 850 (1992).
25. A. W. Weiss, *Phys. Rev. A* **51**, 1067 (1995).
26. J. E. Lawler, in *Lasers, Spectroscopy, and New Ideas: A Tribute to Arthur L. Schawlow* (W. M. Yen and M. D. Levenson, eds.), Springer-Verlag, New York (1988).
27. C. Goldbach, M. Martin, and G. Nollez, *Astron. Astrophys.* **221**, 155 (1989).
28. (a) Q. Zhu, J. M. Bridges, T. Hahn, and W. L. Wiese, *Phys. Rev. A* **40**, 3721 (1989); (b) J. Musielok, W. L. Wiese, and G. Veres, *Phys. Rev. A* **51**, 3588 (1995).
29. S. Glenzer, H. J. Kunze, J. Musielok, Y.-K. Kim, and W. L. Wiese, *Phys. Rev. A* **49**, 221 (1994).
30. E. A. Den Hartog, D. W. Duquette, and J. E. Lawler, *J. Opt. Soc. Am. B* **4**, 48 (1987).
31. T. R. O'Brian, M. E. Wickliffe, J. E. Lawler, W. Whaling, and J. W. Brault, *J. Opt. Soc. Am. B* **8**, 1185 (1991).
32. W. N. Lennard, W. Whaling, J. M. Scalo, and L. Testerman, *Astrophys. J.* **197**, 517 (1975).

33. D. W. Duquette, E. A. Den Hartog, and J. E. Lawler, *J. Quant. Spectrosc. Radiat. Transfer* **35**, 281 (1986).
34. W. Whaling and J. W. Brault, *Phys. Scr.* **38**, 707 (1988).
35. P. Hannaford and R. M. Lowe, *Opt. Eng.* **22**, 532 (1983).
36. W. Schade, L. Wolejko, and V. Helbig, *Phys. Rev. A* **47**, 2099 (1993).
37. A. Musgrove and R. Zalubas, *Natl. Bur. Stand. (U.S.) Special Publ.* **363**, Supplement 3 (1985) and earlier editions, cited therein.
38. J. R. Fuhr, B. J. Miller, and G. A. Martin, *Natl. Bur. Stand. (U.S.) Special Publ.* **505**, (1978) and Supplement 1 (1980).
39. *International Bulletin on Atomic and Molecular Data for Fusion*, International Atomic Energy Agency, Vienna, Nos. 1–48 (1977–1994).
40. J. Reader, C. H. Corliss, W. L. Wiese, and G. A. Martin, Wavelengths and Transition Probabilities for Atoms and Atomic Ions, *Natl. Stand. Ref. Data Ser., Natl. Bur. Stand. (U.S.)* **68**, U.S. Government Printing Office, Washington, D.C. (1980).
41. J. Reader and C. H. Corliss, in *CRC Handbook of Chemistry and Physics*, 73rd and following editions (D. R. Lide, ed.), CRC Press, Boca Raton, Florida (1992, 1993, 1994).
42. A. N. Zaidel', V. K. Prokof'ev, S. M. Raiskii, V. A. Slavnyi, and E. Ya. Shreider, *Tables of Spectral Lines*, IFI/Plenum Press, New York (1970).
43. R. L. Kelly, *J. Phys. Chem. Ref. Data* **16**, Suppl. 1 (1987).
44. A. R. Striganov and N. S. Sventitskii, *Tables of Spectral Lines of Neutral and Ionized Atoms*, IFI/Plenum Press, New York (1968).
45. D. E. Kelleher, Database for Atomic Spectroscopy, NIST Standard Reference Database 61, National Institute of Standards and Technology (NIST), (1995).
46. C. E. Moore, *Tables of Spectra for Hydrogen, Carbon, Nitrogen and Oxygen Atoms and Ions* (J. W. Gallagher, ed.), CRC Press, Boca Raton, Florida (1993).
47. G. A. Odintzova and A. R. Striganov, *J. Phys. Chem. Ref. Data* **8**, 63 (1979).
48. V. Kaufman and W. C. Martin, *J. Phys. Chem. Ref. Data* **20**, 83 (1991).
49. V. Kaufman and W. C. Martin, *J. Phys. Chem. Ref. Data* **20**, 775 (1991).
50. C. E. Moore, Selected Tables of Atomic Spectra, *Natl. Stand. Ref. Data Ser., Natl. Bur. Stand. (U.S.)* **3**, Sections 1 and 2 (1965, 1967).
51. K. Mori, W. L. Wiese, T. Shirai, Y. Nakai, K. Ozawa, and T. Kato, *At. Data Nucl. Data Tables* **34**, 79 (1986).
52. T. Shirai, T. Nakagaki, J. Sugar, and W. L. Wiese, *J. Phys. Chem. Ref. Data* **21**, 273 (1992).
53. T. Shirai, Y. Nakai, T. Nakagaki, J. Sugar, and W. L. Wiese, *J. Phys. Chem. Ref. Data* **22**, 1279 (1993).
54. T. Shirai, Y. Funatake, K. Mori, J. Sugar, and W. L. Wiese, *J. Phys. Chem. Ref. Data* **19**, 127 (1990).
55. T. Shirai, K. Mori, J. Sugar, W. L. Wiese, Y. Nakai, and K. Ozawa, *At. Data Nucl. Data Tables* **37**, 235 (1987).
56. T. Shirai, T. Nakagaki, Y. Nakai, J. Sugar, K. Ishii, and K. Mori, *J. Phys. Chem. Ref. Data* **20**, 1 (1991).
57. T. Shirai, Y. Nakai, K. Ozawa, K. Ishii, J. Sugar, and K. Mori, *J. Phys. Chem. Ref. Data* **16**, 327 (1987).
58. C. E. Moore, Atomic Energy Levels, *Natl. Bur. Stand. (U.S.), Circ.* **467**, Vol. I (1949); Vol. II (1952); Vol. III (1958); reprinted as *Natl. Stand. Ref. Data Ser., Natl. Bur. Stand. (U.S.)* **35** (1971), U.S. Government Printing Office, Washington, D.C.
59. J. Sugar and C. H. Corliss, Atomic Energy Levels of the Iron-Period Elements: Potassium through Nickel, *J. Phys. Chem. Ref. Data* **14**, Suppl. 2 (1985).
60. W. C. Martin, *J. Phys. Chem. Ref. Data*, **2**, 257 (1973); *Phys. Rev. A* **36**, 3575 (1987).
61. W. C. Martin and R. Zalubas, *J. Phys. Chem. Ref. Data* **9**, 1 (1980).

62. W. C. Martin and R. Zalubas, *J. Phys. Chem. Ref. Data* **8**, 817 (1979).
63. W. C. Martin and R. Zalubas, *J. Phys. Chem. Ref. Data* **12**, 323 (1983).
64. J. Sugar and A. Musgrove, *J. Phys. Chem. Ref. Data* **19**, 527 (1990).
65. J. Sugar and A. Musgrove, *J. Phys. Chem. Ref. Data* **20**, 859 (1991).
66. J. Sugar and A. Musgrove, *J. Phys. Chem. Ref. Data* **17**, 155 (1988).
67. J. R. Fuhr and W. L. Wiese, in *CRC Handbook of Chemistry and Physics*, 71st and following editions (D. R. Lide, ed.), CRC Press, Boca Raton, Florida (1990, 1991, 1992, 1993, 1994).
68. W. L. Wiese, M. W. Smith, and B. M. Glennon, Atomic Transition Probabilities—Hydrogen through Neon, *Natl. Stand. Ref. Data Ser., Natl. Bur. Stand. (U.S.)* **4**, U.S. Government Printing Office, Washington, D.C. (1966).
69. W. L. Wiese, M. W. Smith, and B. M. Miles, Atomic Transition Probabilities—Sodium through Calcium, *Natl. Stand. Ref. Data Ser., Natl. Bur. Stand. (U.S.)* **22**, U.S. Government Printing Office, Washington, D.C. (1969).
70. G. A. Martin, J. R. Fuhr, and W. L. Wiese, Atomic Transition Probabilities—Scandium through Manganese, *J. Phys. Chem. Ref. Data* **17**, Suppl. 3 (1988).
71. J. R. Fuhr, G. A. Martin, and W. L. Wiese, Atomic Transition Probabilities—Iron through Nickel, *J. Phys. Chem. Ref. Data* **17**, Suppl. 4 (1988).
72. L. S. Savanov, J. Huovelin, and I. Tuominen, *Astron. Astrophys. Suppl. Ser.* **86**, 531 (1990).
73. G. Nave, S. E. Johansson, R. C. M. Learner, A. P. Thorne, and J. W. Brault, *Astrophys. J. Suppl. Ser.* **94**, 221 (1994).
74. F. M. J. Sawey and K. A. Berrington, *J. Phys. B* **25**, 1451 (1992).
75. S. N. Nahar and A. K. Pradhan, *J. Phys. B* **26**, 1109 (1993).
76. J. O. Ekberg, *Astron. Astrophys. Suppl. Ser.* **101**, 1 (1993).
77. W. H. Parkinson, in *Atomic and Molecular Data for Space Astronomy*, (P. L. Smith and W. L. Wiese, eds.), *Lecture Notes in Physics*, Vol. 407, Springer-Verlag, Berlin (1992), p. 149.
78. U. G. Jorgensen, in *Astrophysical Applications of Powerful New Atomic Databases* (S. J. Adelman and W. L. Wiese, eds.), Astronomical Society of the Pacific Conference Series, **78** (1995).

Chapter 3

Elastic and Excitation Electron Collisions with Atoms

Sandor Trajmar and Isik Kanik

1. INTRODUCTION

Electron–atom (molecule) collision physics is concerned with the many processes (elastic scattering, momentum transfer, excitation, ionization, dissociation, etc.) that can take place as a result of such collisions. Instrumentation and experimental techniques for such studies have improved significantly in recent years, allowing more accurate electron collision measurements to be carried out. This progress has been motivated by the need for electron collision data for a variety of reasons, ranging from understanding the fundamental nature of the interactions to modeling the behavior of several physical systems (various discharge and laser systems, fusion plasmas) and environments (astrophysical plasmas, ionospheric and auroral processes of planetary atmospheres). At the same time, a great deal of progress has also been made in the theoretical area of collision physics, and the interplay between theory and experiments has further enhanced developments in this field.

The edge plasma contains neutral atoms, molecules, and ions which originate either from atomic collisions and plasma interaction with the reactor walls or from

---

SANDOR TRAJMAR and ISIK KANIK • Jet Propulsion Laboratory, California Institute of Technology, Pasadena, California 91109.

Atomic and Molecular Processes in Fusion Edge Plasmas, edited by R. K. Janev. Plenum Press, New York, 1995.

injection for diagnostic and fueling purposes. Electron collision processes involving atomic, molecular, and ionic species play an important role in the plasma edge for a wide variety of conditions and determine, in part, the edge plasma conditions, that is, plasma properties, parameters, and dynamics.[1] The most important collision processes that take place in the edge plasma are excitation, dissociation, ionization, and recombination. Knowledge of quantitative characteristics of these processes is of critical importance for interpretation of experimental observations, modeling, and diagnostics of edge plasmas. Design of the next generation of fusion devices, particularly design of the impurity control, power, and particle exhaust systems, is essentially determined by collision processes taking place in the plasma edge.

In general, the atomic and molecular data base needs for studies of fusion edge plasmas have been discussed extensively.[1-4] Here we are concerned with the cross section measurement techniques and status of the available cross section data for electron collisions with atoms relevant to plasma edge studies. Differential and integral cross sections for elastic scattering and excitation processes as well as momentum transfer and total electron scattering cross sections will be reviewed. The atomic species considered here are the primary species (H, He), common impurities (C, O), metallic impurities (Be, Al, Ti, Cr, Fe, Ni, Cu, Ga, Mo, Ta, W, V, Zr), and diagnostic species (Li, Ne, Ar, Kr, Xe). Electron collision processes involving molecules and atomic and molecular ions are treated separately elsewhere in this volume. Excitations to the electronic continuum (ionization) are also treated by others in this volume.

2. DEFINITION OF CROSS SECTIONS

The quantity that characterizes a scattering process is the cross section. It represents the time-independent probability for the occurrence of a particular process. The interaction between an electron and an atom is a function of electron velocity, the type of the process, and the scattering angle. Consequently, the cross sections are functions of the same parameters. The cross sections of interest to us are the differential, integral, momentum transfer, and total electron scattering cross sections. We shall take this occasion to define these cross sections.

In experiments where only scattered electrons are detected for a particular process n as a function of scattering polar angles $\Omega(\theta, \phi)$, impact energy E_0, and energy loss E, we must, in general, define the doubly differential cross section $\partial^2 \sigma_n(E_0, E, \Omega)/\partial E\, \partial\Omega$. In conventional electron scattering experiments, several conditions may apply which reduce the complexity of these cross sections. When the incident energy of electrons is far from threshold and resonance regions, the cross section does not change appreciably over the energy distribution of the primary electron beam, and it may, therefore, be well represented by an average doubly differential cross section that is evaluated at the mean energy of the incident energy distribution. Similarly, if the cross section does not change appreciably with angle over the range of the angular resolution of the apparatus, the scattering angle can be taken as the median of the scattering angles. Therefore, E_0 and θ represent nominal values. For discrete excitation processes, this cross section may be integrated over the energy-loss profile to obtain the differential (in angle) cross section $d\sigma_n(E_0, \Omega)/d\Omega$. Finally, since the target atoms and molecules can be considered, under normal conditions, as spherically symmetric or randomly oriented, the cross section will be independent of the azimuthal angle ϕ. The differential cross section (DCS) in beam–beam experiments is then expressed as

$$\mathrm{DCS}_n(E_0, \theta) = \overline{d\sigma_n(E_0, \Omega)/d\Omega} \qquad (1)$$

$\mathrm{DCS}_n(E_0, \theta)$ is in units of area per unit solid angle. The bar over the right-hand side of Eq. (1) indicates integration over the energy-loss profile, averaging over instrumental energy and angular resolutions, and the usual statistical averaging over experimentally indistinguishable channels.

Integration of the DCS over all the scattering angles yields the experimental integral and momentum transfer cross sections,

$$Q_n(E_0) = 2\pi \int_0^\pi \mathrm{DCS}_n(E_0, \theta) \sin\theta\, d\theta \qquad (2)$$

$$Q_0^M(E_0) = 2\pi \int_0^\pi \mathrm{DCS}_n(E_0, \theta)(1 - \cos\theta) \sin\theta\, d\theta \qquad (3)$$

(We use Q and σ for experimental and theoretical results, respectively.)

The total scattering cross section represents the sum of integral cross sections for all energetically accessible processes:

$$Q_T(E_0) = \sum_n Q_n(E_0) \qquad (4)$$

Theoretical calculations yield the complex scattering amplitude, which is related to the differential cross sections as

$$\frac{d\sigma_n(E_0, \Omega)}{d\Omega} = \frac{k_n}{k_0} \mid f_n(E_0, \Omega) \mid^2 \qquad (5)$$

where k_0 and k_n are the initial and final momenta of the free electron.

3. EXPERIMENTAL METHODS

The experimental methods for cross section measurements are only briefly discussed here. More detailed descriptions have been given in Refs. 5 and 6 and other references cited therein.

3.1. Differential Cross Sections

Differential cross section measurements are carried out by crossing the target molecular beam (or static gas cell) with a nearly monoenergetic beam of electrons, usually at 90°, and determining the energy and angular distribution of the scattered electrons. These distributions contain the information on the nature of the electron collision processes, the energy level scheme of the target, and the corresponding cross sections.

One customary way to represent the scattering data is by the energy-loss spectra, which are plots of the scattering intensity as a function of the energy lost by the electron at fixed E_0 and θ values. A typical energy-loss spectrum for He, as an example, is shown in Fig. 1. The location of the features appearing in an energy-loss spectrum determines the energy lost by the electron and corresponds to the energy level scheme of the target. The scattering intensities are related to the corresponding DCS_n. As we can see from the figure, not only the optically allowed ^1P level excitation but the symmetry-forbidden ^1S, the spin-forbidden ^3P, and the both symmetry- and spin-forbidden ^3S excitation occur under electron impact. At high impact energies and low scattering angles, the energy-loss spectra are equivalent to photoabsorption spectra; that is, optical selection rules hold, and the energy-loss spectra can be converted to photoabsorption spectra. At low impact energies (and large scattering angles), optical selection rules do not apply, and optically forbidden processes readily occur (as demonstrated in Fig. 1). One can study any of these processes as a function of impact energy by setting the detector to any of these scattering channels and varying the impact energy.

In order to obtain the absolute differential cross section directly from the measured scattering signal, one has to know the electron flux distribution, the target density distribution, the exact scattering geometry, and the overall response function of the apparatus. A straightforward approach to determine all these parameters can be applied successfully at high energies (>200 eV). However, at low electron energies, this approach is not feasible. A number of methods have been devised and used to derive relative DCS from the measured scattering intensities and then to normalize the DCS to the absolute scale. We briefly outline here only the most commonly used procedures.

The most practical and reliable method of obtaining absolute elastic cross sections is by normalizing the relative data to a secondary standard. In this method, which is known as the relative flow technique, the elastic scattering signals of two targets are compared at each energy and angle where the cross section of one of them (standard gas) is known and the other (test gas) is to be determined.[5,7-9] The natural choice for the standard gas is He since elastic cross sections for it are known accurately over a wide energy and angular range and it is experimentally easy to handle. In such measurements, the absolute electron flux and molecular beam density or distributions need not be known. The electron beam should be unchanged

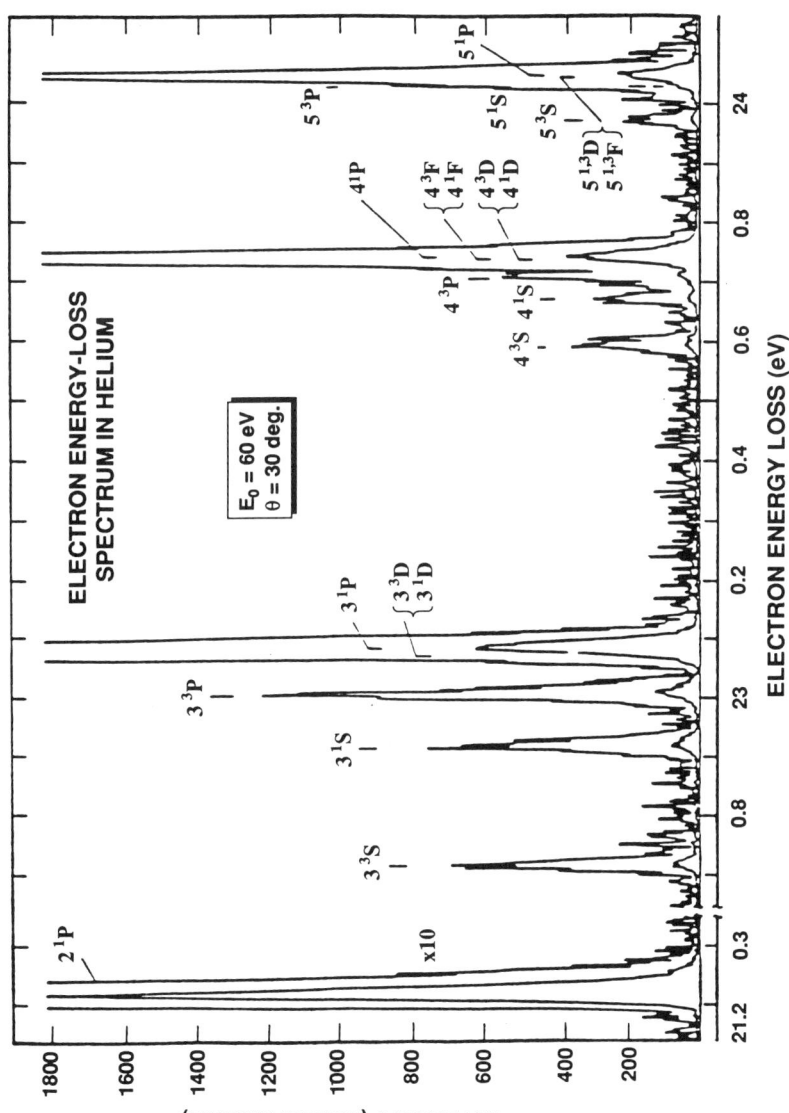

Figure 1. Electron-impact-energy-loss spectrum of He, showing the inelastic scattering region at 60-eV impact energy and 30° scattering angle.

in the two measurements so that the electron flux distributions cancel out. The ratio of the flow rate of the test gas to that of the standard (He) has to be adjusted in such a way that the density distributions of the two gases from the capillary array are identical. The ratio of the flow rates then represents the ratio of the gas densities throughout the scattering volume and leads us to a determination of the ratio of cross sections. No further normalization or fitting is necessary since each absolute differential cross section measurement at a given energy and angle is totally independent of the others. Detailed discussions of this technique have been given by Trajmar and Register,[5] Trajmar and McConkey,[6] and Nickel et al.[9]

Perhaps the most straightforward and efficient way of determining absolute inelastic cross sections is to measure relative inelastic to elastic scattering intensities and achieve normalization through the elastic DCS. The test gas elastic DCS is obtained by utilizing the DCS of He and the relative flow technique as discussed above. Normalization of inelastic to elastic signal in the test gas can be accomplished in two ways. In the first method, the angular distribution of the inelastic process (the relative DCS) is first measured and then, in a subsequent measurement, the inelastic angular distribution is normalized to the elastic cross section at one or two angles. In the second method, the energy-loss spectra, containing both the elastic and the inelastic features, are obtained for each scattering angle at a given impact energy, and the inelastic to elastic DCS calibration is performed at each angle (and energy). In both methods, care must be taken that the elastic and inelastic signals are proportional to the corresponding DCS with the same proportionality factor. In order for elastic and inelastic signals to be proportional to the corresponding DCS, with the same proportionality constant, variation of the detector response throughout the spectrum must be eliminated, or it must be accounted for. A detailed discussion of analyzer response calibration and associated problems can be found in papers by Nickel et al.[9] and Trajmar and McConkey.[6]

3.2. Integral Cross Sections

As discussed earlier, one can obtain the integral and, in the case of elastic scattering, the momentum transfer cross sections by integrating the DCS_n over all scattering angles. Integral cross sections for excitation processes can also be obtained from measurements of the photon emission subsequent to the electron impact excitation. The line emission (or optical excitation) cross sections Q_n^E derived from the measurements can be converted, with appropriate branching ratios, into apparent excitation cross sections Q_n^A, which contain both the direct electron impact excitation and all cascade contributions. The apparent excitation cross sections can be converted to direct electron impact (or level) excitation cross sections (Q_n) if cascade can be properly accounted for. For a more detailed discussion, see Heddle and Gallagher.[10]

3.3. Total Scattering Cross Sections

Total electron scattering cross sections (Q_T) represent the sum of all integral cross sections, $Q_T = Q_0 + Q_E + Q_I$, where Q_0 is the elastic cross section, Q_E is the sum of all excitation cross sections, and Q_I is the ionization cross section. Q_T values are useful for checking the validity of scattering theory, for checking the consistency of available data, for normalizing integral and differential cross sections, and for finding the solution to the Boltzmann equation characterizing various types of plasmas. At low impact energies, the elastic scattering process makes the major contribution to the total scattering cross section. At intermediate and high impact energies, the electronic excitations and ionization become major contributors to the total cross sections.

Basically, two methods are commonly used for measuring Q_T: the transmission method (with or without time of flight) and the target recoil method (for details, see, e.g., Bederson and Kieffer[11] and Trajmar and Register[5]). The total scattering cross sections measured by these techniques are, in general, accurate to within a few percent, and data obtained by various investigators agree with each other within this error limit.

4. REVIEW OF CROSS SECTION DATA

In the plasma edge region, a large number of collision processes involving electrons and atomic (and molecular) species take place. Quantitative information about the collisional properties of the plasma edge constituents is needed for plasma diagnostics and modeling. In this regard, a comprehensive coverage of the present situation regarding differential and integral collision cross sections for elastic and inelastic processes, momentum transfer, and total electron scattering, relevant to plasma edge studies, is given in this section. The two most important elements in the edge plasma are H and He. Fortunately, they are also the ones for which the cross section data base is the most extensive. We are going to give a rather comprehensive review on these two elements.

4.1. Primary Species: H, He

4.1.1. Atomic Hydrogen

The interactions of electrons with hydrogen have been of continuous interest for over half a century because of their role in the behavior of such diverse physical systems as planetary and stellar atmospheres, interstellar clouds, and fusion plasmas. In the plasma edge region, the $H + e^-$ collisions determine the level of hydrogen recycling and energy losses due to radiation. As pointed out in a recent review by King et al.,[12] the available cross section data base is quite limited, and there exist significant discrepancies in the differential and integral cross sections. Experimen-

Table I. Summary of Cross Section Measurements for H

Author(s)	Impact energy (eV)	Angular range (deg)	Type of measurement	Reference
Teubner et al.	50	15–130	$DCS_0(H)/DCS_0(H_2)$	19
Teubner et al.	9.4, 12, 20	15–135	$DCS_0(H)/DCS_0(H_2)$	18
Lloyd et al.	9.4, 12, 20, 30, 50, 100, 200	15–35	$DCS_0(H)/DCS_0(H_2)$	20
Williams	3.4, 8.7	30–150	DCS_0	15
Williams	0.5–8.7	10–150	DCS_0	16
Williams	20–680	10–140	DCS_0	22
Callaway and Williams	12–30	10–140	DCS_0	24
van Wingerden et al.	100, 2000	5–50	DCS_0	21
	100, 200	15–130	DCS_0, Q_0	21
Shyn and Cho	5–30	12–156	$DCS_0 (Q_0, Q^M)$	17
Shyn and Grafe	40–200	12–156	$DCS_0 (Q_0, Q^M)$	23
Williams and Willis	54–680	20–140	$DCS_{(2s+2p)}$	31
Williams	13.87, 16.46, 19.58	15–150	$DCS_{(2s+2p)}$	32
Frost and Weigold	54.4	10–130	DCS_{2s}/DCS_{2p}	138
Williams	54.4	10–140	DCS_{2s}, DCS_{2p}	30
Doering and Vaughan	100	5–120	$DCS_{(2s+2p)}$	33
Lower et al.	100–200	30, 45, 60	$DCS_{(2s+2p)}/DCS_{2s}$	35
Fite and Brackmann	Tr^a–500	—	Q_{2p}^A	51
Fite et al.	Tr–50	—	Q_{2p}^A	39
Stebbings et al.	Tr–600	—	Q_{2s}^A	139
Long et al.	Tr–200	—	Q_{2p}^A	40
McGowan et al.	Tr–200	—	Q_{2p}^A	41
Kauppila et al.	Tr–1000	—	$Q_{2s}^A, Q_{2p}^A, Q_{2s}^A$	43
Williams	Tr–13	—	Q_{2p}^A, Q_{2s}^A	32
Mahan et al.	Tr–500	—	$Q_{3s}^A, Q_{3p}^A, Q_{3d}^A$	47
Neynaber et al.	3.1–12.3	—	Q_T	50
de Heer et al.[b]	0.136–400	—	$Q_0, Q_{exc}, Q_{ion}, Q_T$	13
Shimamura[b]	Tr–10,000	—	$Q_0, Q_{2s}, Q_{2p}, Q_{exc}, Q_{ion}, Q_T$	14

[a] Tr, threshold.
[b] Recommended cross sections.

tal DCS data are limited to elastic scattering and to the $1s \to 2s$ and $1s \to 2p$ excitation processes. An extensive coverage, in energy and angle, is available only for elastic DCS. Inelastic DCS data for the individual $2s$ and $2p$ excitations are available only at 54.4 eV. DCS measurements for the combined $1s \to (2s + 2p)$ process have been carried out from near-threshold to 200-eV impact energies. No DCS data have been reported for excitation of higher manifolds at any energy. Integral and momentum transfer cross sections have been derived from the elastic DCS. For inelastic processes, only integral apparent excitation cross sections are available from optical measurements. Total electron scattering cross sections have been reported only at low impact energies. A summary of cross section measurements is shown in Table I. Based on all available experimental and theoretical information, de Heer et al.[13] and, more recently, Shimamura[14] gave recommended integral and total electron scattering cross sections at impact energies ranging from near threshold to 10 keV. A more detailed discussion is given below.

4.1.1a. Elastic and Momentum Transfer Cross Sections. Measurements of elastic DCS for H atoms at low (below 20 eV) electron impact energies were reported by Williams[15,16] and, more recently, Shyn and Cho.[17] These two measurements are consistent with each other within experimental error limits. Teubner et al.[18] determined the ratios of elastic DCS for atomic and molecular hydrogen at 9.4, 12, and 20 eV. In the intermediate (20 to about 200 eV) energy region, elastic scattering has been extensively studied. There are large deviations (up to 40%) in the DCS data reported by Teubner et al.,[19] by Lloyd et al.,[20] and by van Wingerden et al.[21] DCS measurements of Williams[22] and those of Shyn and Cho[17] are in good agreement at the energies of overlap (20 and 30 eV) up to about 100° scattering angles. Serious discrepancies exist at the higher scattering angles. Later, Shyn and Grafe[23] extended these measurements up to 200 eV (at 40-, 60-, 100-, and 200-eV impact energies). Their data are again in agreement with those of Williams[22] in the forward scattering angles, but substantial discrepancies exist in the backward scattering angles.

Integral elastic and momentum transfer cross sections, obtained via integration of the measured and extrapolated DCSs over all scattering angles, were reported by van Wingerden et al.[21] at impact energies of 100 and 200 eV (elastic only), by Shyn and Cho[17] at impact energies of 5, 7, 15, 20, and 30 eV, and by Shyn and Grafe[23] at impact energies of 40, 60, 100, and 200 eV. We extrapolated and integrated the DCSs measured by Williams,[16,22] Callaway and Williams,[24] and van Wingerden et al.[21] Theoretical elastic scattering results have been reported by Fon et al.[25] (from 1 to 200 eV), Bray et al.[26] (from 13.87 to 200 eV), Bray and Stelbovics[27] (at 35, 54.5, and 100 eV), and Madison and Bubelev[28] (from 15 to 999 eV). We obtained integral elastic cross sections by integrating the DCSs calculated by Scholz et al.[29] (16.51–50 eV). Theoretical momentum transfer cross sections were calculated by Madison and Bubelev[28] (15 to 999 eV), and we obtained these cross sections by integrating the DCS results of Fon et al.[25] (1.22–54.4 eV) and Scholz et al.[29] (16.51–50 eV). De Heer et al.[13] and Shimamura[14] gave a set of recommended integral elastic cross sections based on all available experimental and theoretical data in the 0.136- to 400-eV and 0.00- to 10,000-eV impact energy ranges, respectively. The integral elastic and momentum transfer cross sections are shown and compared in Figs. 2 and 3 together with the lines representing our recommended values (see Table II). The general picture that emerges from these two figures can be briefly summarized as follows. For atomic hydrogen, theoretical methods for calculating integral elastic and momentum transfer cross sections have

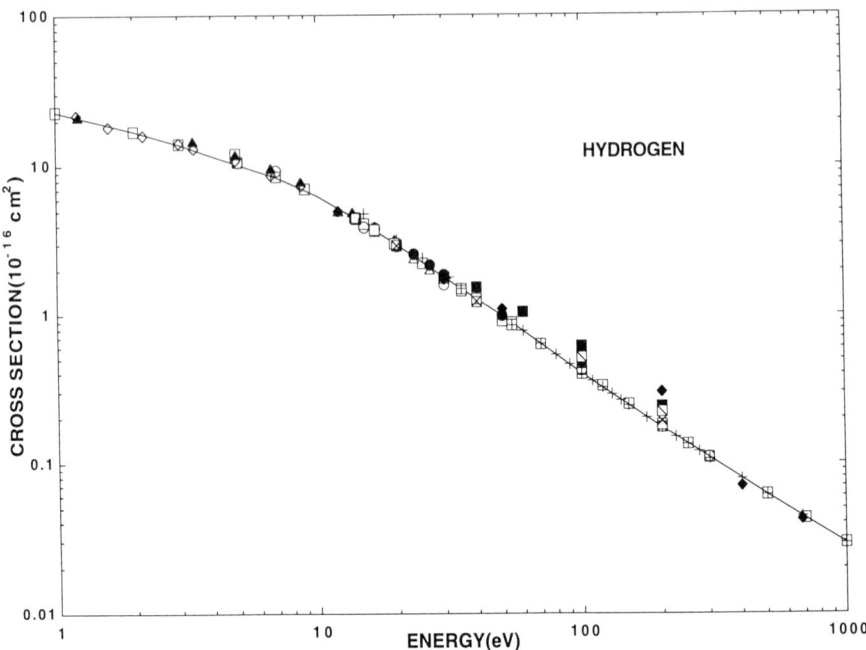

Figure 2. Integral elastic scattering cross sections for H. Experimental results: ◇) Williams[16]; ◆) Williams[22]; △) Callaway and Williams[24]; ⊙) Shyn and Cho[17]; ■) Shyn and Grafe.[23] Theoretical results: ⊠) van Wingerden et al.[21]; ▲) Fon et al.[25]; ○) van Wyngaarden and Walters[34]; ●) Scholz et al.[29]; ⊡) Bray et al.[26]; ⊞) Bray and Stelbovics[27]; +) Madison and Bubelev.[28] Recommended values: ×) de Heer et al.[13]; □) Shimamura[14]; ____) present.

Table II. Summary of Recommended Cross Sections for H (in Units of 10^{-16} cm^2)

E_0 (eV)	Q_0	Q_0^M	Q_E	Q_I	Q_T	Q_{2s}	Q_{2p}	Q_{2s+2p}
1.0	23.0	24.7	—	—	23.0	—	—	—
2.0	17.1	19.5	—	—	17.1	—	—	—
3.0	14.0	15.0	—	—	14.0	—	—	—
5.0	10.2	9.30	—	—	10.2	—	—	—
7.0	8.40	6.50	—	—	8.40	—	—	—
10	6.30	4.10	—	—	6.30	—	—	—
15	4.15	2.27	0.69	—	4.84	0.12	0.42	0.54
20	3.00	1.40	0.81	0.29	4.10	0.090	0.56	0.65
30	1.80	0.68	0.93	0.52	3.25	0.077	0.65	0.73
50	0.98	0.28	0.93	0.68	2.59	0.059	0.68	0.74
100	0.40	0.080	0.73	0.63	1.77	0.036	0.56	0.6
200	0.172	0.024	0.53	0.43	1.13	0.022	0.39	0.41
500	0.061	0.0048	0.28	0.21	0.55	0.010	0.21	0.22
1000	0.029	0.0014	0.17	0.11	0.31	0.005	0.12	0.13

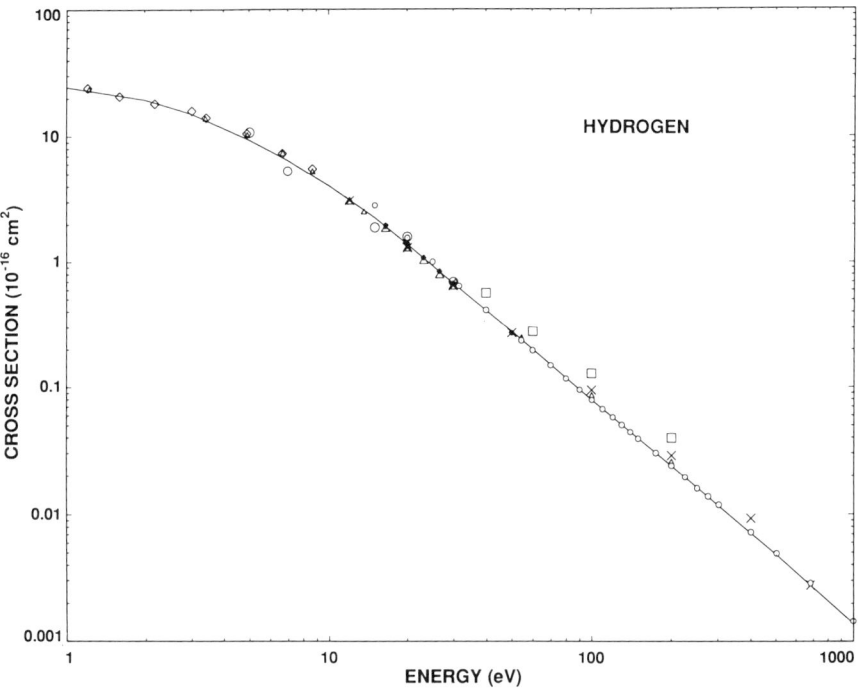

Figure 3. Momentum transfer cross sections for H. Experimental results: ◇) Williams[16]; ×) Williams[22]; △) Callaway and Williams[24]; ○) Shyn and Cho[17]; □) Shyn and Grafe.[23] Theoretical results: ▲) Fon et al.[25]; ◆) Scholz et al.[29]; ⊙) Madison and Bubelev.[28] Recommended values: ____) present.

reached a stage where they are more reliable than experimental approaches. One has to rely on close-coupling methods at near-threshold energies, and these methods now can be extended to the intermediate energy range (convergent close-coupling method[27]). The exact second-order distorted-wave method[28] yields reliable cross sections above about 20 eV. These two approaches and experiment yield consistent results, and we estimate that the integral elastic and momentum transfer cross section values of Table II are accurate to within 20%, in general. Above 100 eV, we relied heavily on the theoretical cross sections. The experimental results would indicate higher values, but they are subject to considerable uncertainty owing to experimental difficulties and extrapolations to near-zero and high scattering angles.

4.1.1b. Inelastic Cross Sections. Experimental DCS data for the individual $1s \rightarrow 2s$ and $1s \rightarrow 2p$ excitations of H are available only at 54.4-eV electron impact energy.[30] The experimental DCS measurements for the $1s \rightarrow (2s + 2p)$ excitation process are those obtained by Williams and Willis[31] at electron impact energies of

54, 100, 136, 200, 300, and 680 eV and by Williams[32] at electron impact energies of 13.87, 16.46 and 19.58 eV. The two measurements at 54.4 and 54 eV are consistent within the stated error limits. Doering and Vaughan[33] reported DCS measurements on the combined $1s \to (2s + 2p)$ excitation at 100 eV. These DCS measurements agree well with those of Williams and Willis[31] at scattering angles below 60°. At larger angles, the agreement between these two measurements gets worse (about 30%). Van Wyngaarden and Walters[34] argued that the values of Williams and Willis[31] are too high in general. Lower et al.[35] reported the ratios of elastic DCS to the $1s \to (2s + 2p)$ excitation DCS at 100 and 200 eV at 30, 45, and 60° scattering angles. These ratios deviate from those calculated from the DCS results of Williams and Willis[31] by a few percent to 20%. DCS measurements for excitation of the $n = 2$ state of H by electron impact, at $E < 40$ eV, and for the $n = 3, 4,$ and 5 states are in progress.[36]

Integral cross sections for the $1s \to 2p$ excitation process are available only from measurements using optical techniques representing apparent excitation cross sections. Little has been done since the review of Callaway and McDowell.[37] (See also the more recent review by Heddle and Gallagher.[10]) The first measurements of $2p$ excitation were carried out by Fite and Brackmann[38] from threshold to 500 eV and by Fite et al.[39] from threshold to 50 eV. Long et al.[40] carried out a similar study from threshold to 200 eV and normalized their data to the Born approximation. McGowan et al.[41] published measurements of these cross sections over the same energy range. All these measurements are in good agreement. Williams[32] reported absolute cross section data for energies between threshold and 13 eV (cascade contributions here can be neglected). From the DCS_{2p} results of Williams,[30] Van Wyngaarden and Walters[34] derived the integral excitation cross section at 54.4 eV. This value (0.78 Å^2) is cascade free, but it is 15% higher than those of Long et al.[40] at 54.4 eV. Since the combined error bars in their measurements were approximately 10%, the results of these experiments would normally be considered to be in agreement. However, this discrepancy is viewed as serious, in part because the measurement of excitation functions of H has fundamental importance for the development of theoretical models, and provoked a sustained debate in the literature. Van Wyngaarden and Walters[34] argued convincingly for the lower values of Long et al.[40] in the intermediate energy region, whereas Whelan et al.[42] and Callaway and McDowell[37] made a strong case for the higher values of Williams.[32] Recent calculations[26,27,29] support the lower values. It is also worth noting the detailed analysis of the Long et al.[40] data carried out by Heddle and Gallagher,[10] which is based on the Bethe approximation. They argued that these experimental data must represent an upper limit of the true cross section.

Integral cross sections for excitation of the $2s$ state are scarce primarily because of the increased experimental difficulties involved in these measurements. There have been only two significant measurements reported in the literature on the integral cross section for the $1s \to 2s$ excitation process. Kauppila et al.[43] obtained apparent excitation cross sections from threshold to 1000 eV, rendering their data absolute by normalizing to the 2p data of Long et al.[40] Williams[32] measured absolute excitation cross sections from the 2s threshold to the n = 3 excitation threshold energy region using electron–photon coincidence techniques. A comparison between these two data sets at overlapping energies of 11.0 and 11.6 eV shows that both experiments agree well at 11.0 eV, but they differ by about 20% at 11.6 eV. At 54.4 eV, we obtained Q_{2s} from the DCS_{2s} values of Williams[30]; this value is about 36% lower than that reported by Kauppila et al.[43]

A large number of calculations have been reported for obtaining the 2s and 2p excitation cross sections. It is not feasible to cite all of them here and to make comparisons among all these results and experiments. We are going to limit our comparisons to only selected recent calculations and make a few general remarks on the status of theory. More detailed summaries of various calculational schemes and comparisons of results have been given by Madison et al.,[44] Bray and Stelbovics,[27] Bray,[45] and Schneider.[46]

The Q_{2p} and σ_{2p} values are shown in Fig. 4. In drawing the curve representing our recommended integral excitation cross sections, we relied on the calculations of van Wyngaarden and Walters,[34] Bray et al.,[26] and Bray and Stelbovics[27] from threshold to 100 eV. Their values are in good agreement with those recommended by Shimamura[14] and slightly lower than the experimental Q_{2p} values of Fite and Brackmann,[38] Fite et al.,[39] Long et al.,[40] McGowan et al.,[41] and Williams.[30,32] The values obtained by Scholz et al.[29] are too low, whereas the results of Madison and Bubelev[28] are too high in this region. Above 100 eV we relied on the results of van Wyngaarden and Walters[34] and the recommendations of Shimamura[14] because of the overall consistency (see Table II and Fig. 4). The results obtained from the second-order distorted-wave method are in good agreement with our recommendations in the 100- to 300-eV range but appear to fall too steeply above this range. It should be noted that the cascade contribution in the apparent excitation cross section is estimated to be a few percent at impact energies higher than 100 eV.[34] Figure 5 summarizes the 2s excitation results. At impact energies below 100 eV, our recommendations are based on the near-threshold measurements of Williams[32] and Kauppila et al.[43] and recent close-coupling calculations. At higher energies we relied on the calculations of van Wyngaarden and Walters[34] and Madison and Bubelev[28] as well as on the recommendations of Shimamura,[14] which are in good agreement with the second-order distorted-wave results. The experimental apparent cross sections of Kauppila et al.[43] have been corrected for cascade contributions by van Wyngaarden and Walters[34] and are in good agreement with our recommended values. In Fig. 6, direct measurements of the combined (2s + 2p) excitation cross sections are shown and compared with cross sections obtained by summing individually measured or calculated 2s and 2p excitation cross sections. Our recommended values correspond closely to those of Shimamura[14] and are in agreement with the recent close-coupling results. The near-threshold cross sections

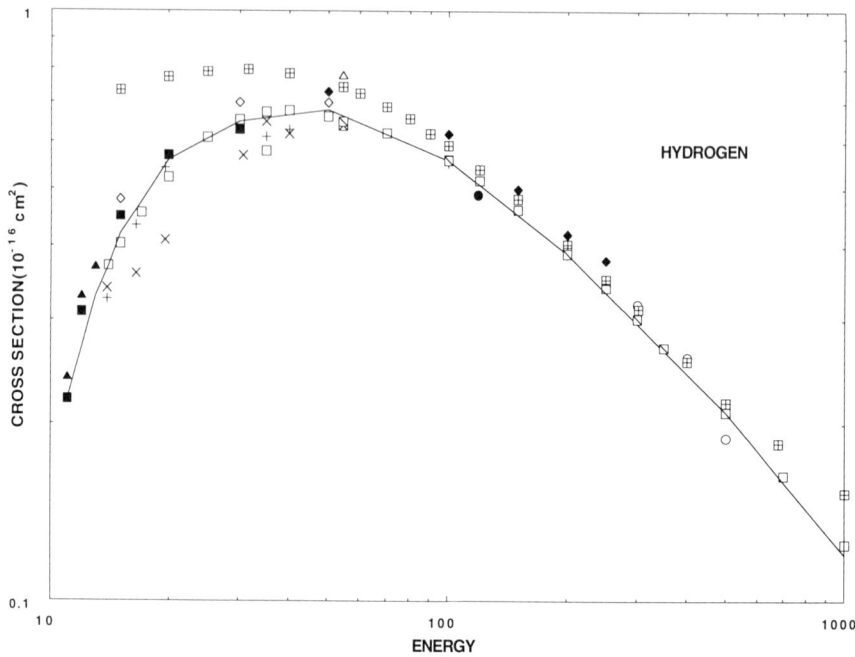

Figure 4. Integral cross sections for excitation of the 2p state in H. Experimental results: ⊙) Fite and Brackman[38]; ■) Fite et al.[39]; ♦) Long et al.[40]; ●) Long et al.[40] (corrected); ◊) McGowan et al.[41]; ▲) Williams[32]; △) Williams.[30] Theoretical results: ⊠) van Wyngaarden and Walters[34]; +) Bray et al.[26]; ×) Scholz et al.[29]; ⊡) Bray and Stelbovics[27]; ⊞) Madison and Bubelev.[28] Recommended values: □) Shimamura[14]; _____) present.

of Williams[32] obtained from optical measurements below 12 eV and by interpreting their DCS results in the 13.87- to 19.59-eV range seem to be somewhat too large. The 54-eV value that we obtained by interpreting the DSC reported by Williams and Willis[31] is in good agreement with our recommended value, whereas a similar procedure applied to the DCS reported by Williams[30] yielded a somewhat larger value. The DCSs for the (2s + 2p) excitation obtained by Williams and Willis[31] at impact energies ranging from 100 to 680 eV at scattering angles ranging from 20° to 140° cannot be reliably integrated because the dominant contributions to the integral cross sections come from angles below 20°. For the second-order distorted-wave approach, the same comment applies as in case of the 2p excitation. Our recommendations for the 2p, 2s, (2s + 2p) and total electronic state excitations are also given in Table II.

The situation for the excitation of $n = 3$ states has remained unchanged since the optical measurements of Mahan et al.[47] In their paper, they reported the individual apparent excitation cross sections for 3s, 3p, and 3d levels with an

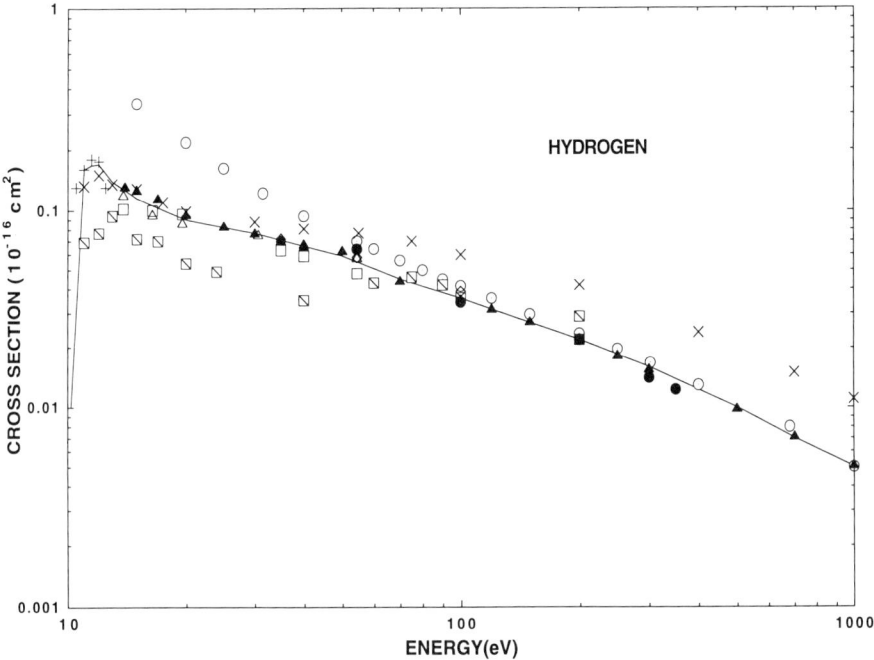

Figure 5. Integral cross sections for excitation of the 2s state in H. Experimental results: ⊠) Stebbings et al.[139]; ×) Kauppila et al.[43]; +) Williams.[32] Theoretical results: ●) van Wyngaarden and Walters[34]; △) Scholz et al.[29]; □) Bray et al.[26]; ◇) Bray and Stelbovics[27]; ⊙) Madison and Bubelev.[28] Recommended values: ▲) Shimamura[14]; ———) present.

assigned error of 10%. The accuracy of these measurements could be further improved, if they were to be repeated with an improved H-atom source.[48] In this regard, electron impact excitation cross section measurements for the $n = 2$ and $n = 3$ states of H employing the Slevin source are being carried out.[49]

Total electron scattering cross sections were measured by Neynaber et al.[50] in the 3.1- to 12.3-eV impact energy range. Theoretical values are available from Bray et al.[26] (13.87–200 eV) and Bray and Stelbovics[27] (35, 54.4, and 100 eV). At impact energies below 10.2 eV, the elastic integral cross sections of Fon et al.[25] (from 1.22 to 8.7 eV) represent total scattering cross sections. Cross section values based on all available information were recommended by de Heer et al.[13] (from 0.136 to 400 eV) and Shimamura[14] (from 0.00 to 10,000 eV). The total electron scattering cross sections are shown in Fig. 7 together with our recommended values. For the purpose of providing an overview, the integral elastic, the total excitation, and the ionization cross sections are also shown. The total excitation cross section data consist of the recommended values given by de Heer et al.,[13] Shimamura,[14] and us. The ionization

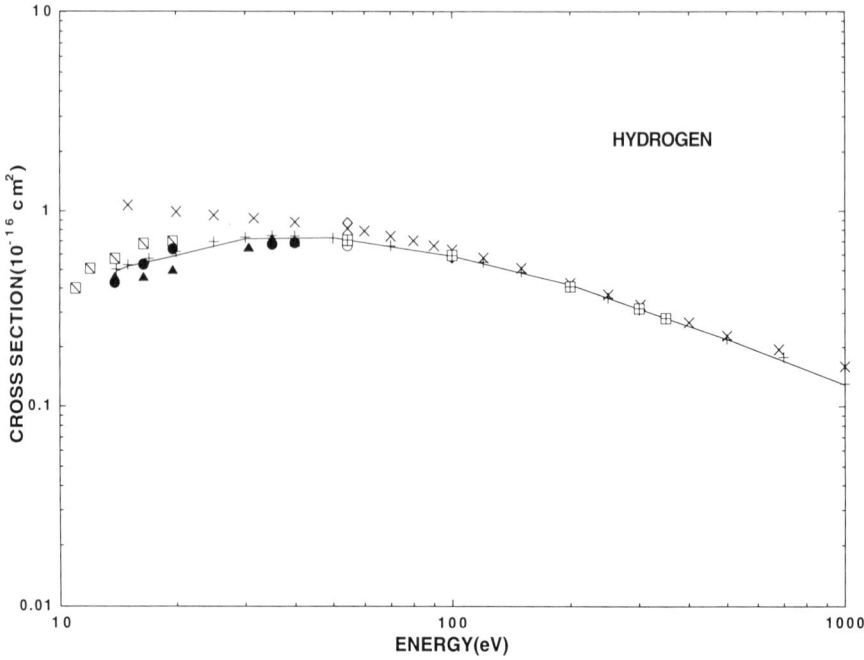

Figure 6. Integral cross sections for excitation of the combined 2s and 2p states in H. Experimental results: ○) Williams and Willis[31]; ⊠) Williams[32]; ◇) Williams[30]; ◆) Doering and Vaughan.[33] Theoretical results: ⊞) van Wyngaarden and Walters[34]; ▲) Scholz et al.[29]; ●) Bray et al.[26]; △) Bray and Stelbovics[27]; ×) Madison and Bubelev.[28] Recommended values: +) Shimamura[14]; _____) present.

cross sections are from Fite and Brackmann[51] from 20 to 100 eV. The recommended values given by de Heer et al.,[13] Shimamura,[14] and us are also shown.

Our recommended values for all these cross sections are listed in Table II. For the total electron scattering cross section, we relied heavily on theoretical results and on consistency in adding up the contributing cross sections. At low energies the close-coupling method of Fon et al.[25] and at intermediate energies the convergent close-coupling method of Bray et al.[26] and Bray and Stelbovics[27] were judged to be most reliable.

4.1.2. Helium

Helium has been the subject of many laboratory and theoretical investigations, partly because it is easy to handle (both experimentally and theoretically) and partly because electron collision processes involving He are important in a large variety of systems and environments (e.g., various discharge and laser systems, planetary and astrophysical environments, and fusion plasmas). Elastic scattering DCSs have been measured and calculated over a wide energy range. Excellent agreement exists

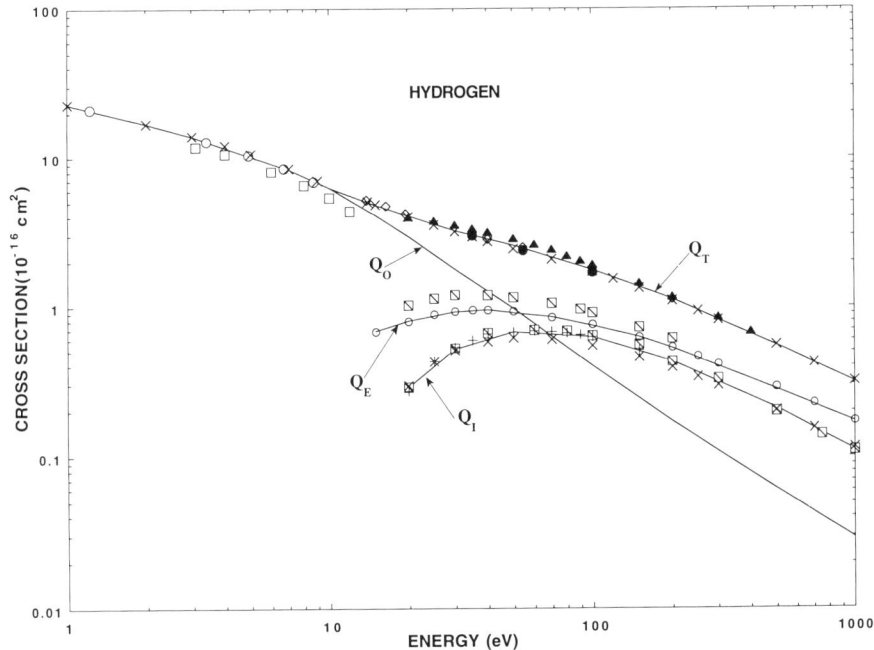

Figure 7. An overview of integral cross sections for H. *Total* scattering cross section (Q_T): □) Neynaber et al.[80]; ⊙) Fon et al.[25]; ◇) Bray et al.[26]; ●) Bray and Stelbovics[27]; ▲) de Heer et al.[13]; ×) Shimamura[14]; ———) present recommended. *Elastic* scattering cross section (Q_0): ———) Present recommended. *Total electronic* state excitation (Q_E): ⊠) de Heer et al.[13]; ○) Shimamura[14]; ———) present recommended. *Ionization* cross section (Q_I): ⊠) Fite and Brackmann[51]; +) de Heer et al.[13]; ×) Shimamura[14]; ———) present recommended.

among the experimental and theoretical results (to within a few percent below the inelastic threshold and to within about 10–30% at higher energies). These are the most reliable DCSs for any atomic species, and they serve as secondary standards for normalizing elastic DCSs for other species. Reliable integral elastic and momentum transfer cross sections were obtained from these DCSs and from swarm measurements. Extensive measurements and calculations have also been carried out for excitation of the $n = 2$ manifold levels. The agreement among the available data is not as good as for elastic scattering, and the typical error limits range from 5 to 50% for the various inelastic processes. Cross section data for excitation to the $n > 2$ manifolds are rather limited. Accurate (to within a few percent) total electron scattering cross sections are available from 0.1 to 2000 eV. It is not possible to list and discuss all the measurements in the present chapter. Fortunately, recent reviews are available. We give references to these review articles and to pertinent recent cross section measurements (see Table III). A summary of experimental and

Table III. Summary of Recent Cross Section Measurements for He

Author(s)	Impact energy (eV)	Angular range (deg)	Notes	Reference
Elastic				
Williams	0.5–20	15–150	Phase shifts	60
Register et al.	5–200	10–140	DCS	61
Shyn	2–400	6–156	DCS, Q, Q^M	62
Golden et al.	2–19	20–120	Phase shifts, Q, Q^M	63
Brunger et al.	1.5–50	10–128	DCS, Q, Q^M	64
Inelastic				
Westerveld et al.	Tr^a–2000	—	$2\,^1P, 3\,^1P, Q^A, Q$	74
van Zyl et al.	50–2000	—	$n\,^1S$ (n = 2, 3, 4, 5, 6), Q^E, Q	77
Phillips and Wong	Tr–24	55, 90	$2\,^3S, 2\,^1S, 2\,^3P, 2\,^1P$, DCS	70
Johnston and Burrow	Tr–20.35	—	$2\,^3S, Q$	140
Shemansky et al.	2.2–2000	—	$n\,^1P$ (n = 2, 3, 4,), Q	75
Brunger et al.	29.6–40.1	2–90	DCS	141
Sakai et al.	200–800	0–12	$2\,^3S$, DCS	73
Allan	Tr–(Tr+4)	20, 60, 90, 120	$m\,^3S, n\,^1S, n\,^3P, n\,^1P$ (m = 2, 3, 4, 5; n = 2, 3), DCS	71
Cartwright et al.	30, 50, 100	10–135	$2\,^1P, (3\,^1P + 3\,^{1,3}D)$, DCS, Q	55
Trajmar et al.	30, 50, 100	10–135	$n\,^3S, n\,^1S, n\,^3P$ (n = 2, 3), DCS, Q	56
Trajmar et al.	25–500	20	$2\,^1P$, DCS	72
Total electron scattering				
Kennerly and Bonham	1–50	—	Q_T	142
Dalba et al.	100–1400	—	Q_T	143
Blaauw et al.	15–75	—	Q_T	144
Charlton et al.	2–50	—	Q_T	145
Dalba et al.	300–2000	—	Q_T	146
Kauppila et al.	30–600	—	Q_T	147
Buckman and Lohmann	0.1–20	—	Q_T	148
Nickel et al.	4–300	—	Q_T	149
Brunger et al.	1.5–50	—	Q_T	64

^{a}Tr, threshold.

theoretical cross section results up to 1977 was given by Bransden and McDowell.[52,53] A compilation of experimental and theoretical integral excitation cross sections was prepared by Kato et al.[54] in graphical form. The latest cross section measurements and other available experimental and theoretical results were presented and compared with each other for the $n\,^1P$ excitation processes by Cartwright et al.[55] and for the $n\,^3S$, $n\,^1S$, and $n\,^3P$ excitation processes by Trajmar et al.[56] A critical review on integral excitation cross section data and a list of preferred data

Elastic and Excitation Electron Collisions with Atoms

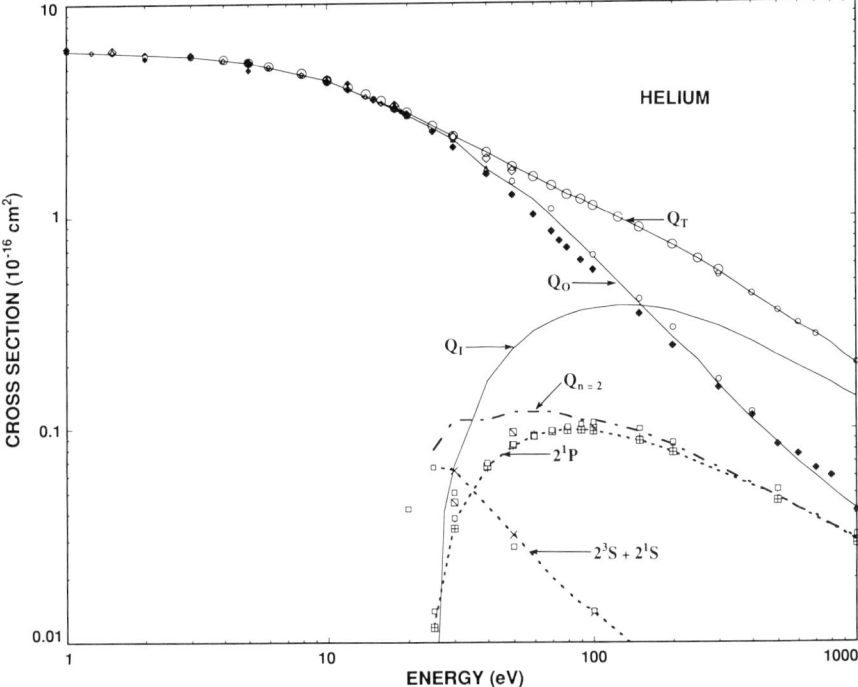

Figure 8. An overview of integral cross sections for He. *Total* scattering cross section (Q_T): ○) Dalba et al.[146]; ◯) Nickel et al.[149]; ◊) Buckman and Lohmann[148]; ◌) Brunger et al.[64]; ———) present recommended. *Elastic* scattering cross section (Q_0): ⊙) Shyn[62]; ♦) Register et al.[61]; ♦) Golden et al.[63]; △) Brunger et al.[64]; ———) present recommended. 2^1P excitation cross section: +) Westerveld et al.[74]; ⊞) Shemansky et al.[75]; ◹) Cartwright et al.[55]; ▫) Hayashi[58]; – – –) present recommended. $2^3S + 2^1S$ excitation cross section: ×) Cartwright et al.[55]; □) recommended, Hayashi[58]; ———) present recommended. $n = 2$ *manifold excitation* cross section: — - —) present recommended. *Ionization* cross section (Q_I): ———) Krishnakumar and Srivastava.[150]

sets based on this review were given recently by de Heer et al.[57] in terms of collision strengths. An exhaustive survey of differential cross section results and a list of recommended sets of cross sections based on all available data and consistency checks are being prepared by Hayashi.[58] In Fig. 8 we present an overview of integral elastic, $n = 2$ manifold excitation, and total electron scattering cross sections (total ionization cross section is also shown for comparison). In Table IV, we list our recommended values for various integral cross sections. A brief summary of the status of the cross section data is given below. Measurements for excitation of high principal quantum number levels are not in general available. A good estimation for these cross sections can be obtained, however, by using sealing laws, as pointed out by Csanak and Cartwright.[59]

Table IV. Recommended Integral Cross Sections for He (in Units of 10^{-16} cm^2)

E_0 (eV)	Q_0	Q_0^M	Q_T	$Q_{2^3S+2^1S}$	Q_{2^1P}	$Q_{n=2}$
1	6.08	6.88	6.08	—	—	—
3	5.70	6.80	5.70	—	—	—
5	5.30	6.20	5.30	—	—	—
10	4.36	4.80	4.36	—	—	—
15	3.55	3.65	3.55	—	—	—
20	3.00	2.70	3.10	—	—	—
25	2.60	2.00	2.68	0.067	0.012	0.08
30	2.30	1.55	2.39	0.063	0.037	0.10
40	1.67	1.00	2.00	0.043	0.067	0.11
50	1.40	0.70	1.72	0.032	0.088	0.12
60	1.20	0.51	1.54	0.025	0.095	0.12
70	1.00	0.40	1.39	0.020	0.097	0.12
80	0.85	0.32	1.27	0.017	0.098	0.115
90	0.74	0.27	1.20	0.015	0.098	0.11
100	0.65	0.22	1.12	0.014	0.098	0.11
150	0.39	0.11	0.88	0.009	0.088	0.097
200	0.27	0.070	0.73	0.006	0.078	0.084
250	0.21	0.050	0.63	0.004	0.072	0.076
300	0.16	0.036	0.55	—	0.064	0.064
400	0.11	0.022	0.43	—	0.056	0.056
500	0.086	0.016	0.36	—	0.048	0.048
600	0.070	0.011	0.31	—	0.042	0.042
700	0.060	0.0088	0.28	—	0.039	0.039
800	0.052	0.0070	0.25	—	0.035	0.035
900	0.046	0.0058	0.22	—	0.033	0.033
1000	0.042	0.0047	0.20	—	0.030	0.030

4.1.2a. Elastic Cross Sections. There have been numerous investigations of differential elastic electron scattering from He. Works prior to 1977 have been summarized by Brandsen and McDowell.[52,53] Later, Williams,[60] Register et al.,[61] Shyn,[62] Golden et al.,[63] and, most recently, Brunger et al.[64] reported elastic DCS measurements. A comparison of this large body of experimental measurements gives us the following picture:

(a) *For electron impact energies below the first inelastic threshold*, the overall agreement among the experimental measurements is very good. Elastic cross sections based on theoretical calculations[65–67] are also in agreement with each other and with experiments. For example, below 20-eV electron impact energy, the agreement between the experimental data of Register et al.[61] and theoretical values of Nesbet[65] is to within 1–3%. One can obtain elastic cross sections with a high degree of confidence from the existing data at any energy and angle below 20 eV.

(b) *For electron energies above the n = 2 inelastic threshold*, there exists some discrepancy among experimental measurements of the elastic cross sections, reaching 30% or more at certain combinations of energy and scattering angle.

Integral elastic and momentum transfer cross sections were obtained from the DCSs and from swarm measurements. Integral elastic cross sections were reported recently by Register et al.,[61] Shyn,[62] Golden et al.,[63] and Brunger et al.[64] The results, in general, agree well with each other. Below the inelastic threshold, where the integral elastic cross sections are identical to total cross sections, the agreement among these data sets is excellent. Register et al.,[61] Shyn,[62] Brunger et al.,[64] Milloy and Crompton,[68] and Ramanan and Freeman[69] reported momentum transfer cross sections. Agreement among these results is good.

4.1.2b. Inelastic Cross Sections. The results for both theoretical and experimental investigations through 1977 were reviewed by Bransden and McDowell.[52,53] More recent cross section measurements are listed in Table III. DCS measurements have been reported for the $2\ ^1P$ and $(3\ ^1P + 3\ ^{1,3}D)$ excitations by Cartwright et al.[55] and for the $n\ ^3S$, $n\ ^1S$, and $n\ ^3P$ excitations ($n = 2, 3$) by Trajmar et al.[56] at intermediate impact energies. These authors also made extensive comparisons with theoretical results and with previous measurements. In the near-threshold region, DCS measurements have been reported by Phillips and Wong[70] for the $n = 2$ manifold levels and by Allan[71] for the $n = 2$ and $n = 3$ manifold levels and for the $4\ ^3S$ and $5\ ^3S$ levels. Trajmar et al.[72] recommended a set of $2\ ^1P$ DCS values at 20° in the 25- to 500-eV impact energy range as secondary standards. Measurements carried out by Sakai et al.[73] at high impact energies revealed a very steep rise in the $2\ ^3S$ excitation cross section near zero scattering angles.

Integral cross sections were obtained from the respective DCSs by Cartwright et al.[55] and Trajmar et al.[56] Westerveld et al.[74] measured optical excitation functions for the $2\ ^1P$ and $3\ ^1P$ levels and also supplied cascade corrections, which allowed us to deduce the direct excitation cross section values. Shemansky et al.[75] measured emission cross sections for the $n\ ^1P$ ($n = 2, 3, 4$) levels at 200 eV and deduced absolute direct excitation cross sections from the measurements. They also applied their analytical and normalization procedures to the optical excitation functions of Donaldson et al.[76] and Westerveld et al.[74] to obtain direct excitation cross sections for the $2\ ^1P$ and $3\ ^1P$ levels from threshold to 2000 eV. Van Zyl et al.[77] measured absolute total emission cross sections for the $n\ ^1S$ ($n = 3, 4, 5, 6$) levels and, by accounting for the proper branching ratios and cascade contributions, deduced absolute direct (or level) excitation cross sections. Their measurements were carried out with great care, and the deduced cross sections can serve as secondary standards.

Fabrikant et al.[78] reviewed the differential and integral cross section data for excitation of the $2\ ^3S$ and $2\ ^1S$ metastable levels.

4.1.2c. Total Electron Scattering Cross Sections. Total electron scattering cross sections for He have been reported over a wide impact energy range. Good agreement is found among the various measurements. Integral elastic scattering cross sections obtained from beam–beam measurements are equivalent to total scattering cross sections below the first inelastic threshold and are consistent with the direct Q_T measurements.

4.2. Common Impurities: C, O

Because of the experimental difficulties of producing atomic carbon, there were no studies of the electron scattering cross sections for this atom prior to 1989. Doering and Dagdigian[79] initiated the the first measurements of the electron energy-loss cross sections for atomic carbon. They reported the relative intensities for excitation of the carbon multiplets at 7.48 eV (165.7 nm) and 7.94 eV (156.1 nm). A summary of their results is included in Table V.

Experimental work on electron impact excitation of atomic oxygen has also been limited because it is somewhat difficult to generate atomic oxygen for laboratory measurements. Total electron scattering cross sections for the 2.3- to 11.6-eV impact energy range were reported by Neynaber et al.,[80] and for the 0.5- to 11.3-eV range by Sunshine et al.[81] The only elastic cross section measurement reported for atomic oxygen was carried out by Dehmel et al.[82] They measured DCSs for the elastic scattering at 5- and 15-eV electron impact energies. Differential and integral excitation cross sections for atomic oxygen were measured by Doering's group[33,83–90] using electron energy-loss spectroscopy. Shyn and Sharp[91] and Shyn et al.[92] measured electron impact differential excitation cross sections of atomic oxygen for the ^3P–^1D and ^3P–^1S transitions, respectively, by employing electron energy-loss spectroscopy. Electron impact-induced optical emission cross sections of various O I lines were reported by several investigators[93–101] (see Table V for summary).

On the theoretical side, there are large numbers of calculations of elastic cross sections for atomic oxygen. Elastic cross section calculations including the polarization effect were performed by Vo Ky Lan et al.,[102] Rountree et al.,[103] Thomas and Nesbet,[104,105] Tambe and Henry,[106,107] and Blaha and Davis.[108] The elastic cross sections, σ_{elas}, reported by Tambe and Henry[107] were found to be in good agreement with the DCSs measured by Dehmel et al.[82] Several electron impact differential excitation and integral excitation cross section calculations for atomic oxygen have also been reported over the past 20 years.[109–112] More recently, Tayal and Henry[113–115] calculated electron impact differential excitation cross sections and integral excitation cross sections using multistate close-coupling theory and compared their results with available experimental data. Serious discrepancies between the theoretical and experimental results exist. Often, various theoretical calculations are not consistent with each other in magnitude and/or shape of the excitation function.

A comprehensive review on cross sections for collisions of electrons with atomic oxygen was given by Itikawa and Ichimura.[116]

Table V. Summary of Electron Impact Cross Section Measurements for C and O

Transition	ΔE (eV)	E_0 (V)	θ (deg)	Author(s)	Reference
Atomic carbon					
$2s^2 2p^2 \, ^3P \to 2s2p3s^3P$		100	2	Doering and Dagdigian	79
$ 2s2p^3 \, ^3D$		100	2	Doering and Dagdigian	79
Atomic oxygen					
$2s^2 2p^4 \, ^3P \to 3s \, ^3S$	9.521	100	5–120	Doering and Vaughan	33
$ 3s \, ^3S$	9.521	16.5–200	5–120	Vaughan and Doering	83
$ 3s \, ^3S$	9.521	16.5–200	5–120	Vaughan and Doering	84
$ 3s' \, ^3D$	12.51	20–200	5–120	Vaughan and Doering	84
$ 3s'' \, ^3P$	14.12	30–200	5–120	Vaughan and Doering	85
$ 2s2p^5 \, ^3P$	15.66	30–200	5–120	Vaughan and Doering	85
$ 4d' \, ^3P$	16.11	30–200	5–120	Vaughan and Doering	85
$ nd \, ^3D^a$	12.07–13.34	30–100	5–120	Vaughan and Doering	85
$ 3p \, ^3P$	10.99	30	2–130	Gulcicek and Doering	86
$ 3p \, ^5P$	10.76	30	25–130	Gulcicek and Doering	86
$ 3p \, ^3P$	10.99	13.9–100	0–135	Gulcicek *et al.*	87
$ 3p \, ^5P$	10.76	13.9–100	0–135	Gulcicek *et al.*	87
$ 2p^4 \, ^1D$	1.97	4–30	10–130	Doering and Gulcicek	88
$ 2p^4 \, ^1S$	4.19	7–30	20–135	Doering and Gulcicek	88
$ 3s \, ^5S$	9.14	13.9–30	20–135	Doering and Gulcicek	89
$ 2p^4 \, ^1D$	1.97	3.5–30	80–134	Doering	90
$ 2p^4 \, ^1D$	1.97	7–30	30–150	Shyn and Sharp	91
$ 2p^4 \, ^1S$	4.19	10–30	30–150	Shyn *et al.*	92
Elastic	0	5, 15	15–150	Dehmel *et al.*	82
Q_T		2.3–11.6		Neynaber *et al.*	80
Q_T		0.5–11.3		Sunshine *et al.*	81
Optical (line emission)[b]					
$3p \, ^5P \to 3s \, ^5S \, (7774)$	1.59	12–17		Germany *et al.*	100
$2s^2 2p^4 \, ^3P \to 3s \, ^3S \, (1304)$	9.521	Tr[c]–100		Wang and McConkey	101
$2s^2 2p^4 \, ^3P \to 3d \, ^3D \, (1027)$	12.07	Tr–100		Wang and McConkey	101
$2s^2 2p^4 \, ^3P \to 3s' \, ^3D \, (989)$	12.54	Tr–100		Wang and McConkey	101
$2s^2 2p^4 \, ^3P \to 3s'' \, ^3P \, (878)$	14.12	Tr–100		Wang and McConkey	101

[a] $n = 3$–7.100. (+) Tr stands for threshold.
[b] For line emission, the wavelength in angstroms is given in parantheses.
[c] Tr, threshold.

4.3. Metallic Impurities: Be, Al, Ti, Cr, Fe, Ni, Cu, Ga, Mo, Ta, W, V, and Zr

Experimental cross section data for metal atoms are scarce. This situation is due partly to experimental difficulties in producing metal vapor beams and partly to problems associated with the normalization of the measurements to the absolute scale. Most of the information is available in the form of electron impact optical excitation functions. Some measurements have been reported for Be, Al, Ti, Cr, Fe, Cu, Ga, Mo, V, and Zr, and this subject was reviewed by Heddle and Gallagher.[10]

Copper is the only metal atom for which elastic scattering and excitation cross sections (nonoptical) have been studied at a few impact energies,[117,118] and the optical excitation function has been reported recently.[119] No cross sections have been reported for Ni, Ta, and W atoms.

4.4. Diagnostic Species: Li, Ne, Ar, Kr, and Xe

Cross sections for electron–Li collision processes, especially differential cross sections, are scarce because of the experimental difficulties in converting the measured intensities to absolute cross sections. Williams et al.[120] reported differential and integral cross sections for elastic scattering and for the excitation of the $2p\,^2P$, $3p\,^2P$, $4p\,^2P$, and $3s\,^2S$ states of Li at electron impact energies of 5.4, 10, 20 and 60 eV. They also reported momentum transfer cross sections at these energies. Shuttleworth et al.[121] measured absolute zero-angle inelastic differential cross sections for Li for electron impact energies from 15 to 190 eV. Later, Vuskovic et al.[122] reported the electron impact differential and integral excitation cross sections of the $2p\,^2P$ state of Li at impact energies of 10, 20, 60, 100, 150, and 200 eV in the angular range 3–120°. Total electron scattering cross sections for Li were measured by Perel et al.[123] and Jaduszliwer et al.[124] in the 1- to 10-eV and 2- to 10-eV impact energy ranges, respectively. Only a few optical excitation function measurements have been carried out for Li. Heddle and Gallagher[10] reviewed this subject.

A large number of experiments have been carried out to measure various differential, integral, and total electron-impact cross sections for rare-gas atoms (Ne, Ar, Kr, Xe). We are going to refer here only to the most recent publications, where references to previous works can be found.

Differential, integral, and momentum transfer cross sections for elastic electron scattering by Ne in the 5- to 100-eV electron impact energy and 10–145° angular ranges were measured by Register and Trajmar.[125] Later, Register et al.[126] reported differential elastic electron scattering cross sections for Xe in the 1- to 100-eV impact energy and 10–146° angular ranges. Wagenaar et al.[127] reported the absolute differential cross sections for elastic scattering of electrons from all rare-gas atoms in the 20- to 200-eV electron impact energy range over small angles (0–10°). More recently, differential elastic scattering cross sections were reported by Weyhreter et al.[128] for Ar, Kr, and Xe in the 0.05- to 2-eV electron impact energy region and by Shi and Burrow[129] for Ne in the 0.25- to 7-eV impact energy region.

Differential excitation cross sections for the lower electronic levels in the near-threshold to 100-eV electron impact energy range were measured by Register et al.[130] for Ne and by Chutjian and Cartwright[131] for Ar. Similar work for Kr was reported by Trajmar et al.[132] and later by Filipovic[133] and more recently by Danjo.[134] In the case of Xe, Filipovic et al.[135] reported inelastic DCSs in the 15- to 80-eV electron impact energy region. In all these measurements, there are large uncertainties associated with the data. Mason and Newell[136] reported integral cross sections for Ne, Ar, Kr, and Xe for excitation to the metastable states. Fabrikant *et al.*[78] reviewed work related to electron impact excitation of metastable levels in the rare gases.

Total electron scattering cross sections for rare gas atoms were summarized by Trajmar and McConkey.[6]

Electron-impact-induced photoemission cross sections for rare-gas atoms in the extreme ultraviolet were reviewed by van der Burgt *et al.*[137] and Heddle and Gallagher.[10]

ACKNOWLEDGMENTS. We wish to express our gratitude to D. H. Madison and V. Bubelev for supplying their theoretical results prior to this publication. The authors' work is, in general, supported by the National Aeronautics and Space Administration.

REFERENCES

1. H. Tawara and R. A. Phaneuf, *Comments At. Mol. Phys.* **21**, 177 (1988).
2. H. Tawara (ed.), Atomic and Molecular Processes in Edge Plasmas Including Hydrocarbon Molecules, Report IPPJ-AM59, Institute of Plasma Physics, Nagoya University, Nagoya (1988).
3. H. W. Drawin, The Application of Atomic and Molecular Physics in Fusion Plasma Diagnostics, Report IPPJ-AM61, Institute of Plasma Physics, Nagoya University, Nagoya (1988).
4. R. K. Janev, M. F. A. Harrison, and H. W. Drawin, *Nucl. Fusion* **29**, 109 (1989).
5. S. Trajmar and D. F. Register, in *Electron Molecule Collisions* (K. Takayanagi and I. Shimamura, eds.), Plenum Press, New York (1984), Chapter VI, pp. 427–493.
6. S. Trajmar and J. W. McConkey, *Advances in Atomic, Molecular and Optical Physics Vol. 33*, (M. Inokuti, ed.), pp. 63–96 (1994).
7. S. Srivastava, A. Chutjian, and S. Trajmar, *J. Chem. Phys.* **63**, 2659 (1975).
8. J. C. Nickel, C. Mott, I. Kanik, and D. C. McCollum, *J. Phys. B* **21**, 1867 (1988).
9. J. C. Nickel, P. Zetner, G. Shen, and S. Trajmar, *J. Phys. E.* **22**, 730 (1989).
10. D. W. O. Heddle and J. W. Gallagher, *Rev. Mod. Phys.* **61**, 221 (1989).
11. B. Bederson and L. J. Kieffer, *Rev. Mod. Phys.* **43**, 601 (1971).
12. G. C. King, S. Trajmar, and J. W. McConkey, *Comments At. Mol. Phys.* **23**, 229 (1989).
13. F. J. de Heer, M. R. C. McDowell, and R. W. Wagenaar, *J. Phys. B* **10**, 1945 (1977).
14. I. Shimamura, *Scientific Papers of the Institute of Physical and Chemical Research (Japan)* **82**, 1 (1989).
15. J. F. Williams, *J. Phys. B* **7**, L56 (1974).
16. J. F. Williams, *J. Phys. B* **8**, 1683 (1975).
17. T. W. Shyn and S. Y. Cho, *Phys. Rev. A* **40**, 1315 (1989).
18. P. J. O. Teubner, C. R. Lloyd, and E. Weigold, *J. Phys. B* **9**, 2552 (1974).
19. P. J. O. Teubner, C. R. Lloyd, and E. Weigold, *J. Phys. B* **6**, L134 (1973).
20. C. R. Lloyd, P. J. O. Teubner, E. Weigold, and B. R. Lewis, *Phys. Rev. A* **10**, 175 (1974).
21. B. van Wingerden, E. Weigold, F. J. de Heer, and K. J. Nygaard, *J. Phys. B* **10**, 1345 (1977).
22. J. F. Williams, *J. Phys. B* **8**, 2191 (1975).
23. T. W. Shyn and A. Grafe, *Phys. Rev. A* **46**, 2949 (1992).
24. J. Callaway and J. F. Williams, *Phys. Rev. A* **12**, 2312 (1975).
25. W. C. Fon, P. G. Burke, and A. E. Kingston, *J. Phys. B* **11**, 521 (1978).

26. I. Bray, D. A. Konovalov, and I. E. McCarthy, *Phys. Rev. A* **44**, 5586 (1991).
27. I. Bray and A. T. Stelbovics, *Phys. Rev. A* **46**, 6995 (1992).
28. D. H. Madison and V. Bubelev, private communication, 1993.
29. T. T. Scholz, H. R. J. Walters, P. G. Burke, and M. P. Scott, *J. Phys. B* **24**, 2097 (1991).
30. J. F. Williams, *J. Phys. B* **14**, 1197 (1981).
31. J. F. Williams and B. A. Willis, *J. Phys. B* **8**, 1641 (1975).
32. J. F. Williams, *J. Phys. B* **9**, 1519 (1976).
33. J. P. Doering and S. O. Vaughan, *J. Geophys. Res.* **91**, 3279 (1986).
34. W. L. van Wyngaarden and H. R. J. Walters, *J. Phys. B* **19**, L53 (1986).
35. J. Lower, I. E. McCarthy, and E. Weigold, *J. Phys. B* **20**, 4571 (1987).
36. T. W. Shyn, private communication, 1994.
37. J. Callaway and M. R. C. McDowell, *Comments At. Mol. Phys.* **31**, 19 (1983).
38. L. W. Fite and R. T. Brackmann, *Phys. Rev.* **112**, 1151 (1958).
39. W. L. Fite, R. F. Stebbings, and R. T. Brackmann, *Phys. Rev.* **116**, 356 (1959).
40. R. L. Long, Jr., D. M. Cox, and S. J. Smith, *J. Res. Natl. Bur. Stand.—A, Phys. Chem.* **72A**, 521 (1968).
41. J. W. McGowan, J. F. Williams, and E. K. Curley, *Phys. Rev.* **180**, 132 (1969).
42. C. T. Whelan, M. R. C. McDowell, and P. W. Edmunds, *J. Phys. B* **20**, 1587 (1987).
43. W. E. Kauppila, W. R. Ott, and W. L. Fite, *Phys. Rev. A* **1**, 1099 (1970).
44. D. H. Madison, I. Bray, and I. E. McCarthy, *J. Phys. B* **24**, 3861 (1991).
45. I. Bray, *Phys. Rev. A* **49**, 1066 (1994).
46. B. I. Schneider, "The role of theory in the evaluation and interpretation of cross section data," in *Advances in Atomic, Molecular and Optical Physics Vol. 33,* (M. Inokuti, ed.), 183–214 (1994).
47. A. H. Mahan, A. Gallagher, and S. J. Smith, *Phys. Rev.* **180**, 132 (1976).
48. J. A. Slevin and W. Sterling, *Rev. Sci. Instrum.* **52**, 1780 (1981).
49. J. M. Ajello, private communication (1994).
50. R. H. Neynaber, L. L. Marino, E. W. Rothe, and S. M. Trujillo, *Phys. Rev.* **124**, 135 (1961).
51. L. W. Fite and R. T. Brackmann, *Phys. Rev.* **112**, 1141 (1958).
52. B. H. Brandsen and M. R. C. McDowell, *Phys. Rep.* **30**, 207 (1977).
53. B. H. Brandsen and M. R. C. McDowell, *Phys. Rep.* **46**, 249 (1978).
54. T. Kato, Y. Itikawa, and K. Sakimoto, Compilation of Excitation Cross Sections for He Atoms by Electron Impact, Research Report ISSN 0915-6364, National Institute for Fusion Science, Nagoya, Japan, March 1992.
55. D. C. Cartwright, G. Csanak, S. Trajmar, and D. F. Register, *Phys. Rev. A* **45**, 1602 (1992).
56. S. Trajmar, D. F. Register, D. C. Cartwright and G. Csanak, *J. Phys. B* **25**, 5233 (1992).
57. F. J. de Heer, R. Hoekstra, A. E. Kingston, and H. P. Summers, in *Atomic and Plasma-Material Interaction Data for Fusion, J. Nucl. Fusion* Suppl., **3**, 19 (1992).
58. M. Hayashi, private communication (1993), to be published as a National Institute of Fusion Science (NIFS) Report.
59. G. Csanak and D. C. Cartwright, *Phys. Rev. A* **34**, 93 (1986).
60. J. F. Williams, *J. Phys. B* **12**, 265 (1979).
61. D. F. Register, S. Trajmar, and S. K. Srivastava, *Phys. Rev. A* **21**, 1134 (1980).
62. T. W. Shyn, *Phys. Rev. A* **22**, 916 (1980).
63. D. E. Golden, J. E. Furst, and M. Maghrefteh, *Phys. Rev. A* **30**, 1247 (1984).
64. M. J. Brunger, S. J. Buckman, L. J. Allen, I. E. McCarthy, and K. Ratnavelu, *J. Phys. B* **25**, 1823 (1992).
65. R. K. Nesbet, *Phys. Rev. A* **20**, 58 (1979).
66. T. Scott and H. S. Taylor, *J. Phys. B* **20**, 3385 (1979).
67. W. C. Fon, K. A. Berrington, and A. Hibbert, *J. Phys. B* **14**, 307 (1981).
68. H. B. Milloy and R. W. Crompton, *Phys. Rev. A* **15**, 1847 (1977).

69. G. Ramanan and G. R. Freeman, *J. Chem. Phys.* **93**, 3120 (1990).
70. J. M. Phillips and S. F. Wong, *Phys. Rev. A* **23**, 3324 (1981).
71. M. Allan, *J. Phys. B* **25**, 1559 (1992).
72. S. Trajmar, J. M. Ratliff, G. Csanak, and D. C. Cartwright, *Z. Phys. D* **22**, 457 (1992).
73. H. Sakai, T. Y. Suzuki, B. S. Min, T. Takayanagi, K. Wakiya, H. Suzuki, S. Ohtani, and H. Takuma, *Phys. Rev. A* **43**, 1656 (1991).
74. W. B. Westerveld, H. G. M. Heideman, and J. van Eck, *J. Phys. B* **12**, 115 (1979).
75. D. E. Shemansky, J. M. Ajello, D. T. Hall, and B. Franklin, *Astrophys. J.* **296**, 774 (1985).
76. F. G. Donaldson, M. A. Hender, and J. W. McConkey, *J. Phys. B* **5**, 1192 (1972).
77. B. van Zyl, G. H. Dunn, G. Chamberlain, and D. W. O. Heddle, *Phys. Rev.* **422**, 1916 (1980).
78. I. I. Fabrikant, O. B. Shpenik, A. V. Snegursky, and A. N. Zavilopulo, *Phys. Rep.* **159**, 1 (1988).
79. J. P. Doering and P. J. Dagdigian, *Chem. Phys. Lett.* **154**, 234 (1989).
80. R. H. Neynaber, L. L. Marino, E. W. Rothe, and S. M. Trujillo, *Phys. Rev.* **123**, 148 (1961).
81. G. Sunshine, B. A. Aubrey, and B. Bederson, *Phys. Rev.* **154**, 1 (1967).
82. R. C. Dehmel, M. A. Fineman, and D. R. Miller, *Phys. Rev. A* **13**, 115 (1976).
83. S. O. Vaughan and J. P. Doering, *J. Geophys. Res.* **91**, 13755 (1986).
84. S. O. Vaughan and J. P. Doering, *J. Geophys. Res.* **92**, 7749 (1987).
85. S. O. Vaughan and J. P. Doering, *J. Geophys. Res.* **93**, 289 (1988).
86. E. E. Gulcicek and J. P. Doering, *J. Geophys. Res.* **92**, 3445 (1987).
87. E. E. Gulcicek, J. P. Doering, and S. O. Vaughan, *J. Geophys. Res.* **93**, 5885 (1988).
88. J. P. Doering and E. E. Gulcicek, *J. Geophys. Res.* **94**, 1541 (1989).
89. J. P. Doering and E. E. Gulcicek, *J. Geophys. Res.* **94**, 2733 (1989).
90. J. P. Doering, *Geophys. Res. Lett.* **19**, 449 (1992).
91. T. W. Shyn and W. E. Sharp, *J. Geophys. Res.* **91**, 1691 (1986).
92. T. W. Shyn, S. Y. Cho, and W. E. Sharp, *J. Geophys. Res.* **91**, 13751 (1986).
93. E. J. Stone and E. C. Zipf, *J. Chem. Phys.* **60**, 4237 (1974).
94. E. C. Zipf, R. W. Melaughlin, and M. R. Gorman, *Planet. Space Sci.* **27**, 719 (1979).
95. E. C. Zipf and W. W. Kao, *EOS Trans. Am. Geophys. Union* **64**, 785, (1983).
96. E. C. Zipf and W. W. Kao, *Chem. Phys. Lett.* **125**, 394 (1986).
97. E. C. Zipf and P. W. Erdman, *J. Geophys. Res.* **90**, 11087 (1985).
98. E. C. Zipf, *J. Phys. B* **19**, 2199 (1986).
99. P. W. Erdman and E. C. Zipf, *J. Chem. Phys.* **87**, 3381 (1987).
100. G. A. Germany, R. J. Anderson, and G. J. Solamo, *J. Chem. Phys.* **89**, 1999 (1988).
101. S. Wang and J. W. McConkey, *J. Phys. B* **25**, 5461 (1992).
102. Vo Ky Lan, N. Feautrier, M. Le Dourneuf, and H. van Regemorter, *J. Phys. B* **5**, 1506 (1972).
103. S. P. Rountree, E. R. Smith, and R. J. W. Henry, *J. Phys. B* **7**, L167 (1974).
104. L. D. Thomas and R. K. Nesbet, *Phys. Rev. A* **11**, 170 (1975).
105. L. D. Thomas and R. K. Nesbet, *Phys. Rev. A* **12**, 1729 (1975).
106. B. R. Tambe and R. J. W. Henry, *Phys. Rev. A* **13**, 224 (1976).
107. B. R. Tambe and R. J. W. Henry, *Phys. Rev. A* **14**, 512 (1976).
108. M. Blaha and J. Davis, *Phys. Rev. A* **12**, 2319 (1975).
109. S. P. Rountree and R. J. W. Henry, *Phys. Rev. A* **6**, 2106 (1972).
110. S. P. Rountree, *J. Phys. B* **10**, 2719 (1977).
111. E. R. Smith, *Phys. Rev. A* **13**, 65 (1976).
112. P. S. Julienne and J. Davis, *J. Geophys. Res.* **81**, 1397 (1976).
113. S. S. Tayal and R. J. W. Henry, *Phys. Rev. A* **38**, 5945 (1988).
114. S. S. Tayal and R. J. W. Henry, *Phys. Rev. A* **39**, 4531 (1989).
115. S. S. Tayal and R. J. W. Henry, *Phys. Rev. A* **42**, 320 (1990).
116. Y. Itikawa and A. Ichimura, *J. Phys. Chem. Ref. Data* **19**, 637 (1990).
117. S. Trajmar, W. Williams, and S. K. Srivastava, *J. Phys. B* **10**, 3323 (1977).

118. K. F. Scheibner, A. U. Hazi, and R. J. Henry, *Phys. Rev. A* **35**, 4869 (1987).
119. C. Flynn, Z. Wei, and B. Stumpf, *Phys. Rev. A* **48**, 1239 (1993).
120. W. Williams, S. Trajmar, and D. Bozinis, *J. Phys. B* **9**, 1529 (1976).
121. T. Shuttleworth, D. E. Burgess, M. A. Hender, and A. C. H. Smith, *J. Phys. B* **12**, 3967 (1979).
122. L. Vuskovic, S. Trajmar, and D. F. Register, *J. Phys. B* **15**, 2517 (1982).
123. J. Perel, P. Englander, and B. Bederson, *Phys. Rev.* **128**, 1148 (1962).
124. B. T. A. Jaduszliwer, B. Bederson, and T. M. Miller, *Phys. Rev. A* **24**, 1249 (1981).
125. D. F. Register and S. Trajmar, *Phys. Rev. A* **29**, 1785 (1984).
126. D. F. Register, L. Vuskovic, and S. Trajmar, *J. Phys. B* **19**, 1685 (1986).
127. R. W. Wagenaar, A. de Boer, T. Van Tubergen, J. Los, and F. J. de Heer, *J. Phys. B* **19**, 3121 (1986).
128. M. Weyhreter, B. Barzick, A. Mann, and F. Linder, *Z. Phys. D* **7**, 333 (1988).
129. X. Shi and P. D. Burrow, *J. Phys. B* **25**, 4273 (1992).
130. D. F. Register, S. Trajmar, G. Steffensen, and D. C. Cartwright, *Phys. Rev. A* **29**, 1793 (1984).
131. A. Chutjian and D. C. Cartwright, *Phys. Rev. A* **23**, 2178 (1981).
132. S. Trajmar, S. K. Srivastava, H. Tanaka, and H. Nishimura, *Phys. Rev. A* **23**, 2167 (1981).
133. D. Filipovic, V. Pejcev, B. Marinkovic, and L. Vuskovic, *Fizika* **20**, 421 (1988).
134. A. Danjo, *J. Phys. B* **22**, 951 (1989).
135. D. Filipovic, B. Marinkovic, V. Pejcev, and L. Vuskovic, *Phys. Rev. A* **37**, 356 (1988).
136. N. J. Mason and W. R. Newell, *J. Phys. B* **20**, 1357 (1987).
137. P. J. M. van der Burgt, W. B. Westerveld, and J. S. Risley, *J. Phys. Chem. Ref. Data* **18**, 1757 (1989).
138. L. Frost and E. Weigold, *Phys. Rev. Lett.* **45**, 247 (1980).
139. R. F. Stebbings, W. L. Fite, D. G. Hummer, and R. T. Brackmann, *Phys. Rev.* **119**, 1939 (1960).
140. A. R. Johnston and P. D. Burrow, *J. Phys. B* **16**, 613 (1983).
141. M. J. Brunger, I. E. McCarthy, K. Ratnavelu, P. J. O. Teubner, A. M. Weigold, Y. Zhou, and L. J. Allen, *J. Phys. B* **23**, 1325 (1990).
142. R. E. Kennerly and R. A. Bonham, *Phys. Rev. A* **17**, 1844 (1978).
143. G. Dalba, P. Fornasini, I. Lazzizzera, G. Ranieri, and A. Zecca, *J. Phys. B* **12**, 3787 (1979).
144. H. J. Blaauw, R. H. Wagenaar, D. H. Berends, and F. J. de Heer, *J. Phys. B* **13**, 359 (1980).
145. M. Charlton, T. C. Griffith, G. R. Heyland, and T. R. Twomey, *J. Phys. B* **13**, L239 (1980).
146. G. Dalba, P. Formasini, R. Grizenti, I. Lazzizzera, G. Ranieri, and A. Zecca, *Rev. Sci. Instrum.* **52**, 979 (1981).
147. W. E. Kauppila, T. S. Stein, J. H. Smart, M. S. Dababneh, Y. K. Ho, J. P. Downing, and V. Pol, *Phys. Rev. A* **24**, 725 (1981).
148. S. J. Buckman and B. Lohmann, *J. Phys. B* **19**, 2547 (1986).
149. J. C. Nickel, K. Imre, D. F. Register, and S. Trajmar, *J. Phys. B* **18**, 125 (1985).
150. E. Krishnakumar and S. K. Srivastava, *J. Phys. B* **21**, 1055 (1988).

Chapter 4

Electron Impact Ionization of Plasma Edge Atoms

T. D. Märk

1. INTRODUCTION

Plasma consists, in general, of electrons, ions, and neutral particles. The most efficient and dominant process for producing these ions is electron impact ionization.[1] Besides plasma physics, many (fundamental and applied) areas of science, such as gas laser physics,[2] quantitative mass spectrometry,[3] and radiobiology,[4] require accurate data sets on the electron impact ionization cross sections σ as a function of incident electron energy E in order to allow an understanding and description of the basic phenomena in the ionized media.

In order to carry out these calculations, cross section functions must be known for these inelastic ionizing electron collisions, including total and partial as well as differential cross sections. In this chapter, available experimental data and theoretical formulas for partial and total electron impact ionization cross sections for atoms present in and of consequence for the plasma edge in magnetic fusion will be presented and discussed critically. Differential cross sections are discussed in Chapter 10 and will therefore not be considered here. The present chapter will start with (i) a short introduction to electron impact ionization mechanisms and processes and (ii) a definition of the concept of total and partial ionization cross sections.

T. D. MÄRK • Institut für Ionenphysik, Leopold Franzens Universität, A-6020 Innsbruck, Austria.

Atomic and Molecular Processes in Fusion Edge Plasmas, edited by R. K. Janev. Plenum Press, New York, 1995.

2. ELECTRON IMPACT IONIZATION: MECHANISMS AND DEFINITIONS

Electrons accelerated through a potential of several tens of volts have a de Broglie wavelength of ~0.1 nm. This wavelength and the atomic dimensions are similar, and mutual quantum effects (distortions) occur; that is, an electron may be promoted from a lower to a higher orbital (excitation) or—if the electron energy is greater than a critical value (the ionization energy or appearance energy)—an electron may be ejected from the target, thus producing a positive ion (cation).

Conversely, direct attachment of the incident electron to an atomic target to give a stable anion is less likely. The reason for this is that the translational energy of the attaching electron and the binding energy (electron affinity) must be taken up (accommodated) in the emerging product (anion). Usually, the excess energy leads to shake-off (detachment) of the electron.

As the electron energy is increased, the variety and abundance of the cations produced from a specific atomic target will increase, because the electron ionization process may proceed via different reaction channels, each of which gives rise to characteristic ionized products. These include the following types of ions: singly charged ions, multiply charged ions, and excited (metastable) ions. For the simple case of an atomic target A, possible reaction channels are:

$$A + e^- \to \begin{cases} A^+ + e_s^- + e_e^- & \text{(single ionization)} \quad (1) \\ A^{2+} + e_s^- + 2e_e^- & \text{(double ionization)} \quad (2) \\ A^{z+} + e_s^- + z \cdot e_e^- & \text{(multiple ionization)} \quad (3) \\ A^{K+} + e_s^- + e_e^- & \text{(K-shell ionization)} \quad (4) \\ A^{**} + e_s^- \to A^+ + e_s^- + e_e^- & \text{(autoionization)} \quad (5) \\ A^{+*} + e_s^- + e_e^- & \text{(ionization plus excitation)} \quad (6) \end{cases}$$

where e_S is a "scattered" electron, and e_e is an "ejected" electron. Other reaction channels may be operative, especially when one is using complex polyatomic targets, that is, molecules[5,6] or clusters.[7–10]

Most of the ionization reactions summarized above [e.g., processes (1) through (4)] can be classified as direct ionization events, in which the ejected and the scattered electron leave the ion within 10^{-16} s. Conversely, there exist alternative ionization channels (competing with direct ionization) in which the electrons are ejected one after the other. For instance, the autoionization process given by Eq. (5) can be described as a two-step reaction. First, a neutral atom is raised to a superexcited state, which can exist for some finite time. Then, radiationless transition into the continuum occurs. Autoionization is a resonance process, and this will complicate the respective ionization cross section dependence (i) at low electron energy (e.g., deviation from the threshold law[5]) but also (ii) at higher electron energies (e.g., see the partial ionization cross section function for the production of Ar^+ shown in Fig. 1). The top curve in Fig. 1 shows the variation of the cross section

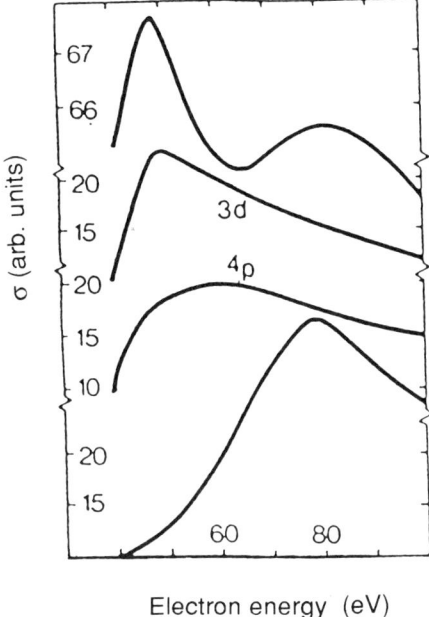

Figure 1. Partial ionization cross section function for $Ar + e^- \rightarrow Ar^+ + 2e^-$ after Crowe et al.[11]. The middle curves illustrate the variation of the strengths of the $3d$ and $4p$ autoionization features (see text). The bottom curve shows the efficiency of production of Ar^+ via direct ionization processes, whereas the top curve is the sum of the two middle curves and the bottom curve.

function obtained by summing over all possible ionization mechanisms. The two curves in the middle illustrate the variation of the strengths of the $3d$ and $4p$ autoionization processes (e.g., $Ar + e^- \rightarrow Ar^*(3s3p^63d) + e^- \rightarrow Ar^+(3s^23p^5) + 2e^-$). The bottom curve shows the behavior of the direct ionization mechanism (Eq. 1).

Quite similarly, multiply charged ions can be formed in a two-step autoionization process. First, a singly charged ion is produced by the ejection of an electron from an inner shell [inner-shell ionization process (Eq. 4)]. This singly ionized atom may then react to form a multiply charged ion by a series of radiationless transitions (Auger effect).

Because of ambiguities in the nomenclature in the literature, some definitions will be given here concerning various ionization cross section terms used.

Consider, as shown in Fig. 2, a parallel, homogeneous, and monoenergetic beam of electrons crossing a semi-infinite medium containing n target particles per cubic meter at rest. If $n(0)_e$ represents the incident electron current, that is, the number of electrons per unit time, the unscattered electron current at depth x (taking into account only the ionizing collisions) is given by the exponential absorption law

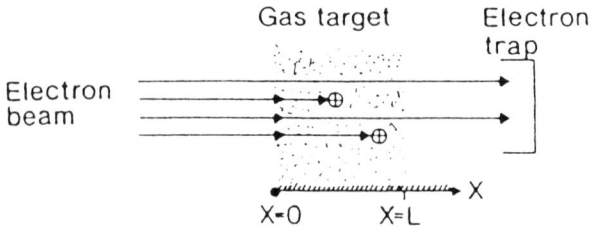

Figure 2. Schematic view of an electron impact ionization experiment.

$$n(x)_e = n(0)_e \exp(-n\sigma x) \tag{7}$$

If $n\sigma x \ll 1$ (single-collision condition†), the number of ions, $n(L)_i$, generated per second along the collision interaction path $x = L$ (over which the ions are collected and analyzed) is

$$n(L)_i = n(0)_e n\sigma_c L \tag{8}$$

where σ_c is the counting ionization cross section in square meters. The total positive-ion current, i_t, produced in this interaction volume is given by

$$i_t = n(0)_e e n\sigma_t L \tag{9}$$

where σ_t is the total ionization cross section, and e is the elemental charge. In the measurement of σ_t, no information is obtained in general about individual partial cross sections. If the produced ions are analyzed with a mass spectrometer according to their mass-to-charge ratio, m/z, however, the respective individual ion currents, i_{ms}, are given by

$$i_{ms} = n(0)_e e n\sigma_{zi} z L \tag{10}$$

where σ_{zi} is the partial ionization cross section for the production of a specific ion i with charge ze.

Total and counting ionization cross sections of a specific target system are the weighted and the simple sum of the various single and multiple partial cross sections, respectively; that is,

$$\sigma_t = \Sigma \sigma_{zi} \cdot z \quad \text{and} \quad \sigma_c = \Sigma \sigma_{zi} \tag{11}$$

---

†This single-collision condition may be achieved by using a low enough number density n. This is also necessary in order to avoid multiple electron collisions and secondary ion–molecule reactions in the ion source, both of which would lead to erroneous cross section results. If this single-collision condition is fullfilled, Eq. (7) may be linearized, allowing the relationship given in Eq. (8) to be derived.

Electron Impact Ionization of Plasma Edge Atoms

Under certain circumstances, single ionization is dominant (e.g., for atoms at electron energies below the double-ionization threshold and for many molecules in the entire energy regime), and then

$$\sigma_t = \sigma_c \tag{12}$$

Sometimes the macroscopic cross section $s = \sigma_t n$ is used; this represents the total effective cross-sectional area for ionization of all target particles in 1 m^3 of the target medium.

For more information and details on the ionization mechanisms and processes, see, for example, Märk,[5,6,9,12,13] Massey et al.,[14] Field and Franklin,[15] Märk and Dunn,[3] and Illenberger and Momigny.[16]

3. TOTAL ELECTRON IMPACT IONIZATION CROSS SECTIONS OF ATOMS AND MOLECULES

3.1. Experimental Methods and Techniques

According to Eq. (9), the determination of the total ionization cross section σ_t requires the measurement of four quantities, namely, i_t, $n(0)_e$, n, and L. Kieffer and Dunn[17] have extensively discussed the problems encountered when trying to measure accurate experimental σ_t values (see also more recent reviews by de Heer and Inokuti[18] and Märk[19]).

One of the earliest and most widely used experimental methods to determine total ionization cross sections is the condenser plate method of Tate and Smith.[20] This is the method that was used very successfully by Rapp and Englander-Golden[21] to produce their benchmark total ionization cross section functions for the rare gases and several small molecules. In this method a magnetically collimated electron beam is directed through a target gas of known density n (see Fig. 3). All ions that are produced in a well-defined region are completely removed and collected. The main limitation of this method is the absolute measurement of the gas density, a

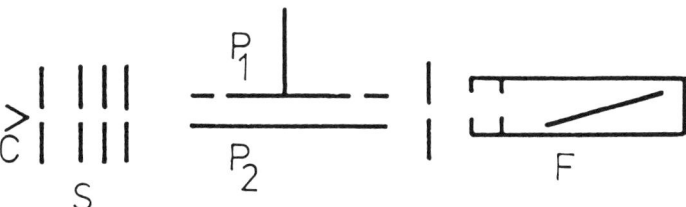

Figure 3. Schematic view of the parallel plate condensor apparatus after Tate and Smith.[20] (C) cathode, (S) collimator slits, (P$_1$ and P$_2$) condensor plates and (F) Faraday cage.

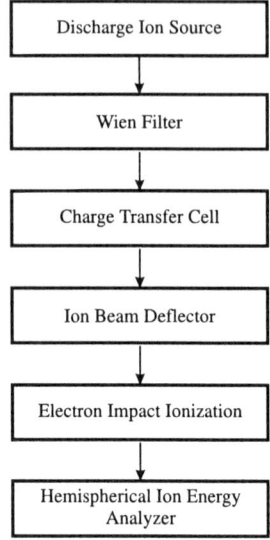

Figure 4. Schematic view of the fast neutral beam apparatus after Freund et al.[22]

difficult matter for many gases, which has been solved by using the capacitance manometer to determine absolute gas pressures. De Heer and Inokuti,[18] in their definitive review on the determination of total electron impact ionization cross sections, discussed and summarized experiments and results up to 1985, including also the Lozier tube, the summation method, gas-filled counters, and various kinds of crossed-beam methods.

Excellent data on total [and partial (see below)] ionization cross sections have been obtained recently by Freund and co-workers[22,23] using a modified crossed-beam method, in which an electron beam is crossed with a fast neutral beam prepared by charge transfer neutralization of a mass-selected ion beam. This approach was first used for ionization cross section measurements by Cook and Peterson,[24] Ziegler et al.,[25] and Smith and co-workers.[26] Extensions by Freund and others[27,28] have made it a powerful method. In the fast neutral beam apparatus of Freund and co-workers, atomic or molecular ions are (i) extracted from a dc discharge, (ii) accelerated to 3 kV, and (iii) mass separated by a Wien filter (see Fig. 4). The ion beam is then neutralized by charge transfer with a gas selected to have an ionization energy resonant with that of the ions. The pressure is adjusted to neutralize several percent of the ions, with the remainder being deflected to a collector. According to Freund, the resulting neutral beam generally has a flux of 10^{10} s^{-1}. Its relative intensity is measured by kinetic electron ejection from a metal surface. For accurate flux measurements, a pyroelectric crystal is used to calibrate the electron ejection coefficient. Ionization is produced by crossing the fast neutral beam with a well-characterized electron beam. The resulting ions are steered and focused with magnetic and electrostatic fields onto a hemispherical energy analyzer. This analyzer separates ions of different charge-to-mass ratios, since all ions retain essentially the same velocity, that of the 3-keV neutral beam. Ions are finally detected by a channel electron multiplier. The biggest advantages of this method are that it permits preparation of a pure beam even of unstable species and that the high beam velocity permits accurate flux measurements.

3.2. Theoretical Considerations

The theoretical treatment of the basic electron impact ionization process (i.e., in the exit channel, a full three-body problem) has received a great deal of attention. Calculations based on quantum-mechanical approximations are difficult; few have been performed, and some of them are not as accurate as necessary.[29] Therefore, other methods have been developed with the goal of obtaining reasonably accurate cross sections. Three different approaches have been used: (i) empirical and semiempirical formulas, (ii) classical theories, and (iii) semiclassical collision theories. Theoretical methods have been reviewed several times in great detail, particularly with regard to the accuracy and reliability of the most widely used formulas.[5,12,13,30,31] Most of these treatments apply only to single ionization. Nevertheless, these formulas are used in general also for the estimation of total ionization cross sections [e.g., see Eq. (12)].

Two very recent theoretical developments (one initiated by Deutsch and Märk[32] and the other by Rudd and Kim[33]) will be presented here in more detail. One of them [the Deutsch–Märk (DM) approach; see details below] allows, in a very general way, the convenient calculation of electron impact (single) ionization cross sections as a function of electron energy for (ground state and excited) atoms. The other one [the binary-encounter-dipole (BED) and binary-encounter-Bethe (BEB) approach; for more detail, see Märk and Rudd[34]] enables the construction of single differential ionization cross sections (and thus, by integration over all secondary electron energies w, also the determination of total ionization cross sections) but, in its more accurate version (BED), requires information on differential oscillator strengths.[33]

Using classical mechanics, Thomson[35] was the first to derive a formula for the electron impact ionization cross section (for single ionization):

$$\sigma = \sum_i 4\pi a_0^2 N_i \left(\frac{R}{B_i}\right)^2 \frac{u-1}{u^2} \tag{13}$$

with a_0 the Bohr radius, N_i the number of electrons in the ith subshell, B_i the ionization energy in the ith subshell, R the ionization energy of H, $u = E/B_i$, and E the energy of the incident electron. This classical treatment has been modified by several authors; however, according to Rudge,[29] none of these formulas represents a substantial improvement over the Thomson formula, because all of them suffer from the same large deviations at high and low energies. A definitive improvement introduced by Gryzinski[36] is the assumption of a continuous velocity distribution for atomic electrons, leading to the following formula for the cross section (integrated over all i shells):

$$\sigma = \sum_i 4\pi a_0^2 N \left(\frac{R}{B_i}\right)^2 f(u) \qquad (14)$$

with

$$f(u) = \frac{1}{u}\left(\frac{u-1}{u+1}\right)^{3/2}\left\{1 + \frac{2}{3}\left(1 - \frac{1}{2u}\right)\ln[2.7 + (u-1)^{1/2}]\right\}$$

Burgess[37] and Vriens[38] have suggested further means of improving the classical theory by incorporating exchange effects. Although all these classical and semiclassical formulas constitute a significant improvement, they still fail to predict the correct magnitude of ionization cross section functions in the case of rather simple atoms such as, for instance, neon, nitrogen, and fluorine (e.g., see examples given by Deutsch et al.[39] and Deutsch and Märk[32]). Moreover, while empirical formulas are helpful in certain cases, many of them are of limited usefulness; that is, they provide good fits to certain classes of known data (see also the approximate analytic formula given by Bell et al.[40]) but will not allow reliable predictions for other unknown systems. Even the widely used Lotz formula[41–44] appears to fail badly in the case of certain atoms (e.g., see examples given in Refs. 32 and 39 and the ionization cross section function of uranium shown in Fig. 5).

Based on a comparison between these classical formulas and the Bethe[45] approximation (which is only accurate at higher energies owing to the use of the Born approximation) in the simplified form of Miller and Platzman,[46]

$$\sigma = \sum_{n,l} \pi a_0^2 \left(\frac{R}{B_{n,l}}\right)^2 N_{n,l} \frac{M_{nl}^2}{u} \ln(4c_{nl}u) \qquad (15)$$

where M_{nl} is a dipole matrix element and c_{nl} is a collisional parameter (to be determined by Fano plot analyses), Deutsch and Märk[32] recently suggested that the Bohr radius in the classical formula (14) be replaced by the radius of the corresponding electron subshell (n, l). This step is in line with (i) a result of Bethe's calculation that the ionization cross section of an atomic electron with quantum numbers (n, l) is approximately proportional to the mean square radius $<r^2>_{nl}$ of the electron shell (n, l)[47] and (ii) the observation of a correlation between the maximum of the atomic cross section and the sum of the mean square radii of all outer electrons.[47,48] Following up this suggestion, Margreiter et al.[49,50] successfully

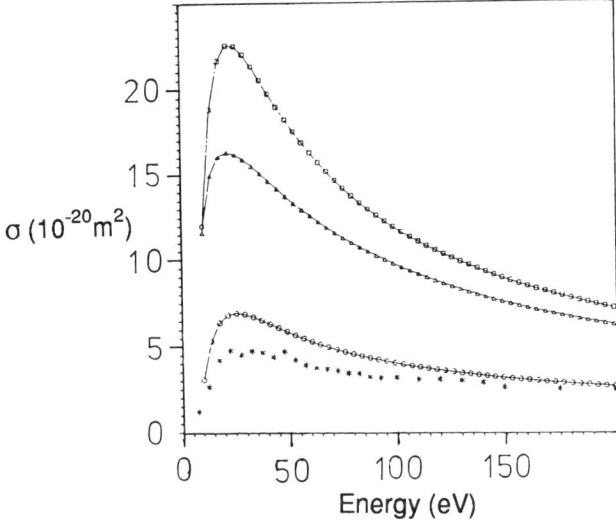

Figure 5. Partial electron impact ionization cross sections function for the reaction $U + e \to U^+ + 2e$:*) experimental results of Halle et al.[51]; ○) DM approach calculations[50]; Δ) calculations using the Lotz formula[41-44]; and □) calculations using the Gryzinski formula.[36]

applied this DM approach to a large number of ground state and excited state atoms using the following expression:

$$\sigma = \sum_{n,l} g_{nl} \pi \, r_{nl}^2 \, N_{nl} f(u) \qquad (16)$$

where r_{nl}^2 is the mean square radius of the shell (n,l) and the g_{nl} are weighting factors. Margreiter et al.[49] determined the necessary generalized weighting factors g_{nl} via a fitting procedure using accurate experimental data for the rare gases and uranium as test cases. It was demonstrated that the formula in Eq. (15) leads, in general, to an improved agreement with the experimental results not only in the case of ground state atoms, but also in the case of excited state atoms.[49] Recently, the same authors[50] derived relationships for the product $g_{nl}B_{nl}$ (given in Table I) that yielded even better overall agreement with the experimental data available for atomic targets[50] (e.g., see Fig. 5) and allowed also the calculation of inner-shell ionization cross sections[52] and outer-shell ionization cross sections.[53]

As already mentioned above, besides this rather general way to calculate total ionization cross sections for atoms, Rudd and Kim[33] recently proposed a method based on the single differential cross section as given in the Mott and binary-encounter formulation. By correcting existing deficiencies, that is, by including the dipole interaction and the quantum-mechanically correct energy dependence as given in

Table I. Weighting Factor Matrix Elements $g_{nl}B_{nl}$ as a Function of Quantum Numbers n and l and the Number of Electrons N_{nl} in the Respective Subshell[a]

	s: N_{nl}		p: N_{nl}						d	f
n	1	2	1	2	3	4	5	6		
1	50	70	—	—	—	—	—	—	—	—
2	12	20	32.5	30	30	30	30	30	—	—
3	14	14	31.5	25	25	25	22	22	13.6	—
4	10	10	31	22.4	22.4	22.4	18.5	17.5	11.2	20 ($N_{nl}=14$)
5	7.5	7.5	30.5	20	20	20	16	13	8.85	1 ($N_{nl}=3$)
6	6	6	30	18	18	18	14.5	7.5	6.5	—
7	5	5	—	—	—	—	—	—	—	—

[a]After Margreiter et al. (Ref. 50).

the Bethe formula, Rudd and Kim developed a single differential cross section formula (see Chapter 10), which can also be successfully applied—after integration over w—to the prediction of total ionization cross sections (e.g., see Figs. 6 and 7). In the simpler version of their model, called the binary-encounter-Bethe (BEB) model, the only input data needed are the values of the binding and kinetic energies and occupation number of each contributing subshell (n, l) and the integrated oscillator strength, whereas in the more accurate version, called the binary-encounter-dipole (BED) model, information on differential oscillator strengths is required.[33,34]

3.3. Consistency Checks

Although theory cannot always predict accurate electron impact ionization cross sections, it can at least identify certain limitations and provide relationships to other collision data and may supply the asymptotic behavior in certain energy regimes.

For instance, in the Bethe approximation the total ionization cross section as given in Eq. (15) represents an asymptotic form of the cross section at high incident energies. This fact can be used for data consistency checks by plotting the cross section multiplied by the reduced energy u versus the logarithm of u. Such a "Fano plot" must approach a straight line at high energies because of the $(1/u) \ln u$ dependence of the leading term in the Bethe equation.[56] If properly normalized, the parameters M_{nl} and c_{nl} in Eq. (15) may be deduced from a least-squares analysis of the Fano plot data points.[18,57] Furthermore, as the quantity M_{nl} is related to the differential dipole optical oscillator strength, df/dt, experimental photoionization results may be used for comparison (e.g., see the results for n-alkanes[58]). This provides an important link between these two types of ionization processes. Another interesting point about the Bethe result is that Eq. (15) is valid for any singly charged particle with $E = m_e v^2/2$ (with m_e the mass of the electron, and v the velocity of the particle). From this follows that, in the high-energy limit, electron impact ionization cross sections and (other) particle (e.g., ion) impact ionization cross sections approach the same values.

Conversely, in the low-energy limit, close to the onset of ionization, the shape of the (partial) electron impact ionization cross section curve is governed by a threshold law, which is usually expressed in the form

$$\sigma_z \sim (E - B_i)^{a \cdot z} \quad (17)$$

with the parameter a close to 1.[5] The accurate shape of the cross section in this region is especially important for the determination (by means of extrapolation techniques) of the respective ionization threshold, a property that can also be derived by other means and therefore constitutes another point of reference.

Finally, another important class of consistency checks derives from the fact that the total ionization cross section is (i) equal to the charge-weighted sum of partial ionization cross sections [see Eq. (11)] and (ii) may be obtained by integrating differential ionization cross sections over all secondary electron energies and angles. The former connection is used in the summation method[18] for calibration purposes, and the latter fact is employed in the frame of Platzman plots.[34] Both relationships provide a basis for checks on the reliability of the absolute magnitude and of the energy dependence of the ionization cross sections under consideration.

3.4. Recommended Total Ionization Cross Sections

As mentioned above, there exist a number of comprehensive reviews on experimental and theoretical cross section determinations including also compilations of recommended electron impact ionization cross section data. It is outside the scope of this chapter to give a detailed account of all available results. Nevertheless, total electron impact ionization cross section functions for some of the important atomic targets in magnetic fusion edge plasmas (e.g., H, He, C, O, N, Ne, Ar, Kr, and Xe) have been assessed and summarized here, and the best data sets available (see also discussion in the figure captions) are presented in graphical form in Figs. 6–12. More details and information on the reliability of these cross sections are available in previous reviews and data compilations; see Refs. 1, 3, 12, 13, 17, 18, 22, 50, and 76–78. Available information about experimental and theoretical cross sections for some of the metallic impurities (i.e., Li, Ti, V, Fe, Ni, Cu, Ga, and Cr) are summarized in two recent reviews by Freund[22] and Margreiter et al.[50] and will not be repeated here. Moreover, cross sections of targets not yet measured may be generated by means of the formula given in Eq. (16); for details, see Margreiter et al.[50]

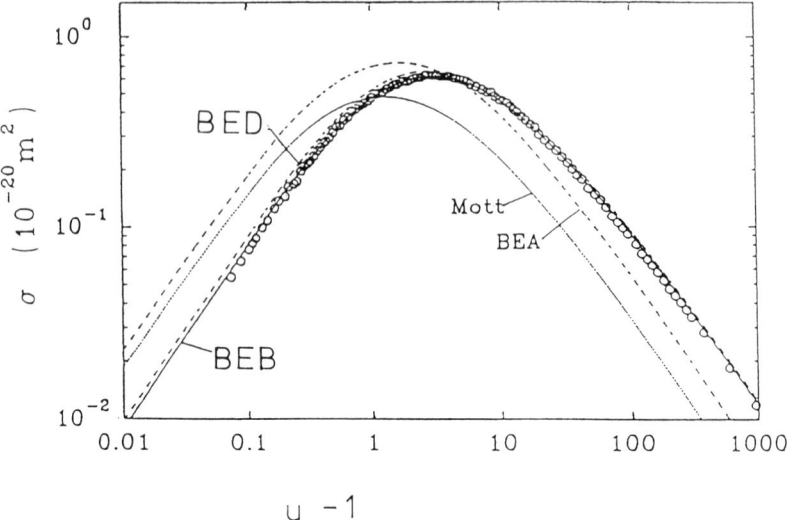

Figure 6. Electron impact ionization cross section function for the reaction $H + e \rightarrow H^+ + 2e$: (○) experimental results of Shah et al.[54] up to 4 keV (above that, the data were generated from the Bethe equation[34]), (—) BEB calculations and dashed line BED calculation[33,34], (···) calculations using the Mott cross section formula and (– – –) (designated BEA) calculations using the BEA formula.[34]

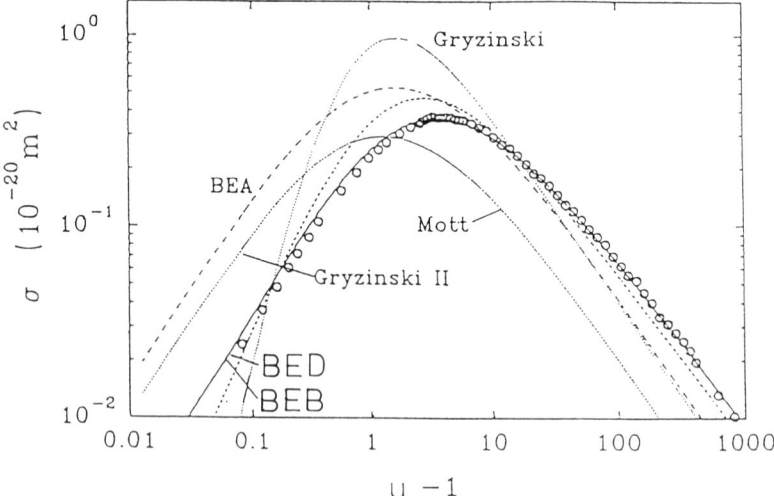

Figure 7. Total electron impact ionization cross section function for helium: (○) experimental data of Shah et al.[55] up to 10 keV (above that, the data were generated[34] from the Bethe equation[34]), (—) BEB and BED calculation[33,34] (both versions give the same result). Also shown for comparison calculated results using the Mott, BEA and Gryzinski formulation.[34]

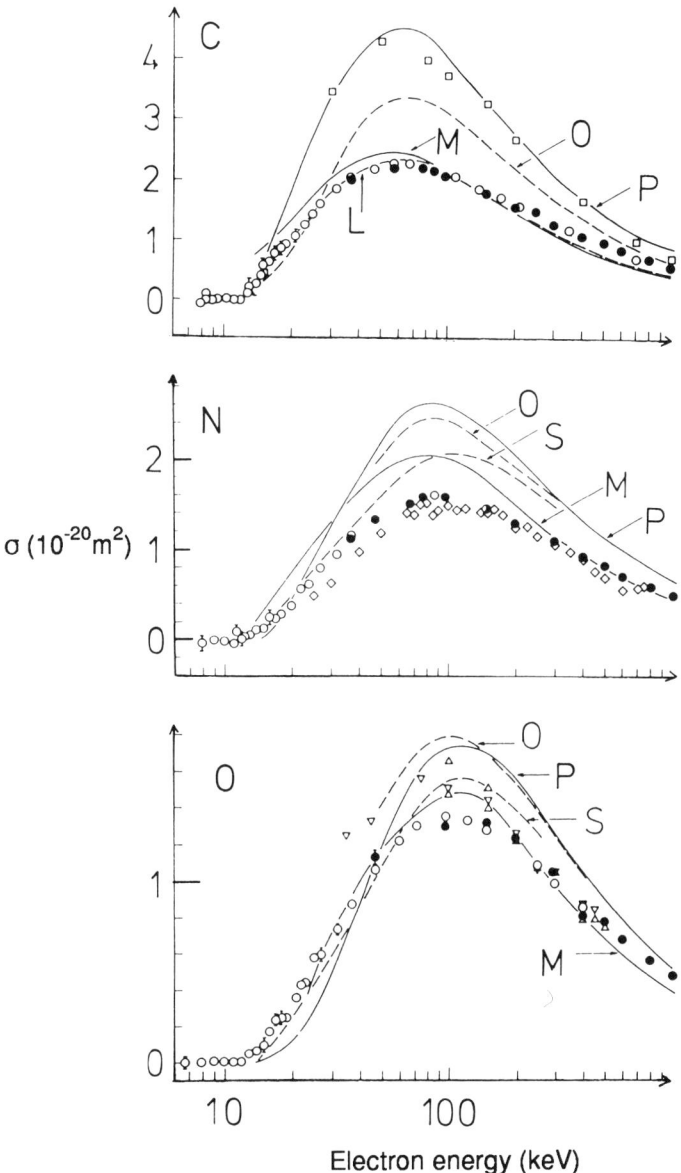

Figure 8. Electron impact ionization cross section function (including only singly-charged products, see below) for atomic carbon, nitrogen and oxygen. (○ and ●) Brook *et al.*[59], (◇) Smith *et al.*[60], (▽) Fite and Brackmann[61], and (△) Rothe *et al.*[62] The theoretical cross section curves are labelled P for Peach[63], O for Omidvar *et al.*[64], M for McGuire[65] and L for Lotz.[41–44]

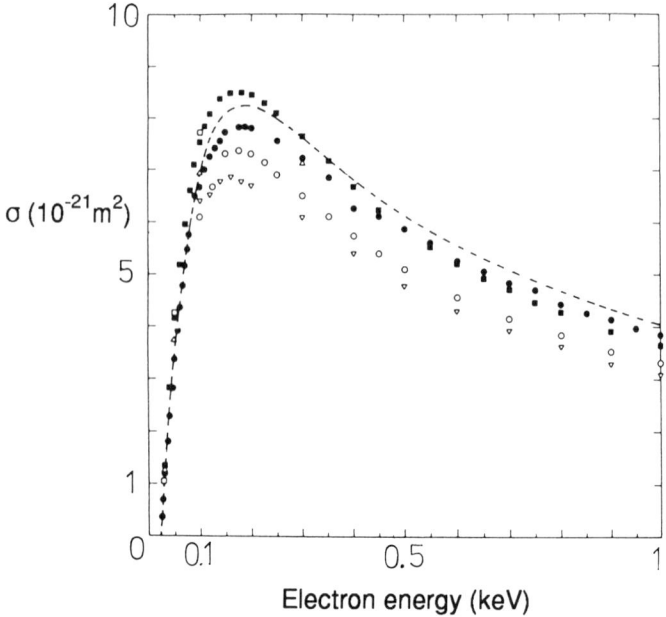

Figure 9. Total electron impact ionization cross section function for neon. (■) Smith[66], (●) Rapp and Englander–Golden[21], (▽) Schram et al.[67,68], (○) Gaudin and Hagemann[69], (△) Fletcher and Cowling[70], (---) Krishnakumar and Srivastava.[71] The recent results of Wetzel et al.[72] (not shown for the sake of clarity) are in good agreement with those of Rapp and Englander–Golden.[21] The data of Schram et al.[67] extend up to an electron energy of 20 keV. Mc Clure[73] summarizes data in the MeV region.

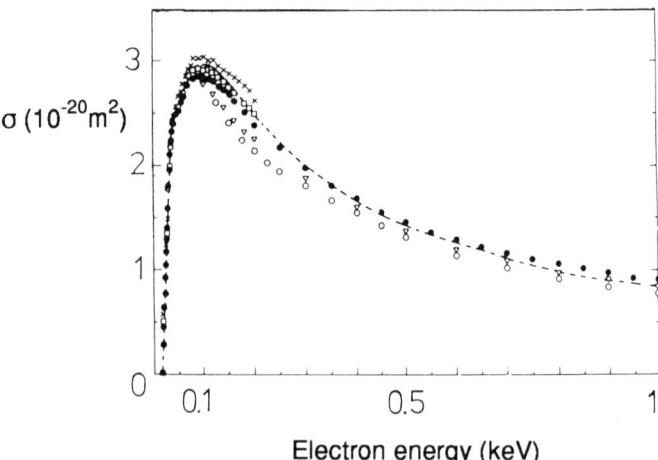

Figure 10. Total electron impact ionization cross section function for argon. (●) Rapp and Englander–Golden[21], (▽) Schram et al.[67,68], (○) Gaudin and Hagemann[69], (□) Kurepa et al.[74], (×) Wetzel et al.[72], (- - -) Krishnakumar and Srivastava.[71] The data of Schram et al.[67] extend up to an electron energy of 20 keV (see also a recent study by Mc Callion et al.[75] extending up to 5.3 keV). McClure[73] summarizes data in the MeV region.

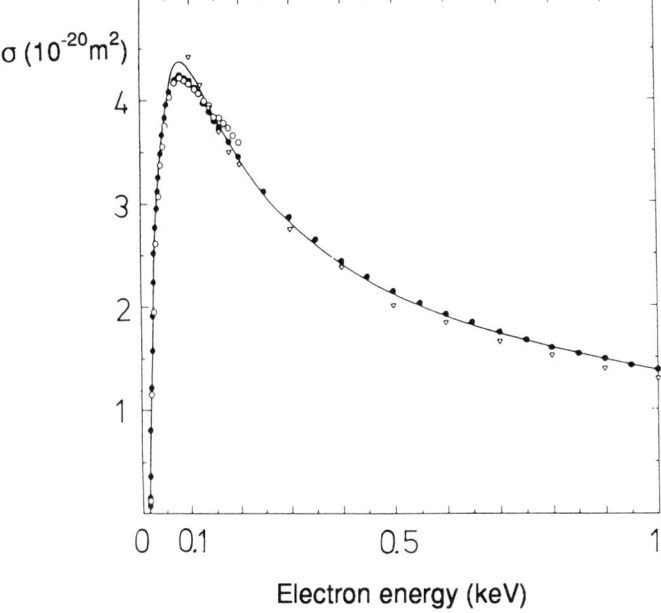

Figure 11. Total electron impact ionization cross section function for krypton. (●) Rapp and Englander–Golden[21], (▽) Schram et al.[67,68], (○) Wetzel et al.[72], (—) Krishnakumar and Srivastava.[71]

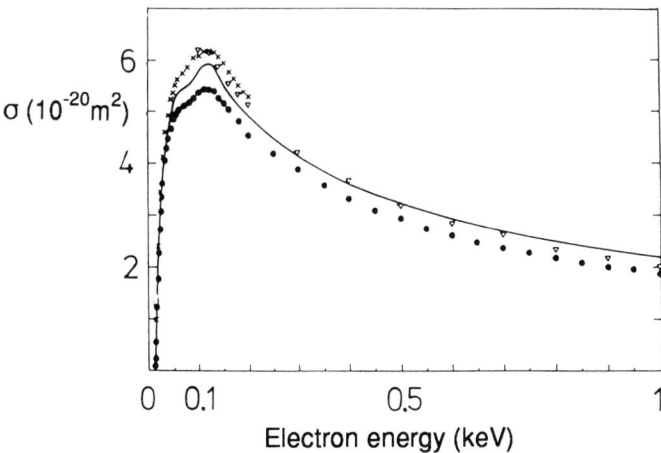

Figure 12. Total electron impact ionization cross section function for xenon. (●) Rapp and Englander–Golden[21], (▽) Schram et al.[67,68], (×) Wetzel et al.[72], (—) Krishnakumar and Srivastava.[71]

4. PARTIAL ELECTRON IMPACT IONIZATION CROSS SECTIONS

4.1. Experimental Methods and Techniques

According to the relation in Eq. (10), the determination of partial electron impact ionization cross sections σ_{zi} requires not only the measurement of $n(0)_e$, n, L, and i_t (as in the case of total cross section measurements), but in addition an accurate analysis of i_t in terms of individual ion currents, i_{ms}, produced. In order to measure individual ion currents, i_{ms}, mass spectrometers must be used in the corresponding experimental setups. Because of the great difficulty in achieving a known and reliable transmission and collection efficiency (independent of the m/z value of the ion to be analyzed) in those instruments, their primary value has been for a long time the identification of the different ions produced and the measurement of approximate partial ionization cross section ratios.

The first mass-spectrometric determinations of partial electron impact ionization cross section functions were made in the 1930s. Some of these studies were repeated later on; however, until recently, large differences (as exemplified by the variability in the cross section ratio between singly and doubly charged Ar ions shown in Fig. 13) existed in both the magnitude and the shape of partial electron impact ionization cross section functions obtained in various studies. As pointed out by many workers (see Refs. 5, 12, 13, 17, 19, and 22), this is due to large

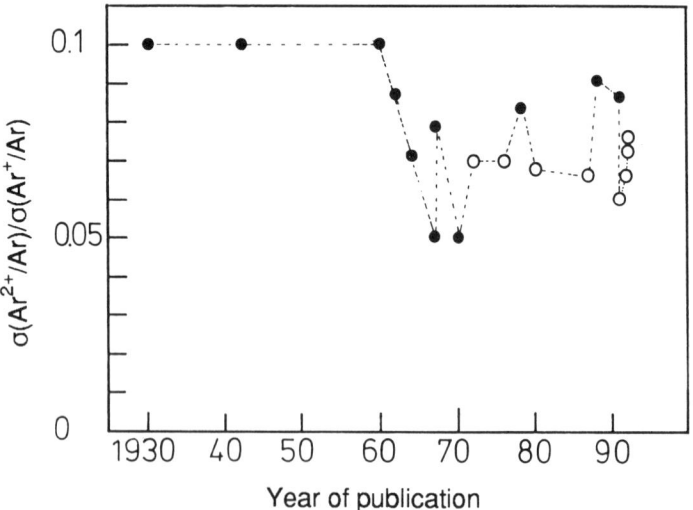

Figure 13. Measured partial ionization cross section ratio σ (Ar^{2+}/Ar) / σ (Ar^{+}/Ar) at 100 eV electron energy versus year of publication of these results (original data references given in Margreiter et al.[79] and Märk[1]). Cross section ratios designated with open circles are measured with improved and controlled experimental conditions (see text) and agree quite well with each other within the experimental error bars.

Electron Impact Ionization of Plasma Edge Atoms 75

discrimination effects occurring at the ion source exit and at the mass spectrometer slits (discussed in more detail below). Moreover, a common problem (never solved satisfactorily) is the absolute calibration. Closely related to this is the fact that discrimination may also occur at the ion detector.[80,81]

There exist, however, some recent experimental studies in which new and sophisticated approaches have been used in order to overcome these difficulties. Some of these new studies come very close to meeting the main prerequisites for measuring accurate partial ionization cross section functions, namely, a constant (known) or complete ion source–mass spectrometer collection efficiency independent of the mass-to-charge ratio of the ion under study, independent of the incident electron energy, and in some cases independent of the initial kinetic ion energy. These studies include the following techniques: improved crossed thermal beam methods, crossed fast atom beam techniques, improved metastable ion detection techniques, trapped ion mass spectrometry, and improved ion extraction and transmission techniques (also in combination with molecular beam techniques).

These methods have been summarized and referenced recently[1,34,82] and therefore will not be reviewed here. However, one of these methods, employing a Nier type ion source in combination with a sector field mass spectrometer system (Fig. 14), will be discussed as an example in detail in the following paragraph, because of (i) the widespread use of this instrumentation in mass spectrometry laboratories and (ii) its recent success in the determination of accurate partial ionization cross section functions for atoms, molecules, and clusters.

The extraction of ions from the ionization region in a Nier type ion source depends under usual experimental conditions on various parameters; the initial energy of the ions, the mass-to-charge ratio, m/z, the guiding magnetic field, the electron beam space charge, and the applied extraction field. Usually, ions are extracted from the ionization region (in which there is a crossed electric and magnetic field) by a weak electric field applied between the collision chamber exit slit and an electrode opposite to the exit slit (i.e., pusher). This extraction, however, is not complete and results in discrimination between ions with different m/z. In an alternative approach, a penetrating field from external electrodes may be used; that is, all electrodes confining the collision chamber are kept at the same potential (e.g., ion acceleration voltage of 3 kV), and ions are drawn out of the collision chamber through the ion source exit slit under the action of an electric field applied to the external electrodes (see Fig. 14). It has been shown that this penetrating field extraction assures saturation of the parent ion current.[84] Ions extracted in this manner are then centered and focused by various elements and reach the end of the acceleration region at the so-called earth slit. Stephan *et al.*[84] additionally introduced two pairs of deflection plates in front of the mass spectrometer entrance slit S_1 (see L_6–L_9 in Fig. 14); these plates serve to sweep the ion beam across the mass spectrometer entrance slit S_1 in the y direction (perpendicular to S_1) and in the z

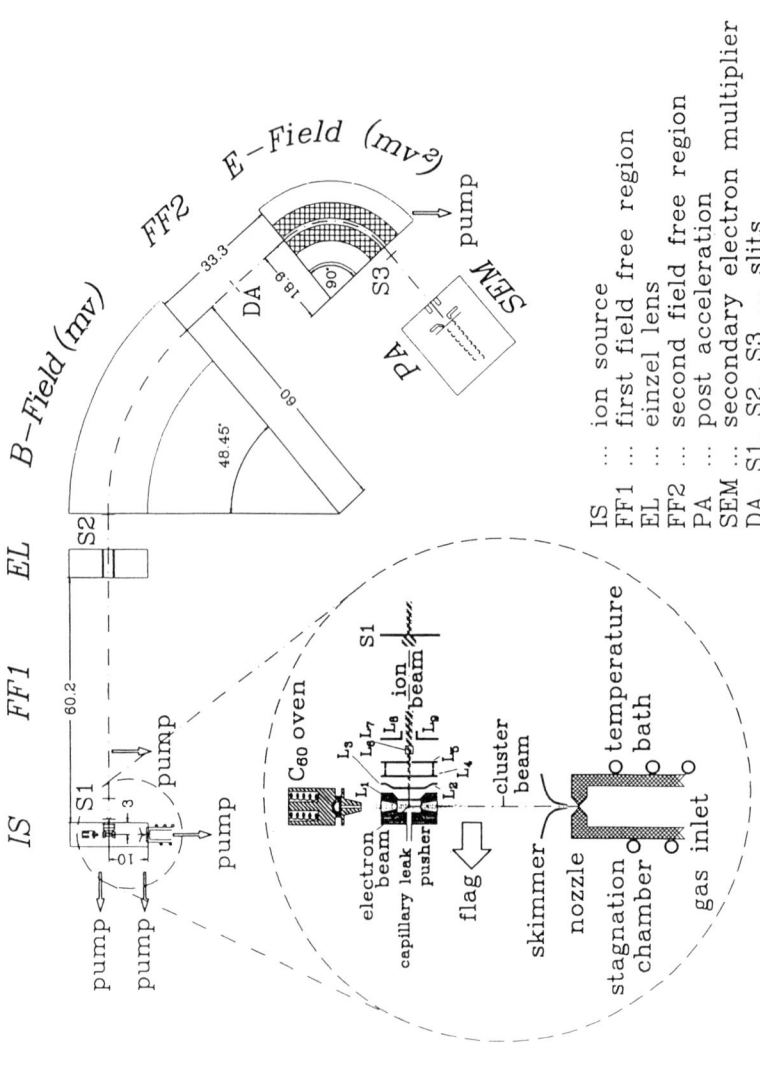

Figure 14. View (to scale, numbers given in cm) of the electron impact ion source and of the double focussing mass spectrometer system after Grill et al.[83] L_1, collision chamber exit slit electrodes; L_2, penetrating field electrodes; L_3 and L_4, focussing electrodes; L_5, ground slit; L_6 to L_9 ion beam deflection electrodes; S_1, mass spectrometer entrance slit; z-direction: perpendicular to the x-y plane (right-hand axes).

IS   ... ion source
FF1   ... first field free region
EL   ... einzel lens
FF2   ... second field free region
PA   ... post acceleration
SEM ... secondary electron multiplier
DA, S1, S2, S3 ... slits

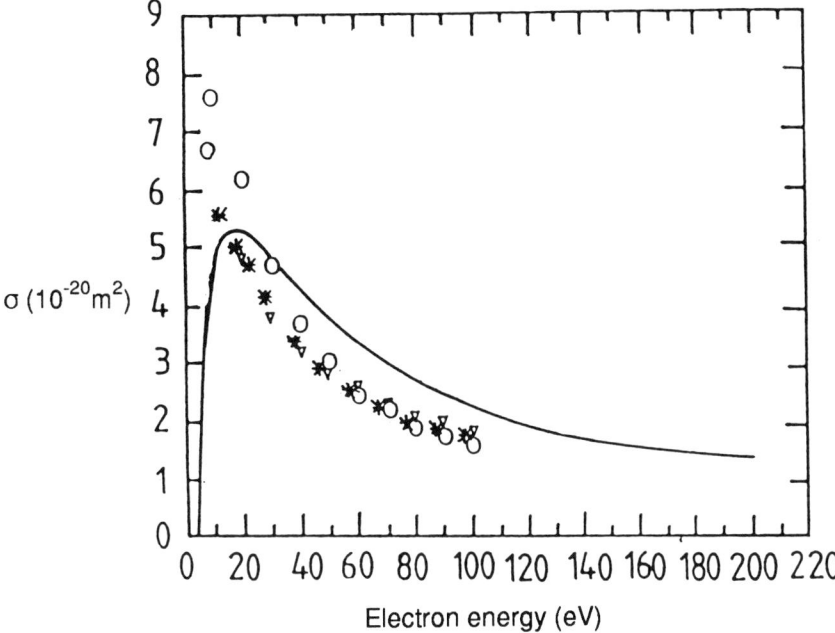

Figure 15. Partial electron impact ionization cross section function for the reaction $Ne^m + e \rightarrow Ne^+ + 2e$ after Dixon et al.[89] designated *. Also shown for comparison theoretical data using the Born approximation (Ton That and Flannery[90]) designated ○, using the scaled Born approximation (Mc Guire[91]) designated ▽, and using the semiclassical DM approach Eq. (16) (Margreiter et al.[49]) designated —.

direction (parallel to S_1). This allows the recording and integration of the ion beam profile, and hence discrimination at S_1 can be avoided. It is of interest to note that this technique has been recently extended to the quantitative detection of fragment ions.[79,83,85–87]

4.2. Theoretical Considerations

In contrast to the situation for total ionization cross sections, almost no theoretical treatments are available for the determination of absolute partial ionization cross sections (except for those cases with atomic targets where the total cross section is about equal to the single ionization cross section; see above). Besides a few quantum-mechanical and classical approximations for the production of doubly charged atoms, which are usually in strong disagreement with the accepted experimental results (e.g., see Figs. 16–19 in Ref. 13, showing results for doubly charged rare-gas atoms), the only approach to the calculation of absolute

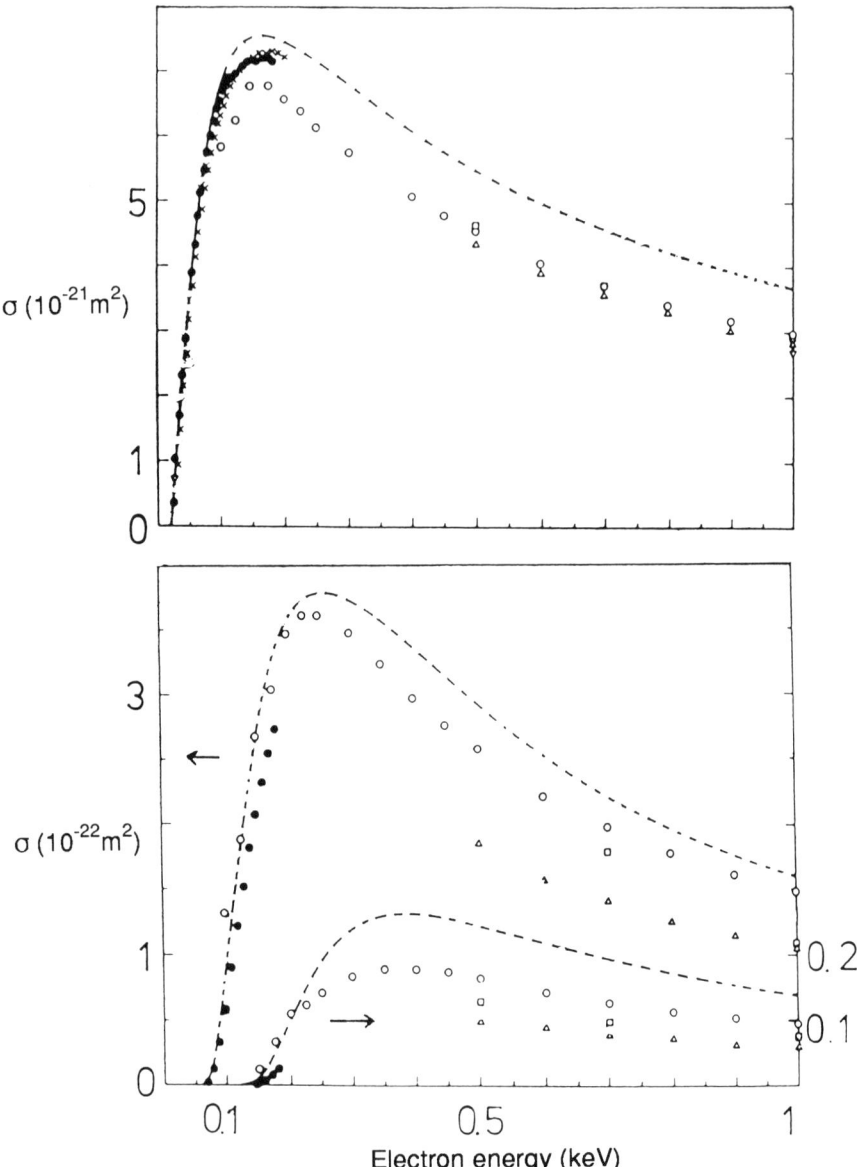

Figure 16. Partial electron impact ionization cross section function for the reactions Ne + e → Ne$^+$ + 2e (upper part), Ne + e → Ne^{2+} + 3e (lower part, left scale) and Ne + e → Ne^{3+} + 4e (lower part, right scale). (△) Schram et al.[92], (○) Gaudin and Hagemann[69], (□) Nagy et al.[93], (●) Stephan et al.[84], (×) Wetzel et al.[72] and (- - -) Krishnakumar and Srivastava.[71] The data of Nagy et al. extend up to 5 keV and those of Schram et al. to 16 keV (including up to fivefold charged ions).

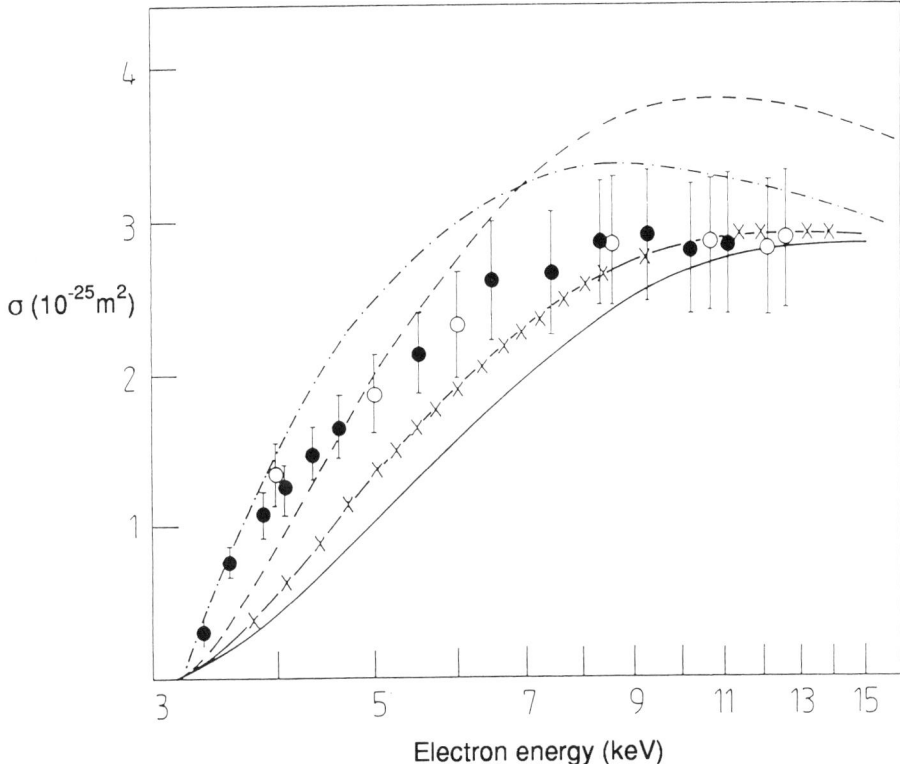

Figure 17. K-shell ionization cross section versus electron energy for argon: (●) experimental results of Hippler et al.[96], (○) experimental results of Tawara et al.[97], (---) PWBA calculation of Hippler and Jitschin[98], (···) PWBA Ochkur calculation of Hippler and Jitschin[98], (-×-×) Coulomb–Born exchange calculation of Moores et al.[99] and (—) DM approach using a relativistic version of Eq. (16) after Deutsch et al.[52]

partial ionization cross section functions is the semiempirical treatment of molecules by Khare and Meath.[88]

4.3. State-Selected Partial Ionization Cross Sections

Most of the partial electron impact ionization cross section functions reported in the literature so far concern ionization of a stagnant (in thermal equilibrium) target system in its electronic ground state into a specific ion distinguished only by charge-to-mass ratio and not due to its electronic state. For a limited number of cases, however, there exist partial cross sections for electron impact ionization reactions taking into account the electronic states of either the neutral target or the ion produced (for a detailed discussion, see Märk[19]). In order to illustrate this point,

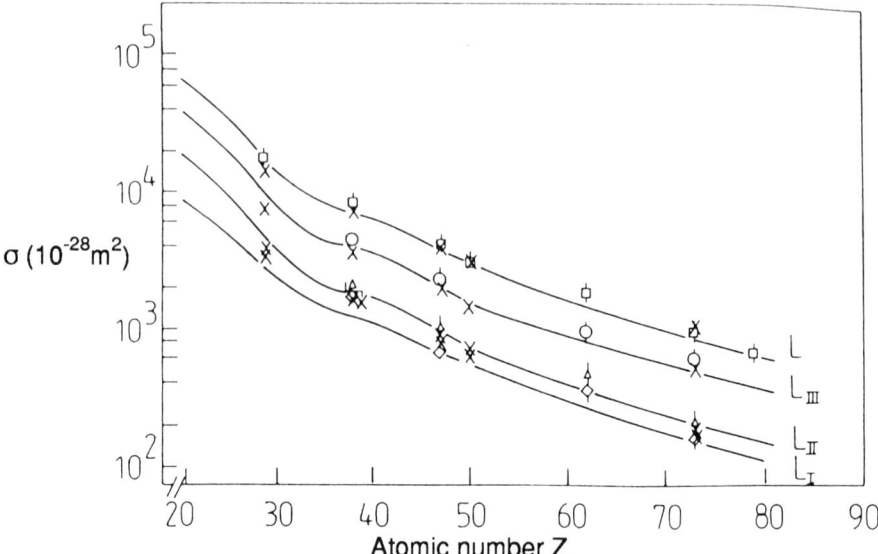

Figure 18. L-shell and L-subshell ionization cross section at an electron energy of 150 keV versus atomic number. Full line relativistic calculation of Scofield[100], (×) DM results obtained using a relativistic version of Eq. (16) after Deutsch et al.[52], and (□, ○, △ and ◇) experimental results obtained by Reusch et al.[101] for L, L_{III}, L_{II} and L_I shell ionization, respectively.

we present in Fig. 15 the corresponding cross section function for the single ionization of metastable neon atoms. It is interesting to note that the cross section for this metastable atom is approximately one order of magnitude larger than that for the ground state atom (shown in Fig. 16). Absolute cross section data for single ionization of metastable atoms of H, He, Ne, Ar, Kr, Xe, and Rn have been summarized recently by Margreiter et al.[49]

Moreover, in certain instances it is of interest also to determine the (state-selected) partial cross section for the removal of electrons from a specific target shell in the course of an electron impact ionization process. Typical examples of such cases are inner-shell ionization reactions, which have been reviewed by Powell[94] and Hippler[95] and recently by Deutsch et al.[52] Figures 17 and 18 show, as an example, theoretical and experimental results for the K-shell ionization cross section function of argon and for the dependence of the L-shell ionization cross section at 150 keV on the atomic number Z, respectively. Furthermore, it is interesting to note that K-shell ionization cross sections may be expressed in a scaled form; that is, the reduced cross section $\sigma_{nl}B_{nl}^2$ does not depend on Z (see discussion and Fig. 6-1 in Ref. 94). This scaling law has been rationalized recently in the frame of the DM approach[52] and extended into the relativistic energy range.[52] Another class of cross sections falling into the category dealing with the removal of a specific electron is

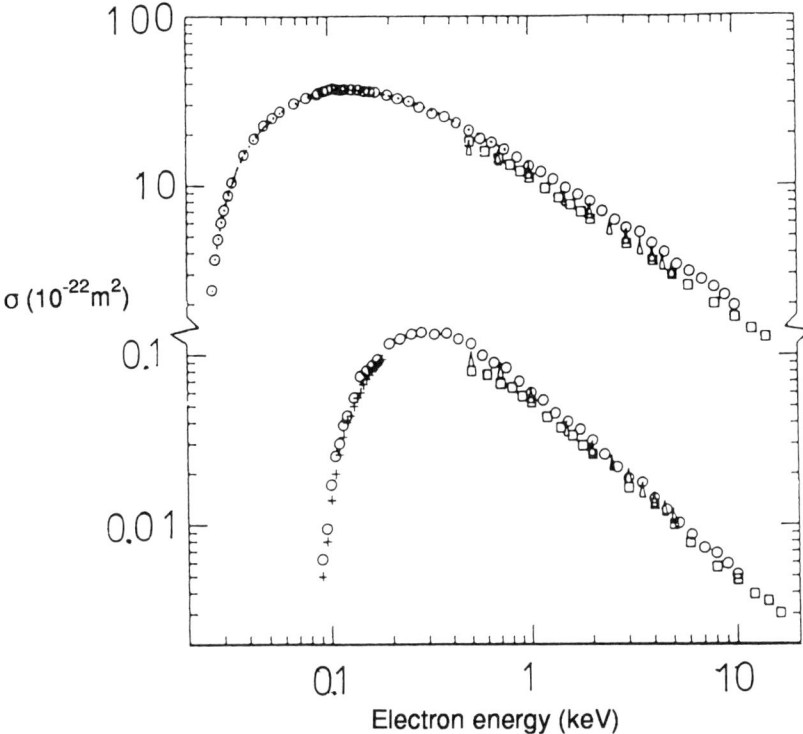

Figure 19. Partial electron impact ionization cross section function for the reactions $He + e \rightarrow He^+ + 2e$ (upper part) and $He + e \rightarrow He^{2+} + 3e$ (lower part). (□) Schram et al.[92], (△) Nagy et al.[93], (+) Stephan et al.[84], (•) Montague et al.[26], (◇) Wetzel et al.[72] and (○) Shah et al.[55]

outer-shell cross sections. As single ionization of rare-gas atoms proceeds mainly by the removal of an outer-shell electron, the determination of these cross sections is of basic importance in the understanding of electron impact ionization. Ionization cross section functions for s-electron ionization of the rare gases, that is, the outer shell reaction $(ns^2, np^6) \rightarrow (ns, np^6)$, have been recently summarized by Deutsch and Märk.[53]

4.4. Recommended Partial Cross Sections

There exist several comprehensive reviews on partial electron impact ionization cross section determinations including also compilations of recommended data (Refs. 1, 3, 5, 6, 12, 13, 17, 19, 22, 76, 78, and 82). Moreover, the available data for partial cross sections are less reliable than those for total cross sections owing to errors introduced by discrimination effects in the mass spectrometer and ion

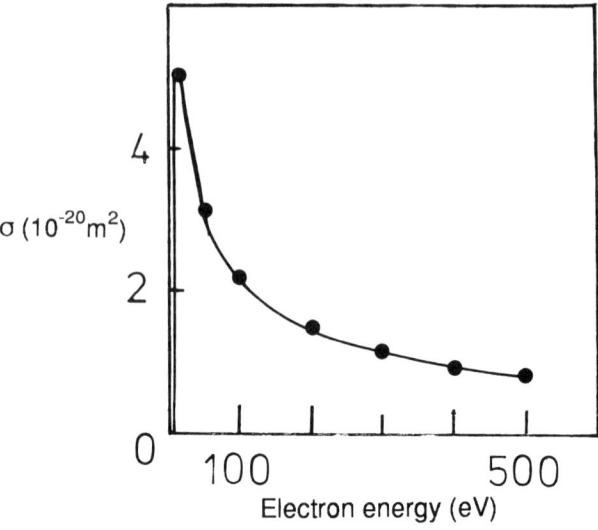

Figure 20. Partial electron impact ionization cross function for the reaction Li + $e \to$ Li$^+$ + 2e after McFarland and Kinney.[102] This cross section curve is according to the authors identical to the total ionization cross section function.

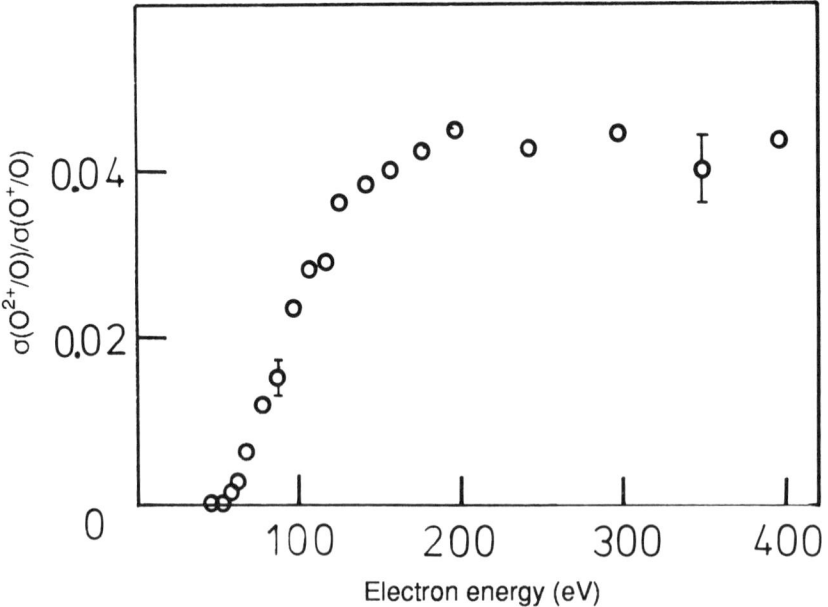

Figure 21. Ratio of the electron impact cross sections for double and single ionization of atomic oxygen after Ziegler et al.[103]

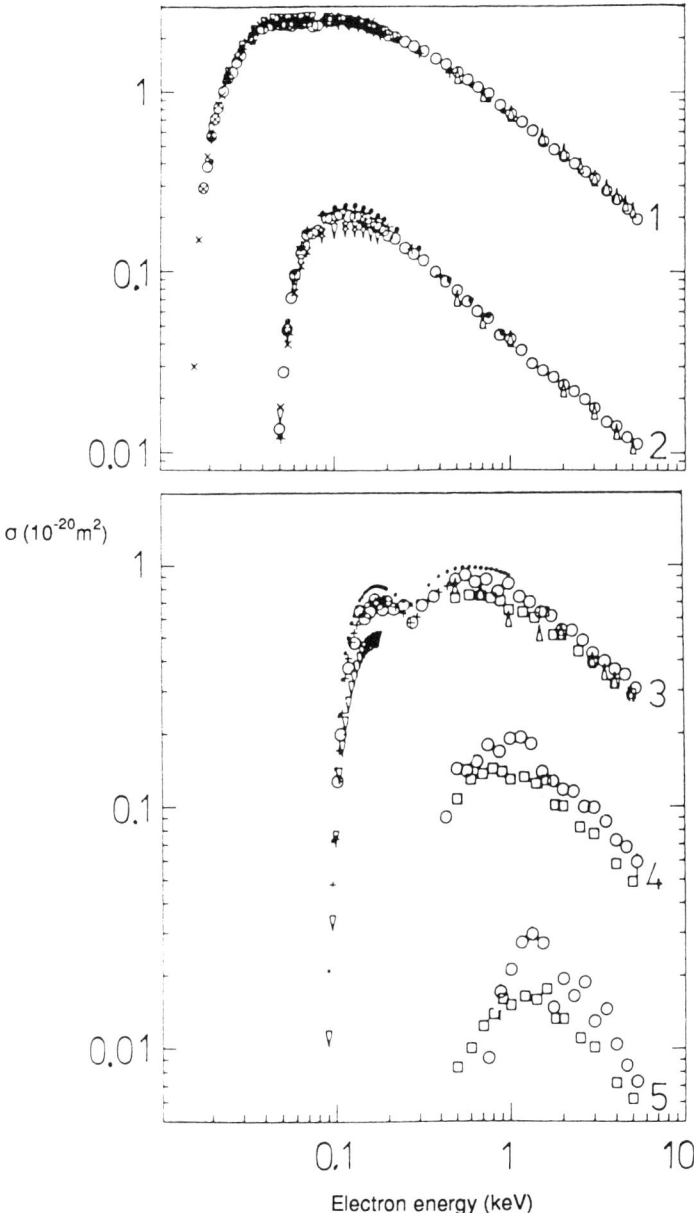

Figure 22. Partial electron impact ionization cross section function for the reactions $Ar + e \rightarrow Ar^+ + 2e$ (designated 1), $Ar + e \rightarrow Ar^{2+} + 3e$ (designated 2), $Ar + e \rightarrow Ar^{3+} + 4e$ (designated 3), $Ar + e \rightarrow Ar^{4+} + 5e$ (designated 4) and $Ar + e \rightarrow Ar^{5+} + 6e$ (designated 5). (□) Schram[104], (△) Nagy et al.[93], (▽) Stephan et al.[84], (×) Wetzel et al.[72], (●) Krishnakumar and Srivastava[71], (+) Ma et al.[105] and (○) McCallion et al.[75] The data of Schram extend up to 18 keV including sevenfold charged ions.

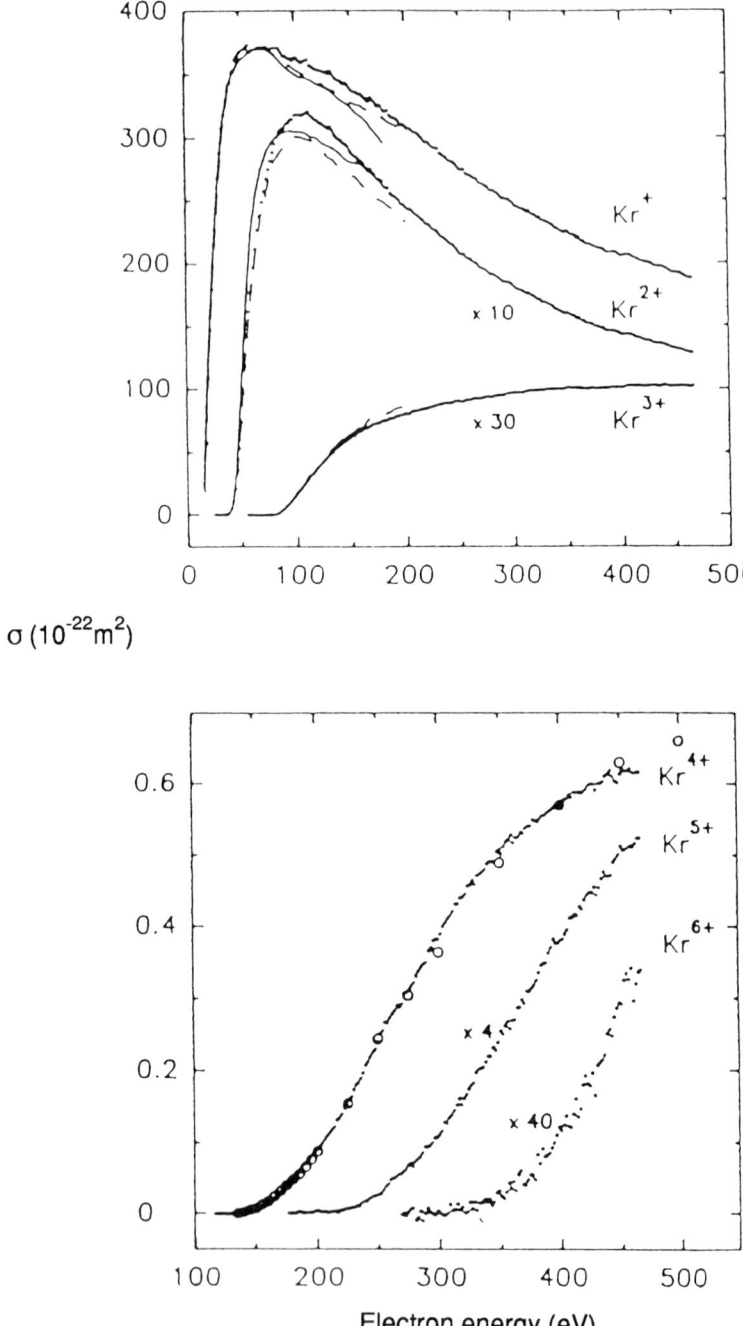

Figure 23. Partial electron impact ionization cross section function for the reactions $Kr + e \rightarrow Kr^{z+} + (z+1)e$ for $z = 1$ to 6. (—) Stephan et al.[84], (- - -) Wetzel et al.[72], (○) Krishnakumar and Srivastava[71] (their data were multiplied by 0.50 to obtain overlap with the data of Syage[106]) and (···) Syage.[106] For data at higher electron energies see Schram.[104]

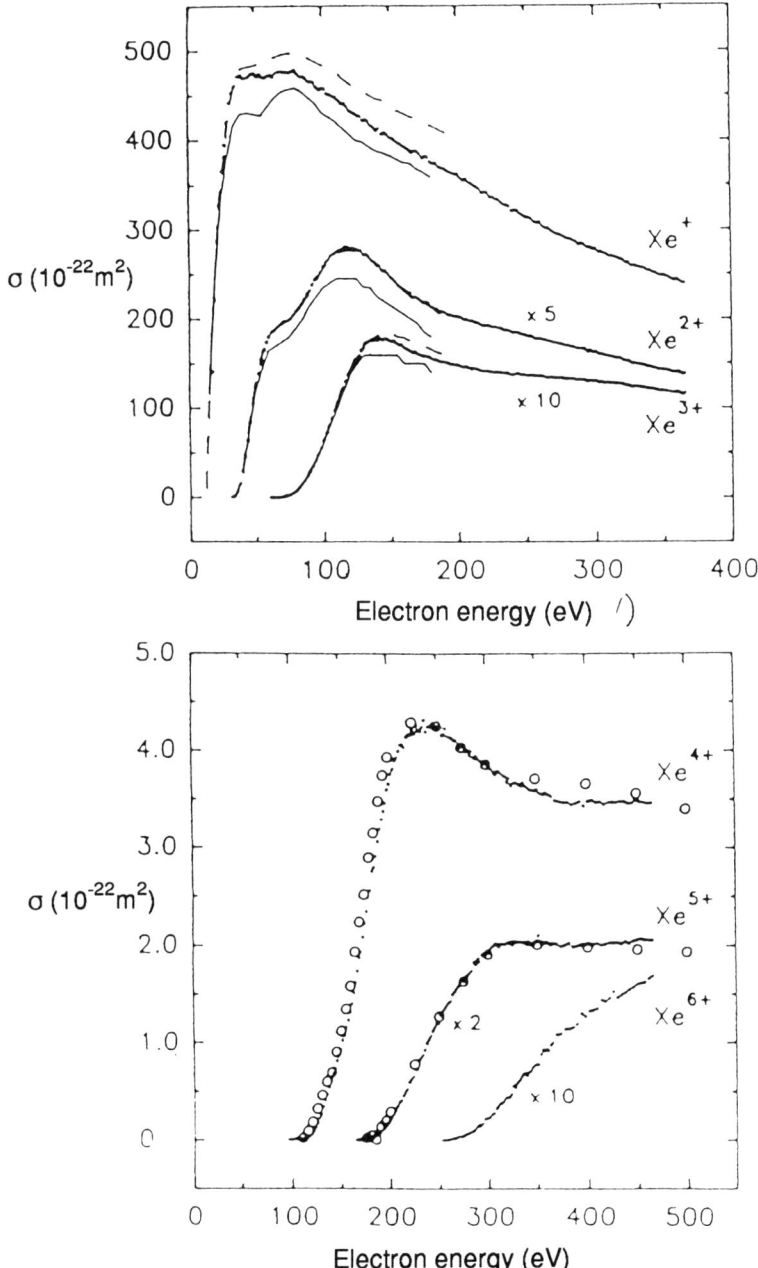

Figure 24. Partial electron impact ionization cross section function for the reaction $Xe + e \rightarrow Xe^{z+} + (z+1)e$ for $z = 1$ to 6. (—) Stephan and Märk[107], (- - -) Wetzel *et al.*[72], (○) Krishnakumar and Srivastava[71] (their data were multiplied by 0.65 to obtain overlap with the data of Syage[106]) and (···) Syage.[106] For data at higher electron energies see Schram.[104]

source and at the detector and because of problems arising in the absolute calibration.[19] These effects have only been taken into account properly in the past few years. Nevertheless, the cross sections for some of the most important targets in magnetic fusion edge plasmas (e.g., He, Li, O, Ne, Ar, Kr, and Xe) have been measured recently with improved methods. The available data on these targets have been assessed and summarized here, and the best data sets available (partial cross section functions or ratios) are presented in graphical form in Figs. 16 and 19–24. In the case of the other targets (in particular, the metallic impurities), much less is known about the cross sections for the production of doubly charged and more highly charged ions. (The single ionization cross section is usually very similar to the total ionization cross section owing to the small contributions from the higher charge states; see Eq. (12) and also the discussion in Section 3.3 and 3.4 and references given therein). The only reliable data available appear to be the recent measurements of Freund et al.[23] on Fe, Cu, and Ga, where the double-to-single ionization cross section ratios at 100 eV are 0.60, 0.055, and 0.065, respectively, and the triple-to-single ratio for Ga at 150 eV is 0.0061 (for details of the measured cross section functions up to 200 eV, see Freund et al.[22]), and of Gilbody and co-workers[108] on Fe and Cu.

ACKNOWLEDGMENTS. This work was partially supported by the Österreichische Fonds zur Förderung der Wissenschaftlichen Forschung and the Bundesministerium für Wissenschaft und Forschung, Vienna, Austria.

REFERENCES

1. T. D. Märk, *Plasma Phys. Controlled Fusion* **34**, 2083–2090 (1992).
2. M. Janossy, L. Csillag, Z. Donko, and K. Rozsa, *Acta Phys. Hung.* **73**, 311–343 (1993).
3. T. D. Märk and G. H. Dunn, *Electron Impact Ionization*, Springer-Verlag, Vienna (1985).
4. M. Inokuti (ed.), *Atomic and Molecular Data for Radiotherapy*, IAEA-TECDOC, **799** (1995).
5. T. D. Märk, in *Electron–Molecule Interactions and Their Applications*, Vol. 1 (L. G. Christophorou, ed.), Academic Press, Orlando (1984), pp. 251–334.
6. T. D. Märk, in *Gaseous Ion Chemistry and Mass Spectrometry* (J. H. Futrell, ed.), John Wiley & Sons, New York (1986), pp. 61–93.
7. T. D. Märk, *Int. J. Mass Spectrom. Ion Processes* **79**, 1–59 (1987).
8. T. D. Märk, *Int. J. Mass Spectrom. Ion Processes* **107**, 143–163 (1991).
9. T. D. Märk, in *Linking the Gaseous and Condensed Phases of Matter: The Behavior of Slow Electrons* (L. G. Christophorou, W. F. Schmidt, and E. Illenberger, eds.), Plenum, New York (1994).
10. T. D. Märk and A. W. Castleman, Jr., *Adv. Atom. Mol. Phys.* **20**, 65–172 (1985).
11. A. Crowe, J. A. Preston, and J. W. McConkey, *J. Chem. Phys.* **57**, 1620–1625 (1972).
12. T. D. Märk, *Int. J. Mass Spectrom. Ion Phys.* **45**, 125–145 (1982).
13. T. D. Märk, *Beitr. Plasmaphysik* **22**, 257–294 (1982).
14. H. S. W. Massey, E. H. S. Burhop, and H. B. Gilbody, *Electronic and Ionic Phenomena*, Clarendon Press, Oxford (1969).
15. F. H. Field and J. L. Franklin, *Electron Impact Phenomena*, Academic Press, New York (1970).

16. E. Illenberger and J. Momigny, *Gaseous Molecular Ions*, Steinkopff, Darmstadt (1992).
17. L. J. Kieffer and G. H. Dunn, *Rev. Mod. Phys.* **38**, 1–35 (1966).
18. F. J. de Heer and M. Inokuti, in *Electron Impact Ionization* (T. D. Märk and G. H. Dunn, eds.), Springer-Verlag, Vienna, (1985), pp. 232–276.
19. T. D. Märk, in *Electron Impact Ionization* (T. D. Märk and G. H. Dunn, eds.), Springer-Verlag, Vienna, (1985), pp. 137–197.
20. J. T. Tate and P. T. Smith, *Phys. Rev.* **39**, 270–277 (1932).
21. D. Rapp and P. Englander-Golden, *J. Chem. Phys.* **43**, 1464–1479 (1965).
22. R. S. Freund, in *Swarm Studies and Inelastic Electron–Molecule Collisions* (L. C. Pitchford, B. V. McKoy, A. Chutjian, and S. Trajmar, eds.), Springer-Verlag, New York (1987), pp. 329–346.
23. R. S. Freund, R. C. Wetzel, R. J. Shul, and T. R. Hayes, *Phys. Rev. A* **41**, 3575–3595 (1990).
24. C. J. Cook and J. R. Peterson, *Phys. Rev. Lett.* **9**, 164–166 (1962).
25. D. L. Ziegler, J. H. Newman, K. A. Smith, and R. F. Stebbings, *Planet. Space Sci.* **30**, 451–456 (1982).
26. R. G. Montague, M. F. A. Harrison, and A. C. H. Smith, *J. Phys. B* **17**, 3295–3310 (1984).
27. V. Tarnovsky and K. Becker, *Z. Phys. D* **22**, 603–610 (1992).
28. K. Becker, in *Proceedings of the XVIIIth International Conference on Physics of Electronic and Atomic Collisions*, Aarhus, 1993, pp. 234–248.
29. M. R. H. Rudge, *Rev. Mod. Phys.* **40**, 564–590 (1968).
30. S. M. Younger, in *Electron Impact Ionization* (T. D. Märk and G. H. Dunn, eds.), Springer-Verlag, Vienna (1985), pp. 1–23.
31. S. M. Younger and T. D. Märk, in *Electron Impact Ionization* (T. D. Märk and G. H. Dunn, eds.), Springer-Verlag, Vienna (1985), pp. 24–41.
32. H. Deutsch and T. D. Märk, *Int. J. Mass Spectrom. Ion Processes* **70**, R1–R8 (1987).
33. Y. K. Kim and M. E. Rudd, *Phys. Rev. A* **50**, 3954–3967 (1994).
34. T. D. Märk and M. E. Rudd, in *ICRU Report on Secondary Electron Spectra*, International Commission on Radiation Units, Bethesda, Maryland (1995), Chapter 3.
35. J. J. Thomson, *Phil. Mag.* **23**, 449–457 (1912).
36. M. Gryzinski, *Phys. Rev. A* **138**, 305–358 (1965).
37. A. Burgess, in *Atomic Collision Processes* (M. R. C. McDowell, ed.), North-Holland, Amsterdam (1964), pp. 237–242.
38. L. Vriens, *Phys. Rev.* **141**, 88–141 (1966).
39. H. Deutsch, P. Scheier, and T. D. Märk, *Int. J. Mass Spectrom. Ion Processes.* **74**, 81–95 (1986).
40. K. L. Bell, H. B. Gilbody, G. G. Hughes, A. E. Kingston, and F. J. Smith, *J. Phys. Chem. Ref. Data* **12**, 891–916 (1983).
41. W. Lotz, *Astrophys. J. Suppl.* **128**, 207–238 (1967).
42. W. Lotz, *Z. Phys.* **206**, 205–211 (1967).
43. W. Lotz, *Z. Phys.* **216**, 241–247 (1968).
44. W. Lotz, *Z. Phys.* **232**, 101–107 (1970).
45. H. Bethe, *Ann. Physik* **5**, 325–400 (1930).
46. W. F. Miller and R. L. Platzman, *Proc. Phys. Soc. A* **70**, 299–303 (1957).
47. J. W. Otvos and D. P. Stevenson, *J. Am. Chem. Soc.* **78**, 546–551 (1956).
48. J. B. Mann, *J. Chem. Phys.* **46**, 1646–1651 (1967).
49. D. Margreiter, H. Deutsch, and T. D. Märk, *Contrib. Plasma Phys.* **30**, 487–495 (1990).
50. D. Margreiter, H. Deutsch, and T. D. Märk, *Int. J. Mass Spectrom. Ion Processes*, **139**, 127–139 (1995).
51. J. C. Halle, H. H. Lo, and W. L. Fite, *Phys. Rev. A* **23**, 1708–1716 (1981).
52. H. Deutsch, D. Margreiter, and T. D. Märk, *Z. Phys. D* **29**, 31–37 (1944).
53. H. Deutsch and T. D. Märk, *Contrib. Plasma Phys.* **34**, 19–24 (1994).
54. M. B. Shah, D. S. Elliott, and H. B. Gilbody, *J. Phys. B* **20**, 3501–3514 (1987).

55. M. B. Shah, D. S. Elliott, P. McCallion, and H. B. Gilbody, *J. Phys. B* **21**, 2751–2761 (1988).
56. Y. K. Kim, in *Physics of Ion–Ion and Electron–Ion Collisions* (F. Brouillard and J. W. McGowan, eds.), Plenum, New York (1983), pp. 101–165.
57. F. F. Rieke and W. Prepejchal, *Phys. Rev. A* **6**, 1507–1519 (1972).
58. B. L. Schram, M. J. van der Wiel, F. J. deHeer, and H. R. Moustafa, *J. Chem. Phys.* **44**, 49–54 (1966).
59. E. Brook, M. F. A. Harrison, and A. C. H. Smith, *J. Phys. B* **11**, 3115–3132 (1978).
60. A. C. W. Smith, E. Caplinger, R. H. Neynaber, E. W. Rothe, and S. M. Trujillo, *Phys. Rev.* **127**, 1647–1649 (1962).
61. W. L. Fite and R. T. Brackmann, *Phys. Rev.* **113**, 815–816 (1959).
62. E. W. Rothe, L. L. Marino, R. H. Neynaber, and S. M. Trujillo, *Phys. Rev.* **125**, 582 (1962).
63. G. Peach, *J. Phys. B* **3**, 328–349 (1970); **4**, 1670–1677 (1971).
64. K. Omdivar, H. L. Kyle, and E. C. Sullivan, *Phys. Rev. A* **5**, 1174–1187 (1972).
65. E. J. McGuire, *Phys. Rev. A* **3**, 267–279 (1971).
66. P. T. Smith, *Phys. Rev.* **36**, 1293–1302 (1930).
67. B. L. Schram, F. J. deHeer, M. J. van der Wiel, and J. Kistemaker, *Physica* **31**, 94–112 (1965).
68. B. L. Schram, H. R. Moustafa, J. Schutten, and F. J. deHeer, *Physica* **32**, 734–740 (1966).
69. A. Gaudin and R. Hagemann, *J. Chim. Phys.* **64**, 1209–1221 (1967).
70. J. Fletcher and I. R. Cowling, *J. Phys. B* **6**, L258–L261 (1973).
71. E. Krishnakumar and S. K. Srivastava, *J. Phys. B* **21**, 1055–1082 (1988).
72. R. C. Wetzel, F. A. Baiocchi, T. R. Hayes, and R. S. Freund, *Phys. Rev. A* **35**, 559–577 (1987).
73. G. W. McClure, *Phys. Rev.* **90**, 796–803 (1953).
74. M. V. Kurepa, I. M. Cadez, and V. M. Pejcev, *Fizika* **6**, 185–209 (1974).
75. P. McCallion, M. B. Shah, and H. B. Gilbody, *J. Phys. B* **25**, 1061–1071 (1992).
76. H. Tawara and T. Kato, *At. Data Nucl. Data Tables* **36**, 167–353 (1987).
77. M. A. Lennon, K. L. Bell, H. B. Gilbody, J. G. Hughes, A. E. Kinston, M. J. Murray, and F. J. Smith, *J. Phys. Chem. Ref. Data* **17**, 1285–1363 (1988).
78. I. Shimamura, *Sci. Papers Inst. Phys. Chem. Res. (Jpn.)* **82**, 1–51 (1989).
79. D. Margreiter, G. Walder, H. Deutsch, H. U. Poll, C. Winkler, K. Stephan, and T. D. Märk, *Int. J. Mass Spectrom. Ion Processes* **100**, 143–156 (1990).
80. J. A. Syage, *J. Phys. B* **24**, L527–L532 (1991).
81. F. Aumayr, T. D. Märk, and H. Winter, *Int. J. Mass Spectrom. Ion Processes* **129**, 17–29 (1993).
82. T. D. Märk, *IAEA-TECDOC* **506**, 179–193 (1989).
83. V. Grill, G. Walder, D. Margreiter, T. Rauth, H. U. Poll, P. Scheier, and T. D. Märk, *Z. Phys. D* **25**, 217–226 (1993).
84. K. Stephan, H. Helm, and T. D. Märk, *J. Chem. Phys.* **73**, 3763–3778 (1980).
85. K. Stephan, H. Deutsch, and T. D. Märk, *J. Chem. Phys.* **83**, 5712–5720 (1985).
86. K. Leiter, P. Scheier, G. Walder, and T. D. Märk, *Int. J. Mass Spectrom. Ion Processes* **87**, 209–224 (1989).
87. H. U. Poll, C. Winkler, D. Margreiter, V. Grill, and T. D. Märk, *Int. J. Mass Spectrom. Ion Processes* **112**, 1–17 (1992).
88. S. P. Khare and W. J. Meath, *J. Phys. B* **20**, 2101–2116 (1987).
89. A. J. Dixon, M. F. A. Harrison, and A. C. H. Smith, in *Proceedings of the VIIIth International Conference on Physics of Electronic and Atomic Collision,* Belgrad, 1973, 405–406.
90. D. Ton-That and M. R. Flannery, *Phys. Rev. A* **15**, 517–526 (1977).
91. E. J. Mc Guire, *Phys. Rev. A* **20**, 445–456 (1979).
92. B. L. Schram, J. H. Boerboom, and J. Kistemaker, *Physica* **32**, 185–196 (1966).
93. N. Nagy, P. Skutlartz, and V. Schmidt, *J. Phys. B* **13**, 1249–1267 (1980).
94. C. J. Powell, in *Electron Impact Ionization* (T. D. Märk and G. W. Dunn, eds.), Springer-Verlag, Vienna (1985), pp. 198–231.

95. R. Hippler, in *Progress in Atomic Spectroscopy* (H. G. Kleinpoppen, ed.), Plenum, New York (1984), pp. 511–575.
96. R. Hippler, K. Saeed, I. McGregor, and H. Kleinpoppen, *Z. Phys. A* **307**, 83–87 (1982).
97. H. Tawara, K. G. Harrison, and F. J. deHeer, *Physica* **63**, 351–367 (1973).
98. R. Hippler and W. Jitschin, *Z. Phys. A* **307**, 287–292 (1982).
99. D. L. Moores, L. B. Golden, and D. H. Samson, *J. Phys. B* **13**, 385–395 (1980).
100. J. H. Scofield, *Phys. Rev. A* **18**, 963–970 (1978).
101. S. Reusch, H. Genz, W. Löw, and A. Richter, *Z. Phys. D* **3**, 379–389 (1986).
102. R. H. McFarland and J. D. Kinney, *Phys. Rev. A* **137**, 1058–1061 (1965).
103. D. L. Ziegler, J. H. Newman, K. A. Smith, and R. F. Stebbings, *Planet. Space Sci.* **30**, 451–456 (1982).
104. B. L. Schram, *Physica* **32**, 197–208 (1966).
105. C. Ma, C. R. Sporleder, and R. A. Bonham, *Rev. Sci. Instrum.* **62**, 909–923 (1991).
106. J. A. Syage, *Phys. Rev. A* **46**, 5666–5679 (1992).
107. K. Stephan and T. D. Märk, *J. Chem. Phys.* **81**, 3116–3117 (1984).
108. M. A. Bolorizadeh, C. J. Patton, M. B. Shah, and H. B. Gilbody, *J. Phys. B* **27**, 175–183 (1994).

Chapter 5

Electron–Ion Recombination Processes in Plasmas

Yukap Hahn

1. INTRODUCTION

As an initially disturbed plasma proceeds toward steady state, recombination processes, together with collisional excitation and ionization, play an important role in determining the relaxation time and final-state ionization balance of the plasma. Recombination of free electrons[1] with ions results in the emission of radiation that carries information about the radiating system. This recombination radiation and radiative decay X rays of excited states provide important tools with which the structure of the radiating system and its environment can be examined. Modeling and diagnostics of laboratory and astrophysical plasmas have relied heavily on the information carried by these radiations.

There are basically three different modes of recombination processes. Radiative recombination (RR) is a one-step process of the electron–ion (e–I) collision system, in which continuum electrons of energy E_c descend to one of the available bound levels by emitting X rays:

YUKAP HAHN • Department of Physics, University of Connecticut, Storrs, Connecticut 06269 and CTR, Storrs, Connecticut 06268.

Atomic and Molecular Processes in Fusion Edge Plasmas, edited by R. K. Janev. Plenum Press, New York, 1995.

$$e^- + A^{Z+}(i) \rightarrow A^{(Z-1)+}(f) + x \quad \text{(RR)} \qquad (1)$$

where f is one of the initially unoccupied excited states of the recombined ion $A^{(Z-1)+}$. The process given by Eq. (1) has been known since the mid-1920s, and various theoretical forms for the cross section and the rate have been available.[2,3] For the bare target ions, with no electrons initially, the theory can be given in an exact form, both nonrelativistic and relativistic. However, for ions with one or more electrons initially, the theory is necessarily approximate.

On the other hand, dielectronic recombination (DR) involves a two-step resonant process, first suggested by Bates and Massey[4] in the 1940s and rediscovered two decades later by Burgess[5,6] in connection with the study of solar corona. The process is schematically described by

$$e^- + A^{Z+}(i) \rightarrow A^{(Z-1)+}(d) \rightarrow A^{(Z-1)+}(f) + x \quad \text{(DR)} \qquad (2)$$

where d represents doubly (or multiply) excited (resonance) states of the recombined ion that are degenerate with one or more continua of the system $A^{(Z-1)+}$. The DR process requires the resonance condition $E_c = E_d - E_i$. In the mid-1970s, the importance of this process in high-temperature tokamak plasmas was realized, and intensive research has followed,[1] especially directed toward the experimental measurement of the cross sections directly,[7–11] with and without external field perturbations.[12–14]

Improved understanding of the intricacies of the recombination processes has been achieved by a combination of much high-precision experimentation and improved theoretical calculations. Thus, the current experimental data have confirmed the validity of the theoretical procedures in calculating the RR and DR cross sections and rate coefficients.

In addition to RR and DR, it is also possible to have recombination in a plasma by a three-body (two continuum electrons and the ion core) collisional recombination (TBR):

$$e^- + e'^- + A^{Z+} \rightarrow e''^- + A^{(Z-1)+}(f) \quad \text{(TBR)} \qquad (3)$$

This process is, of course, important at high electron density. Note that this process does not involve radiation emission, unless states f further decay radiatively to lower states. The TBR is effective in populating high Rydberg states (HRS) of ions in plasmas, rapidly reaching the Saha equilibrium condition.

It has also been proposed recently[15] that the processes in Eqs. (1) and (2) may be accompanied by a simultaneous emission of X rays:

$$e^- + A^{Z+}(i) \rightarrow A^{(Z-1)+}(d') + x' \quad \text{(RDR)}$$
$$\rightarrow A^{(Z-1)+}(f') + x'' \qquad (4)$$
$$\rightarrow A^{Z+}(f'') + e'$$

A process roughly the inverse of that in Eq. (4) was considered earlier,[16–18] but with an X ray in the "final" state (i). The process in Eq. (4), the radiative DR (RDR), does not require that the resonance condition be satisfied; instead, $E_x = E_c + E_i - E_d$. Therefore, the RDR may effectively compete with the RR of Eq. (1) at all E_c, and its contribution to the total recombination rate can be sizable. Incidentally, the signature of the RDR, as distinct from the DR and RR, is the X-ray (x') spectrum, which is red-shifted from that of the RR by the excitation energies, and the follow-up line emission or Auger emission. A process analogous to RDR should also occur in ion–atom (I–A) collisions, as with resonant transfer excitation followed by radiative decay (RTEX), which is related to DR. We denote this new process as XTEX. This aspect of the recombination problem in the e–I and I–A collision systems has not yet been fully investigated.[19]

The resonance enhancement in the DR through the intermediate states (d) is, of course, not unique to DR. Such resonance phenomena in collisional excitation and ionization have been known experimentally[20,21] from the early 1960s, and a preliminary theoretical analysis of indirect multiple ionization for high-Z ions was presented[22] in 1977. Much progress has been made since then.[23,24] A unified picture of all these resonant processes in both e–I and I–A collisions has been presented,[24] where the common link between these processes is the resonant intermediate states (d) and the resulting decay branchings (Auger and fluorescence yields). A compact summary of these relationships is given[24] in terms of a "resonance cube," and they have been successfully exploited in a number of experiments and theories. In the case of electron–ion collisions, we thus have

RR = radiative electron capture, in e–I collisions;
~ inverse photoionization;
~ radiative electron capture in I–A collisions (REC).
DR = complement of resonant excitation in e–I collisions;
~ inverse photo-Auger ionization;
~ resonant transfer excitation and radiative decay (RTEX).
RDR = radiative dielectronic recombination in e–I collisions;
~ inverse radiative autoionization;
~ radiative transfer with X-ray emission (XTEX).

We first briefly review the theory of recombination processes in Section 2, including RR, DR, RDR, and TBR. Recent attempts to incorporate the plasma field effects are also discussed. Currently available recombination data on RR and DR are then summarized in Sections 3 and 4, in compact forms using the approximate scaling properties and empirically fitted formulas, all in the zero-density, zero-field limit. The data on recombination are far from complete, in terms of their scope and reliability, and we discuss some of the outstanding problems in Section 5.

2. THEORY OF ELECTRON–ION RECOMBINATION IN PLASMAS

The general theory of electron–ion recombination in plasmas has been given previously[25,26] in various formulations, and we briefly review the basic formulas, following the work of Ref. 26, with an important extension to incorporate the effect of external perturbations[27–29] and plasma field effects (PFE).

The total Hamiltonian for a system of plasma particles (electrons and ions) inside a small volume is given by

$$H_{tot} = (H_A + H_X + H_P + D) + H_{AP}$$
$$\equiv H_0 + H_{AP} \qquad (5)$$

where H_A denotes the Hamiltonian for the particular atomic system selected for study and H_{AP} represents the atom–plasma interaction. H_X and H_P are the radiation field and plasma perturber Hamiltonians, respectively, and D is the radiation–atom coupling. The many-body problem, $[H_{tot} - E_{tot}]\Psi_{tot} = 0$, may be expressed in terms of the plasma state projections $\mathcal{T}_P = \Pi + \Sigma$, as

$$\Pi[H_0 + (H_{AP} + H_{AP}G^\Sigma H_{AP}) - E_{tot}]\Pi\Psi_{tot} = 0 \qquad (6)$$

where the interaction inside the parentheses is the effective plasma potential (EPP) which distorts the atomic Hamiltonian. With $H_{AP} = H_{AP,e} + H_{AP,I}$ for the electron and ion perturbers, we define the EPP as follows:

$$V_P^{tot} = \langle \Pi(H_{AP,e} + H_{AP,I})\Pi \rangle_P + \langle \Pi(H_{AP,e} + H_{AP,I})G^\Sigma(H_{AP,e} + H_{AP,I})\Pi \rangle_P \qquad (7)$$

where $\langle \Pi \ldots \Pi \rangle_P$ denotes the various averaging processes over the plasma particles in their "ground state." Depending on the strength of the coupling and plasma conditions (density, temperature, etc.), the averaging procedures[27,28] have to be adjusted accordingly for each term in V_P^{tot}.

The entire problem posed by H_{tot} is then approximately reduced to that in terms of EPP, defined by

$$[H_A + H_X + D + V_P^{tot} - E]\Psi = 0 \qquad (8)$$

In general, V_P^{tot} contains all the plasma particle perturbations, and it has been used extensively in the pressure broadening theory (PBT)[30–32] in various approximations. In fact, the spectral profile function in the PBT may be expressed compactly in terms of the operator

$$D[E - H_A - V_P^{tot}]^{-1} D \qquad (8')$$

The entire V_P^{tot} is treated here in terms of the modified Hamiltonian; it contains the level shifts and widths. This is a reasonable approach when the effect of plasma perturbations is small. However, when the plasma potential strongly mixes HRS and the cascade transitions caused by fast plasma particles are important, the PBT procedure becomes cumbersome. In a recent formulation of the PFE,[28] the part of the EPP that corresponds to electronic perturbations which are "fast" relative to the atomic relaxation time is omitted from V_P^{tot} and is treated by the rate equations (REQ) approach, while the "slow" component of the EPP is retained in the modified atomic Hamiltonian in Eq. (8') as a distortion potential. We denote the latter part as V_P, which contains most of the (slow) ionic perturbations. Obviously, this separation of V_P^{tot} shows that a close interrelationship must be maintained between the slow plasma field distortion of the atomic states that are included in the rate calculation and the fast cascade transitions to be included in the rate equations.

The RR and DR rates are evaluated in terms of the states distorted by V_P. We write the Hamiltonian for the plasma-distorted ionic system A as $H_A = H_C + K_0 + V$, where H_C is for the target ion, K_0 is for the kinetic energy of the incoming electron, and V is for the interaction of the incoming electron with core electrons. Omitting the couplings D and V in Eq. (8), we have then a new EPP distorted basis generated by

$$[H_C + K_0 + H_X + V_P - i\partial_t]\Phi_p = 0 \qquad (9)$$

Following Ref. 26, we further define the projection operators P and Q for the target ground and excited states, respectively, of H_C, with $[H_C, P] = 0$. Then, the recombination amplitudes are given in the distorted-wave approximation by

$$M_{fi}^{recomb} = \langle \Phi_f | D | \Phi_i \rangle + \langle \Phi_f | V | \Phi_{i'} \rangle + \langle \Phi_f | DG^Q V | \Phi_i \rangle + \cdots$$
$$= M_{fi}^{RR} + M_{fi'}^{TBR} + M_{fi}^{DR} + M_{fi}^{RDR} + \cdots \qquad (10)$$

where some of the amplitudes may vanish depending on the particular final states chosen, owing to symmetries, selection rules, etc. The projected Green's function G^Q describes the intermediate excited states of the ion plus the captured electron and gives rise to DR in the on-shell limit[26] and to RDR in the off-shell limit, as $M_{fi}^{RDR} = \langle \Phi_f | D\overline{G}^Q V | \Phi_i \rangle$, with $\overline{G}^Q$ the principal part of G^Q. The generalization of this formulation that includes the important multiple cascade effects is given in Ref. 26. The cross sections thus contain the RR and DR components plus the interference between the RR and DR amplitudes, as

$$\sigma_{fi}^{recomb} = \sigma_{fi}^{RR} + \sigma_{fi'}^{TBR} + \sigma_{fi}^{DR} + \sigma_{fi}^{RDR} + \sigma_{fi}^{interf} \qquad (11)$$

For most cases of physical interest, the interference terms between RR and DR, and between RR and RDR, are found to be small, except when the core charge is very large. This situation may change in the presence of the plasma perturbations, however, and this question is being investigated.

In the isolated resonance approximation, the intermediate states d for the (on-shell part of the) Green's function G^Q may be expressed as a sum over the doubly excited states d, and we have

$$\sigma_{fi}^{DR} \simeq \sum_d \sigma_{fdi}^{DR}, \quad \text{where } \sigma_{fdi}^{DR} \propto V_a(i \to d) A_r(d-f)/\Gamma(d) \tag{12}$$

and where $V_a(i \to d)$ is the inverse of the Auger process $A_a(d \to i)$. The total widths of states d are given by

$$\Gamma(d) = \sum_{i'} A_a(d \to i') + \sum_{f'} A_r(d-f') + \Gamma^{sk}(d) + \Gamma_F(d) \tag{13}$$

in terms of the Auger and radiative transition probabilities. The width corresponding to the shake-off, $\Gamma^{sk}(d)$, is added to the total width, as the shake-off contribution is found[33] to be important when strong correlations among the electrons in A are present during the V_a and A_a processes that involve inner-shell electrons. The field ionization probabilities Γ_F are also added, where the ionizing field can be external or plasma microfields.

The rates are defined by averaging the cross sections over the Maxwellian distribution of the continuum electrons, under the assumption of local thermal equilibrium, as

$$\alpha_{fi} = \int v \sigma_{fi} \, d\phi_v(T, v) \tag{14}$$

$$d\phi_v = [4\pi/(2\pi k_B T_e)^{3/2}] \exp(-E_c/k_B T_e) v dE_c,$$

where $E_c = mv^2/2$ for the continuum electrons which are assumed to be in Maxwell distribution.

The rates thus generated are put into the rate equations for determination of the excited state population densities and dominant charge state distributions. As noted earlier, however, the structure of the REQ is no longer arbitrary; it must correctly reflect the contents of the V_P used in calculating the rates, and vice versa. In particular, when some of the states (d) are included in the DR rates, the same states (d) must not appear explicitly in the rate equations. In cases in which all the d's are to be included in the rate equations, the DR rates then should be simply the $V_a(i \to d)$ without the branching ratios A_r/Γ, rather than the full rates as given by Eqs. (14) and (12). This point is obvious but has been overlooked in most of the modeling calculations.

A plasma, when disturbed externally, relaxes to its equilibrium state provided the plasma conditions are "stable." During the very early stage of this relaxation, the electron distribution may not be a simple Maxwellian. In fact, by definition of a time-evolving plasma, the distribution must be non-Maxwellian during the initial period, although the continuum electrons reach equilibrium much faster than the electron ion system A. The effect of the nonequilibrium nature of the electron distribution was studied recently,[34] and it was found that the ionization rates are especially sensitive to the form of the distribution in Eq. (14), whereas the radiative capture rates were relatively insensitive. This will be further discussed in Section 5.

Finally, the RDR contribution to the total recombination is difficult to estimate, because of the appearance of the correlated vertex $VGD + DGV$, where G is a principal part of G^Q. This problem requires extensive numerical computation and is being investigated.

3. RADIATIVE RECOMBINATION AND SCALED RATES

Radiative recombination is the simplest capture process, especially when the ionic target is completely stripped of electrons so that the final recombined ion is hydrogenic. For target ions with one or more electrons initially, several theoretical approaches are available, such as the single-configuration Hartree–Fock (HF) approximation and the distorted-wave method. More elaborate multiconfiguration HF and close-coupling methods have also been used. However, extensive numerical calculations using these refined methods showed that a simple hydrogenic treatment with a judicious choice of an effective charge works reasonably well in many cases. Because the RR cross sections and rate coefficients scale in the hydrogenic ion limit, it is convenient to document the rates using the effective charge approach. The cross sections involve integrals that may be evaluated exactly for the hydrogenic initial and final states. Detailed tabulation of the cross sections and rate coefficients is available[35,36] in terms of the scaled variables. Therefore, we have the RR cross sections, given in atomic units,

$$\sigma_{nl}^{RR} = (16\pi/3)\alpha_0^3(\hbar\omega)^3[l|R_{l-}|^2 + (l+1)|R_{l+}|^2]/E_c(\pi a_0^2) \quad (15)$$

where α_0 is the fine-structure constant, $l_\pm$ means $l_c = l \pm 1$, and $|R|$ are the radial dipole matrix elements between the bound and continuum orbitals. The continuum wave function is normalized here as

$$u_c \to \sqrt{2/\pi p_c} \sin(p_c r + \cdots)/r \quad (16)$$

The cross section in the hydrogenic model scales in $\eta = Z/p_c$, such that

$$\sigma_{fi}^{RR}(\eta, f = nl) = 2.06 \times 10^{-8}\eta K(\eta, nl)(\pi a_0^2) \quad (17)$$

where K is expressed by a series of analytic functions.[35] Table I contains the scaled RR cross sections.[36] The σ_{fi}^{RR} are given in the scaled parameter $\eta = Z_{eff}/(pa_0)$, and $f = nl$ and i is the continuum with energy p_c^2 Ry. Fully filled subshells for $nl \le n_0 l_0$ are assumed. For partially filled shells, the cross sections given in Table I are to be multiplied by $N_f/(4l + 2)$, where N_f is the number of holes in state f before capture. The numbers in parentheses are powers of 10, and the cross sections are given in units of πa_0^2. Table II contains the total RR cross sections σ_i^{RR}, given in the scaled variables, for capture to all the states that lie above, but exclusive of, the states

Table I. RR Cross Sections ($i \to f$) for $f = nl$ (in Units of πa_0^2)[a]

nl	\multicolumn{7}{c}{η}						
	0.50	1.0	2.0	5.0	10.0	20.0	40.0
1s	1.19 (–7)	1.13 (–6)	6.53 (–6)	4.64 (–5)	1.78 (–4)	7.27 (–4)	2.91 (–3)
2s	1.56 (–8)	1.59 (–7)	9.72 (–7)	6.87 (–6)	2.81 (–5)	1.13 (–4)	4.51 (–4)
2p	2.71 (–9)	9.16 (–8)	1.36 (–6)	1.63 (–5)	7.37 (–5)	3.04 (–4)	1.23 (–3)
3s	4.68 (–9)	4.89 (–8)	3.13 (–7)	2.31 (–6)	9.56 (–6)	3.83 (–5)	1.53 (–4)
3p	9.62 (–10)	3.32 (–8)	5.10 (–7)	6.24 (–6)	2.83 (–5)	1.16 (–4)	4.69 (–4)
3d	2.88 (–11)	3.79 (–9)	1.82 (–7)	5.32 (–6)	2.99 (–5)	1.30 (–4)	5.38 (–4)
4s	1.98 (–9)	2.09 (–8)	1.38 (–7)	1.06 (–6)	4.49 (–6)	1.80 (–5)	7.21 (–5)
4p	4.30 (–10)	1.50 (–8)	2.35 (–7)	2.96 (–6)	1.36 (–5)	5.58 (–5)	2.25 (–4)
4d	1.75 (–11)	2.30 (–9)	1.12 (–7)	3.33 (–6)	1.88 (–5)	8.23 (–5)	3.37 (–4)
4f	0.00 (00)	1.01 (–10)	1.67 (–8)	1.52 (–6)	1.20 (–5)	5.89 (–5)	2.48 (–4)
5s	1.02 (–9)	1.08 (–8)	7.23 (–8)	5.79 (–7)	2.51 (–6)	1.01 (–5)	4.05 (–5)
5p	2.26 (–10)	7.89 (–9)	1.25 (–7)	1.63 (–6)	7.62 (–6)	3.14 (–5)	1.27 (–4)
5d	1.03 (–11)	1.35 (–9)	6.64 (–8)	2.02 (–6)	1.15 (–5)	5.05 (–5)	2.07 (–4)
5f	0.00 (00)	8.65 (–11)	1.45 (–8)	1.33 (–6)	1.06 (–5)	5.19 (–5)	2.18 (–4)
5g	0.00 (00)	2.06 (–12)	1.11 (–9)	3.67 (–7)	4.68 (–6)	2.68 (–5)	1.18 (–4)
6	7.27 (–10)	1.18 (–8)	1.70 (–7)	4.11 (–6)	2.84 (–5)	1.39 (–4)	5.92 (–4)
7	4.59 (–10)	7.51 (–9)	1.10 (–7)	2.93 (–6)			
8	3.08 (–10)	5.06 (–9)	7.56 (–8)	2.15 (–6)			
9	2.17 (–10)	3.57 (–9)	5.38 (–8)	1.61 (–6)			
10	1.58 (–10)	2.61 (–9)	3.97 (–8)	1.23 (–6)			
>10	7.15 (–10)	1.18 (–8)	1.80 (–7)	5.57 (–6)			

[a] Numbers in parentheses are powers of 10.

$n_0 l_0$. All the other notations are as in Table I. For simplicity, the HRS contributions are estimated using the n^{-3} scaling. (This is valid for $E_c > Z^2/n^2$; otherwise, $\sigma \simeq 1/n$.)

The corresponding RR rates are given by

$$\alpha_{fi}^{RR} = \sqrt{2\pi} Z \Theta^{-3/2} \int d\eta \; \eta^{-5} \exp[-1/(2\Theta \eta^2)] \; \sigma_{fi}^{RR}(\eta) \tag{18}$$

where $\Theta = k_B T_e / Z^2$, in atomic units. The total RR rates summed over all the states above $n_0 l_0$ are given in Table III, where for simplicity the n^{-3} scaling was used again to estimate the high-n contribution. (Evidently, this depends on temperature as well.) All the states that lie below and inclusive of the states $n_0 l_0$ are assumed filled before the capture. The temperature $k_B T_e$ in Table III is given in rydbergs, and the rates are in units of cubic centimeters per second.

As mentioned earlier, the scaled cross sections and rate coefficients are useful also for ions with one or more electrons before capture. In terms of the effective charge Z_{eff}, we can immediately obtain the desired cross sections and rates from the tables. A simple form of the effective charge is given by

Table II. Total RR Cross Sections (in Units of πa_0^2)[a]

$n_0 l_0$	\multicolumn{7}{c}{η}						
	0.50	1.0	2.0	5.0	10.0	20.0	40.0
0	1.49 (−7)	1.56 (−6)	1.13 (−5)	1.16 (−4)	5.74 (−4)	2.40 (−3)	9.93 (−3)
1s	3.03 (−8)	4.38 (−7)	4.75 (−6)	6.95 (−5)	3.96 (−4)	1.68 (−3)	7.02 (−3)
2s	1.46 (−8)	2.78 (−7)	3.78 (−6)	6.26 (−5)	3.68 (−4)	1.56 (−3)	6.57 (−3)
2p	1.19 (−8)	1.87 (−7)	2.41 (−6)	6.26 (−5)	2.95 (−4)	1.26 (−3)	5.34 (−3)
3s	7.25 (−9)	1.38 (−7)	2.10 (−6)	4.40 (−5)	2.85 (−4)	1.22 (−3)	5.19 (−3)
3p	6.29 (−9)	1.05 (−7)	1.59 (−6)	3.77 (−5)	2.57 (−4)	1.11 (−3)	4.72 (−3)
3d	6.26 (−9)	1.01 (−7)	1.41 (−5)	3.24 (−5)	2.27 (−4)	9.76 (−4)	4.18 (−3)
4s	4.28 (−9)	7.98 (−8)	1.27 (−6)	3.14 (−5)	2.22 (−4)	9.58 (−4)	4.11 (−3)
4p	3.85 (−9)	6.49 (−8)	1.04 (−6)	2.84 (−5)	2.09 (−4)	9.02 (−4)	3.89 (−3)
4d	3.84 (−9)	6.26 (−8)	9.25 (−7)	2.51 (−5)	1.90 (−4)	8.20 (−4)	3.55 (−3)
4f	3.84 (−9)	6.25 (−8)	9.09 (−7)	2.35 (−5)	1.78 (−4)	7.61 (−4)	3.30 (−3)
5s	2.82 (−9)	5.17 (−8)	8.36 (−7)	2.30 (−5)	1.76 (−4)	7.51 (−4)	3.26 (−3)
5p	2.59 (−9)	4.38 (−8)	7.11 (−7)	2.13 (−5)	1.68 (−4)	7.19 (−4)	3.13 (−3)
5d	2.58 (−9)	4.25 (−8)	6.46 (−7)	1.93 (−5)	1.56 (−4)	6.69 (−4)	2.93 (−3)
5f	2.58 (−9)	4.24 (−8)	6.30 (−7)	1.80 (−5)	1.46 (−4)	6.17 (−4)	2.71 (−3)
5g	2.58 (−9)	4.24 (−8)	6.29 (−7)	1.76 (−5)	1.41 (−4)	5.90 (−4)	2.59 (−3)
6	1.91 (−9)	3.05 (−8)	4.59 (−7)	1.35 (−5)	1.13 (−4)	4.51 (−4)	2.00 (−3)
7	1.40 (−9)	2.30 (−8)	3.49 (−7)	1.06 (−5)	8.99 (−5)	3.45 (−4)	1.55 (−3)
8	1.09 (−9)	1.80 (−8)	2.73 (−7)	8.43 (−6)	7.24 (−5)	2.71 (−4)	1.21 (−3)
9	8.70 (−10)	1.44 (−8)	2.19 (−7)	6.82 (−6)	5.80 (−5)	2.17 (−4)	9.72 (−4)
10	7.13 (−10)	1.18 (−8)	1.80 (−7)	5.58 (−6)	4.75 (−5)	1.78 (−4)	7.96 (−4)

[a]Numbers in parentheses are powers of 10.

$$Z_{\text{eff}} \simeq (Z_C + Z_I)/2 \equiv Z_{\text{eff}}^{(0)}, \quad (19)$$

for low Rydberg states where Z_C is the nuclear core charge, and $Z_I (\equiv Z)$ is the ionic charge before capture. The number of electrons N in the target ion before capture is then $N = Z_C − Z_I$. For $Z_I < Z_C/2$, another form, $Z_{\text{eff}} \simeq (Z_C Z_I)^{1/2}$, was suggested. For HRS with $n \gg 1$, a slightly improved form, $Z_{\text{eff}} \simeq Z_{\text{eff}}^{(0)} − (N/2)(n − n_0 − 1)/n$, may work better. There are more refined forms available, which depend on the continuum energy E_c, l and n, etc., but we simply refer to the original report[37] for such refinements.

There have been several accurate measurements of the RR cross sections reported recently[38] for the fully stripped and single-electron target ions. The results agree well with the theory with Z_{eff}, within the experimental uncertainty.

Alternatively, approximate formulas for the RR cross sections and rate coefficients have been given in the past, the most prominent of which is Kramer's formula[2]:

$$\sigma_{fi}^{\text{RR-K}} = 6.1 \alpha_0^3 v^{-3} Z^4 / [E_c(E_c + Z^2/v^2)](\pi a_0^2) \quad (20)$$

Table III. The RR Rates $\alpha_i^{RR}/Z_{\text{eff}}$ (in Units of cm^3/s)^a

$\log_{10}\theta' \setminus n_0 l_0$	0	1s	2s	2p	3s	3p	3d
−2.0	1.23 (−12)	8.38 (−13)	7.58 (−13)	6.24 (−13)	5.97 (−13)	5.49 (−13)	4.72 (−13)
−1.5	6.18 (−13)	3.97 (−13)	3.57 (−13)	2.82 (−13)	2.69 (−13)	2.20 (−13)	2.08 (−13)
−1.0	2.86 (−13)	1.64 (−13)	1.43 (−13)	1.08 (−13)	1.01 (−13)	8.73 (−14)	7.54 (−14)
−0.5	1.21 (−13)	5.93 (−14)	4.93 (−14)	3.51 (−14)	3.22 (−14)	2.67 (−14)	2.33 (−14)
0.0	4.60 (−14)	1.86 (−14)	1.45 (−14)	9.97 (−15)	8.75 (−15)	7.10 (−15)	6.32 (−15)
0.5	1.52 (−14)	5.11 (−15)	3.66 (−15)	2.55 (−15)	2.11 (−15)	1.71 (−15)	1.55 (−15)
1.0	4.33 (−15)	1.26 (−15)	8.29 (−16)	5.93 (−16)	4.66 (−16)	3.80 (−16)	3.51 (−16)
1.5	1.05 (−15)	2.79 (−16)	1.74 (−16)	1.27 (−16)	9.64 (−17)	7.95 (−17)	7.42 (−17)
2.0	2.22 (−16)	5.65 (−17)	3.42 (−17)	2.50 (−17)	1.88 (−17)	1.56 (−17)	1.47 (−17)
2.5	4.27 (−17)	1.07 (−17)	6.38 (−18)	4.62 (−18)	3.49 (−18)	2.91 (−18)	2.74 (−19)
3.0	7.81 (−18)	1.94 (−18)	1.15 (−18)	8.32 (−19)	6.30 (−19)	5.30 (−19)	4.97 (−19)

	4s	4p	4d	4f	5s	5p	5d
−2.0	4.64 (−13)	4.32 (−13)	3.95 (−13)	3.68 (−13)	3.63 (−13)	3.46 (−13)	3.22 (−13)
−1.5	2.02 (−13)	1.86 (−13)	1.68 (−13)	1.58 (−13)	1.55 (−13)	1.46 (−13)	1.35 (−13)
−1.0	7.25 (−14)	6.54 (−14)	5.85 (−14)	5.51 (−14)	5.37 (−14)	4.98 (−14)	4.61 (−14)
−0.5	2.19 (−19)	1.93 (−14)	1.72 (−14)	1.64 (−14)	1.57 (−14)	1.42 (−14)	1.33 (−14)
0.0	5.71 (−15)	4.97 (−15)	4.45 (−15)	4.29 (−15)	3.99 (−15)	3.57 (−15)	3.39 (−15)
0.5	1.34 (−15)	1.16 (−15)	1.06 (−15)	1.03 (−15)	9.26 (−16)	8.24 (−16)	7.90 (−16)
1.0	2.92 (−16)	2.54 (−16)	2.34 (−16)	2.29 (−16)	2.00 (−16)	1.78 (−16)	1.72 (−16)
1.5	6.00 (−17)	5.26 (−17)	4.89 (−17)	4.80 (−17)	4.08 (−17)	3.67 (−17)	3.56 (−17)
2.0	1.16 (−17)	1.03 (−17)	9.60 (−18)	9.43 (−18)	7.91 (−18)	7.15 (−18)	6.95 (−18)
2.5	2.16 (−18)	1.91 (−18)	1.79 (−18)	1.76 (−18)	1.47 (−18)	1.33 (−18)	1.29 (−18)
3.0	3.90 (−19)	3.45 (−19)	3.24 (−19)	3.19 (−19)	2.65 (−19)	2.40 (−19)	2.34 (−19)

	5f	5g	6	7	8	9	10
−2.0	2.98 (−13)	2.87 (−13)	2.23 (−13)	1.74 (−13)	1.37 (−13)	1.11 (−13)	9.01 (−14)
−1.5	1.25 (−13)	1.21 (−13)	9.41 (−14)	7.37 (−14)	5.76 (−14)	4.71 (−14)	3.81 (−14)
−1.0	4.25 (−14)	4.15 (−14)	3.18 (−14)	2.48 (−14)	1.94 (−14)	1.59 (−14)	1.29 (−14)
−0.5	1.22 (−14)	1.20 (−14)	9.15 (−15)	7.08 (−15)	5.55 (−15)	4.51 (−15)	3.67 (−15)
0.0	3.13 (−15)	3.09 (−15)	2.34 (−15)	1.80 (−15)	1.41 (−15)	1.14 (−15)	9.29 (−16)
0.5	7.38 (−16)	7.29 (−16)	5.48 (−16)	4.19 (−16)	3.28 (−16)	2.64 (−16)	2.16 (−16)
1.0	1.62 (−16)	1.61 (−16)	1.20 (−16)	9.15 (−17)	7.16 (−17)	5.76 (−17)	4.71 (−17)
1.5	3.39 (−17)	3.36 (−17)	2.48 (−17)	1.90 (−17)	1.48 (−17)	1.19 (−17)	9.76 (−18)
2.0	6.64 (−18)	6.60 (−18)	4.82 (−18)	3.71 (−18)	2.90 (−18)	2.33 (−18)	1.91 (−18)
2.5	1.24 (−18)	1.23 (−18)	8.93 (−19)	6.90 (−19)	5.40 (−19)	4.35 (−19)	3.56 (−19)
3.0	2.24 (−19)	2.23 (−19)	1.61 (−19)	1.25 (−19)	9.77 (−20)	7.87 (−20)	6.45 (−20)

^{a}Numbers in parentheses are in powers of 10.

where E_c is the continuum energy in rydbergs, and v is the effective principal quantum number for the final state f with the quantum defects. The initial and final l's are summed. At small $E_c \to 0$, σ diverges, whereas for finite E_c, σ decreases rapidly with increasing v, as

$$\sigma \to E_c^{-1} \ (E_c \to 0) \quad \text{and} \quad \sigma - v^{-3} \ (v \to \infty) \tag{21}$$

However, when $E_c << Z^2/v^2$, $\sigma \to v^{-1}$. The corresponding rates are given by

$$\alpha_{fi}^{RR-K} = 1.3 \times 10^{-14} E_n^2 g_n (k_B T_e)^{-3/2} \exp[-E_n/k_B T_e] \, \mathscr{E}_1(E_n/k_B T_e) \tag{22}$$

in units of cubic centimeters per second. $\mathscr{E}_1$ is the elliptic integral of the first kind. As noted above, these formulas may be used with the effective charge for ions with one or more electrons initially.

There have been some theoretical attempts to include the plasma field effects in the RR rate calculation, especially when low-energy electrons and capture to HRS are involved, as these cases are most sensitive to field perturbations. Preliminary results indicate that the field effect is very large when a special momentum coherence due to field-induced dressing occurs.[39] However, the situations have not been fully clarified.

Summarizing the discussion on the RR, the following problems require further study:

1. The RR rates for the excited ionic targets are not available; they are needed in the rate equations.
2. The plasma effects on the RR rates, especially for low energies and HRS capture, are still very poorly understood. Further experimental and theoretical work is needed.

4. DIELECTRONIC RECOMBINATION FOR THE GROUND STATES

Much theoretical work[1,26] on DR cross sections and rate coefficients has been carried out in the distorted-wave approximation, using the single-configuration HF wave functions. The configuration mixing and relativistic effects can be important, and some specific cases have been analyzed using more elaborate theories. However, for the purpose of generating a large number of rates for plasma modeling, such approaches are prohibitive, but they provide valuable tests of crude approximations. The DR cross sections exhibit the resonant behavior as, in atomic units,

$$\sigma_{fdi}^{DR} = (2\pi/E_c) V_a(i \to d) \, \omega(d \to f) \tilde{\delta}(E_c)(\pi a_0^2) \tag{23}$$

where $V_a(i \to d) = A_a(d \to i)(g_d/2g_i)$ by detailed balance, the g's are the statistical factors, and

$$\tilde{\delta} = [\Gamma(d)/2\pi]/[(E_i - E_d)^2 + \Gamma(d)^2/4] \tag{24}$$

is a Lorentzian line profile, with $\int \tilde{\delta} \, dE_d = 1$, indicating the resonance property of DR. The partial and total fluorescence yields are defined by

$$\omega(d \to f) = A_r(d \to f)/\Gamma(d) \tag{25}$$

and

$$\omega(d) = \sum_f \omega(d \to f) = \Gamma_r(d)/\Gamma(d) \qquad (26)$$

In these expressions, the total widths are given as in Eq. (13),

$$\Gamma(d) = \Gamma_a(d) + \Gamma_r(d) + \Gamma^{sk}(d) + \Gamma_F(d) \qquad (27)$$

in terms of the Auger and radiative contributions to the doubly excited levels d. The shake-off contribution may be neglected in general, except for some special cases in which it can be very important.[33] The above formulas are valid in the isolated resonance approximation, and in the zero-field/zero-density limit. The DR rates, in units of cubic centimeters per second, are defined by Eq. (14) with Eq. (23) as

$$\alpha_{fdi}^{DR} = (4\pi \, Ry/k_B T_e)^{3/2} V_a(i \to d) \, \omega(d \to f) \qquad (28)$$

where Ry = rydbergs.

As noted earlier, there are still large gaps[40] in the available DR rates, especially for ions in the regions $N = 5$–8 and $N > 12$, even when i is restricted to the ground states. A careful fit to the existing data helps to extrapolate into the gap regions, but much more work is needed to test and improve such extrapolations, even in the absence of all the above complications. Experimental data are sometimes available for comparison. Nevertheless, we have recently attempted a general fit to the available data, with some assessment of their accuracy. Instead of the usual fitting procedure along the isoelectronic sequences, it was decided that fitting for each excitation mode may be more appropriate. Previous work[40,41] indicated that the DR rates are not always smooth functions of the electron temperature T_e, because different processes become important at different temperatures; that is, the $\Delta n = 0$ and $\Delta n \neq 0$ excitation modes, represented by V_a corresponding to intra- and intershell transitions, respectively, give contributions at two different temperatures. The DR rates for the degenerate multiplet excitations are large at even lower energies than in the intrashell case. In addition, when the fit is carried out along the isoelectronic sequences, some of the transitions that contribute a small but significant fraction are buried in the phenomenological formulas. This aspect has been improved upon in the new fit.

We obtained a preliminary fit[41] for a small set of ions using the new approach, which showed relatively smooth variations in N and Z. The new rate formulas that we obtained[42] for i = ground states are simple to use and, more importantly, do not depend on the external inputs, such as the level energies and oscillator strengths. The excitation modes are specified in the definition of the rates α:

$$\alpha(1s) = (A_1 e^{-A_2/T} T^{-3/2}) \exp[-A_3(N-2)^2][6/(4+N)]^{0.9} \qquad (29)$$

$$\alpha(2s) = (B_1 e^{-B_2/T} T^{-3/2})(N-2)(10-N)(N+B_3)^{-5/2}(1+0.3T^{-0.21}) \qquad (30)$$

$$\alpha(2p) = F_C C_1 \exp\{-C_2[1 + 0.0001/(N+1)]/T\} T^{-3/2} \quad (31)$$

where

$$F_C = \exp[-C_3|N - 9.6|[10/(N+1)]^{0.5}$$

$$\alpha(3s) = (D_1 \exp\{-D_2[1 - 0.15/(N-10)^{1.5}]/T\} T^{-3/2})$$
$$\cdot (N-10)(Z_C - N)(N - D_3)^{-1} \quad (32)$$

$$\alpha(3p) = (E_1 e^{-E_2/T}/T^{3/2}) \exp^{-E_3(N-12)^2}(N-10)(Z_C - 10)/8 \quad (33)$$

where the 2s and 3s formulas are for the $\Delta n = 0$ transitions, whereas the 1s, 2p, and 3p formulas are for the $\Delta n \neq 0$ case and include the $\Delta n \neq 0$ excitations of the 2s and 3s electrons. The rates are given in units of 10^{-13} cm³/s, and the temperature is in keV.

The coefficients introduced in the above formulas are in general Z-dependent:

$$A_1 = 1230 \exp[-44/(Z_C + 2.86)] Z_C^{-0.14}$$

$$A_2 = 0.0075(Z_C + 1/N)^2 \quad \text{and} \quad A_3 = 0.0222 Z_C \quad (34)$$

where, unlike the other parameters, A_2 depends on N. Now, let $Z = Z_C - 2$, and

$$B_1 = 52 \exp[-18/(Z+1)][Z_C(10 + 0.011 Z_C^2)^{-1}]^{0.65}$$

$$B_2 = 0.0023 Z(1 + 0.0015 Z^2)^{-1} \quad \text{and} \quad B_3 = 0.8 \quad (35)$$

$$C_1 = 2.15 \exp[-0.004(Z - 35)^2] Z^{1.8}$$

$$C_2 = 0.00115 Z^2(1 - 0.003 Z) \quad \text{and} \quad C_3 = 0.17 \quad (36)$$

With $Z' = Z_C - 10$, we have

$$D_1 = 0.16 Z'^2 \exp(-0.11 Z')$$

$$D_2 = 0.0024 Z'(1 - 0.01 Z') \quad \text{and} \quad D_3 = 6 \quad (37)$$

Furthermore, with $Z'' = Z' + 3 = Z_C - 7$,

$$E_1 = 0.45 \exp[Z'/(4 + 0.02 Z')]/Z'$$

$$E_2 = 0.0003 Z''^2 (1 - 0.003 Z'') \quad \text{and} \quad E_3 = 0.02 \quad (38)$$

We note that all the parameters above are independent of N, except A_2. Table IV contains the DR rates for C ions that are generated by the new rate formulas, in units of 10^{-13} cm³/s, and the electron temperature T_e in keV. N is the number of electrons in the target ion before capture. For $\Delta n > 0$, intershell excitations occur during the radiationless capture (V_a). As stated earlier, the rates for the 2s and 3s

Table IV. DR Rates for C Ions (in Units of 10^{-13} cm^3/s)

N	T (keV)	1S	2S	2P	3S	3P	$\Delta n > 0$	$\Delta n = 0$	Total
1	0.01	—	—	—	—	—	—	—	—
1	0.02	—	—	—	—	—	—	—	—
1	0.05	0.4	—	—	—	—	0.4	—	0.4
1	0.10	5.5	—	—	—	—	5.5	—	5.5
1	0.20	12.2	—	—	—	—	12.2	—	12.2
1	0.50	9.3	—	—	—	—	9.3	—	9.3
1	1.00	4.8	—	—	—	—	4.8	—	4.8
1	2.00	2.0	—	—	—	—	2.0	—	2.0
1	5.00	0.6	—	—	—	—	0.6	—	0.6
1	10.00	0.2	—	—	—	—	0.2	—	0.2
2	0.01	—	—	—	—	—	—	—	—
2	0.02	—	—	—	—	—	—	—	—
2	0.05	1.1	—	—	—	—	1.1	—	1.1
2	0.10	8.9	—	—	—	—	8.9	—	8.9
2	0.20	15.3	—	—	—	—	15.3	—	15.3
2	0.50	10.0	—	—	—	—	10.0	—	10.0
2	1.00	4.9	—	—	—	—	4.9	—	4.9
2	2.00	2.0	—	—	—	—	2.0	—	2.0
2	5.00	0.6	—	—	—	—	0.6	—	0.6
2	10.00	0.2	—	—	—	—	0.2	—	0.2
3	0.01	—	180.1	46.7	—	—	46.7	180.1	226.7
3	0.02	—	93.8	40.9	—	—	40.9	93.8	134.7
3	0.05	1.1	28.9	17.9	—	—	19.0	28.9	47.8
3	0.10	7.9	10.6	7.6	—	—	15.5	10.6	26.1
3	0.20	12.6	3.8	2.9	—	—	15.6	3.8	19.3
3	0.50	7.9	0.9	0.8	—	—	8.7	0.9	9.6
3	1.00	3.8	0.3	0.3	—	—	4.0	0.3	4.4
3	2.00	1.5	0.1	0.1	—	—	1.6	0.1	1.8
3	5.00	0.4	—	—	—	—	0.5	—	0.5
3	10.00	0.2	—	—	—	—	0.2	—	0.2
4	0.01	—	172.1	49.5	—	—	49.5	172.1	221.6
4	0.02	—	89.7	43.4	—	—	43.4	89.7	133.1
4	0.05	0.8	27.6	18.9	—	—	19.7	27.6	47.3
4	0.10	5.1	10.2	8.0	—	—	13.1	10.2	23.3
4	0.20	7.8	3.6	3.1	—	—	10.9	3.6	14.5
4	0.50	4.8	0.9	0.8	—	—	5.6	0.9	6.5
4	1.00	2.3	0.3	0.3	—	—	2.6	0.3	2.9
4	2.00	0.9	0.1	0.1	—	—	1.0	0.1	1.1
4	5.00	0.3	—	—	—	—	0.3	—	0.3
4	10.00	0.1	—	—	—	—	0.1	—	0.1
5	0.01	—	134.1	53.5	—	—	53.5	134.1	187.6
5	0.02	—	69.8	47.0	—	—	47.0	69.8	116.8
5	0.05	0.4	21.5	20.5	—	—	20.9	21.5	42.4
5	0.10	2.5	7.9	8.7	—	—	11.2	7.9	19.1
5	0.20	3.7	2.8	3.4	—	—	7.1	2.8	9.8
5	0.50	2.2	0.7	0.9	—	—	3.1	0.7	3.8
5	1.00	1.0	0.2	0.3	—	—	1.4	0.2	1.6
5	2.00	0.4	0.1	0.1	—	—	0.5	0.1	0.6
5	5.00	0.1	—	—	—	—	0.1	—	0.2
5	10.00		—	—	—	—	0.1	—	0.1

Table V. DR Rates for O Ions (in Units of 10^{-13} cm^3/s)

N	T (keV)	1S	2S	2P	3S	3P	$\Delta n > 0$	$\Delta n = 0$	Total
1	0.01	—	—	—	—	—	—	—	—
1	0.02	—	—	—	—	—	—	—	—
1	0.05	—	—	—	—	—	—	—	—
1	0.10	1.1	—	—	—	—	1.1	—	1.1
1	0.20	8.5	—	—	—	—	8.5	—	8.5
1	0.50	13.2	—	—	—	—	13.2	—	13.2
1	1.00	8.6	—	—	—	—	8.6	—	8.6
1	2.00	4.1	—	—	—	—	4.1	—	4.1
1	5.00	1.2	—	—	—	—	1.2	—	1.2
1	10.00	0.5	—	—	—	—	0.5	—	0.5
2	0.01	—	—	—	—	—	—	—	—
2	0.02	—	—	—	—	—	—	—	—
2	0.05	—	—	—	—	—	—	—	—
2	0.10	2.2	—	—	—	—	2.2	—	2.2
2	0.20	11.9	—	—	—	—	11.9	—	11.9
2	0.50	15.3	—	—	—	—	15.3	—	15.3
2	1.00	9.3	—	—	—	—	9.3	—	9.3
2	2.00	4.3	—	—	—	—	4.3	—	4.3
2	5.00	1.3	—	—	—	—	1.3	—	1.3
2	10.00	0.5	—	—	—	—	0.5	—	0.5
3	0.01	—	395.1	16.5	—	—	16.5	395.1	411.6
3	0.02	—	252.8	44.6	—	—	44.6	252.8	297.4
3	0.05	—	88.0	38.2	—	—	38.2	88.0	126.2
3	0.10	2.0	33.7	20.3	—	—	22.3	33.7	56.0
3	0.20	9.6	12.2	8.8	—	—	18.4	12.2	30.6
3	0.50	11.6	3.0	2.5	—	—	14.1	3.0	17.2
3	1.00	6.9	1.0	0.9	—	—	7.8	1.0	8.9
3	2.00	3.2	0.4	0.3	—	—	3.5	0.4	3.9
3	5.00	0.9	0.1	0.1	—	—	1.0	0.1	1.1
3	10.00	0.3	—	—	—	—	0.4	—	0.4
4	0.01	—	377.7	17.5	—	—	17.5	377.7	395.2
4	0.02	—	241.6	47.3	—	—	47.3	241.6	288.9
4	0.05	—	84.1	40.5	—	—	40.5	84.1	124.6
4	0.10	1.2	32.2	21.5	—	—	22.7	32.2	54.9
4	0.20	5.3	11.6	9.3	—	—	14.6	11.6	26.2
4	0.50	6.2	2.9	2.7	—	—	8.8	2.9	11.7
4	1.00	3.6	1.0	1.0	—	—	4.6	1.0	5.6
4	2.00	1.7	0.3	0.4	—	—	2.0	0.3	2.4
4	5.00	0.5	0.1	0.1	—	—	0.6	0.1	0.7
4	10.00	0.2	—	—	—	—	0.2	—	0.2
5	0.01	—	294.2	19.0	—	—	19.0	294.2	313.1
5	0.02	—	188.2	51.2	—	—	51.2	188.2	239.4
5	0.05	—	65.5	43.8	—	—	43.8	65.5	0.4
5	0.10	0.5	25.1	23.3	—	—	23.7	25.1	48.8
5	0.20	2.0	9.1	10.1	—	—	12.1	9.1	21.2
5	0.50	2.3	2.3	2.9	—	—	5.2	2.3	7.5
5	1.00	1.4	0.8	1.1	—	—	2.4	0.8	3.2
5	2.00	0.6	0.3	0.4	—	—	1.0	0.3	1.3
5	5.00	0.2	0.1	0.1	—	—	0.3	0.1	0.3
5	10.00	0.1	—	—	—	—	0.1	—	0.1

Table VI. DR Rates for Ar Ions (in Units of 10^{-13} cm^3/s)

N	T (keV)	1S	2S	2P	3S	3P	$\Delta n > 0$	$\Delta n = 0$	Total
1	0.01	—	—	—	—	—	—	—	—
1	0.02	—	—	—	—	—	—	—	—
1	0.05	—	—	—	—	—	—	—	—
1	0.10	—	—	—	—	—	—	—	—
1	0.20	—	—	—	—	—	—	—	—
1	0.50	1.0	—	—	—	—	1.0	—	1.0
1	1.00	5.2	—	—	—	—	5.2	—	5.2
1	2.00	7.2	—	—	—	—	7.2	—	7.2
1	5.00	4.1	—	—	—	—	4.1	—	4.1
1	10.00	1.9	—	—	—	—	1.9	—	1.9
2	0.01	—	—	—	—	—	—	—	—
2	0.02	—	—	—	—	—	—	—	—
2	0.05	—	—	—	—	—	—	—	—
2	0.10	—	—	—	—	—	—	—	—
2	0.20	—	—	—	—	—	—	—	—
2	0.50	1.7	—	—	—	—	1.7	—	1.7
2	1.00	7.6	—	—	—	—	7.6	—	7.6
2	2.00	9.8	—	—	—	—	9.8	—	9.8
2	5.00	5.3	—	—	—	—	5.3	—	5.3
2	10.00	2.4	—	—	—	—	2.4	—	2.4
3	0.01	—	675.4	—	—	—	—	675.4	675.4
3	0.02	—	848.4	—	—	—	—	848.4	848.5
3	0.05	—	442.8	12.6	—	—	12.6	442.8	455.4
3	0.10	—	194.3	73.7	—	—	73.7	194.3	267.9
3	0.20	—	75.0	105.7	—	—	105.7	75.0	180.7
3	0.50	1.1	19.5	62.0	—	—	63.1	19.5	82.6
3	1.00	4.7	6.8	29.0	—	—	33.7	6.8	40.5
3	2.00	5.8	2.4	11.8	—	—	17.6	2.4	20.0
3	5.00	3.1	0.6	3.2	—	—	6.4	0.6	7.0
3	10.00	1.4	0.2	1.2	—	—	2.6	0.2	2.8
4	0.01	—	645.6	—	—	—	—	645.6	645.6
4	0.02	—	811.1	—	—	—	—	811.1	811.1
4	0.05	—	423.3	13.4	—	—	13.4	423.3	436.7
4	0.10	—	185.7	78.1	—	—	78.1	185.7	263.8
4	0.20	—	71.7	112.1	—	—	112.1	71.7	183.8
4	0.50	0.3	18.6	65.8	—	—	66.0	18.6	84.7
4	1.00	1.3	6.5	30.8	—	—	32.0	6.5	38.6
4	2.00	1.6	2.3	12.5	—	—	14.1	2.3	16.4
4	5.00	0.8	0.6	3.4	—	—	4.3	0.6	4.8
4	10.00	0.4	0.2	1.3	—	—	1.6	0.2	1.8
5	0.01	—	502.8	—	—	—	—	502.8	502.8
5	0.02	—	631.7	—	—	—	—	631.7	631.7
5	0.05	—	329.6	14.5	—	—	14.5	329.6	344.1
5	0.10	—	144.6	84.5	—	—	84.5	144.6	229.1
5	0.20	—	55.8	121.3	—	—	121.3	55.8	177.1
5	0.50	—	14.5	71.1	—	—	71.2	14.5	85.7
5	1.00	0.2	5.1	33.3	—	—	33.4	5.1	38.5
5	2.00	0.2	1.8	13.5	—	—	13.7	1.8	15.5
5	5.00	0.1	0.4	3.7	—	—	3.8	0.4	4.3
5	10.00	0.0	0.2	1.4	—	—	1.4	0.2	1.6

Table VI. (continued)

N	T (keV)	1S	2S	2P	3S	3P	Δn > 0	Δn = 0	Total
10	0.01	—	—	—	—	—	—	—	—
10	0.02	—	—	—	—	—	—	—	—
10	0.05	—	—	21.9	—	—	21.9	—	21.9
10	0.10	—	—	127.4	—	—	127.4	—	127.4
10	0.20	—	—	183.0	—	—	183.0	—	183.0
10	0.50	—	—	107.3	—	—	107.3	—	107.3
10	1.00	—	—	50.2	—	—	50.2	—	50.2
10	2.00	—	—	20.4	—	—	20.4	—	20.4
10	5.00	—	—	5.6	—	—	5.6	—	5.6
10	10.00	—	—	2.0	—	—	2.0	—	2.0
11	0.01	—	—	—	1324.9	11.3	11.3	1324.9	1336.2
11	0.02	—	—	—	992.4	23.1	23.1	992.4	1015.4
11	0.05	—	—	17.7	393.9	16.7	34.4	393.9	428.3
11	0.10	—	—	102.9	161.8	8.4	111.3	161.8	273.2
11	0.20	—	—	147.8	61.7	3.5	151.3	61.7	213.0
11	0.50	—	—	86.7	16.3	1.0	87.7	16.3	104.0
11	1.00	—	—	40.6	5.9	0.4	40.9	5.9	46.8
11	2.00	—	—	16.5	2.1	0.1	16.6	2.1	18.7
11	5.00	—	—	4.5	0.5	—	4.6	0.5	5.1
11	10.00	—	—	1.7	0.2	—	1.7	0.2	1.9
12	0.01	—	—	—	1594.8	23.0	23.0	1594.8	1617.8
12	0.02	—	—	—	1301.3	47.0	47.1	1301.3	1348.4
12	0.05	—	—	14.3	543.8	34.1	48.4	543.8	592.2
12	0.10	—	—	83.4	227.3	17.1	100.6	227.3	327.8
12	0.20	—	—	119.8	87.4	7.2	127.0	87.4	214.4
12	0.50	—	—	70.3	23.2	2.0	72.3	23.2	95.5
12	1.00	—	—	32.9	8.4	0.7	33.6	8.4	42.0
12	2.00	—	—	13.4	3.0	0.3	13.6	3.0	16.6
12	5.00	—	—	3.7	0.8	0.1	3.7	0.8	4.5
12	10.00	—	—	1.3	0.3	—	1.4	0.3	1.6

shells include only the $\Delta n = 0$ mode; the $1s$, $2p$, and $3p$ modes are for the $\Delta n > 0$ excitations and include the contributions of the $2s$ and $3s$ electron excitations as well. Tables V–VII contain the rates for O, Ar, and Fe ions, respectively. These sample rates may be used as a check in generating rates with the formulas presented in Eqs. (29)–(33).

As noted in Section 1, the DR process is intrinsically related[24] to the other resonant electron–ion collision processes, such as resonant excitation and excitation autoionization.[22] It is also related to the resonant transfer excitation in ion–atom collisions, with X-ray decay (RTEX). In fact, some of the earliest experiments carried out on DR with $\Delta n \neq 0$ were in the RTEX[11,23] mode.

Table VII. DR Rates for Fe Ions (in Units of 10^{-13} cm^3/s)

N	T (keV)	1S	2S	2P	3S	3P	$\Delta n > 0$	$\Delta n = 0$	Total
1	0.01	—	—	—	—	—	—	—	—
1	0.02	—	—	—	—	—	—	—	—
1	0.05	—	—	—	—	—	—	—	—
1	0.10	—	—	—	—	—	—	—	—
1	0.20	—	—	—	—	—	—	—	—
1	0.50	—	—	—	—	—	—	—	—
1	1.00	0.5	—	—	—	—	0.5	—	0.5
1	2.00	2.6	—	—	—	—	2.6	—	2.6
1	5.00	3.4	—	—	—	—	3.4	—	3.4
1	10.00	2.1	—	—	—	—	2.1	—	2.1
2	0.01	—	—	—	—	—	—	—	—
2	0.02	—	—	—	—	—	—	—	—
2	0.05	—	—	—	—	—	—	—	—
2	0.10	—	—	—	—	—	—	—	—
2	0.20	—	—	—	—	—	—	—	—
2	0.50	—	—	—	—	—	—	—	—
2	1.00	0.9	—	—	—	—	0.9	—	0.9
2	2.00	4.3	—	—	—	—	4.3	—	4.3
2	5.00	5.3	—	—	—	—	5.3	—	5.3
2	10.00	3.2	—	—	—	—	3.2	—	3.2
3	0.01	—	755.6	—	—	—	—	755.6	755.6
3	0.02	—	1104.1	—	—	—	—	1104.1	1104.1
3	0.05	—	630.9	0.1	—	—	0.1	630.9	631.0
3	0.10	—	285.3	14.1	—	—	14.1	285.3	299.4
3	0.20	—	111.8	107.6	—	—	107.6	111.8	219.4
3	0.50	—	29.3	172.2	—	—	172.2	29.3	201.5
3	1.00	0.5	10.3	112.5	—	—	113.0	10.3	123.3
3	2.00	2.2	3.6	54.1	—	—	56.3	3.6	59.9
3	5.00	2.6	0.9	16.5	—	—	19.1	0.9	20.0
3	10.00	1.6	0.3	6.2	—	—	7.7	0.3	8.1
4	0.01	—	722.3	—	—	—	—	722.3	722.3
4	0.02	—	1055.5	—	—	—	—	1055.5	1055.5
4	0.05	—	603.1	0.1	—	—	0.1	603.1	603.2
4	0.10	—	272.7	14.9	—	—	14.9	272.7	287.7
4	0.20	—	106.9	114.1	—	—	114.1	106.9	221.0
4	0.50	—	28.0	182.5	—	—	182.5	28.0	210.5
4	1.00	0.1	9.8	119.3	—	—	119.4	9.8	129.2
4	2.00	0.3	3.4	57.4	—	—	57.7	3.4	61.1
4	5.00	0.4	0.8	17.5	—	—	17.9	0.8	18.7
4	10.00	0.2	0.3	6.6	—	—	6.8	0.3	7.1

Table VII. (continued)

N	T (keV)	1S	2S	2P	3S	3P	$\Delta n > 0$	$\Delta n = 0$	Total
5	0.01	—	562.5	—	—	—	—	562.5	562.5
5	0.02	—	822.1	—	—	—	—	822.1	822.1
5	0.05	—	469.7	0.1	—	—	0.1	469.7	469.8
5	0.10	—	212.4	16.2	—	—	16.2	212.4	228.6
5	0.20	—	83.2	123.5	—	—	123.5	83.2	206.7
5	0.50	—	21.8	197.5	—	—	197.5	21.8	219.3
5	1.00	—	7.7	129.1	—	—	129.1	7.7	136.8
5	2.00	—	2.7	62.1	—	—	62.1	2.7	64.8
5	5.00	—	0.7	18.9	—	—	18.9	0.7	19.6
5	10.00	0.0	0.2	7.1	—	—	7.1	0.2	7.3
10	0.01	—	—	—	—	—	—	—	—
10	0.02	—	—	—	—	—	—	—	—
10	0.05	—	—	0.1	—	—	0.1	—	0.1
10	0.10	—	—	24.4	—	—	24.4	—	24.4
10	0.20	—	—	186.2	—	—	186.2	—	186.2
10	0.50	—	—	297.9	—	—	297.9	—	297.9
10	1.00	—	—	194.7	—	—	194.7	—	194.7
10	2.00	—	—	93.6	—	—	93.6	—	93.6
10	5.00	—	—	28.5	—	—	28.5	—	28.5
10	10.00	—	—	10.7	—	—	10.7	—	10.7
11	0.01	—	—	—	1362.7	0.1	0.1	1362.7	1362.8
11	0.02	—	—	—	1897.6	4.8	4.8	1897.6	1902.4
11	0.05	—	—	0.1	1092.8	26.0	26.1	1092.8	1118.8
11	0.10	—	—	19.7	508.2	25.5	45.2	508.2	553.4
11	0.20	—	—	150.4	206.1	15.0	165.4	206.1	371.5
11	0.50	—	—	240.6	56.6	5.2	245.8	56.6	302.4
11	1.00	—	—	157.3	20.6	2.0	159.3	20.6	179.9
11	2.00	—	—	75.6	7.4	0.8	76.4	7.4	83.7
11	5.00	—	—	23.0	1.9	0.2	23.2	1.9	25.1
11	10.00	—	—	8.6	0.7	0.1	8.7	0.7	9.4
12	0.01	—	—	—	1550.4	0.2	0.2	1550.4	1550.6
12	0.02	—	—	—	2524.5	9.8	9.8	2524.5	2534.3
12	0.05	—	—	0.1	1596.8	53.0	53.1	1596.8	1649.8
12	0.10	—	—	16.0	766.2	52.0	68.0	766.2	834.2
12	0.20	—	—	121.9	315.6	30.6	152.6	315.6	468.2
12	0.50	—	—	195.0	87.5	10.5	205.5	87.5	293.0
12	1.00	—	—	127.5	31.9	4.1	131.6	31.9	163.5
12	2.00	—	—	61.3	11.5	1.5	62.8	11.5	74.3
12	5.00	—	—	18.6	2.9	0.4	19.0	2.9	22.0
12	10.00	—	—	7.0	1.0	0.1	7.2	1.0	8.2

Finally, the effects of shake-off and the new mode of capture, RDR, have not been included in the result presented above. Theoretical calculation of these effects is difficult, as they are the result of strong electron–electron correlations. Since RDR is an off-shell part of the DR type amplitude, its contribution to the total recombination will be significant only for energies away from the DR resonances. We expect its effect to be sizable near $E_c \simeq 0$, at higher energies $E_c > E_{DR}$ for the DR resonance energies, and for processes in which DR is forbidden by selection rules.

Insofar as the availability of DR rates is concerned, we note several critical points that have yet to be addressed:

1. The DR rates for the ground states of ions with the number of electrons $N = 5, 6, 7,$ and 8 and $N > 11$ are needed, as pointed out earlier. For open-shell ions, the configuration mixing can be strong.

2. DR rates for excited states, including the metastable states, are almost nonexistent, although they are not any harder to evaluate than those for ground states, at least in the distorted-wave approximation.

3. The effects of plasma environment on the DR rates, especially for the intrashell excitation mode, have been totally neglected in all the available rates, although the seriousness of the problem is well known.

4. The DR process is essentially a two-step process, so that steps $i \to d$ and $d \to f$ steps are separately completed, in the definition (23) and (28). Therefore, in using these rates in the rate equations, the states d should not appear. Otherwise, double counting results. Alternatively, these states may be included explicitly in the REQ, in which case the corresponding DR rates should simply be $V_a(i \to d)$.

5. Nearly all the available data on DR are for the ground final (f) states, which are presumably both radiatively and Auger stable. In actual cases, however, all the final states, which are only Auger stable, have been summed. This can again lead to inconsistencies. The DR rates for all the excited states f included in the REQ are then needed, where the choice of the set $\{f\}$ will affect the rates themselves.

5. PLASMA FIELD EFFECTS AND RATE EQUATIONS

We have already mentioned some of the potentially serious problems caused by the presence of plasma particles in a strongly perturbing environment.[43] The plasma perturbations cause essentially two effects: (a) distortions of the ionic states involved in the various transitions and (b) cascade transitions among the ionic levels. The RR cross sections do not seem to be affected much by the presence of the plasma field unless $E_c \simeq 0$ or captures to high Rydberg states are involved. The momentum coherence may then take place,[39] which generally enhances the RR cross sections. Even then, the total sums over all the allowed final states become less sensitive to the effective plasma potential (EPP), and the resulting rates are further desensitized to the EPP when the cross sections are averaged over the continuum energy distribution. On the other hand, the plasma cascade effect (b) can be sizable, as seen, for example, in the effective collisional radiative recombination and ionization rates obtained by solving the appropriate set of rate equations.[44,45]

The DR cross sections are similarly affected by the plasma field effects,[46,47] especially when the intermediate states d contain one or more electrons in high Rydberg states. In the (dominant) dipole approximation for the electron–electron correlation involving a high Rydberg electron and a low-lying electron in state d, the transition V_a in DR is very much like that in RR in a PFE; a similar momentum coherent effect is therefore expected for DR. On the other hand, in DR the fluorescence yields are also affected by the PFE, so that an alteration in the delicate balance between the two effects produces changes in DR rates. When the cascade transitions induced by the plasma particles are included, the effective DR rates may be further modified. This problem has only begun to be treated.[48,49]

Recently, as sketched in Section 2, we have formulated a simple and workable theory of PFE, based on previously developed pressure broadening theory (PBT)[30–32,43,50–53] and the rate equations approach (REQ). It involves essentially three interrelated steps. The first step is construction of a simple EPP, as in Eq. (7), which carries all the "slowly varying" part of the plasma atom interactions. The "fast" component is deleted in the EPP. In the course of simplifying the EPP, many drastic approximations are introduced, and they are to be tested and improved. In the second step, the distorted atomic states are generated using this EPP, as in Eq. (9), and the distorted state amplitudes (DSA) and rates involving plasma electron perturbations are estimated. They correspond to the "fast varying" part of the PFE. The dressing of atomic states is a nontrivial mathematical problem, and the evaluation of the rates also requires extensive numerical work. In the third step, the appropriate rate equations are set up that reflect correctly the contents of the EPP and the resulting rates. The cascade transitions caused by the PFE can be naturally included by solving the rate equations. The double-counting problem and the effects of the truncation of the states in the REQ must be correctly treated. Finally, many upper states in the REQ may be converted into a "continuum" and to a Fokker–Planck type structure.[54]

As a preliminary to a more extensive calculation based on the improved theory of PFE, we reanalyzed the collisional radiative model for the hydrogen plasma. It was found that:

(a) The collisional cascade effect dominates[44,45] both for the collisional ionization rates and the capture rates at densities higher than 10^{15} cm^{-3}, as shown in Fig. 1. At lower densities, the radiative transitions are important.

(b) The upper Rydberg states are almost always in Saha equilibrium population, thus allowing a cutoff of an otherwise infinite set of coupled rate equations.

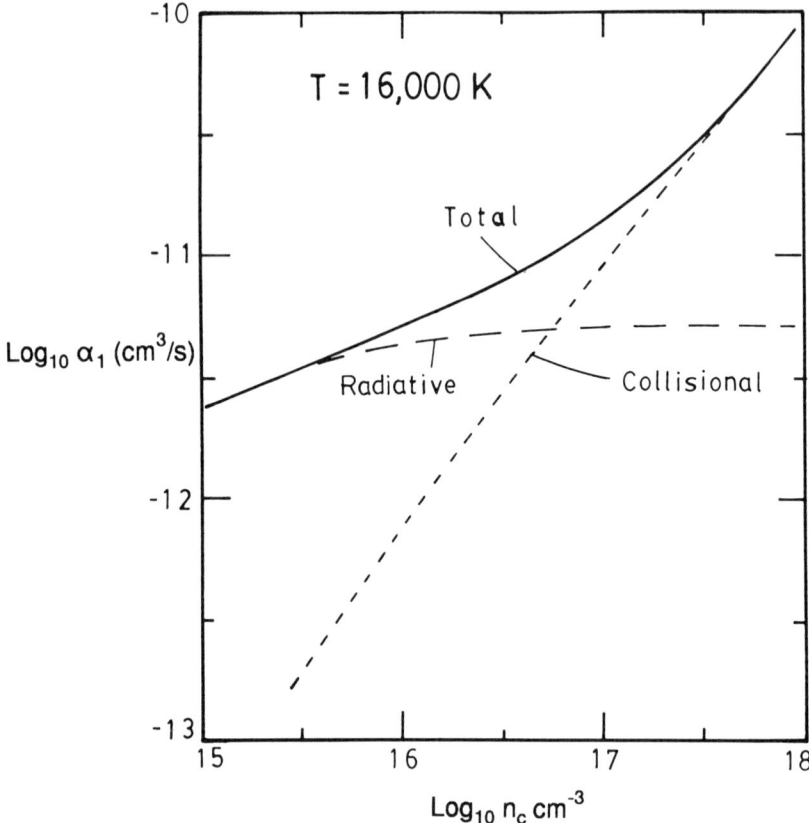

Figure 1. The collisional radiative recombination rate for the ground state as a function of the electron density, for the hydrogen plasma.

(c) More importantly, these high Rydberg states serve as an instant source (and sink) of electrons to which the lower states are coupled. This is illustrated in Fig. 2, where the free electron density is held fixed during the relaxation.

(d) The nearest-neighbor approximation for the collisional transitions among the upper states is a good approximation, enabling the application of a continuum Fokker–Planck diffusion picture.

(e) Finally, the nonequilibrium nature of the plasma electrons has a profound effect on the rates, especially for the collisional excitation and ionization rates.[34] For the purposes of a simple demonstration, a sum of two Maxwellian distributions was tried:

$$\phi_v = \mu_a \phi_a(T_a) + \mu_b \phi_b(T_b), \quad \text{with} \quad \mu_a + \mu_b = 1$$

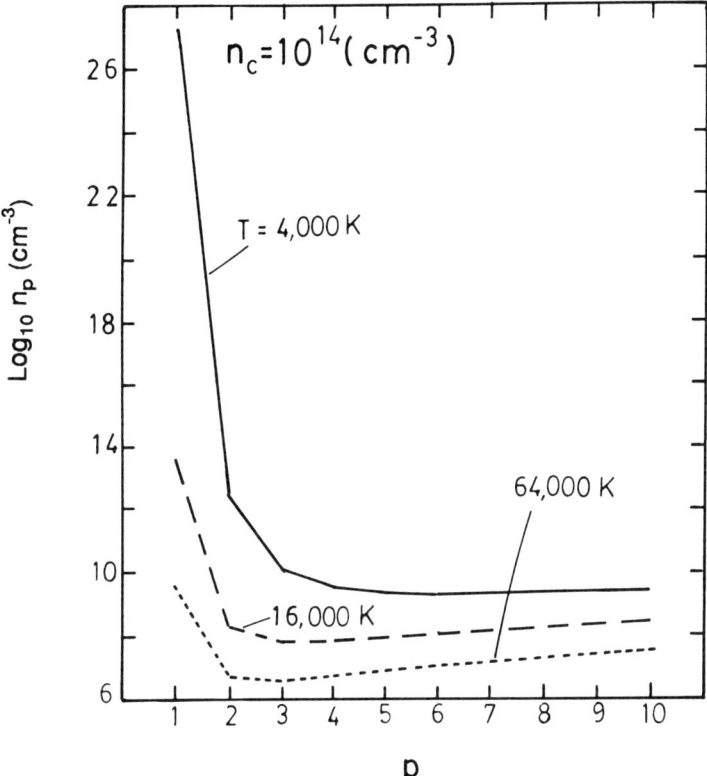

Figure 2. The excited state population densities n_p for the states p in hydrogen plasma.

and where $\mu_a \gg \mu_b$ with $T_a < T_b$. Plots of the collisional radiative ionization rate versus T_b for a 1% mixture ($\mu_b = 0.01$) of the higher temperature component at several values of the electron density, n_c, are shown in Fig. 3. It was found that the collisional radiative ionization rate for the ground state S_1 is extremely sensitive to the nonequilibrium mixture, whereas the collisional radiative recombination rates are much less affected.

In connection with the use of the rates and its relationship to the actual form of the rate equations, we stress several critical points:

1. Consistency of the rate calculation and the structure of the rate equations. This was pointed out already in connection with DR.

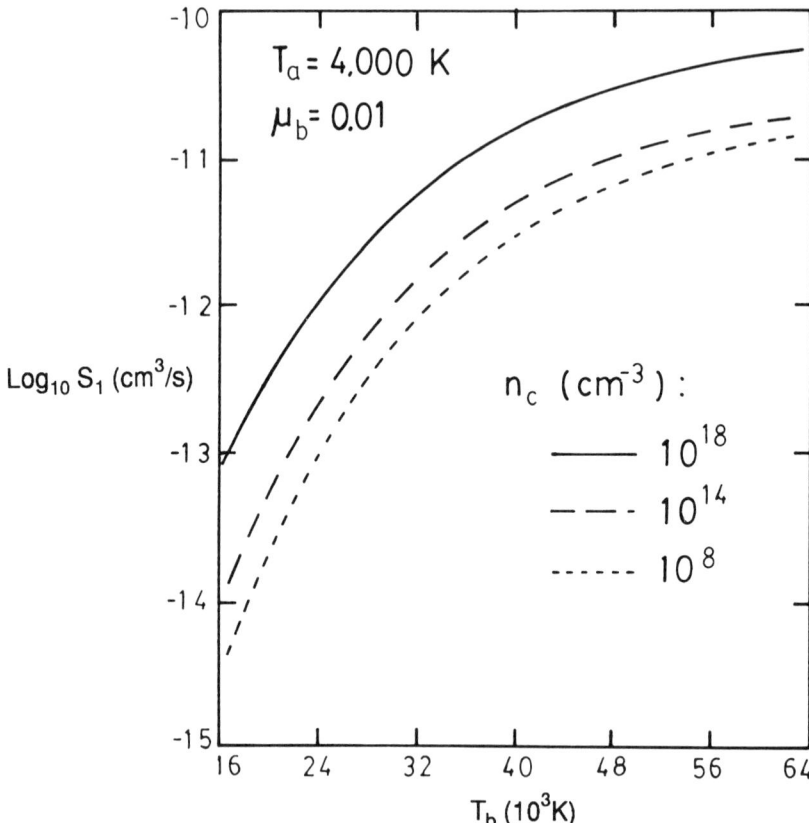

Figure 3. The effect of nonequilibrium electron distribution on the collisional ionization rate S.

2. Validity of the rate equation approach, with regard to the various relaxation times involved with the ionic excited states and plasma particle collision times.
3. Sensitivity of the rates to plasma fluctuations and nonequilibria. This is important not only during the initial stage of stabilization, but eventually will affect the population densities of excited states.

6. SUMMARY AND CONCLUSIONS

In this chapter, we have summarized the results obtained from the experimental and theoretical efforts of the past 15 years on direct radiative recombination and resonant dielectronic recombination. The cross sections and rate coefficients for the RR process are summarized in a compact form in terms of the effective charge (Eq. 19) for low-lying Rydberg states. For high Rydberg states with $n >> 1$, other n-dependent forms of the effective charge given earlier may be useful. In the limited number of cases in which accurate RR rates are required, it is always possible to carry out refined calculations; there are many simple programs available for this purpose. On the other hand, in view of the vast amount of rates that are needed for plasma modeling, with all possible ionic targets and many excited states being considered, the mode of presentation chosen here seems to be reasonable.

Insofar as the DR rates are concerned, we presented a set of parametrized rate formulas that can readily be used to generate actual rates for the initial ground states of the ions. The validity of these formulas has been checked against the available data, but the situation is far from satisfactory. No data are given that involve excited states of ions initially; they can readily be calculated, but their tabulation becomes an unmanageable bookkeeping problem. A limited amount of such data may be generated as needed. Again, a set of simple rate formulas for the excited initial states is desirable.

As noted above, there are large gaps in the available data on the RR and DR rates for (i) some of the ions of open-shell isoelectronic sequences and (ii) the excited states of the ions. Furthermore, the following problems still remain to be treated: (iii) plasma-field-distorted rates, (iv) adjustment for the contributions of the states omitted in the rate equations, (v) nonequilibrium effect of the plasma, and (vi) the effects of radiatively unstable final states that are included in the currently available DR rates. Note that item (iv), for example, depends on processes other than recombination, such as the collisional excitation and ionization rates. Item (ii) is, of course, related to item (iii) since the field mixing in (iii) involves many of the excited states. Item (iv) also depends on items (ii) and (iii) as well. A great deal more remains to be done in the way of testing, improving, and extending the data before we have a good understanding of the entire recombination process.

The most serious general conclusion reached in this work is that, aside from the many additional important rates that are yet to be generated, a consistent analysis and modeling of plasma may be carried out conveniently only if the proper rates are generated with the full knowledge of the structure of the REQ into which these rates are to be inserted.[28] In fact, almost all the rates that have been generated thus far are only of very limited applicability at best, and even then extreme caution is required as to their applicability.

ACKNOWLEDGMENTS. This work was supported in part by a grant from the Basic Interactions Branch, Division of Chemical Sciences, Office of Energy Research, U.S. Department of Energy. Most of the results presented here are the result of collaboration with many people during the past 15 years.

REFERENCES

1. W. G. Graham, W. Frisch, Y. Hahn, and J. Tanis (eds.), *Recombination of Atomic Ions*, NATO ASI Series B: Physics, No. 296 Plenum (1992).
2. H. A. Kramers, *Phil. Mag.* **46**, 836 (1923).
3. H. A. Bethe and E. E. Salpeter, *Quantum Mechanics of One and Two Electron Atoms*, Springer-Verlag, Berlin (1957).
4. D. R. Bates and H. S. W. Massey, *Phil. Trans. R. Soc.* **A239**, 269 (1943).
5. A. Burgess, *Astrophys. J.* **139**, 776 (1964).
6. A. Burgess, *Astrophys. J.* **141**, 1588 (1965).
7. D. S. Belic, G. H. Dunn, T. J. Morgan, D. W. Muller, and C. Timmer, *Phys. Rev. Lett.* **50**, 339 (1983); A. Muller, D. S. Belic, B. D. DePaola, N. Djuric, G. H. Dunn, D. W. Muller, and C. Timmer, *Phys. Rev. Lett.* **56**, 127 (1986).
8. B. A. Mitchell, C. T. Ng, J. L. Forand, D. P. Levac, R. E. Mitchell, A. Sen, D. B. Miko, and J. W. McGowan, *Phys. Rev. Lett.* **50**, 335 (1983).
9. P. F. Dittner, S. Datz, P. D. Miller, C. D. Moak, P. H. Stelson, C. Bottcher, W. B. Dress, G. D. Alton, and N. Neskovic, *Phys. Rev. Lett.* **51**, 31 (1983).
10. A. R. Young, L. D. Gardner, D. W. Savin, G. P. Lafyatis, A. Chutjian, S. Bliman, and J. L. Kohl, *Phys. Rev. A* **49**, 357 (1994).
11. J. A. Tanis, *Nucl. Instrum. Methods, Sect. A* **262**, 52 (1987).
12. V. L. Jacobs, J. Davis, and P. C. Kepple, *Phys. Rev. Lett.* **37**, 1390 (1976).
13. K. LaGattuta and Y. Hahn, *Phys. Rev. Lett.* **51**, 558 (1983).
14. K. LaGattuta, I. Nasser, and Y. Hahn, *J. Phys. B* **20**, 1565, 1577 (1987).
15. Y. Hahn, *Physica Scripta*, submitted.
16. T. Aberg and J. Utriiainen, *Phys. Rev. Lett.* **22**, 1346 (1969).
17. T. Aberg, K. Reinikainen, and O. Keski-Rahkonen, *Phys. Rev. A* **23**, 153 (1981).
18. E. M. Bernstein, M. W. Clark, C. S. Oglesby, J. A. Tanis, W. G. Graham, R. H. McFarland, T. J. Morgan, B. M. Johnson, and K. W. Jones, *Phys. Rev. A* **41**, 2594 (1990).
19. Y. Hahn, *Phys. Rev. A* **40**, 950 (1989).
20. See, for example, T. D. Mark and G. H. Dunn (eds.), *Electron Impact Ionization*, Springer-Verlag, Vienna (1985).
21. D. H. Crandall, in *Physics of Ion–Ion and Electron–Ion Collisions* (F. Brouillard and J. W. McGowan, eds.), Plenum, New York (1983), p. 201.
22. Y. Hahn, *Phys. Rev. Lett.* **39**, 82 (1977).
23. Y. Hahn and K. LaGattuta, *Phys. Reports* **166**, 195 (1988).
24. Y. Hahn, *Comments Atom. Molec. Phys.* **19**, 99 (1987).
25. B. W. Shore, *Astrophys. J.* **158**, 1205 (1969).
26. Y. Hahn, in *Advances in Atomic and Molecular Physics, Vol. 21* (D. R. Bates and B. Bederson, eds.), Academic Press, New York, p. 123 (1985).
27. Y. Hahn and P. Krstic, *Physica Scripta*, **48**, 340 (1993).
28. Y. Hahn, (unpublished).
29. Y. Hahn, P. Krstic and J. Li, in *9th Int. Conf. on Atomic Processes in Plasmas* (W. L. Rowan, ed.), AIP, New York, p. 69–83 (1995).
30. M. Baranger, *Phys. Rev.* **111**, 481 (1958).
31. A. C. Kolb and H. Griem, *Phys. Rev.* **111**, 514 (1958).
32. H. Griem, *Spectral Line Broadening by Plasmas*, Academic Press, New York (1974).
33. Y. Hahn, *Phys. Rev. A* **46**, 4433 (1992).
34. J. Li and Y. Hahn, *Phys. Rev. E* **49**, 927 (1994).
35. Y. Hahn and D. W. Rule, *J. Phys. B* **10**, 2689 (1977).
36. D. J. McLaughlin and Y. Hahn, *Phys. Rev. A* **43**, 1313 (1991) and Errata.

37. A. Abdel-Hady, I. Nasser, and Y. Hahn, *J. Quant. Spectrosc. Radiat. Transfer* **39**, 197 (1988).
38. L. Andersen, Electron–Ion Recombination at Low Energies, Aarhus, 1993. XVIII ICPEAC.
39. P. Krstic and Y. Hahn, *Phys. Lett. A*, **192**, 47 (1994).
40. Y. Hahn, *J. Quant. Spectrosc. Radiat. Transfer* **41**, 315 (1989).
41. A. H. Moussa and Y. Hahn, *J. Quant. Spectrosc. Radiat. Transfer* **43**, 45 (1990).
42. Y. Hahn, *J. Quant. Spectrosc. Radiat. Transfer* **49**, 81 (1993) and Errata; *JQSRT* **51**, 663 (1994).
43. G. Peach, *Adv. Phys.* **30**, 367 (1981).
44. D. R. Bates, A. E. Kingston, and R. W. P. McWhirter, *Proc. R. Soc. London, Ser. A* **267**, 297 (1962).
45. J. Li and Y. Hahn, *Phys. Rev. E* **48**, 2934 (1993).
46. P. Krstic and Y. Hahn, *Phys. Rev. A* **48**, 4515 (1993).
47. Y. Hahn and P. Krstic, *J. Phys. B* **26**, L291 (1993).
48. D. B. Riesenfeld, *Astrophys. J.* **398**, 386 (1992).
49. D. B. Riesenfeld, J. R. Raymond, A. R. Young, and J. L. Kohl, *Astrophys. J.* **389**, L37 (1992).
50. C. F. Hooper, *Phys. Rev.* **149**, 77 (1966).
51. C. F. Hooper, *Phys. Rev.* **165**, 215 (1968).
52. S. Ichimaru (ed.), *Strongly Coupled Plasma Physics,* North-Holland, Amsterdam (1990).
53. F. J. Rogers and H. E. Dewitt (eds.), *Strongly Coupled Plasma Physics*, NATO ASI Series B, No. 154, Plenum, New York (1986).
54. J. Li and Y. Hahn, *Phys. Rev. E*, in press.

Chapter 6

Excitation of Atomic Ions by Electron Impact

Swaraj S. Tayal, Anil K. Pradhan, and Michael S. Pindzola

1. INTRODUCTION

Recently, a review of theoretical methods and data for low-energy electron impact excitation of atomic ions was completed by Pradhan and Gallagher.[1] It includes a comprehensive annotated bibliography of articles giving data on this subject. In this chapter we focus on three aspects of a continuing worldwide effort to better understand electron–ion excitation processes for application to astrophysical and laboratory plasma research. The first part is concerned with the evaluation and recommendation of the best of a large body of atomic data available for carbon and oxygen ions. The second is a theoretical effort, called the Iron Project, to calculate collision cross sections and rate coefficients for the electron impact excitation of iron ions. The third is a joint effort to benchmark theory against experiment for the electron impact excitation of selected rare-gas ions.

Direct excitation (DE) from an initial state i to a final state f is given by

---

SWARAJ S. TAYAL • Department of Physics and Center for Theoretical Studies of Physical Systems, Clark Atlanta University, Atlanta, Georgia 30314.   ANIL K. PRADHAN • Department of Astronomy, Ohio State University, Columbus, Ohio 43210.   MICHAEL S. PINDZOLA • Department of Physics, Auburn University, Auburn, Alabama 36849.

Atomic and Molecular Processes in Fusion Edge Plasmas, edited by R. K. Janev. Plenum Press, New York, 1995.

$$e^- + A_i^{q+} \rightarrow A_f^{q+} + e^- \tag{1}$$

while a second process called resonant excitation (RE), involving dielectronic capture followed by autoionization, is given by

$$e^- + A_i^{q+} \rightarrow \left[A_j^{(q-1)+} \right] \rightarrow A_f^{q+} + e^- \tag{2}$$

where q is the charge on the atomic ion A, and the brackets indicate a doubly excited resonance state. Many of the distorted-wave calculations in the literature only include the DE process and thus are not as accurate as the close-coupling calculations which include both processes. There are examples where the RE process is over an order of magnitude more probable than the DE process in the near-threshold region. A more general formulation of the distorted-wave method,[2] however, includes both the direct and resonant excitation processes.

In electron collisional excitation of ions, it is convenient to present results in the form of a symmetric and dimensionless collision strength Ω rather than the collision cross sections. The collision strength for excitation from level i to level f is related to the cross section σ, in units of πa_0^2, by the relation

$$\Omega(i \rightarrow f) = \omega_i k_i^2 \sigma(i-f) \tag{3}$$

where ω_i is the statistical weight of level i, and k_i^2 is the energy, in rydbergs, of the electron relative to level i.

For many applications in astrophysics and fusion plasmas, effective collision strengths are needed rather than the collision strengths, which vary rapidly as a function of energy of the incident electron in the threshold region. The effective collision strengths are obtained by integrating collision strengths over a Maxwellian velocity distribution for the electron. The effective collision strength Υ for a transition from state i to state f is given by

$$\Upsilon(i \rightarrow f) = \int_0^\infty \Omega(i \rightarrow f) \exp\left(-\frac{\varepsilon_f}{kT_e}\right) d\left(\frac{\varepsilon_f}{kT_e}\right) \tag{4}$$

where T_e is the electron temperature, ε_f is the energy of the incident electron with respect to upper level f, and k is the Boltzmann constant. The electron excitation rate $C(i \rightarrow f)$ and deexcitation rate $C(f \rightarrow i)$ can simply be determined from the effective collision strength by the relations

$$C(i \rightarrow f) = \frac{8.6 \times 10^{-6}}{\omega_i T_e^{1/2}} \Upsilon(i \rightarrow f) \exp\left(-\frac{\Delta E_{if}}{kT_e}\right) \text{ cm}^3/\text{s} \tag{5}$$

and

$$C(f \rightarrow i) = \frac{8.6 \times 10^{-6}}{\omega_j T_e^{1/2}} \Upsilon(i \rightarrow f) \text{ cm}^3/\text{s} \tag{6}$$

where ω_f is the statistical weight of level f, and ΔE_{if} is the excitation energy.

We present sources of the best currently available data on all ionization stages of carbon and oxygen ions in Section 2. In Section 3 we report on progress on the iron ions, in Section 4 we report on progress on the rare-gas ions, and Section 5 contains a brief summary.

2. CARBON AND OXYGEN IONS

2.1. General Considerations

The singly and multiply charged ions of Be, B, C, and O atoms are usually present as impurities in controlled thermonuclear fusion plasmas, particularly in the edge plasma region. The electron collisional excitation and ionization of ions play an important role in the determination of energy loss in the plasmas. The impurity ions can also be used for diagnostic purposes. Berrington and Clark[3] recently reviewed the electron impact excitation data for the Be and B ions. They have presented the recommended data for these ions. In this section we evaluate the available electron collisional excitation cross sections for all charge states of C and O atoms and make recommendation of the best currently available data.

The review by Pradhan and Gallagher[1] provides a list of articles along with the accuracy ratings published up to mid-1990. Since then, several papers on the electron impact excitation of C and O ions have appeared in the literature. The updated list of references from 1970 to mid-1994 is given in Tables I and II. In addition, we have relied on the previously published compilation of the recommended data for the C and O ions by Itikawa et al.,[4] Aggarwal et al.,[5] Shevelko et al.,[6] and Phaneuf et al.[7,8]

The available experimental measurements on the electron impact excitation of these ions are still very few. Electron collision excitation rate coefficients were measured for Li-like O^{5+} by Kunze and Johnson[9] and Datla and Kunze[10] for several transitions from $n = 2$ to $n = 3$, 4, and 5 by observing photons emitted by ions in plasmas. Similar plasma rate coefficient measurements have been reported for the Be-like O^{4+} ion[11] and the He-like C^{4+} ion.[12,13] The quoted uncertainties of these measurements are typically 40% or more. There are three crossed-beams measurements available for absolute electron impact excitation cross sections of C and O ions. One of these measurements is on the B-like C^+,[14] and the other two are on the Li-like C^{3+}.[15,16] The total systematic uncertainties in these measurements are quoted in the range 16–19%. There is good agreement between the measurements and the close-coupling calculations[17,18] in the near-threshold region. The measurements of absolute excitation cross sections of multiply charged ions using the crossed-beams technique is difficult because of low detection efficiency and low signal rates. These difficulties of the crossed-beams technique are not present in the merged-beams experiments recently developed at the Joint Institute for Laboratory Astrophysics

Table I. Electron-Impact Excitation Data for Carbon Ions

Ion	N^a	Author(s)	Year	Method^b	Coupling	Resonances included	Data^c	Rating^d	Reference
C^+	5	Jackson	1972	CC	LS	✓	Ω	B	95
		Jackson	1973	CC	LS	✓	Ω,Υ	B	82
		Kato	1976	Other	LS		Ω, q	D	96
		Tambe	1977	CC	IC^e		Ω	B	97
		Brandus	1980	B	LS		Q	E	98
		Hayes and Nussbaumer	1984	CC	LS, IC	✓	Ω,Υ	A	42
		Hayes and Nussbaumer	1984	CC	LS, IC	✓	Ω,Υ	A	43
		Keenan et al.	1986	CC	IC	✓	Ω,Υ	A	44
		Luo and Pradhan	1990	CC	IC	✓	Ω	A	39
		Blum and Pradhan	1991	CC	IC	✓	Ω	A	40
		Blum and Pradhan	1992	CC	IC	✓	Ω,Υ	A	41
C^{2+}	4	Osterbrock	1970	CC	LS		Ω	C	104
		Flower and Launay	1972	CC, DW	LS		Ω	C	105
		Nussbaumer	1972	Other	LS			C	106
		Flower and Launay	1973	DW, CC	LS		Ω	C	107
		Hershkowitz and Seaton	1973	DW, CC	LS	✓		C	108
		Banyard and Taylor	1974	B	LS		Q	D	109
		Kato	1976	Other	LS		Ω,q	D	96
		Berrington et al.	1977	CC	LS	✓	Ω	A	53
		Nakazaki and Hashino	1977	CB	LS		Q	D	110
		Osterbrock and Wallace	1977	CC, DW, CB	LS		Υ	C	56
		Dufton et al.	1978	CC	IC	✓	q	B	111
		Ganas and Green	1979	B	LS		Q	E	112
		Brandus	1980	B	LS		Q	E	98
		Allouche and Marinelli	1981	Other	LS		Q	D	113
		Berrington et al.	1981	CC	LS	✓	Ω	B	54
		Burke et al.	1981	CC	LS	✓	Ω	B	114
		Peek and Mann	1982	DW	LS		Ω	C	59
		Nakazaki and Hashino	1982	CB	LS		Q	C	58
		Berrington	1985	CC	LS	✓	Υ	A	49
		Berrington et al.	1985	CC	LS	✓	Ω	A	55
		Burke and Seaton	1986	CC	LS	✓	Ω	A	115
		Berrington et al.	1989	CC	LS	✓	Υ	A	50
C^{3+}	3	Mathur et al.	1971	Other	LS		q	E	122
		Petrini	1972	CB	LS		Ω	E	123
		Kato	1976	Other	LS		Ω,q	D	96
		Callaway et al.	1977	CC		✓	Ω	A	124
		Gau and Henry	1977	CC	LS		Ω	B	125
		Osterbrock and Wallace	1977	DW, CB	LS		Υ	C	56

Table I. (continued)

Ion	N^a	Author(s)	Year	Method[b]	Coupling	Resonances included	Data[c]	Rating[d]	Reference
C^{3+}	3	Callaway et al.	1979	CB	IC		Q	C	61
		Henry	1979	CC	LS		Ω	C	126
		Brandus	1980	B	LS		Q	E	98
		Bazylev and Chibisov	1981	Other	IC		Q	C	118
		Peek and Mann	1982	DW	LS		Ω	C	59
		Bhadra and Henry	1982	CC		✓	Ω	B	63
		Soh et al.	1982	B	IC		Q	U	127
		Sampson et al.	1985	CB	IC		Ω	D	128
		Sampson et al.	1985	CB	IC		Ω	D	129
		Glyamzha et al.	1987	CB	LS		Q	D	130
		Glyamzha et al.	1987	CB	LS		Q	D	131
		Itikawa	1987	DW	LS		Ω	C	132
		Badnell et al.	1991	DW, CC	LS	✓	Q	A	219
		Burke	1992	CC	LS	✓	Ω, Υ	A	60
C^{4+}	2	Kato	1976	Other	LS		Ω, q	D	96
		Nakazaki	1976	CB	LS		Q	D	138
		Bhatia and Temkin	1977	DW	LS		Q	C	139
		McDowell et al.	1977	DW	LS		Q	C	140
		Davis et al.	1978	DW	LS		Ω, q	D	141
		Tully	1978	Other	LS		Q	D	142
		van Wyngaarden and Henry	1979	CC	LS		Ω	C	143
		Baluja and Doyle	1980	DW	LS		Q, Υ	C	144
		Goett et al.	1980	CB	IC		Ω	C	145
		Sampson and Clark	1980	CB	IC		Ω	C	146
		Tully	1980	Other	LS		Q	E	147
		Ganas	1981	Other	LS		Q	D	148
		Pradhan	1981	DW	LS	✓	Ω, Υ	C	67
		Pradhan et al.	1981	DW	LS	✓	Υ	C	68
		Peek and Mann	1982	DW	LS		Ω	C	59
		Goett et al.	1983	CB	IC		Ω	C	149
		Sampson et al.	1983	CB	IC		Ω	C	70
		Sampson et al.	1983	CB	IC		Ω	C	71
		Nakazaki	1983	Other	LS		Ω	D	150
		Singh et al.	1983	Other	LS		Q	D	151
		Tayal and Kingston	1984	CC	LS	✓	Υ	A	64
		Badnell	1985	Other	LS		Ω	D	152
		Tayal	1986	CC	LS	✓	Ω, Υ	A	65
		Glyamzha et al.	1987	CB	LS		Q	D	130
		Glyamzha et al.	1987	CB	LS		Q	D	131
		Saxena and Mathur	1987	DW	LS		Ω	C	153
		Srivastava and Katiyar	1987	DW	LS		Q	C	154
		Katiyar and Srivastava	1988	DW	LS		Q	C	155
		Kato and Nakazaki	1989	Other			q	U	156

(continued)

Table I. (continued)

Ion	N^a	Author(s)	Year	Method[b]	Coupling	Resonances included	Data[c]	Rating[d]	Reference
C^{5+}	5	Jacobs	1971	CB	LS		Q	D	164
		Kato	1976	Other	LS		Ω,q	D	96
		Mitra and Sil	1976	CB, other	LS		Q	D	165
		Baluja and McDowell	1977	DW	LS		Ω	D	166
		Hayes and Seaton	1977	DW, CC	LS		Ω	B	167
		McDowell et al.	1977	DW	LS		Q	C	140
		Davis et al.	1978	DW	LS		Ω,q	D	141
		Hayes and Seaton	1978	CC	LS	✓	Ω	A	168
		Mitra and Sil	1978	CB, other	LS		Q	C	169
		Nakazaki	1978	CB	LS		Q	Q	170
		Morgan	1980	CC	LS		Q	A	73
		Abu-Salbi and Callaway	1981	CC, CB	LS	✓	Ω,Q	A	74
		Das and Sil	1984	Other	LS		Q	E	171
		Zygelman and Dalgarno	1987	CC	jj		Q	A	172
		Ritchie	1989	Other	jj		Ω	C	173
		Srivastava et al.	1991	DW	LS			C	157
		Zou and Shirai	1991	CC	LS			B	176
		Aggarwal and Kingston	1991	CC	LS		Ω,Υ	A	72

[a] N = number of electrons.
[b] Method: CC, close coupling; DW, distorted wave; CB, Coulomb–Born; CBe, Coulomb–Bethe approximation; DRM, Dirac R-matrix; B, Born approximation; BP, Breit–Pauli.
[c] Data: Ω, collision strength; Υ, Maxwellian averaged collision strength; Q, cross section; q, rate coefficient; GF, Gaunt factor.
[d] Rating: A, uncertainties within 10%; B, uncertainties within 20%; C, uncertainties within 30%; D, uncertainties within 50%; E, uncertainties greater than 50%; U, uncertain.
[e] Coupling scheme: IC, intermediate coupling

(JILA),[19,20] at the Jet Propulsion Laboratory (JPL),[21] and in other laboratories. The merged-beams electron energy loss experiments normally have very good resolution and can measure absolute excitation cross sections of multiply charged ions in the near-threshold energy region, where they can provide a benchmark for testing various theoretical methods.

Theoretical methods for electron impact excitation of ions are well developed, and several fairly extensive calculations have been reported in the literature for these ions. Most of these calculations are primarily based on the close-coupling (CC), distorted-wave (DW), and Coulomb–Born (CB) approximations. The reliability of these methods basically depends on the electron impact energy, type of transition, and charge state of the ion.

Excitation of Atomic Ions by Electron Impact

Table II. Electron-Impact Excitation Data for Oxygen Ions[a]

Ion	N	Author(s)	Year	Method	Coupling	Resonances included	Data	Rating	Reference
O^+	7	Ormonde et al.	1973	CC	LS	✓	Ω,Q	C	75
		Davis	1974	Other	LS		GF^b	D	76
		Davis et al.	1975	DW	LS		q	D	77
		Pradhan	1976	CC	LS	✓	Ω	A	27
		Pradhan	1976	CC, DW	IC	✓	Ω,Υ	B	78
		Strobel and Davis	1980	DW			q	D	79
		Ho and Henry	1983	CC	LS	✓	Ω,q	B	26
		McLaughlin et al.	1987	CC	LS	✓	Ω	A	25
		McLaughlin and Bell	1993	CC	LS	✓	Ω,Υ	A	28
		McLaughlin and Bell	1993	CC	LS	✓	Ω,Υ	A	29
		McLaughlin and Bell	1993	CC	LS	✓	Ω,Υ	A	30
		McLaughlin and Bell	1994	CC	LS	✓	Ω,Υ	A	31
O^{2+}	6	Poshyunaite et al.	1970	B, other	LS		Q	D	80
		Jackson	1973	CC		✓	Ω,Υ	B	81
		Jackson	1973	CC	LS	✓	Ω,Υ	B	82
		Ormonde et al.	1973	CC	LS	✓	Ω,Q	C	75
		Davis	1974	Other	LS		GF	D	76
		Eissner and Seaton	1974	CC	LS	✓	Ω,Υ	A	83
		Davis et al.	1975	DW	LS		q	D	77
		Seaton	1975	CC	LS	✓	Ω	A	84
		Pindzola	1977	DW	LS	✓	Q	B	85
		Doschek et al.	1978	DW	IC		Ω	C	86
		Bhatia et al.	1979	DW	IC	✓	Ω	C	37
		Baluja et al.	1980	CC	LS	✓	Ω,Υ	A	87
		Ganas	1980	B	LS		Q	E	89
		Strobel and Davis	1980	DW			q	D	79
		Baluja et al.	1981	CC	LS	✓	Ω,Υ	A	88
		Tully and Baluja	1981	CB, CC	LS	✓	Ω	B	35
		Aggarwal et al.	1982	CC	LS	✓	Ω,Υ	A	90
		Aggarwal	1983	CC	IC	✓	Ω,Υ	A	91
		Ho and Henry	1983	CC	LS	✓	Ω,Υ	B	26
		Aggarwal	1985	CC	LS	✓	Ω,Υ	A	92
		Aggarwal	1986	CC	LS	✓	Ω	A	93
		Itikawa and Sakimoto	1986	DW	LS		Ω	C	36
		Burke et al.	1988	CC	IC	✓	Ω,Υ	A	94
		Aggarwal and Hibbert	1991	CC	LS	✓	Ω	A	33
		Aggarwal	1993	CC	LS	✓	Y	A	34
		Bhatia and Kastner	1993	DW	IC		Ω	C	38
O^{3+}	5	Davis	1974	Other	LS		GF	D	76
		Davis et al.	1975	DW	LS		q	D	77
		Flower and Nussbaumer	1975	DW	LS		Ω	C	99
		Kato	1976	Other	LS		Ω,q	D	96
		Osterbrock and Wallace	1977	CC, DW, CB	LS		Υ	C	56
		Ganas	1979	B	LS		Q	E	100

(continued)

Table II. (continued)

Ion	N	Author(s)	Year	Method	Coupling	Resonances included	Data	Rating	Reference
O^{3+}	5	Hayes	1980	CC, DW	LS		Ω	B	101
		Kastner	1982	CBe	LS		q	D	102
		Hayes	1982	CC	IC	✓	Ω,Υ	A	45
		Hayes	1983	CC	LS	✓	Ω,Υ	B	103
		Hayes and Nussbaumer	1983	CC	IC	✓	Ω,Υ	A	46
		Luo and Pradhan	1990	CC	IC	✓	Ω,Υ	A	39
		Blum and Pradhan	1992	CC	IC	✓	Ω,Υ	A	41
O^{4+}	4	Osterbrock	1970	CC	LS		Ω	C	104
		Hershkowitz and Seaton	1973	DW, CC	LS	✓	Ω	C	108
		Banyard and Taylor	1974	B	LS		Q	D	109
		Davis	1974	Other	LS		GF	D	76
		Malinovsky	1975	DW	LS		Ω	C	116
		Davis et al.	1975	DW	LS		q	D	77
		Kato	1976	Other	LS		Ω,q	D	96
		Berrington et al.	1977	CC	LS	✓	Ω	A	53
		Nakazaki and Hashino	1977	CB	LS		Q	D	110
		Osterbrock and Wallace	1977	CC, DW, CB	LS		Υ	C	56
		Dufton et al.	1978	CC	IC	✓	q	B	111
		Berrington et al.	1979	CC	LS	✓	Ω,q	A	51
		Ganas and Green	1979	B	LS		Q	E	112
		Cowan	1980	DW	LS	✓	q	B	117
		Younger	1980	DW, CB, B	LS		Ω	C	57
		Allouche and Marinelli	1981	Other	LS		Q	C	113
		Bazylev and Chibisov	1981	Other	IC		Q	C	118
		Berrington et al.	1981	CC	LS	✓	Ω	B	54
		Nakazaki and Hashino	1982	CB	LS		Q	C	58
		Widing et al.	1982	CC	LS	✓	Υ	A	52
		Berrington et al.	1985	CC	LS	✓	Ω	A	55
		Kato and Safronova	1991	Other	LS		q	D	119
		Zhang and Sampson	1993	DW	IC		Ω	C	120
		Zhang and Sampson	1992	DW	IC		Ω	C	121

Table II. (continued)

Ion	N	Author(s)	Year	Method	Coupling	Resonances included	Data	Rating	Reference
O^{5+}	3	Mathur et al.	1971	Other	LS		Q	E	122
		Petrini	1972	CB	LS		Ω	E	123
		Davis	1974	Other	LS		GF	D	76
		Davis et al.	1975	DW	LS		GF	C	77
		Presnyakov and Urnov	1975	Other	LS	✓	q	D	133
		Presnyakov and Urnov	1975	Other	LS	✓	Q	D	134
		Kato	1976	Other	LS		Ω,q	D	96
		Osterbrock and Wallace	1977	DW, CB	LS		Υ	C	56
		Henry	1979	CC	LS		Ω	C	126
		Bely-Dubau et al.	1981	DW	LS		Ω,q	C	136
		Bhadra and Henry	1982	CC	LS	✓	Ω	B	63
		Goett and Sampson	1983	CB	IC		Ω	C	135
		Sampson et al.	1985	CB	IC		Ω	D	128
		Sampson et al.	1985	CB	IC		Ω	D	129
		Itikawa	1987	DW	LS		Ω	C	132
		Zhang et al.	1990	DW	IC		Ω	C	137
		Badnell et al.	1991	CC	LS	✓	q	A	219
O^{6+}	2	Kato	1976	Other	LS		Ω,q	D	96
		Nakazaki	1976	CB	LS		Q	D	138
		Bhatia and Temkin	1977	DW	LS		Q	C	139
		McDowell et al.	1977	DW	LS		Q	C	140
		Tully	1978	Other	LS		Q	E	142
		Clark and Sampson	1978	CB	IC		Ω	C	158
		Pindzola et al.	1979	DW	LS	✓	Q	C	159
		van Wyngaarden and Henry	1979	CC	LS		Ω,q	C	143
		Baluja and Doyle	1980	DW	LS		Ω,Υ	C	144
		Goett et al.	1980	B	IC		Ω	C	145
		Tully	1980	Other	LS		Q	E	147
		Bely-Dubau et al.	1981	DW	LS		Ω,q	C	136
		Ganas	1981	Other	LS		Q	D	148
		Pradhan	1981	DW	LS	✓	Ω	B	67
		Pradhan et al.	1981	DW	LS	✓	Ω,Υ	C	68
		Pradhan et al.	1981	DW	LS	✓	Υ	C	69
		Goett and Sampson	1983	CB	IC		Ω	C	135
		Kingston and Tayal	1983	CC	LS	✓	Υ	A	160
		Kingston and Tayal	1983	CC	LS	✓	Ω,Υ	A	161
		Nakazaki	1983	Other	LS		Ω	D	150
		Sampson et al.	1983	CB	IC		Ω	C	70
		Sampson et al.	1983	CB	IC		Ω	C	71
		Singh et al.	1983	Other	LS		Q	D	151
		Steenman-Clark and Faucher	1984	DW	IC	✓	Υ	D	162
		Tayal and Kingston	1984	CC	LS	✓	Υ	A	66

(continued)

Table II. (continued)

Ion	N	Author(s)	Year	Method	Coupling	Resonances included	Data	Rating	Reference
O^{6+}	2	Badnell	1985	Other	LS		Ω	D	152
		Itikawa and Sakimoto	1985	DW	LS		Ω	D	163
		Saxena and Mathur	1987	DW	LS		Q	C	153
		Srivastava and Katiyar	1987	DW	LS		Q	C	154
		Katiyar and Srivastava	1988	DW	LS		Q	C	155
		Kato and Nakazaki	1989	Other			q	U	156
		Srivastava et al.	1991	DW	LS			C	157
O^{7+}	1	Jacobs	1971	CB	LS		Q	D	164
		Tully	1973	CB	LS		Q	D	174
		Bransden and Noble	1976	Other	LS		Q	C	177
		Kato	1976	Other	LS		Ω,q	D	96
		Baluja and McDowell	1977	DW	LS		Ω	D	166
		McDowell et al.	1977	DW	LS		Q	C	140
		Nakazaki	1978	CB	LS		Q	C	170
		Singh et al.	1979	Other	LS		Q	D	175
		Abu-Salbi and Callaway	1981	CC, CB	LS	✓	Q,Ω	A	74
		Das and Sil	1984	Other	LS		Q	E	171

[a]Notation as in Table I.
[b]GF, Gaunt factor.

The CC methods are believed to be the most reliable, particularly at low energies in the near-threshold region. These are the most elaborate of the theoretical methods and need the most computational power. During the past two decades, several computer programs have been developed to perform CC calculations. Burke and co-workers at the Queen's University of Belfast developed the RMATRX packages,[22] Seaton and co-workers at the University College London developed the IMPACT computer codes,[23] and Henry and co-workers at Louisiana State University developed the NIEM computer codes.[24] The accuracy of the CC collision calculations depends directly to first order on the accuracy of the target-state wave functions, on the convergence of the CC expansion, and on whether the autoionizing resonances are included in the calculations. The number of terms needed for convergence of the CC expansion varies, depending on the initial and final states of a transition of interest. In principle, all the states coupled to initial and final states should be included. Among these three well-known methods, the **R**-matrix method has been used most in recent years because it properly and most efficiently takes into account low-energy collision effects such as channel coupling, autoionizing resonances, electron correlation, and electron exchange. Relativistic effects are not significant for ions with low nuclear charge.

Several variants of the DW method have been used in theoretical studies of electron impact excitation of ions. The DW approximation includes the distortion of the target by the incident electron and is suitable for intermediately and highly charged ions. Resonance effects can be included in the DW method. In the CB approximation, the distortion of the target ion by the incident electron is ignored, and the radial wave functions are represented by the Coulomb waves. The CB method can provide reasonable results for ions of high charge states and for high impact energies. Thus, in many cases accurate low-energy CC results can be supplemented by the high-energy DW and/or CB results to provide excitation cross sections over a wide energy range for applications in astrophysics and fusion energy research.

The recommended data at low energies (except for O^{5+} and O^{7+}) have been produced by the Belfast **R**-matrix computer packages. Because of the very extensive amounts of data on these ions, it does not seem feasible to present the recommended data in this chapter. Fortunately, for many cases the data have been stored in the Belfast/Daresbury Atomic Data Bank and are freely available on request from the authors. We have adopted the recommendations of Pradhan and Gallagher[1] except for cases where new data have become available. We have used the criteria of Pradhan and Gallagher[1] for the accuracy assessment of the new data. The recommended data for all ions except O^{5+} have been rated "A" with an accuracy estimate of 10. We present a brief discussion of the recommended data sources for each ionization stage of the ions. We also give effective collision strengths for C^{4+} and O^{6+} ions in tabular form for transitions from the ground to the excited states.

2.2. O^+ Cross Sections

McLaughlin et al.[25] performed 9- and 34-state **R**-matrix calculations to obtain collision strengths for the $2s^22p^3\ {}^4S^O \to 2s^22p^23s\ {}^4P$, $2s^22p^3\ {}^2D^O \to 2s^22p^23s\ {}^2P$, and $2s^22p^3\ {}^2P^O \to 2s^22p^23s\ {}^2P$ transitions in O^+, and Ho and Henry[26] carried out a two-state CC calculation for the $2s^22p^3\ {}^4S^O \to 2s2p^4\ {}^4P$ and $2s^22p^3\ {}^4S^O \to 2s^22p^23s\ {}^4P$ transitions. Pradhan[27] performed a five-state ($2s^22p^3\ {}^4P^O,\ {}^2D^O,\ {}^2P^O,\ 2s2p^4\ {}^4P,\ {}^2D$) CC calculation to obtain excitation rate coefficients for transitions among these states. Because McLaughlin et al.[25] used very limited correlation in the target wave functions in their 34-state calculation, this calculation is less accurate than their 9-state calculation. More recently, McLaughlin and Bell[28–31] carried out an 11-state **R**-matrix calculation, where they represented the target eigenstates by fairly extensive configuration-interaction (CI) wave functions and included the closed-channel

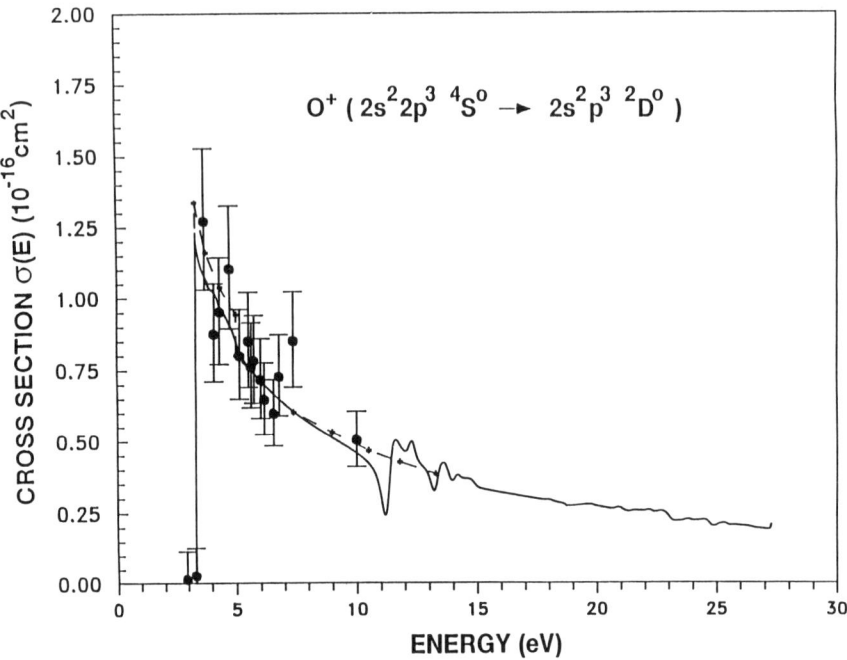

Figure 1. Cross sections for electron impact excitation of the forbidden $^4S^O \to 2s^22p^3\,^2D^O$ transition in O^+. •) Merged-beams electron energy loss experiment[32]; —) 11-state **R**-matrix calculation[32]; ---) two-state close-coupling calculation.[178]

resonances converging to the target excited state thresholds. They calculated collision strengths and effective collision strengths for all possible 55 inelastic transitions. Their calculation is the most accurate to date. However, they have ignored higher $n = 3$ excited states that may show significant channel coupling to lower excited states included in their calculation. The effect of neglected states on the collision strengths presented in their work needs to be examined.

The merged-beams measurements for electron impact excitation of the resonance $2s^22p^3\,^4S^O \to 2s2p^4\,^4P$ transition and the forbidden $2s^22p^3\,^4S^O \to 2s^22p^3\,^2D^O$ transition have been carried out at the JPL[32] and agree very well with the recent **R**-matrix calculation, providing a support to theory. In Figs. 1 and 2 we have presented a comparison between the measured cross sections and the 11-state and 2-state CC calculations. The 11-state theoretical results have been convoluted with the experimental energy spread of 250-meV full width at half maximum (FWHM).

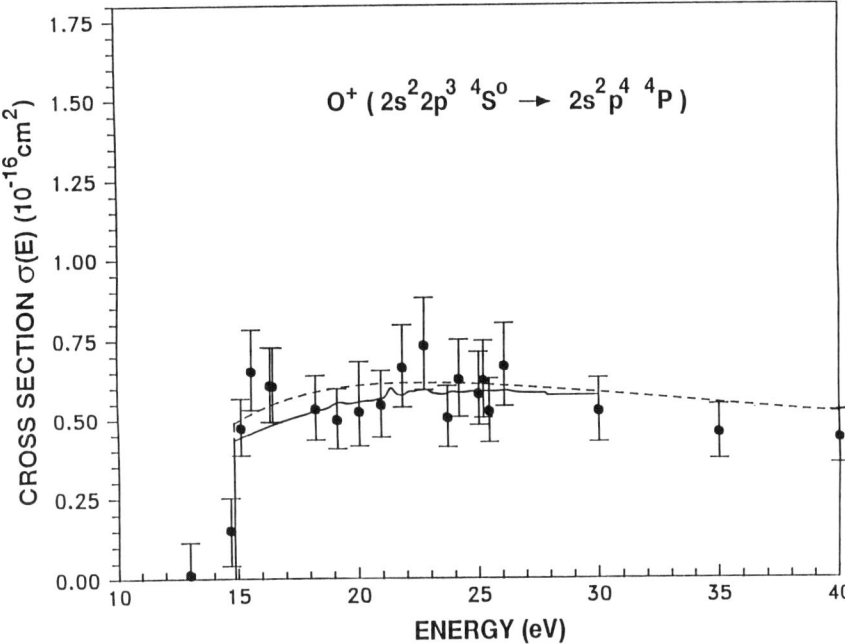

Figure 2. Cross sections for electron impact excitation of the resonance $^4S^o \rightarrow 2s2p^4\ ^4P$ transition in O^+. ●) Merged-beams electron energy loss experiment[32]; —) 11-state **R**-matrix calculation[32]; ---) two-state close-coupling calculation.[26]

2.3. O^{2+} Cross Sections

The O^{2+} ion is one of the most theoretically studied ionic species. There are about 38 calculations reported so far for O^{2+}. Aggarwal and Hibbert[33] and Aggarwal[34] have recently reported extensive CC calculations for electron impact excitation of O^{2+} using the **R**-matrix method. They included 26 LS target eigenstates with $2s^22p^2$, $2s2p^3$, $2p^4$, $2s^22p3s$, $2s^22p3p$, and $2s^22p3d$ configurations in the CC expansion of the total wave function. These states are represented by the CI wave functions constructed from physical and correlation orbitals. Aggarwal and Hibbert[33] reported collision strengths for all 325 possible inelastic transitions among these 26 states in the energy region (4–12 Ry) above the highest excitation threshold in their calculation. Aggarwal[34] calculated collision strengths in the threshold region, where they vary rapidly with energy of the incident electron owing to the presence of autoionizing resonances. He integrated the collision strengths obtained in the threshold and above-threshold energy regions over a Maxwellian distribution of electron energies to obtain effective collision strengths. He assessed the accuracy of his results to about 10% for transitions involving the lower excited states and about 20% for transitions involving higher excited states.

The **R**-matrix results can be supplemented by the CB calculation of Tully and Baluja[35] for the electric quadrupole transition $2p^2(^1D \rightarrow {}^1S)$ and by the DW calculation of Itikawa and Sakimoto[36] for the dipole-allowed $2p^2\,{}^3P \rightarrow 2s2p^3\,{}^3D^O, {}^3S^O$ and forbidden $2p^2\,{}^3P \rightarrow 2p^2\,{}^1S, 2s2p^3\,{}^1P^O$ transitions. The DW calculation of Bhatia et al.[37] can also be combined with the **R**-matrix collision strengths for available transitions to extend the energy range. Bhatia and Kastner[38] performed a DW calculation using the same terms as Aggarwal and Hibbert[33] and calculated collision strengths for fine-structure transitions in an intermediate coupling scheme.

2.4. C$^+$ and O^{3+} Cross Sections

The **R**-matrix calculations by Luo and Pradhan[39] and Blum and Pradhan[40,41] are the most extensive calculations published for B-like C$^+$ and O^{3+} ions so far. They included 10 eigenstates ($2s^22p\,{}^2P^O, 2s2p^2\,{}^4P, {}^2D, {}^2S, {}^2P, 2s^23s\,{}^2S, 2s^23p\,{}^2P^O, 2p^3\,{}^4S^O, {}^2D^O, {}^2P^O$) of C$^+$ and 8 eigenstates ($2s^22p\,{}^2P^O, 2s2p^2\,{}^4P, {}^2D, {}^2S, {}^2P, 2p^3\,{}^4S^O, {}^2D^O, {}^2P^O$) of O^{3+} in the CC expansion and represented these states by the CI wave functions. They included closed-channel resonances in their calculations. They have also reported collision strengths for transitions between the fine-structure levels obtained in an intermediate coupling scheme. The experimental data for electron impact excitation of the resonance ${}^2P^O \rightarrow {}^2D$ transition in C$^+$ reported by Lafyatis and Kohl[14] are in very good agreement with the results of Luo and Pradhan,[39] providing an experimental support for the validity and accuracy of the theory. Several other CC calculations[42–46] with an "A" rating are available for these ions. These give collision strengths for transitions among the low-lying states.

The low-energy **R**-matrix data can be augmented by the Coulomb–Born-exchange calculation of Mann[47] for C$^+$ and by the DW calculation of Mann[48] for O^{3+} for the available transitions.

2.5. C^{2+} and O^{4+} Cross Sections

The 12-state **R**-matrix calculations for electron impact excitation of the C^{2+} ion[49,50] and the O^{4+} ion[51,52] are the most accurate of the currently available calculations on these ions in the low- and intermediate-energy regions ($E \leq 8.0$ Ry for C^{2+} and $E \leq 12.0$ Ry for O^{4+}). These calculations include the 12 lowest eigenstates with configurations $2s^2, 2s2p, 2p^2, 2s3s, 2s3p,$ and $2s3d$ in the CC expansion and replace earlier 6-state **R**-matrix calculations of Berrington et al.[53–55] Sophisticated CI wave functions are used to represent the target states, and resonance series converging to the $n = 2$ and $n = 3$ thresholds are explicitly included in these calculations.

Several DW and CB calculations of comparable accuracy are available for the higher energy region and can be combined with the **R**-matrix data to extend their

Excitation of Atomic Ions by Electron Impact

energy range. Osterbrock and Wallace[56] reported DW and CB calculations for C^{2+} and O^{4+} ions whereas Younger[57] used these perturbative methods for O^{4+}. Nakazaki and Hashino[58] used CB to calculate electron collisional excitation cross sections of C^{2+} and O^{4+}. Peek and Mann[59] performed DW calculation for C^{2+}.

2.6. C^{3+} and O^{5+} Cross Sections

C^{3+} and O^{5+} are among the few ions that have been studied experimentally. Taylor et al.[15] and Lafyatis and Kohl[14] measured absolute excitation cross sections for the resonance $2s \rightarrow 2p$ transition in C^{3+} using the crossed-beams technique. Datla and Kunze[10] measured relative excitation rate coefficients of O^{5+} using the plasma rate method. Recently, Burke[60] calculated accurate collision strengths by taking full account of resonances for all transitions between the $n = 2$ to 4 states of C^{3+} using the **R**-matrix method. The measured cross sections reported by Taylor et al.[15] and Lafyatis and Kohl[14] for the resonance $2s \rightarrow 2p$ transition and the **R**-matrix calculation of Burke[60] are currently the best available data for C^{3+}. These data can be supplemented by the DW calculation of Peek and Mann[59] and by the CB calculation of Callaway et al.[61] in the high-energy region for the available transitions.

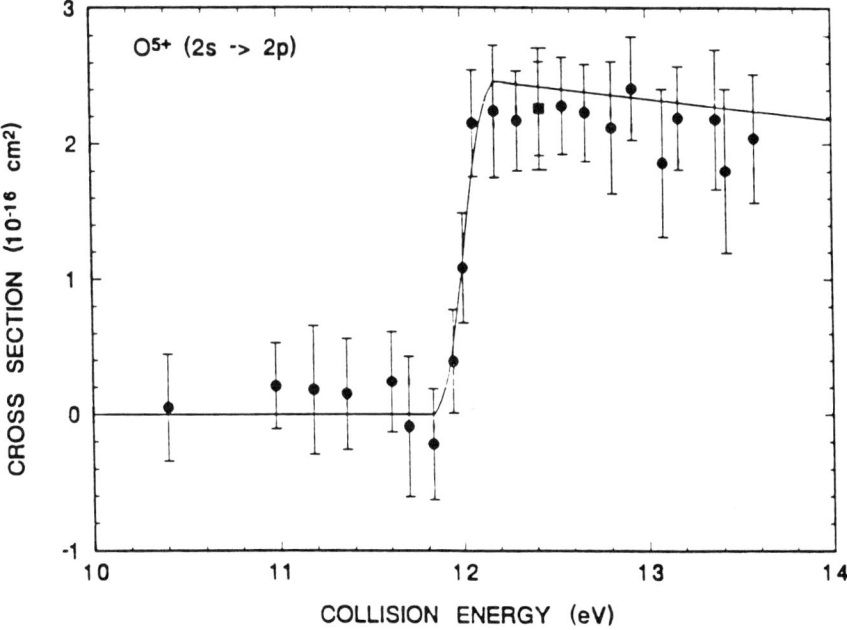

Figure 3. Excitation cross section for the $2s \rightarrow 2p$ transition in O^{5+}. —) Seven-term close-coupling calculation after convolution with an electron-energy distribution of 0.2-eV FWHM; ●) merged-beams experiment.[62]

Merged-beams measurements[62] for the $2s \rightarrow 2p$ excitation in O^{5+} are compared with a seven-state CC calculation in Fig. 3. After convoluting the calculations with a 0.20-eV energy spread for the electron beam, the agreement between theory and experiment is seen to be very good. The five-state CC calculation of Bhadra and Henry[63] is recommended for other excitations in O^{5+}. There is a need for fairly extensive accurate calculation for O^{5+} over the entire energy region, particularly for excitation of the $n = 4$ states.

2.7. C^{4+} and O^{6+} Cross Sections

The **R**-matrix calculations by Tayal and Kingston[64] and Tayal[65] for C^{4+} and by Tayal and Kingston[66] for O^{6+} are the most elaborate and accurate of the available calculations to date for the near-threshold region. They included the 11 lowest target states ($1s^2$ ^1S, $1s2s$ ^3S, ^1S, $1s2p$ $^3P^O$, $^1P^O$, $1s3s$ ^3S, ^1S, $1s3p$ $^3P^O$, $^1P^O$, $1s3d$ ^3D, ^1D) in the CC expansion of the total wave function and took full account of all the resonance Rydberg series converging to the $n = 2$ and $n = 3$ excited states. They used sophisticated CI wave functions to represent the target states. They calculated collision strengths and effective collision strengths for all 55 possible inelastic transitions in these ions.

The DW calculations[67–69] and CB calculations[70,71] can be used to extend the range of **R**-matrix calculations to intermediate and high energies, respectively. There is normally good agreement between the **R**-matrix and DW calculations in the above-threshold region at intermediate energies.

The effective collision strengths for transitions from the ground ($n = 1$) to the excited ($n = 2$ and $n = 3$) states from the 11-state **R**-matrix calculations[64–66] are given in Table III at four temperatures.

Table III. Effective Collision Strengths for Transitions from the Ground State to Excited States in C^{4+} and O^{6+} at Four Temperatures

Transition	C^{4+} (K)				O^{6+} (K)			
	1.0×10^5	6.0×10^5	1.0×10^6	6.0×10^6	1.0×10^5	6.0×10^5	1.0×10^6	6.0×10^6
1^1S-2^3S	0.0207	0.0107	0.0089	0.0051	0.0178	0.0083	0.0067	0.0039
1^1S-2^3P	0.0395	0.0341	0.0309	0.0193	0.0263	0.0224	0.0206	0.0129
1^1S-2^1S	0.0111	0.0113	0.0116	0.0139	0.0149	0.0087	0.0081	0.0083
1^1S-2^1P	0.0304	0.0412	0.0465	0.0810	0.0189	0.0235	0.0263	0.0446
1^1S-3^3S	0.0035	0.0022	0.0020	0.0012	0.0017	0.0012	0.0011	0.0009
1^1S-3^3P	0.0104	0.0079	0.0070	0.0043	0.0056	0.0049	0.0046	0.0033
1^1S-3^1S	0.0023	0.0021	0.0022	0.0028	0.0016	0.0013	0.0013	0.0017
1^1S-3^3D	0.0027	0.0018	0.0015	0.0009	0.0013	0.0011	0.0010	0.0007
1^1S-3^1D	0.0013	0.0009	0.0009	0.0010	0.0006	0.0006	0.0006	0.0008
1^1S-3^1P	0.0085	0.0090	0.0099	0.0180	0.0044	0.0048	0.0053	0.0089

2.8. C^{5+} and O^{7+} Cross Sections

Aggarwal and Kingston[72] presented collision strengths and effective collision strengths for all transitions between the states with $n = 1$–5 in C^{5+} in the energy region from threshold to 50 Ry using the **R**-matrix method. They included resonances in their calculation. This is the most extensive and detailed calculation for C^{5+} currently available. The other two calculations of comparable accuracy are those of Morgan[73] and Abu-Salbi and Callaway.[74] Both of these calculations used exact and pseudostates in the CC expansion and have given results for the $1s \rightarrow 2s$ and $1s \rightarrow 2p$ transitions only. Abu-Salbi and Callaway[74] used the CB approximation with exchange to calculate high partial waves and high electron impact energy results. They found that the resonance contribution to the effective collision strengths for the $1s \rightarrow 2s$ transition is about 6% whereas for the $1s \rightarrow 2p$ transition it is about 3%. The three calculations agree with each other normally within 10%.

For O^{7+}, the data reported by Abu-Salbi and Callaway[74] are the best of the available calculations for $1s \rightarrow 2s$ and $1s \rightarrow 2p$ transitions. There is a need for data for the transitions to higher excited states.

3. IRON IONS

3.1. General Considerations

All ionization stages of iron are expected to be observed and used for diagnostic purposes in fusion plasmas. There have been relatively few calculations for the iron ions isoelectronic with the fourth-row elements of the periodic table, Fe^0–Fe^{7+}. In the compilations on electron impact excitation data by Pradhan and Gallagher[1] and by Kingston and Lennon,[179] no entries are listed for Fe^0 and Fe^{4+}. The situation, however, is expected to improve considerably with the new Iron Project[180] calculations currently under way for Fe^+–Fe^{5+} using the **R**-matrix method, which has been widely employed for low-energy electron scattering. The primary goal of the Iron Project is to calculate accurate and extensive collision data for electron impact excitation of iron ions and ions of other iron peak elements. For example, Pradhan and Zhang[181,182] have recently reported collision strengths and Maxwellian averaged rate coefficients for 10,011 transitions in Fe^+. Although these calculations were motivated by astrophysical needs, the data have been tabulated in the temperature range accessible to fusion plasmas as well. In this brief review we discuss the available data for the heavy iron ions and present a discussion of the associated uncertainties in the data.

Close-coupling calculations for the fourth-row elements are complicated on account of the extensive electron correlation required to adequately represent the open $3d$ shell target states. Experience shows[183,184] that in addition to the ground configuration $3d^n$, it is also necessary to include coupling to excited configurations $3d^{n-1} 4s$, $3d^{n-1} 4p$, and preferably $3d^{n-1} 4d$. Collision strengths for transitions among the low-lying states show a considerable amount of resonance structure owing to coupling to higher levels, particularly for transitions among the low-lying levels of the even-parity configurations that couple strongly to the odd-parity configuration $3d^{n-1}\,4p$. However, the number of *LS* terms dominated by the four configurations is very large, and a judicious choice is necessary to make the calculations feasible even on the Cray Y-MP or the Cray-2.

In the Iron Project calculations the eigenfunction basis includes the following *LS* terms for Fe^+–Fe^{5+}:

Fe^+ (145 *LS* terms; $3d^64s$, $3d^7$, $3d^64p$)
Fe^{2+} (83 *LS* terms; $3d^6$, $3d^54s$, $3d^54p$)
Fe^{3+} (104 *LS* terms; $3d^5$, $3d^44s$, $3d^44p$)
Fe^{4+} (38 *LS* terms; $3d^4$, $3d^34s$, $3d^34p$)
Fe^{5+} (43 *LS* terms; $3d^3$, $3d^24s$, $3d^24p$)

Not all the terms dominated by the configurations listed could be included in the calculations. It is implied that in each case electron scattering cross sections are to be calculated among all the *LS* terms, *and associated fine-structure levels*, included in the target ion. The fine-structure calculations may be carried out in three different ways: (i) nonrelativistic *LS* coupling collision calculations followed by a purely algebraic transformation to a pair-coupling scheme, (ii) quasi-relativistic calculations including term coupling between the target terms in the algebraic transformation, and (iii) the Breit–Pauli **R**-matrix calculations, which account exactly for the intermediate coupling. The three methods are discussed in some detail in a paper by Hummer *et al.*,[180] where a description of the codes is given. In the future, one also expects to be able to carry out Dirac **R**-matrix calculations using codes under development by P. H. Norrington and I. P. Grant.

In Table IV we summarize the data sources for the iron ions. *The transitions listed in the footnotes are only from the recommended sources in the immediately preceding reference(s)*. Below, we describe the available data for the various iron ions.

3.2. Fe^+ Cross Sections

The first calculations, by Nussbaumer and Storey,[185] were carried out in the distorted-wave (DW) approximation with no resonance effects included. They considered the four lowest states of Fe^+: $3d^64s(^6D, {}^4D)$, $3d^7({}^4F, {}^4P)$. Subsequent work by the Belfast group for the same transitions[186,187] yielded significant differences with respect to the earlier work. The calculations by Berrington *et al.*[186] were also carried out in the Breit–Pauli approximation, including resonances, for the forbidden transitions. A comparison of rate coefficients by Berrington *et al.* shows that the Nussbaumer and Storey DW collision strengths underestimate the excita-

Table IV. Electron Impact Excitation Data for Fe^+–Fe^{7+} Ions[a]

Ion	N	Author(s)	Year	Method	Coupling	Resonances included	Data	Rating	Reference
Fe^{+b}	25	Nussbaumer and Storey	1980	DW	IC		Ω	D	185
		Berrington et al.	1988	CC	BP	✓	Ω	B	186
		Keenan et al.	1988	CC	BP	✓	Y	B	187
		Pradhan and Berrington	1993	CC	LS	✓	Ω	B	183
		Zhang and Pradhan	1994	CC	IC	✓	Ω, Υ	B	182
Fe^{2+c}	24	Garstang et al.	1978	CC	IC		Ω	D	189
		Berrington et al.	1991	CC	BP	✓	Υ	C	190
Fe^{3+d}	23	Berrington et al.	1992	CC	LS	✓	ü	C	191
Fe^{5+e}	21	Garstang et al.	1978	CC	IC		Ω	D	189
		Nussbaumer and Storey	1978	DW	IC		Ω	C	192
Fe^{6+}	20	Norrington and Grant	1987	DRM	jj	✓	Ω	B	193
		Nussbaumer and Osterbrock	1970	DW	IC		Ω	C	194
		Nussbaumer and Storey	1982	DW	IC		Ω	C	195
Fe^{7+}	19	Czyzak and Krueger	1966	CB	IC		Ω	D	196
		Blaha	1969	CB, other	IC		Q	D	197
		Datla et al.	1975	CB	IC		q	D	198
		Kato	1976	Other	LS		Ω, q	D	199
		Pindzola et al.	1989	CC, DW	LS		Q	C	200

[a] Notation as in Table I.
[b] Transitions: All transitions between 142 fine-structure levels of the 38 LS terms: $3d^64s(a^6D, a^4D, b^4P, a^4H, b^4F, a^4G, b^4D, c^4P, c^4F)$, $3d^7(a^4F, a^4P)$, and $3d^64p(z^6D^O, z^6F^O, z^6P^O, z^4F^O, z^4D^O, z^4P^O, z^4S^O, y^4P^O, z^4G^O, z^4H^O, z^4I^O, y^4D^O, y^4F^O, x^4D^O, y^4G^O, x^4F^O, y^4H^O, {}^4P^O, w^4F^O, w^4D^O, v^4D^O, {}^4G^O, {}^4S^O, {}^4P^O, {}^4D^O, u^4F^O)$; Ref. 182.
[c] Transitions: Transitions between fine-structure levels of LS terms: $3d^6(^5D, {}^3P, {}^3F, {}^3G, {}^3H)$; Ref. 190.
[d] Transitions: $3d^6$: ${}^6S$-${}^4G$, ${}^4P$, ${}^4D$, ${}^4F$; Ref. 191.
[e] Transitions: $3d^3(^4F_{3/2,5/2,7/2,9/2}, {}^4P_{1/2,3/2,5/2}, {}^2G_{7/2,9/2}, {}^2P_{1/2,3/2}, {}^2D_{3/2,5/2}, {}^2H_{9/2,11/2}, {}^2F_{7/2,9/2}, {}^2D_{3/2,5/2})$, all $(J \rightarrow J')$; Ref. 192.
[f] Transitions: $3d^2$: ${}^3F_2$-${}^1D_2$, ${}^3P_1$, ${}^1G_4$, ${}^1S_0$; Ref. 193.

tion rate as the low-energy resonances are not included in the forbidden transitions (see below). Pradhan and Berrington[183] carried out close-coupling (CC) calculations for collision strengths and rate coefficients in 38-state LS coupling and 41-level Breit–Pauli (BP) approximations. The BP calculations include fine-structure levels dominated by 10 LS states. In general, it is found that the new CC results differ considerably (the collision strengths for the forbidden transitions being larger by factors of up to 2–3) from the earlier DW data of Nussbaumer and Storey[185] and Nussbaumer et al.[188] Also, there are significant differences up to about 30%, for the allowed transitions, for example, $3d^64s(^6D) \rightarrow 3s^64p(^6P^O, {}^6D^O, {}^6F^O)$.

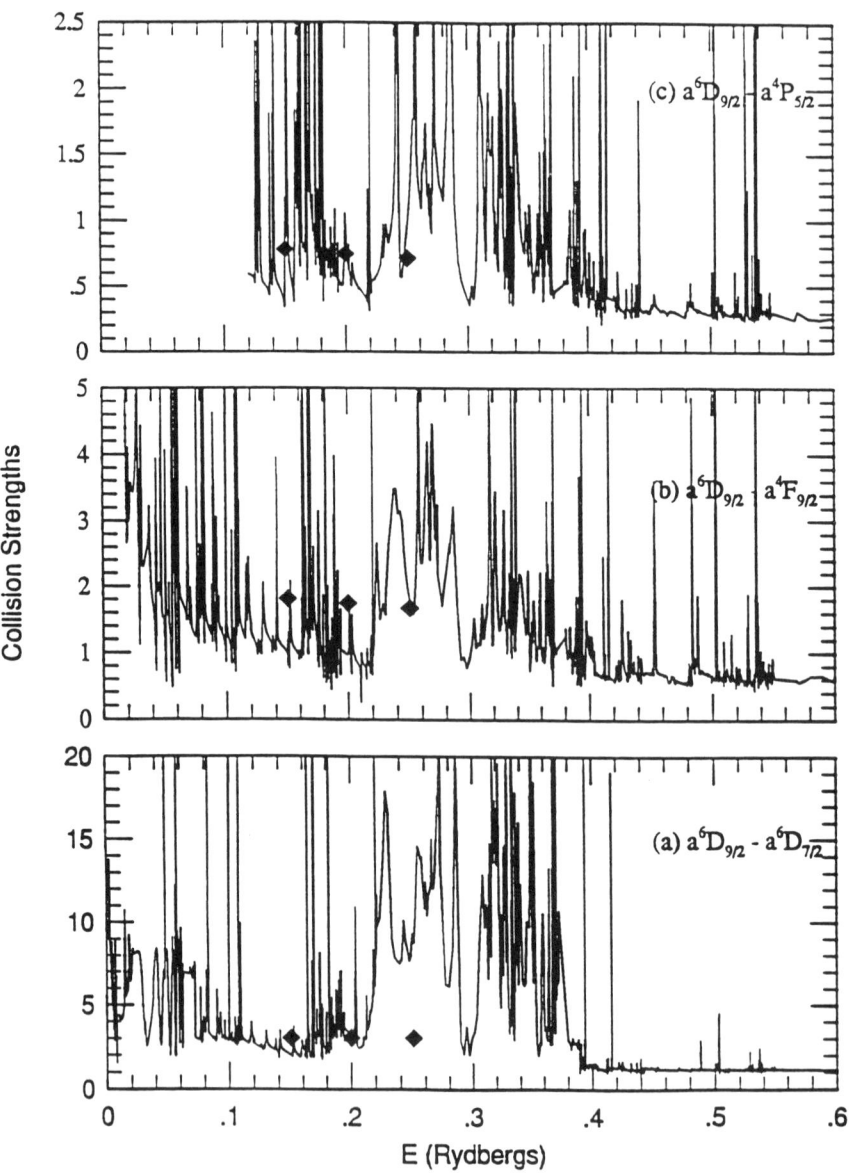

Figure 4. Collision strengths for three forbidden transitions in Fe$^+$. —) Close-coupling calculation[182]; ◆) distorted-wave calculations.[188]

In Figure 4 the recent CC results of Zhang and Pradhan[182] for the collision strengths for three forbidden transitions in Fe^+ are compared with the earlier DW results of Nussbaumer et al.[188] The Maxwellian integrated collision strengths from Ref. 182 are generally higher than the corresponding DW values by factors of up to 3 or 4. The four-state CC results by Berrington et al.[186] and Keenan et al.[187] are in reasonable agreement with the recent work of Zhang and Pradhan.[182] For example, the collision strength at 0.15 Ry for the ^6D–^4F multiplet, summed over all the fine-structure components, is given as 6.82 in Ref. 185 and as 5.81 in Ref. 183, a difference of 17%. This indicates an uncertainty better than 20%, *i.e.* a "B" rating as in Table IV.

The Zhang and Pradhan calculations[182] (which are an extension of the work in Ref. 183) provide the collision strengths for all 10,011 transitions among the 142 fine-structure levels of the 38 *LS* terms. These results are based on approximation (i) for the fine structure (see Section 3.1) and should be accurate as the relativistic effects for Fe^+ are expected to be small. Breit–Pauli calculations for Fe^+ are also in progress.

3.3. Fe^{2+} Cross Sections

Garstang, Robb, and Rountree (GRR)[189] carried out several sets of close-coupling calculations for selected iron ions. Their calculations included a very limited number of states, and the resonance structures were not delineated. Recently, Berrington et al.[190] completed **R**-matrix calculations in the Breit–Pauli approximation for 17 fine-structure levels; the transitions are listed in Table IV. Berrington et al.[190] also carried out a detailed comparison of their calculations with the GRR calculations, as well as with a five-state *LS* coupling **R**-matrix calculation followed by an algebraic transformation based on approximation (i) in Section 3.1. They reported good agreement with the latter work, but their results disagreed by up to a factor of 3 with the GRR results. It is concluded that the GRR work contains errors and the more recent data should be employed. The Berrington et al. work[190] includes collision strengths; Maxwellian averaged effective collision strengths are also given in their Table 4.

The target wave functions employed in the Berrington et al. work[190] are as used in the Opacity Project calculations for radiative data for the $(e + Fe^{2+}) \rightarrow Fe^+$ system. However, the target states included are only the ones dominated by the $3d^6$ configuration, and higher states are not considered. In particular, the coupling to the terms dominated by the $3d^54p$ configuration is not included; this is likely to be of importance in terms of strong resonance series converging onto the odd-parity terms and contributing to the effective collision strengths, as seen in the case of Fe^+. Therefore, although these are CC calculations, the data are rated "C" in Table IV.

Further calculations that are being carried out by Pradhan and co-workers include many of the 136 terms from the $3d^6$, $3d^54s$, and $3d^54p$ configurations (up to 83 selected terms in the eigenfunction expansion).

3.4. Fe^{3+} Cross Sections

New calculations have now been done by Berrington et al.[191] for a few transitions, as shown in Table IV. Again, the data are rated "C" for essentially the same reason as in the case of the data for Fe^{2+}, namely, an eigenfunction expansion that is perhaps too restrictive. The calculations are in the LS coupling approximation, and Maxwellian averaged collision strengths have been calculated. The data may be obtained from the Belfast data bank.

3.5. Fe^{5+} Cross Sections

The excitation data available come from two sources, the CC calculations by Garstang et al.[189] and the DW calculations by Nussbaumer and Storey.[192] As shown in Table IV, the CC values are rated lower than the DW results. The reason, as explained by Nussbaumer and Storey, is that the CC calculations employed an incomplete partial wave expansion up to only $l \leq 2$, whereas the DW calculations included up to $l \leq 5$. Consequently, the DW collision strengths are often higher than the CC ones by a factor of about 2. However, neither set of calculations included resonances or the coupling effects due to higher configurations. Therefore, even the recommended DW results may not be very accurate.

The collision strength data for the forbidden transitions in Fe^{5+} were given by Nussbaumer and Storey.[192]

3.6. Fe^{6+} Cross Sections

The most accurate relativistic calculations on Fe^{6+} have been carried out by Norrington and Grant[193] using the Dirac **R**-matrix method. A detailed comparison shows large differences from the results of the earlier work by Nussbaumer and Osterbrock[194] and Nussbaumer and Storey[195] using the DW approximation and intermediate coupling coefficients from the SUPERSTRUCTURE code. Norrington and Grant[193] found very different energy behavior of collision strengths at higher energies that can affect the rate coefficients considerably. The **R**-matrix values are seen to decrease with increasing energy above all thresholds, whereas the DW values show an increasing trend. Although the target expansion employed in the **R**-matrix work is smaller than in the DW work, the authors of the former work[193] believe that the differences at higher energies are not due to that fact but due to redistribution of flux among inelastic channels accounted for in the **R**-matrix calculations (but not in the DW work) which yields decreasing collision strengths. Norrington and Grant also state that at even higher energies they expect the DW

Excitation of Atomic Ions by Electron Impact

and the **R**-matrix calculations to agree better as higher partial waves make more contribution.

The transitions listed in Table IV are the same ones as considered in both sets of calculations.

3.7. Fe^{7+} Cross Sections

Although several workers[196–200] have carried out the collisional calculations, the accuracy is not sufficient to recommend uncritical use in applications. The calculations by Pindzola et al.[200] for the allowed transitions between the $3d$, $4s$, and $4p$ states should be of acceptable accuracy, although they do not include the terms dominated by the $3p^5 3d^2$ configuration that lie between the $3p^6 4s$, $4p$, and $4d$ terms. The uncertainty in the data may therefore be somewhat higher than the "C" rating suggests, although these data are much better than the other data listed. Pindzola et al.[200] presented the collision strengths in graphical form for the transitions between these states. The data for some other transitions may be obtained from the earlier references by Czyzak and Krueger[196] and Blaha[197]; however, those data are not considered reliable.

3.8. Further Considerations

The situation regarding collisional data for the heavy iron ions, Fe^+–Fe^{7+}, has recently improved primarily owing to new calculations from the Iron Project. However, many of the calculations are still in progress. As yet, no data are available for Fe^0 and Fe^{4+}. Very sparse data exist for Fe^{3+} and Fe^{7+}. Collision strengths have recently been calculated for Fe^+ for a large number of transitions; however, the relativistic calculations for fine-structure transitions are yet to be completed. The current data for Fe^{2+}, Fe^{5+}, and Fe^{6+} are likely to be of less than desirable accuracy and have been computed only for the lowest transitions.

4. RARE-GAS IONS

4.1. General Considerations

A knowledge of atomic collision processes involving rare-gas ions has been found to be quite useful in understanding the physics at the plasma edge and in divertors. Rare gases are introduced as both a diagnostic impurity and a source of electrons for studies of the edge plasma. Rare gases also make excellent radiation "coolants" for divertors. In this section we review the atomic data for the electron impact excitation of argon, krypton, and xenon in low ionization stages, present detailed calculations for three of the rare-gas ions, and compare theory and experiment for both Ar^{7+} and Kr^{6+}.

From the review by Pradhan and Gallagher[1] we constructed Table V, which lists a number of theoretical calculations of electron impact excitation cross sections

Table V. Electron Impact Excitation Cross Section Calculations for Argon in Low Ionization Stages[a,b]

Ion	N	Author(s)	Year	Method	Coupling	Data	Rating	Reference
Ar^+	1	Blaha	1968	CB, DW	IC	$\Omega\,(x=1)$	D	201
		Blaha	1969	Other	IC	$\Omega\,(x=1)$	D	202
		Krueger and Czyzak	1970	DW	IC	Ω	D	203
		Brandi and Koster	1973	DW, CC	LS	Q	C	204
Ar^{2+}	2	Czyzak et al.	1967	CB, DW	IC	$\Omega\,(x=1)$	D	205
		Blaha	1968	CB, DW	IC	$\Omega\,(x=1)$	D	201
		Blaha	1969	Other	IC	$\Omega\,(x=1)$	D	202
		Krueger and Czyzak	1970	DW	IC	Ω	D	203
Ar^{3+}	3	Czyzak and Krueger	1967	Other	LS	Ω	E	206
		Blaha	1969	Other	IC	$\Omega\,(x=1)$	D	202
		Conneely et al.	1970	CC	LS	Ω, Q	B	207
		Czyzak et al.	1970	DW	LS	Ω	D	208
		Krueger and Czyzak	1970	DW	IC	Ω	D	203
		Zeippen et al.	1987	CC	IC	Ω, γ	A	209
Ar^{4+}	4	Czyzak and Krueger	1963	Other		Ω	D	210
		Bely	1966	Other	LS	Ω	E	211
		Bely	1967	Other	LS	$\Omega\,(x=1)$	D	212
		Czyzak and Krueger	1967	DW	LS	$\Omega\,(x=1)$	D	206
		Blaha	1968	CB, DW	IC	$\Omega\,(x=1)$	D	201
		Blaha	1969	Other	IC	$\Omega\,(x=1)$	D	202
		Krueger and Czyzak	1970	DW	IC	Ω	D	203
Ar^{5+}	5	—						
Ar^{6+}	6	Blaha	1968	CB, DW	IC	$\Omega\,(x=1)$	D	201
		Younger	1980	DW, CB, B	LS	Ω	C	213
		Christensen et al.	1986	DW	IC	Ω	C	214
		Pradhan	1988	DW, CBE	IC	γ	B	215
Ar^{7+}	7	Datla et al.	1972	CB	LS	q	D	216
		Kim and Cheng	1978	Other	IC	Q	U	217

[a] From Ref. 1.
[b] Notation as in Table I.

and rate coefficients for Ar^+ through Ar^{7+}. A similar list for Kr^+ through Kr^{7+} would only include entries for the Kr^{6+} ion[213] and the Kr^{7+} ion,[218] while the list for Xe^+ through Xe^{7+} would be completely blank.

4.2. Ar^{7+} Cross Sections

Electron impact excitation cross sections for the dipole-allowed $3s \to 3p$ transition in Ar^{7+} obtained by the close-coupling and the distorted-wave methods[219] are compared in Fig. 5. The most significant differences between the results are the small window resonance features in the close-coupling results, which are not present in the distorted-wave results. The windows are caused by interference between the direct and resonant excitation processes. The close-coupling calcula-

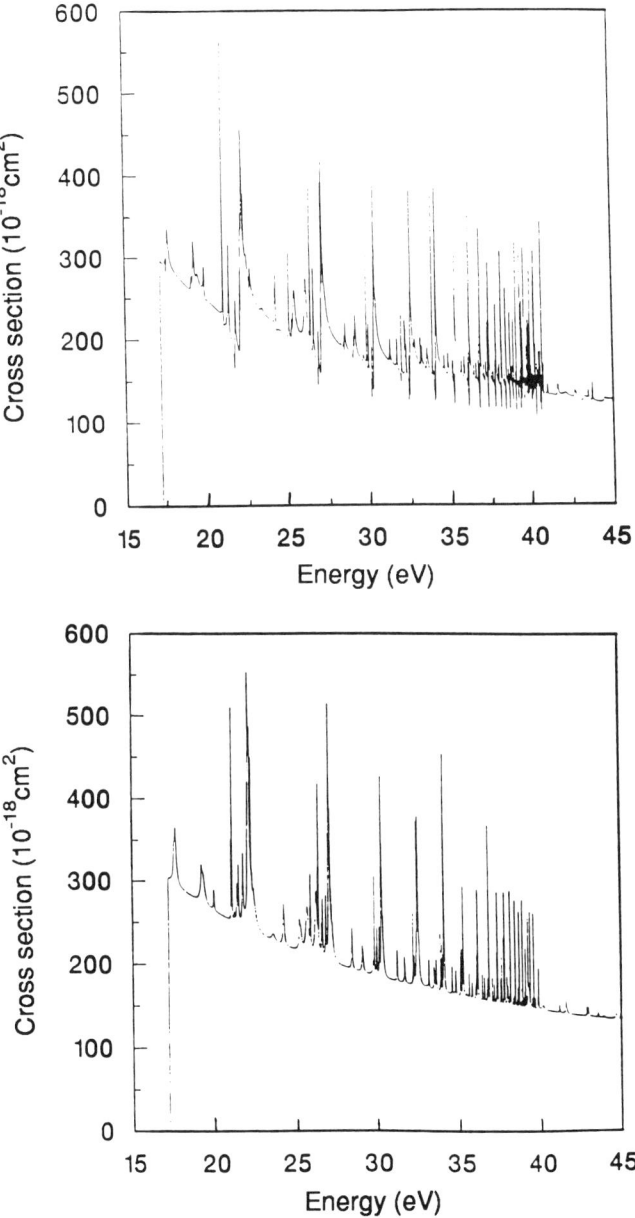

Figure 5. Excitation cross section for the 3s→3p transition in Ar[7+]. *Top*: Seven-term close-coupling calculation[219]; *bottom*: distorted-wave calculation.[219]

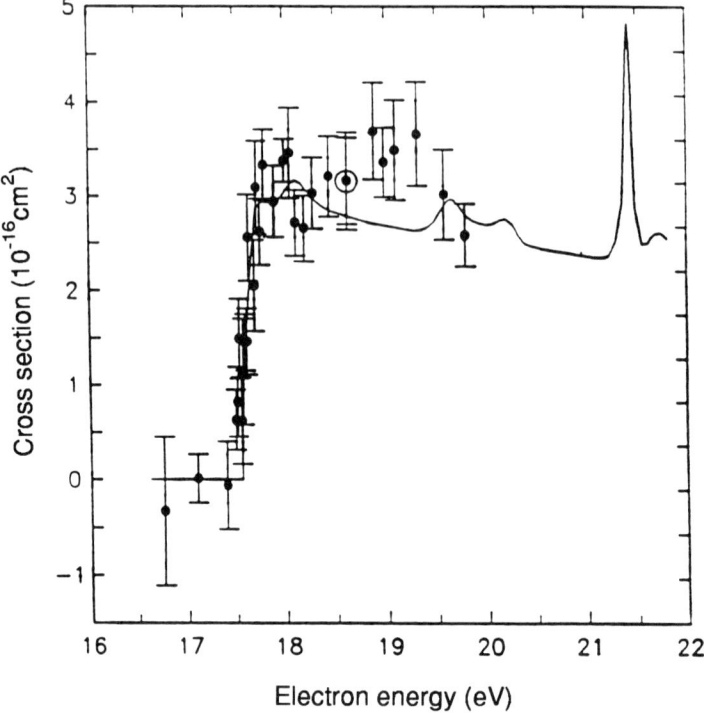

Figure 6. Excitation cross section for the $3s \to 3p$ transition in Ar^{7+}. —) seven-term close-coupling calculation[219] after convolution with an electron-energy distribution of 0.2-eV FWHM; ●) merged-beams experiment.[20]

tions convoluted with a 0.2-eV FWHM Gaussian to simulate the electron energy distribution are compared with the data from a merged-beams experiment[20] in Fig. 6. The agreement between theory and experiment is very good over the small energy range investigated, but many of the resonance features predicted by theory are beyond the energy range of the currently available experimental data.

4.3. Ar^{6+} Cross Sections

In support of current experiments, electron impact excitation cross sections for Ar^{6+} were calculated by the close-coupling method.[220] The dipole-allowed $3s^2\ {}^1S \to 3s3p\ {}^1P$ transition is shown in Fig. 7, while the spin-forbidden $3s^2\ {}^1S \to 3s3p\ {}^3P$ transition is shown in Fig. 8. The resonant excitation process dominates the direct excitation process for the spin-forbidden transition. Merged-beams experiments for the spin-forbidden transition are in progress.

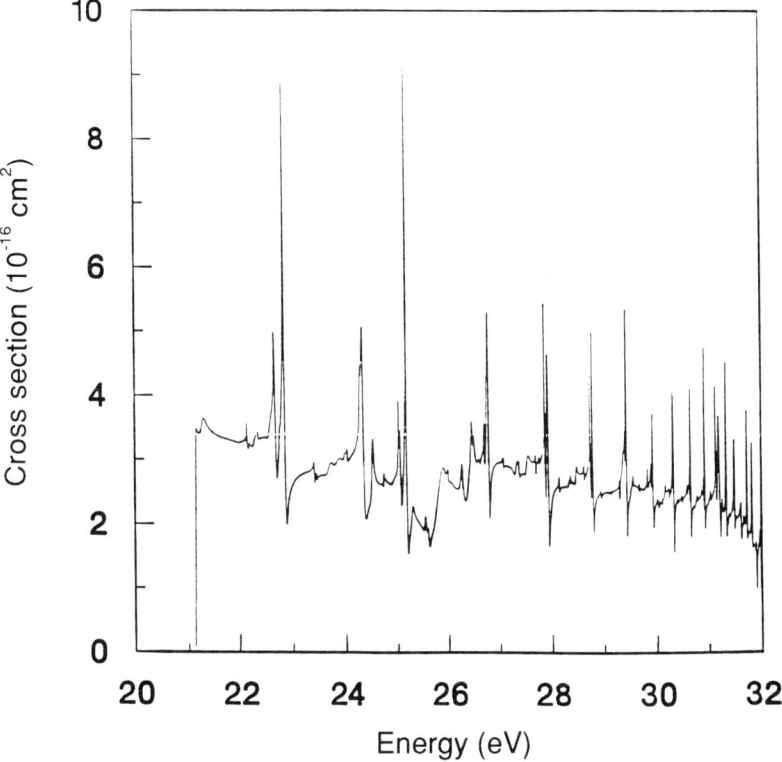

Figure 7. Excitation cross section for the $3s^2\ ^1S \to 3s3p\ ^1P$ transition in Ar^{6+} from an eight-term close-coupling calculation.[220]

4.4. Kr^{6+} Cross Sections

The electron impact excitation cross section for the spin-forbidden $4s^2\ ^1S \to 4s4p\ ^3P$ transition in Kr^{6+} has recently been measured.[221] In this work the first measurements of absolute total cross sections for electron–ion excitation of an intercombination transition have been reported. Theoretical calculations using an intermediate-coupling **R**-matrix method are in progress.

5. CONCLUSIONS

For carbon and oxygen ions, accurate excitation cross sections for all but O^{5+} and O^{7+} ions are now available in the near-threshold region for many transitions. There is a need for data at intermediate and high energies for excitation to higher states to augment the low-energy close-coupling results. The Iron Project's work

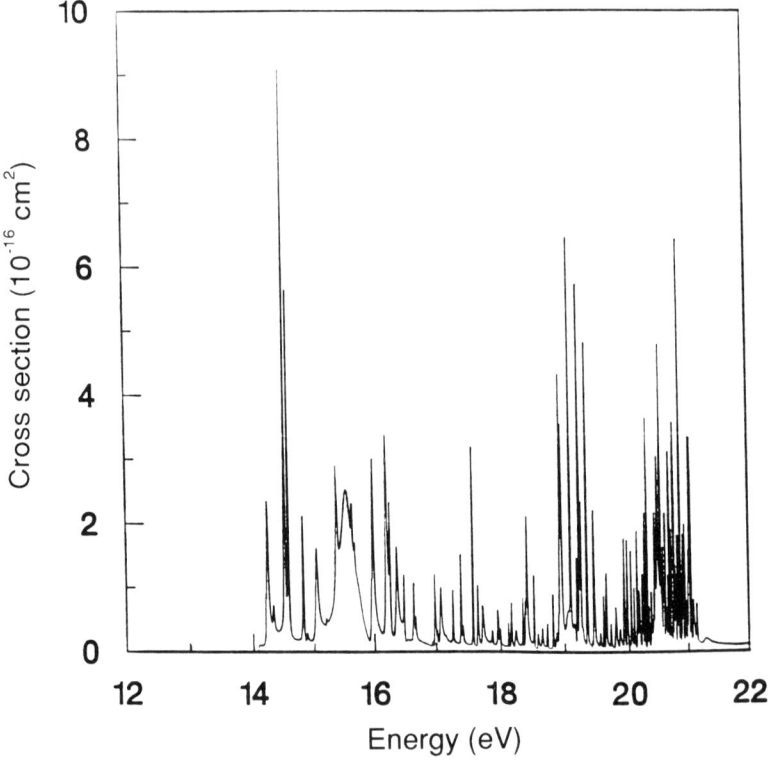

Figure 8. Excitation cross section for the $3s^2\ ^1S \to 3s3p\ ^3P$ transition in Ar^{6+} from an eight-term close-coupling calculation.[220]

on the heavy iron ions exemplifies the complexity involved in doing accurate electron scattering physics for atomic systems with unfilled d shells. In order to include most of the expected electron correlation in the target ion, the **R**-matrix scattering calculation must include a very large number of close-coupled channels. The solution of large numbers of coupled channel equations pushes at the frontier of current high-performance computing machines. A preliminary conclusion of the joint theoretical and experimental project on the rare-gas ions is that the strong dipole-allowed transitions, with a relatively small contribution from resonance structures, are well predicted by theory. The spin-forbidden transitions, with relatively large resonance features, may be more problematic for theory.

ACKNOWLEDGMENTS. We would like to thank Drs. Sultana Nahar, Hong Lin Zhang, Thomas W. Gorczyca, Mark E. Bannister, and A. Chutjian for contributions and Dr.

Keith Berrington for sending the Fe^{2+} and Fe^{3+} data from the Belfast data bank. This work was supported in part by a grant from the NASA Planetary Atmospheres Program (NAGW-4447), the U.S. National Science Foundation (PHY-9115057), the NASA Long Term Space Astrophysics (LTSA) program (NAGW-3315), and the Office of Fusion Energy, U.S. Department of Energy (DE-FG05-86ER53217). SST was also supported by the Oak Ridge Institute for Science and Education, Office of Fusion Energy, Faculty/Student Team Research Participation Program and the Oak Ridge National Laboratory Controlled Fusion Atomic Data Center, where a part of this chapter was completed.

REFERENCES

1. A. K. Pradhan and J. W. Gallagher, *At. Data Nucl. Data Tables* **52**, 227 (1992).
2. D. C. Griffin, M. S. Pindzola, F. Robicheaux, T. W. Gorczyca, and N. R. Badnell, *Phys. Rev. Lett.* **72**, 3491 (1994).
3. K. A. Berrington and R. E. H. Clark, Atomic and Plasma–Material Interaction Data for Fusion, *Nucl. Fusion, Suppl.* **3**, 87 (1992).
4. Y. Itikawa, S. Hara, T. Kato, S. Nakazaki, M. S. Pindzola, and D. H. Crandall, in Recommended Data on Excitation of Carbon and Oxygen Ions by Electron Collisions, Institute of Plasma Physics, Nagoya University Report IPPJ-AM-27, 1983; *At. Data Nucl. Data Tables* **33**, 149 (1985).
5. K. M. Aggarwal, K. A. Berrington, W. B. Eissner, and A. E. Kingston, Report on Recommended Data, Atomic Data Workshop, Daresbury Laboratory, Daresbury, United Kingdom, March 1986.
6. V. P. Shevelko, L. A. Vainshtein, and E. A. Yukov, *Phys. Scr.* **T28**, 39 (1989).
7. R. A. Phaneuf, R. K. Janev, and M. S. Pindzola, Collisions of Carbon and Oxygen Ions with Electrons, H, H_2 and He, Atomic Data for Fusion, Volume 5, Oak Ridge National Laboratory Report ORNL-6090, 1987.
8. R. A. Phaneuf, P. Deferance, D. C. Griffin, Y. Hahn, M. S. Pindzola, L. Roszman, and W. L. Weise, *Phys. Scr.* **T28**, 5 (1987).
9. H.-J. Kunze and W. D. Johnson III, *Phys. Rev. A* **3**, 1384 (1971).
10. R. U. Datla and H.-J. Kunze, *Phys. Rev. A* **37**, 4614 (1988).
11. W. D. Johnson III and H.-J. Kunze, *Phys. Rev. A* **4**, 962 (1971).
12. H.-J. Kunze, A. H. Gaberiel, and H. R. Griem, *Phys. Rev.* **165**, 267 (1968).
13. A. N. Prasad and M. F. El-Menshawy, *J. Phys. B* **1**, 471 (1968).
14. G. P. Lafyatis and J. L. Kohl, *Phys. Rev. A* **36**, 59 (1987).
15. P. O. Taylor, D. C. Gregory, G. H. Dunn, R. A. Phaneuf, and D. H. Crandall, *Phys. Rev. Lett.* **39**, 1256 (1977).
16. G. P. Lafyatis, G. L. Kohl, and L. D. Gardner, *Rev. Sci. Instrum.* **58**, 383 (1987).
17. N. H. Magee, Jr., J. B. Mann, A. L. Merts, and W. D. Robb, Electron Impact Excitation of Carbon and Oxygen Ions, Los Alamos Scientific Laboratory Report No. LA-6691-MS, April 1977 (unpublished).
18. A. L. Merts, J. B. Mann, W. D. Robb, and N. H. Magee, Electron Excitation Collision Strengths for Positive Atomic Ions: A Collection of Theoretical Data, Los Alamos Scientific Laboratory Report No. LA-8267-MS, 1980 (unpublished).
19. E. K. Wahlin, J. S. Thompson, G. H. Dunn, R. A. Phaneuf, D. C. Gregory, and A. C. H. Smith, *Phys. Rev. Lett.* **66**, 157 (1991).
20. X. Q. Guo, E. W. Bell, J. S. Thompson, G. H. Dunn, M. E. Bannister, R. A. Phaneuf, and A. C. H. Smith, *Phys. Rev. A* **47**, R9 (1993).

21. S. J. Smith, K.-F. Man, R. J. Mawhorter, I. D. Williams, and A. Chutjian, *Phys. Rev. Lett.* **67**, 30 (1991).
22. K. A. Berrington, P. G. Burke, M. LeDourneuf, W. D. Robb, K. T. Taylor, and Vo Ky Lan, *Comput. Phys. Commun.* **14**, 367 (1978).
23. M. A. Crees, M. J. Seaton, and P. M. H. Wilson, *Comput. Phys. Commun.* **15**, 23 (1978).
24. R. J. W. Henry, S. P. Rountree, and E. R. Smith, *Comput. Phys. Commun.* **23**, 233 (1980).
25. B. M. McLaughlin, P. G. Burke, and A. E. Kingston, *J. Phys. B* **20**, L55, (1987).
26. Y. K. Ho and R. J. W. Henry, *Astrophys. J.* **264**, 733 (1983).
27. A. K. Pradhan, *J. Phys. B* **9**, 433 (1976).
28. B. M. McLaughlin and K. L. Bell, *J. Phys. B* **26**, 1797 (1993).
29. B. M. McLaughlin and K. L. Bell, *J. Phys. B* **26**, 3313 (1993).
30. B. M. McLaughlin and K. L. Bell, *Astrophys. J.* **408**, 753 (1993).
31. B. M. McLaughlin and K. L. Bell, *Mon. Not. R. Astron. Soc.* **267**, 231 (1994).
32. M. Zou, S. J. Smith, A. Chutjian, I. D. Williams, S. S. Tayal, and B. M. McLaughlin, *Astrophys. J.* **440**, 421 (1995).
33. K. M. Aggarwal and A. Hibbert, *J. Phys. B* **24**, 3445 (1991).
34. K. M. Aggarwal, *Astrophys. J. Suppl.* **85**, 197 (1993).
35. J. A. Tully and K. L. Baluja, *J. Phys. B* **14**, L831 (1981).
36. Y. Itikawa and K. Sakimoto, *Phys. Rev. A* **33**, 2320 (1986).
37. A. K. Bhatia, G. A. Doschek, and U. Feldman, *Astron. Astrophys.* **76**, 359 (1979).
38. A. K. Bhatia and S. O. Kastner, *At. Data Nucl. Data Tables* **54**, 133 (1993).
39. D. Luo and A. K. Pradhan, *Phys. Rev. A* **41**, 165 (1990).
40. R. D. Blum and A. K. Pradhan, *Phys. Rev. A* **44**, 6123 (1991).
41. R. D. Blum and A. K. Pradhan, *Astrophys. J. Suppl.* **80**, 425 (1992).
42. M. A. Hayes and H. Nussbaumer, *Astron. Astrophys.* **134**, 193 (1984).
43. M. A. Hayes and H. Nussbaumer, *Astron. Astrophys.* **139**, 233 (1984).
44. F. P. Keenan, D. J. Lennon, C. T. Johnson, and A. E. Kingston, *Mon. Not. R. Astron. Soc.* **220**, 571 (1986).
45. M. A. Hayes, *Mon. Not. R. Astron. Soc.* **200**, 49p (1982).
46. M. A. Hayes and H. Nussbaumer, *Astron. Astrophys.* **124**, 279 (1983).
47. J. B. Mann (1977), quoted in Ref. 4.
48. J. B. Mann (1981), quoted in Ref. 4.
49. K. A. Berrington, *J. Phys. B* **18**, L395 (1985).
50. K. A. Berrington, V. M. Burke, P. G. Burke, and S. Scialla, *J. Phys. B* **22**, 665 (1989).
51. K. A. Berrington, P. G. Burke, P. L. Dufton, and A. E. Kingston, *J. Phys. B* **12**, L275 (1979).
52. K. G. Widing, J. G. Doyle, P. L. Dufton, and A. E. Kingston, *Astrophys. J.* **257**, 913 (1982).
53. K. A. Berrington, P. G. Burke, P. L. Dufton, and A. E. Kingston, *J. Phys. B* **10**, 1465 (1977).
54. K. A. Berrington, P. G. Burke, P. L. Dufton, and A. E. Kingston, *At. Data Nucl. Data Tables* **26**, 1 (1981).
55. K. A. Berrington, P. G. Burke, P. L. Dufton, and A. E. Kingston, *At. Data Nucl. Data Tables* **33**, 195 (1985).
56. D. E. Osterbrock and R. K. Wallace, *Astrophys. Lett.* **19**, 11 (1977).
57. S. M. Younger, *J. Quant. Spectrosc. Radiat. Transfer* **23**, 489 (1980).
58. S. Nakazaki and T. Hashino, *J. Phys. B* **15**, 2767 (1982).
59. J. M. Peek and J. B. Mann, *Phys. Rev. A* **25**, 749 (1982).
60. V. M. Burke, *J. Phys. B* **25**, 4917 (1992).
61. J. Callaway, R. J. W. Henry, and A. Z. Msezane, *Phys. Rev. A* **19**, 1416 (1979).
62. E. W. Bell, X. Q. Guo, J. L. Forand, K. Rinn, D. R. Swenson, J. S. Thompson, G. H. Dunn, M. E. Bannnister, D. C. Gregory, R. A. Phaneuf, A. C. H. Smith, A. Muller, C. A. Timmer, E. K. Wahlin, B. D. DePaola, and D. S. Belic, *Phys. Rev. A* **49**, 4585 (1994).

63. K. Bhadra and R. J. W. Henry, *Phys. Rev. A* **26**, 1848 (1982).
64. S. S. Tayal and A. E. Kingston, *J. Phys. B* **17**, L145 (1984).
65. S. S. Tayal, *Phys. Rev. A* **34**, 1847 (1986).
66. S. S. Tayal and A. E. Kingston, *J. Phys. B* **17**, 1383 (1984).
67. A. K. Pradhan, *Phys. Rev. Lett.* **47**, 79 (1981).
68. A. K. Pradhan, D. W. Norcross, and D. G. Hummer, *Astrophys. J.* **246**, 1031 (1981).
69. A. K. Pradhan, D. W. Norcross, and D. G. Hummer, *Phys. Rev. A* **23**, 619 (1981).
70. D. H. Sampson, S. J. Goett, and R. E. H. Clark, *At. Data Nucl. Data Tables* **28**, 299 (1983).
71. D. H. Sampson, S. J. Goett, and R. E. H. Clark, *At. Data Nucl. Data Tables* **29**, 467 (1983).
72. K. M. Aggarwal and A. E. Kingston, *J. Phys. B* **24**, 4583 (1991).
73. L. A. Morgan, *J. Phys. B* **13**, 3703 (1980).
74. N. Abu-Salbi and J. Callaway, *Phys. Rev. A* **24**, 2372 (1981).
75. S. Ormonde, K. Smith, K. W. Torres, and A. R. Davies, *Phys. Rev. A* **8**, 262 (1973).
76. J. Davis, *J. Quant. Spectrosc. Radiat. Transfer* **14**, 549 (1974).
77. J. Davis, P. C. Kepple, and M. Blaha, *J. Quant. Spectrosc. Radiat. Transfer* **15**, 1145 (1975).
78. A. K. Pradhan, *Mon. Not. R. Astron. Soc.* **177**, 31 (1976).
79. D. F. Strobel and J. Davis, *Astrophys. J.* **238**, L49 (1980).
80. N. P. Poshyunaite, A. V. Lyash, and A. B. Bolotin, *Opt. Spectrosc. (USSR)* **29**, 424 (1970).
81. A. R. G. Jackson, *J. Phys. B* **6**, 2325 (1973).
82. A. R. G. Jackson, *Mon. Not. R. Astron. Soc.* **165**, 53 (1973).
83. W. Eissner and M. J. Seaton, *J. Phys. B* **7**, 2533 (1974).
84. M. J. Seaton, *Mon. Not. R. Astron. Soc.* **170**, 475 (1975).
85. M. S. Pindzola, *Phys. Rev. A* **15**, 2238 (1977).
86. G. A. Doschek, U. Feldman, A. K. Bhatia, and H. E. Mason, *Astrophys. J.* **226**, 1129 (1978).
87. K. L. Baluja, P. G. Burke, and A. E. Kingston, *J. Phys. B* **13**, 829 (1980).
88. K. L. Baluja, P. G. Burke, and A. E. Kingston, *J. Phys. B* **14**, 119 (1981).
89. P. S. Ganas, *Astron. Astrophys. Suppl.* **40**, 259 (1980).
90. K. M. Aggarwal, K. L. Baluja, and J. A. Tully, *Mon. Not. R. Astron. Soc.* **201**, 923 (1982).
91. K. M. Aggarwal, *Astrophys. J. Suppl.* **52**, 387 (1983).
92. K. M. Aggarwal, *Astron. Astrophys.* **146**, 149 (1985).
93. K. M. Aggarwal, *Mon. Not. R. Astron. Soc.* **218**, 123 (1986).
94. V. M. Burke, D. Lennon, and M. J. Seaton, *Mon. Not. R. Astron. Soc.* **236**, 353 (1988).
95. A. R. G. Jackson, *J. Phys. B* **5**, L83 (1972).
96. T. Kato, *Astrophys. J. Suppl.* **30**, 397 (1976).
97. B. R. Tambe, *J. Phys. B* **10**, L249 (1977).
98. L. Brandus, *Rev. Roum. Phys.* **25**, 121 (1980).
99. D. R. Flower and H. Nussbaumer, *Astron. Astrophys.* **45**, 145 (1975).
100. P. S. Ganas, *Phys. Lett.* **74**, 307 (1979).
101. M. A. Hayes, *J. Phys. B* **13**, 819 (1980).
102. S. O. Kastner, *Astron. Astrophys.* **108**, 361 (1982).
103. M. A. Hayes, *J. Phys. B* **16**, 285 (1983).
104. D. E. Osterbrock, *J. Phys. B* **3**, 149 (1970).
105. D. R. Flower and J. M. Launay, *J. Phys. B* **5**, L207 (1972).
106. H. Nussbaumer, *Astron. Astrophys.* **16**, 77 (1972).
107. D. R. Flower and J. M. Launay, *Astron. Astrophys.* **29**, 321 (1973).
108. M. D. Hershkowitz and M. J. Seaton, *J. Phys. B* **6**, 1176 (1973).
109. K. E. Banyard and G. K. Taylor, *Phys. Rev. A* **10**, 1019 (1974).
110. S. Nakazaki and T. Hashino, *J. Phys. Soc. Jpn.* **43**, 281 (1977).
111. P. L. Dufton, K. A. Berrington, P. G. Burke, and A. E. Kingston, *Astron. Astrophys.* **62**, 111, (1978).
112. P. S. Ganas and A. E. S. Green, *Phys. Rev. A* **19**, 2197 (1979).

113. A. Allouche and F. Marinelli, *J. Phys. B* **14**, 2069 (1981).
114. P. G. Burke, K. A. Berrington, and C. V. Sukumar, *J. Phys. B* **14**, 289 (1981).
115. V. M. Burke and M. J. Seaton, *J. Phys. B* **19**, L527 (1986).
116. M. Malinovsky, *Astron. Astrophys.* **43**, 101 (1975).
117. R. D. Cowan, *J. Phys. B* **13**, 1471 (1980).
118. V. A. Bazylev and M. I. Chibisov, *Opt. Spektrosk.* **50**, 833 (1981) [*Opt. Spectrosc. (USSR)* **50**, 457 (1981)].
119. T. Kato and V. I. Safronova, *Opt. Spectrosc.* **70**, 291 (1991).
120. H. L. Zhang and D. H. Sampson, *Phys. Rev. A* **47**, 208 (1993).
121. H. L. Zhang and D. H. Sampson, *At. Data Nucl. Data Tables* **52**, 143 (1992).
122. K. C. Mathur, A. N. Tripathi, and S. K. Joshi, *Int. J. Mass Spectrom. Ion Phys.* **7**, 167 (1971).
123. D. Petrini, *Astron. Astrophys.* **17**, 410 (1972).
124. J. Callaway, J. N. Gau, R. J. W. Henry, D. H. Oza, V. K. Lan, and M. LeDourneuf, *Phys. Rev. A* **16**, 2288 (1977).
125. J. N. Gau and R. J. W. Henry, *Phys. Rev. A* **16**, 986 (1977).
126. R. J. W. Henry, *J. Phys. B* **12** L309 (1979).
127. D. S. Soh, B. H. Cho, and Y.-K. Kim, *Phys. Rev. A* **26**, 1357 (1982).
128. D. H. Sampson, S. J. Goett, G. V. Petrou, H. Zhang, and R. E. H. Clark, *At. Data Nucl. Data Tables* **32**, 343 (1985).
129. D. H. Sampson, G. V. Petrou, S. J. Goett, and R. E. H. Clark, *At. Data Nucl. Data Tables* **32**, 403 (1985).
130. K. K. Glyamzha, A. V. Kuplyauskene, and Z. I. Kuplyauskis, *Opt. Spectrosc. (USSR)* **62**, 14 (1987).
131. K. K. Glyamzha, A. V. Kuplyauskene, and Z. I. Kuplyauskis, *Opt. Spectrosc. (USSR)* **63**, 261 (1987).
132. Y. Itikawa, *Phys. Rev. A* **36**, 1088 (1987).
133. L. P. Presnyakov and A. M. Urnov, *J. Phys. B* **8**, 1280 (1975).
134. L. P. Presnyakov and A. M. Urnov, *Sov. Phys. JETP* **41**, 31 (1975).
135. S. J. Goett and D. H. Sampson, *At. Data Nucl. Data Tables* **29**, 535 (1983).
136. F. Bely-Dubau, J. Dubau, P. Faucher, and L. Steenman-Clark, *J. Phys. B* **14**, 3313 (1981).
137. H. L. Zhang, D. H. Sampson, and C. J. Fontes, *At. Data Nucl. Data Tables* **44**, 31 (1990).
138. S. Nakazaki, *J. Phys. Soc. Jpn.* **41**, 2084 (1976).
139. A. K. Bhatia and A. Temkin, *J. Phys. B* **10**, 2893 (1977).
140. M. R. C. McDowell, L. A. Morgan, V. P. Myerscough, and T. Scott, *J. Phys. B* **10**, 2727 (1977).
141. J. Davis, K. G. Whitney, and J. P. Apruzese, *J. Quant. Spectrosc. Radiat. Transfer* **20**, 353 (1978).
142. J. A. Tully, *J. Phys. B* **11**, 2923 (1978).
143. W. L. van Wyngaarden and R. J. W. Henry, *Phys. Rev. A* **20**, 1409 (1979).
144. K. L. Baluja and J. G. Doyle, *Phys. Lett. A* **77**, 153 (1980).
145. S. J. Goett, R. E. H. Clark, and D. H. Sampson, *At. Data Nucl. Data Tables* **25**, 185 (1980); Erratum, **27**, 617 (1982).
146. D. H. Sampson and R. E. H. Clark, *Astrophys. J. Suppl.* **44**, 169 (1980); Erratum, **49**, 593 (1982).
147. J. A. Tully, *J. Phys. B* **13**, 3023 (1980).
148. P. S. Ganas, *J. Appl. Phys.* **52**, 6482 (1981).
149. S. J. Goett, D. H. Sampson, and R. E. H. Clark, *At. Data Nucl. Data Tables* **28**, 279 (1983).
150. S. Nakazaki, *J. Phys. Soc. Jpn.* **52**, 1555 (1983).
151. C. S. Singh, R. Srivastava, and D. K. Rai, *Can. J. Phys.* **61**, 981 (1983).
152. N. R. Badnell, *J. Phys. B* **18**, 955 (1985).
153. S. Saxena and K. C. Mathur, *Z. Phys. D* **5**, 57 (1987).
154. R. Srivastava and A. K. Katiyar, *Phys. Rev. A* **35**, 1080 (1987).
155. A. K. Katiyar and R. Srivastava, *Phys. Rev. A* **38**, 5415 (1988).

156. T. Kato and S. Nakazaki, *At. Data Nucl. Data Tables* **42**, 313 (1989).
157. R. Srivastava, Y. Itikawa, and K. Sakimoto, *Phys. Rev. A* **43**, 4736 (1991).
158. R. E. H. Clark and D. H. Sampson, *At. Data Nucl. Data Tables* **22**, 527 (1978).
159. M. S. Pindzola, A. Temkin, and A. K. Bhatia, *Phys. Rev. A* **19**, 72 (1979).
160. A. E. Kingston and S. S. Tayal, *J. Phys. B* **16**, L53 (1983).
161. A. E. Kingston and S. S. Tayal, *J. Phys. B* **16**, 3465 (1983).
162. L. Steenman-Clark and P. Faucher, *J. Phys. B* **17**, 73 (1984).
163. Y. Itikawa and K. Sakimoto, *Phys. Rev. A* **31**, 1319 (1985).
164. A. Jacobs, *J. Quant. Spectrosc. Radiat. Transfer* **11**, 143 (1971).
165. C. Mitra and N. C. Sil, *Phys. Rev. A* **14**, 1009 (1976).
166. K. L. Baluja and M. R. C. McDowell, *J. Phys. B* **10**, L673 (1977).
167. M. A. Hayes and M. J. Seaton, *J. Phys. B* **10**, L573 (1977).
168. M. A. Hayes and M. J. Seaton, *J. Phys. B* **11**, L79 (1978).
169. C. Mitra and N. C. Sil, *Phys. Rev. A* **18**, 1758 (1978).
170. S. Nakazaki, *J. Phys. Soc. Jpn.* **45**, 225 (1978).
171. A. K. Das and N. C. Sil, *J. Phys. B* **17**, 4001 (1984).
172. B. Zygelman and A. Dalgarno, *Phys. Rev. A* **35**, 4085 (1987).
173. B. Ritchie, *Phys. Rev. A* **40**, 1310 (1989).
174. J. A. Tully, *Can J. Phys.* **51**, 2047 (1973).
175. S. N. Singh, S. Kumar, and M. K. Srivastava, *J. Phys. B* **12**, 2351 (1979).
176. Y. Zou and T. Shirai, *Phys. Rev. A* **45**, 6902 (1991).
177. B. H. Bransden and C. J. Noble, *J. Phys. B* **9**, 1507 (1976).
178. R. J. W. Henry, P. G. Burke, A.-L. Sinfailam, *Phys. Rev. A* **178**, 218 (1969).
179. A. E. Kingston and M. A. Lennon, *Nucl. Fusion, Special Supplement* **1987**, 43.
180. D. G. Hummer, K. A. Berrington, W. Eissner, A. K. Pradhan, H. E. Saraph, and J. A. Tully, *Astron. Astrophys.* **279**, 298 (1993).
181. A. K. Pradhan and H. Zhang, *Astrophys. J. Lett.* **409**, L77 (1993).
182. H. Zhang and A. K. Pradhan, *Astron. Astrophys.* **293**, 953 (1995).
183. A. K. Pradhan and K. A. Berrington, *J. Phys. B* **26**, 157 (1993).
184. M. Le Dourneuf, S. N. Nahar, and A. K. Pradhan, *J. Phys. B* **26**, L1 (1993).
185. H. Nussbaumer and P. J. Storey, *Astron. Astrophys.* **89**, 308 (1980).
186. K. A. Berrington, P. G. Burke, A. Hibbert, M. Mohan, and K. L. Baluja, *J. Phys. B* **21**, 339 (1988).
187. F. P. Keenan, A. Hibbert, P. G. Burke, and K. A. Berrington, *Astrophys. J.* **332**, 539 (1988).
188. H. Nussbaumer, M. Petrini, and P. J. Storey, *Astron. Astrophys.* **102**, 351 (1981).
189. R. H. Garstang, W. D. Robb, and S. P. Rountree, *Astrophys. J.* **222**, 384 (1978).
190. K. A. Berrington, C. J. Zeippen, M. Le Dourneuf, W. Eissner, and P. G. Burke, *J. Phys. B* **24**, 3467 (1991).
191. K. A. Berrington and J. C. Pelan, *Astron. Astrophys. Suppl.*, in press.
192. H. Nussbaumer and P. J. Storey, *Astron. Astrophys.* **70**, 37 (1978).
193. P. H. Norrington and I. P. Grant, *J. Phys. B* **20**, 4869 (1987).
194. H. Nussbaumer and D. E. Osterbrock, *Astrophys. J.* **161**, 811 (1970).
195. H. Nussbaumer and P. J. Storey, *Astron. Astrophys.* **113**, 21 (1982).
196. S. J. Czyzak and T. K. Krueger, *Astrophys. J.* **144**, 381 (1966).
197. M. Blaha, *Astrophys. J.* **157**, 473 (1969).
198. R. U. Datla, M. Blaha, and M. J. Kunze, *Phys. Rev. A* **12**, 1076 (1975).
199. T. Kato, *Astrophys. J. Suppl. Ser.* **30**, 397 (1976).
200. M. S. Pindzola, D. C. Griffin, and C. Bottcher, *Phys. Rev. A* **39**, 2385 (1989).
201. M. Blaha, *Ann. Astrophys.* **31**, 311 (1968).
202. M. Blaha, *Astron. Astrophys.* **1**, 42 (1969).
203. T. K. Krueger and S. J. Czyzak, *Proc. R. Soc. London, Ser. A* **318**, 531 (1970).

204. H. S. Brandi and G. F. Koster, *Phys. Rev. A* **8**, 1303 (1973).
205. S. J. Czyzak, T. K. Krueger, H. E. Saraph, and J. Shemming, *Proc. Phys. Soc. London* **92**, 1146 (1967).
206. S. J. Czyzak and T. K. Krueger, *Proc. Phys. Soc. London* **90**, 623 (1967).
207. M. J. Conneely, K. Smith, and L. Lipsky, *J. Phys. B* **3**, 493 (1970).
208. S. J. Czyzak, T. K. Krueger, P. de A. P. Martins, H. E. Saraph, and M. J. Seaton, *Mon. Not. R. Astron. Soc.* **148**, 361 (1970).
209. C. J. Zeippen, K. Butler, and J. LeBourlet, *Astron. Astrophys.* **188**, 251 (1987).
210. S. J. Czyzak and T. K. Krueger, III International Conference on the Physics of Electronic and Atomic Collisions Proceedings, 1963.
211. O. Bely, *Proc. Phys. Soc. London* **87**, 1010 (1966).
212. O. Bely, *Nuovo Cimento* **49**, 66 (1967).
213. S. M. Younger, *J. Quant. Spectrosc. Radiat. Transfer* **23**, 489 (1980).
214. R. B. Christensen, D. W. Norcross, and A. K. Pradhan, *Phys. Rev. A* **34**, 4704 (1986).
215. A. K. Pradhan, *At. Data Nucl. Data Tables* **40**, 335 (1988).
216. R. U. Datla, H. J. Kunze, and D. Petrini, *Phys. Rev. A* **6**, 38 (1972).
217. Y. K. Kim and K. T. Cheng, *Phys. Rev. A* **18**, 36 (1978).
218. S. M. Younger, *J. Quant. Spectrosc. Radiat. Transfer* **22**, 155 (1979).
219. N. R. Badnell, M. S. Pindzola, and D. C. Griffin, *Phys. Rev. A* **43**, 2250 (1991).
220. D. C. Griffin, M. S. Pindzola, and N. R. Badnell, *Phys. Rev. A* **47**, 2871 (1993).
221. M. E. Bannister, X. Q. Guo, T. M. Kojima, and G. H. Dunn, *Phys. Rev. Lett.* **72**, 3336 (1994).

Chapter 7

Ionization of Atomic Ions by Electron Impact

P. Defrance, M. Duponchelle, and D. L. Moores

I. INTRODUCTION: TYPES OF IONIZATION PROCESSES

Electron impact ionization of a positive ion may occur directly, when one or more electrons are ejected from an outer or inner shell by direct impact with the incident electron, or indirectly, via an intermediate autoionizing state of the target system. The most important indirect ionization process is excitation–autoionization (EA), in which the incident electron excites the target ion usually, but not always, from an inner shell into a short-lived state with energy in excess of that required to ionize the system, which subsequently decays by autoionization (an Auger transition). Alternatively, the incident electron may be captured by the target ion into a short lived state of a compound ion which then decays by two successive autoionizing transitions. This sequence of processes is called resonant-recombination–double autoionization (REDA) although the words "capture" or "recombination" may be more appropriate than "excitation" in this context. If the compound state decays

---

P. DEFRANCE and M. DUPONCHELLE • Department of Physics, Université Catholique de Louvain, B-1348 Louvain-la-Neuve, Belgium.   D. L. MOORES • Department of Physics and Astronomy, University College London, London WC1E6BT, United Kingdom.

Atomic and Molecular Processes in Fusion Edge Plasmas, edited by R. K. Janev. Plenum Press, New York, 1995.

directly by simultaneous ejection of two electrons, the process is called resonant-excitation–auto-double ionization (READI). These three processes all lead to single ionization of the original target ion.

Indirect multiple ionization events (which, by analogy, could be assigned the acronyms EMA, REMA, and REAMI, where M stands for "multiple") can also result if the state into which excitation or capture occurs lies above the energy required to eject more than one electron. The principal indirect mechanism for multiple ionization is, however, ionization–autoionization (IA), in which direct inner-shell ionization (primary ionization) by the incident electron is followed by single or multiple autoionization, either by successive, cascading transitions or by simultaneous ejection of two or more electrons.

In all processes involving intermediate autoionizing states, there will be competition between autoionization and radiative decay. An autoionizing state is also able to make a radiative transition to a pure bound state, the final step in the dielectronic recombination process. This radiative stabilization will reduce the overall ionization cross section by an amount depending on the relative probabilities for autoionization and radiative decay. In the limit that the radiative decay probability is much the larger of the two, this process dominates, and the indirect contribution to ionization will be very small.

If X^{z+} denotes a z-times ionized ion, * represents an excited state, and ** a doubly excited state, the principal mechanisms for ionization can be summarized as follows:

A. Direct Ionization

$$X^{z+} + e^- \rightarrow X^{(z+n)+} + (n+1)e^- \qquad (1)$$

The ionized ion $X^{(z+n)+}$ may be left in either the ground or an excited state.

B. Excitation–Autoionization (EA)

$$X^{z+} \rightarrow (X^{z+})^* + e^-$$

This process is then followed by either

$$(X^{z+})^* \rightarrow X^{(z+1)+} + e^-$$

or

$$(X^{z+})^* \rightarrow X^{z+} + e^- + h\nu \qquad (2)$$

C. Resonant-Excitation (Capture) Processes
    REDA:

$$X^{z+} + e^- \rightarrow (X^{(z-1)+})^{**}$$

$$(X^{(z-1)+})^{**} \rightarrow (X^{z+})^* + e^- \rightarrow X^{(z+1)+} + e^- + e^- \qquad (3)$$

READI:
$$X^{z+} + e^- \rightarrow (X^{(z-1)+})^{**} \rightarrow X^{(z+1)+} + e^- + e^- \qquad (4)$$

D. Ionization–Autoionization (IA)
$$X^{z+} + e^- \rightarrow (X^{(z+1)+})^{**} + e^- + e^- \rightarrow X^{(z+n)+} + (n+1)e^- \qquad (5)$$

In processes (3) and (4), any of the autoionizing states may also undergo radiative decay as in process (2), but for the sake of clarity this has not been shown here. In process (5), steps involving intermediate stages of ionization (cascades) have likewise been omitted.

A schematic cross section for single ionization, showing contributions from A, B, and C above, is plotted in Fig. 1.

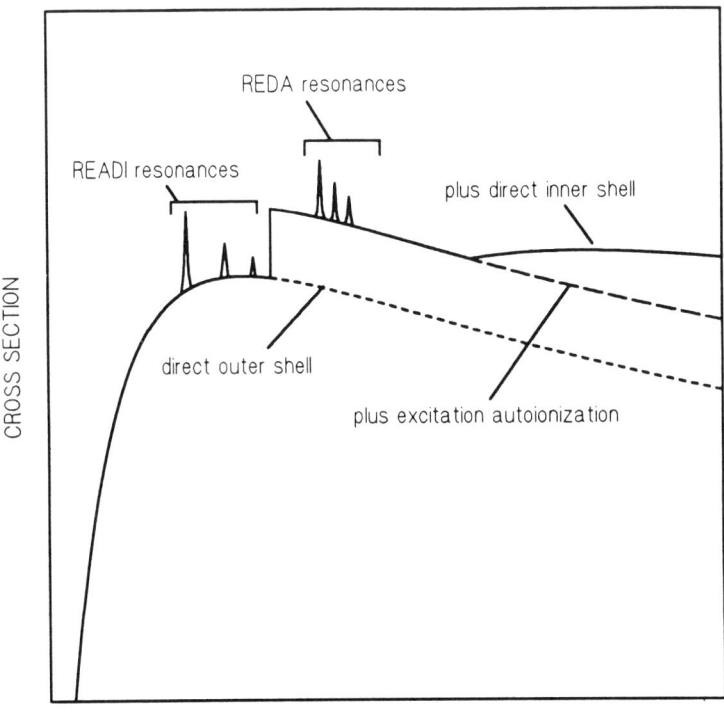

Figure 1. Schematic electron impact ionization cross section showing contributions from direct ionization, excitation–autoionization, REDA, and READI.

2. ELECTRON IMPACT IONIZATION: THEORETICAL METHODS

2.1. Direct, Single Ionization

Methods for calculating cross sections for direct, single ionization of positive ions by electron impact may be divided into three categories: *ab initio* quantum-mechanical approximations, classical and semiclassical methods, and semiempirical formulas.

2.1.1. Quantal Approximations

Quantal approximations are based on the calculation of cross sections starting from the integral expression for the amplitudes,

$$f^S(\mathbf{k}_1, \mathbf{k}_2) = (2\pi)^{-5/2} M^S_{\beta\alpha} \tag{6}$$

where $\mathbf{k}_1$ and $\mathbf{k}_2$ are the momenta of ejected and scattered electrons, respectively, S is the total spin, and $M^S_{\beta\alpha}$ is proportional to the on-shell transition matrix element

$$\langle 0\beta | T | 1\alpha \rangle$$

The ionization cross section is then

$$Q^{\text{ion}}(E_0) = \frac{1}{8\pi k_0} \sum_S (2S+1) \int_0^{E/2} k_1 k_2 \sigma^S(\mathbf{k}_1, \mathbf{k}_2) \, dk_2^2 \tag{7}$$

where

$$\sigma^S(\mathbf{k}_1, \mathbf{k}_2) = \frac{1}{4\pi} \int d\hat{\mathbf{k}}_0 \, d\hat{\mathbf{k}}_1 \hat{\mathbf{k}}_2 \left| f^S(\mathbf{k}_1, \mathbf{k}_2)^2 \right| \tag{8}$$

$\mathbf{k}_0$ being the incident wave vector and E_0 the incident energy.

The general expression for the **T**-matrix element[1] provides a starting point for different types and orders of approximation. To date, few calculations of order higher than the first order in the interaction have been made. All methods that have been used are essentially high-energy or high-charge approximations—no reliable theoretical technique capable of assured accuracy for ions of low charge at low energies (up to a few times threshold) has so far been developed.

In the distorted-wave (DW) method, the incident and scattered electrons are assumed to move in a central potential that takes into account distortion due to the field of the target, exchange with the target, and possibly polarization. The Coulomb–Born (CB) method is an approximation to the DW method in which only the asymptotic Coulomb interaction between projectile and target is taken into account.

Both methods should be fairly accurate for highly charged ions, provided exchange is taken into account, except close to threshold since they take full account of the dominant part of the interaction.

In the plane-wave Born approximation, this interaction is dropped, the incident and scattered electrons being described by plane waves: accuracy can only be expected at very high energies and low effective charge.

At low energies, exchange between ejected and scattered electrons is most important: its inclusion requires an additional approximation for the relative phase of direct and exchange amplitudes. No rigorous theoretical justification has been given for any particular choice.

In all these methods, an accurate theoretical description of the target—initial bound state and final continuum state—is essential. This is particularly true for heavy ions, in which the ejected orbitals can depend sensitively on the term value of the final ionized ion, a phenomenon known as term dependence.

In ionization of complex ions with incomplete shell structure, both initial and final configurations may have a very large number of terms. In the configuration-average DW method, a single cross section for transition between initial and final configurations is determined, resulting in enormous savings in computational effort over level-to-level calculations that calculate a separate cross section for each of the large number of pairs of initial and final levels. In some cases, however, term dependence may make this procedure insufficiently accurate.

In summary, one might expect the distorted-wave exchange method to give the most accurate results followed by the Coulomb–Born exchange method. Also, these approximations should improve with increasing charge on the ion and increasing incident energy. Scattering exchange is usually unimportant for incident energies greater than 4 to 5 times the ionization energy.

However, the accuracy of any of these dynamical approximations can be severely limited by a crude description of the target ion.

At very high energies the ionization cross section goes over to the Bethe form,

$$XQ(X) = A\ln X + B \tag{9}$$

where $X = E_0/I$, I being the ionization energy; A and B are independent of X, with A given by

$$A = 4 \int_I^\infty \frac{df}{d\varepsilon} \frac{d\varepsilon}{\varepsilon} \tag{10}$$

where $df/d\varepsilon$ is the differential dipole continuum oscillator strength of the ion.

Estimates of unknown cross sections can be obtained by making use of the fact that for a given isoelectronic sequence the reduced cross section

$$Q_R(X) = I^2 Q^{\text{ion}}(X) \tag{11}$$

varies slowly as a function of the nuclear charge Z. This technique is referred to as scaling.

At energies above a few kilo-electron-volts, relativistic kinematics apply, and Eq. (9) should be replaced by

$$XQ(X) = A\left[\ln\frac{X}{(1-\beta^2)} - \beta^2\right] + B \tag{12}$$

where $X = mv^2/2$ and $\beta = v/c$, v being the incident electron velocity in the laboratory frame. If a nonrelativistic approximation to the target wave functions is valid, the constants A and B in Eq. (12) have the same values as in Eq. (9).

At higher energies, and for highly stripped ions with nuclear charge greater than about 25, a fully relativistic calculation in which the wave functions satisfy Dirac equations is required.

2.1.2. Classical and Semiclassical Methods

2.1.2a. The ECIP Method. The exchange–classical impact parameter (ECIP) method[2] treats close collisions as classical binary encounters while distant collisions are handled by an impact parameter technique. Electron exchange is taken into account.

The method was developed as an efficient way of generating rate coefficients for use in ionization balance calculations and designed to give reliable results for ionization from highly excited states. In practice, however, the ECIP method has been found in many cases to give reasonable agreement with experimental data for ground state ionization of ions of low charge, H-like and He-like ions.[3] It may be extended to take approximate account of excitation–autoionization by lowering the inner-shell ionization threshold and simultaneously increasing the outer-shell occupancy number by an amount equal to the number of active inner-shell electrons.

2.1.2b. Wannier Threshold Law. On the basis of classical mechanics, Wannier[4] has derived the threshold law for single ionization

$$Q(E_0) \propto (E_0 - I)^\nu \tag{13}$$

where

$$\nu = \frac{1}{4}\left[\left(\frac{100z + 91}{4z + 3}\right)^{1/2} - 1\right] \tag{14}$$

This law is consistent with experimental data for $z = 0$ and $z = 1$. For larger z, $\nu \to 1$. A different threshold law is obtained for multiple ionization.

2.1.3. Semiempirical Formulas

2.1.3a. The Lotz Formula. The Lotz formula[5,6] has been the most successful and extensively used of the semiempirical formulas for estimating direct ionization cross sections. Its form is based on the Coulomb–Born approximation for hydrogenic ions and thus has an underlying quantum-mechanical basis. It follows the correct (Bethe) behavior (Eq. 9) at high energy and at the same time its linear threshold behavior corresponds to the prediction for highly charged ions given by Eqs. (13) and (14). The general form is

$$Q^{ion}(E_0) = \sum_j a_j \xi_j \frac{\ln(E_0/I_j)}{E_0 I_j} \{1 - b_j \exp[-c_j(E/I_j - 1)]\} \tag{15}$$

where the sum is over subshells j, E_0 is the incident energy, I_j is the ionization energy of subshell j, ξ_j is the occupation number of shell j, and a_j, b_j, and c_j are constants obtained by fitting to experimental or theoretical data that have been judged to be reliable. The one-parameter form is obtained by setting $b_j = c_j = 0$ and $a_j = 4.5 \times 10^{-14}$, which gives the cross section $Q^{ion}(E_0)$ in square centimeters.

The Seaton formula[7] was designed for use only in the near-threshold region and should not be used at higher energies, as it can yield highly inaccurate results. In all circumstances, it is far safer and just as easy to use the Lotz formula.

2.1.3b. The Burgess and Chidichimo Formula. Burgess and Chidichimo[8] have given the formula

$$Q^{ion} = C \sum_j \xi_j \left(\frac{I_H}{I_j}\right)^2 \frac{1}{x_j} \ln x_j \, w(x_j) \, \pi a_0^2 \tag{16}$$

where the sum is over subshells j, ξ_j is the effective occupation number of shell j, and I_j is the effective ionization energy; $x_j = E_0/I_j$ and $I_H = 13.6058$ eV. The function $w(x_j)$ is evaluated as follows:

$$w(x_j) = \begin{cases} (\ln x_j)^{v/x_j}, & \text{if } E > I_j \\ 0, & \text{if } E \leq I_j \end{cases} \tag{17}$$

(where v is given by Eq. (14) and ensures the Wannier threshold behavior).

C is a constant obtained from analysis of experimental data whose value may be revised as the results of more measurements become available. A prescription is given for including excitation–autoionization contributions in this formula by redefining I_j and ξ_j as discussed under the ECIP method.

2.1.3c. Gryzinski's Formulas. These are based on a purely classical binary-encounter formalism. For single ionization the Lotz formula is more soundly based and reliable, but for direct multiple ionization the Gryzinski formula is the only theory available. We refer the reader to Ref. 9 for the explicit expressions.

2.1.3d. The Method of Sampson. This may be regarded as occupying a position midway between the *ab initio* quantal methods and the semiempirical formulas. It is based on the use of Eq. (6) for hydrogenic ions. The scaled hydrogenic reduced cross section for initial state nl,

$$Q_R^H(nl, X) = \left(\frac{Z}{n}\right)^4 \frac{Q^H(nl, X)}{\pi a_0^2} \tag{18}$$

calculated in the Coulomb-Born–exchange approximation, tends to a finite limit as the nuclear charge $Z \to \infty$. The ionization cross section is approximated[10] by

$$Q_j(nl, X) = \frac{\pi a_0^2}{I_{nl}^j} \left[\frac{n}{Z_{\text{eff}}(nl)}\right]^2 \xi_{nl} Q_R^H(nl, X) \tag{19}$$

The effective charge parameter Z_{eff} depends on the structure of the ion, but $Q_R^H(nl, X)$ depends only on nl. An alternative form

$$Q_j(nl, X) = \pi a_0^2 \left[\frac{n}{Z_{\text{eff}}(nl)}\right]^4 \xi_{nl} Q_R^H(nl, X) \tag{20}$$

has also been used. The formula is applicable to N-electron ions such that

$$Z \gg N$$

but can give 20–30% accuracy for $Z/N \sim 2$.

2.2. Indirect Ionization

2.2.1. Excitation–Autoionization

It is usually assumed that the interference between direct and indirect ionization can be neglected and the two processes treated as independent. In many cases this assumption is an excellent one, and we shall refer to it as the independent-processes model. The ionization cross section for an incident electron energy E_0 above the first autoionization threshold may then be written

$$Q^{\text{tot}}(E_0) = Q^{\text{ion}}(E_0) + \sum_i Q_i^{\text{ex}}(E_0) B_i \tag{21}$$

where the sum is over all energetically accessible autoionizing states i. $Q^{\text{tot}}(E_0)$ is the total cross section, $Q^{\text{ion}}(E_0)$ is the direct ionization cross section, and $Q_i^{\text{ex}}(E_0)$ is the excitation cross section of autoionizing state i, regarded for this purpose as if it were a pure bound state. B_i is the branching ratio,

$$B_i = \frac{\sum_j A_{ij}^a}{\sum_j A_{ij}^a + \sum_k A_{ik}^r} \tag{22}$$

where A_{ij}^a is the probability of autoionization to state j of the ionized ion, and A_{ik}^r is the probability of radiative decay of state i to a bound state k of the target ion. The size of the EA contribution thus depends on the relative values of the excitation and ionization cross sections and also on the relative values of the autoionization and radiative decay probabilities. For ions of low charge z, the autoionization probabilities tend to be much larger than the radiative decay probabilities, i.e.,

$$A^r \ll A^a \quad \text{and} \quad B \to 1 \quad \text{(low charge)} \tag{23}$$

Thus, for ions of low charge we have

$$Q^{\text{tot}}(E_0) = Q^{\text{ion}}(E_0) + \sum_i Q_i^{\text{ex}}(E_0) \tag{24}$$

The EA contribution is then obtained merely by adding the excitation cross sections for all contributing states to the direct ionization cross section. Knowledge of A^a and A^r is not required.

Systems for which we might expect EA to be particularly significant will be those with filled or almost-filled subvalent shells (of p, d, or even f electrons) immediately below a smaller number of valence electrons. In these systems, the probability of excitation of an inner electron is relatively higher owing to the larger occupation number of the subvalent shell.

Along an isoelectronic sequence, the direct ionization cross section scales as z^4 while the excitation cross section scales as z^2 so we expect the size of the indirect contribution to increase relative to the direct contribution with increasing charge on the ion.

The same effect is observed along an isonuclear sequence, although the changing atomic structure as valence electrons are successively removed also affects the shape of the cross sections. For sufficiently highly charged ions, the EA contribution may dominate direct ionization.

Another effect must be considered, however, in very highly charged ions. As z increases, the radiative probability A^r for a given state increases as z^4 while the autoionization probability remains roughly constant. Thus, B_i decreases, and as a result EA is reduced, along an isoelectronic sequence; the excited states prefer to stabilize rather than autoionize. In the limit,

$$A^r \gg A^a \quad \text{and} \quad B \to 0 \quad \text{(high charge)} \tag{25}$$

For sufficiently large z, radiative decay can destroy the EA contribution completely, and only direct ionization will be important:

$$Q^{\text{tot}}(E_0) = Q^{\text{ion}}(E_0) \tag{26}$$

It should be emphasized, however, that Eqs. (24) and (26) just mark general trends. For all ionic charges, there will be some individual transitions for which neither of the inequalities may be assumed, and excitation–autoionization contributions can persist even for very highly stripped ions such as U^{81+}.[11] Also, not all radiative decays lead to stabilization, since an autoionizing state may make a radiative transition to another autoionizing state with a branching ratio close to unity that promptly autoionizes. This should be taken into account when making calculations.

The detailed behavior of the ionization cross section depends on the structure of the target ion. At each autoionization threshold, an additional term will be added to the sum in Eq. (21). Since the excitation cross section of a positive ion is finite at its threshold, this will lead to an abrupt jump in the total cross section. As more and more states are excited, more and more jumps will occur, their height decreasing as the excitation cross sections decrease. They eventually merge into a semicontinuum converging on an inner-shell ionization threshold. If the lowest energy configuration attainable by excitation of an inner-shell electron lies well above the outer-shell ionization threshold, the cross section will display a series of fairly well-defined jumps well above threshold, starting in the region of or above the peak in the direct ionization. The schematic cross section shown in Fig. 1 is an example of this type of behavior, which corresponds to category (i) distinguished by Burgess et al.[3] and Burgess and Chidichimo.[8] If the autoionization threshold lies well below the outer-shell ionization threshold [category (ii) of Burgess et al.], the EA contribution will manifest itself as a large number of small jumps, usually not resolved, immediately above the outer-shell ionization threshold and up to the inner-shell threshold. As z increases, the principal quantum number n_0 of the first autoionizing state increases.

For such systems, one would not expect to observe structure in the cross section, the EA giving instead a smooth overall enhancement detectable as a faster rise from threshold than that due to direct ionization alone. The cross sections for many ions combine the features of both categories. In other configurations such as the $2p^6$ ground state of Ne-like ions, the EA contribution is small, and very little structure is seen.

As z increases along either an isoelectronic or an isonuclear sequence, the binding energy of a given autoionizing state will increase faster than the level separations in the $(z + 1)$-times ionized ion, leading to a progressive switch from autoionizing to bound state at some value of z. This brings about sudden variations in the size of the EA contributions at points along an isoelectronic sequence, since when a state becomes bound, it no longer appears in the sum in Eq. (21). In such circumstances, estimates of the EA contribution by isoelectronic extrapolation or interpolation ("scaling") should only be attempted with extreme caution, after a careful study of the energy level structure of the systems of interest.

2.2.2. Resonant Capture Processes

The resonant capture processes can be regarded as a special case of EA in which the intermediate state is a resonance in the scattering of an electron by the target. Whenever EA is important, it is likely that it will be accompanied by a corresponding REDA contribution at slightly lower incident energies. The effects of REDA appear as peaked structures in the total cross section below each autoionization threshold. Each state into which capture occurs is a member of a Rydberg series of autoionizing states that converges onto a series limit that is itself an autoionizing state. The large possible number of states may mean that high experimental resolution is required to observe the structure, and it can be difficult to distinguish between resonant capture and the onset of higher EA processes.

REDA may be included in a theoretical treatment by using some approximation that includes resonances, such as the close-coupling approximation, to calculate the excitation cross sections appearing in Eq. (21). If allowance is to be made for the fact that the resonances may also decay radiatively, it may be easier to include the resonant capture process by adding to Eq. (21) terms of the form

$$\sum_k Q_k^{\text{cap}}(E_0) B_k^{\text{da}} \quad (27)$$

where $Q_k^{\text{cap}}(E_0)$, the capture cross section, is obtained on application of the principle of detailed balance:

$$Q_k^{\text{cap}}(E_0) = \frac{2.6741 \times 10^{-32}}{E_0 \Delta E} I_H^2 \frac{g_k}{2g_i} A_{ki}^a \ (\text{cm}^2) \quad (28)$$

where ΔE the energy bin width, E_0 is the energy released in the inverse process, I_H is the ionization energy of hydrogen in the same units as ΔE and E_0, and g_k and g_i are the statistical weights of the intermediate state k and the initial state of the ion, respectively. The branching ratio for REDA is given by

$$B_k^{\text{da}} = \frac{\sum_{k'} A_{kk'}^a \sum_f A_{k'f}^a}{(\sum_m A_{km}^a + \sum_n A_{kn}^r)(\sum_m A_{k'm}^a + \sum_n A_{k'n}^r)} \quad (29)$$

If this procedure is followed, an approximation that does not allow for resonances should be used to obtain the excitation cross sections in Eqs. (21).

In the case of READI, a different formula must be used. Instead of Eq. (27), one must add

$$\sum_f Q_f^{\text{cap}}(E_0) B_f^{\text{ad}} \quad (30)$$

where $Q_f^{\text{cap}}(E_0)$ is the capture cross section into the intermediate state f and depends on the single autoionization rate of state f via an expression of the form of Eq. (28). The capture cross sections depend on the energy bin width ΔE, which is to some extent arbitrary. It is often allowed for by convoluting each resonance with a

Gaussian of width commensurate with the typical experimental energy resolution. The branching ratio in Eq. (30) is

$$B_f^{\text{ad}} = \frac{\sum_i A_{fi}^{a2}}{\sum_j A_{fj}^a + \sum_i A_{fi}^{a2} + \sum_n A_{fn}^r} \tag{31}$$

The auto-double ionization probability A_{fi}^{a2} depends critically on the electron–electron correlation and is best calculated by means of many-body perturbation theory.[12] The calculated value is very sensitive to the basis set chosen and to the continuum phase selected in the matrix elements.

2.3. Multiple Ionization

The ejection of two or more electrons may take place directly or via the indirect inner-shell ionization–autoionization process, in which primary ionization creates a hole state that decays either by a single Auger transition or by a cascade of such transitions to give net multiple ionization. Multiple ionization may also occur via multiple primary ionization, multiple Auger decay of a hole state, or any other energetically allowed combination of these processes. It may also occur via inner-shell excitation or even resonant capture followed by multiple Auger transitions.

Direct multiple ionization is in general of higher order in the interaction than the corresponding indirect process, and the only theory available that is suitable for low impact energies is the classical binary-encounter approximation of Gryzinski.[9] Fortunately, the direct cross sections decrease with increasing ionic charge and also with the number of electrons ejected, and for complex ions ionized more than once, multiple ionization is usually dominated by the indirect processes, which, since they generally involve one-electron transitions, are easier to deal with theoretically.

If the only important contributors are direct ionization and IA, the total double ionization cross section is given by

$$Q_{+2}^{\text{ion}}(E_0) = Q_{+2}^{\text{ion,dir}} + \sum_i Q_i^{\text{ion}}(E_0) B_i \tag{32}$$

where $Q_{+2}^{\text{ion,dir}}$ is the direct double ionization cross section, $Q_i^{\text{ion}}(E_0)$ is the single ionization cross section corresponding to hole state i, and the sum is over all hole states that can be created which contribute to double ionization. B_i is the branching ratio,

$$B_i = \sum_j A_{ij}^a / \Gamma \tag{33}$$

where Γ is the sum of all possible decay probabilities of state i. The threshold for double ionization corresponds to the minimum energy required to eject two electrons, the onset of direct double ionization. Below the IA threshold, only direct double ionization can occur. The onset of the IA process is marked by a sudden change of slope in the cross section. From Eq. (32) it is seen that if the direct cross section is small and the branching ratio unity, the total cross section is just determined by the sum of the cross sections for single ionization from the inner shells.

If excitation followed by double autoionization (EDA) is also important, a term

$$\sum_j Q_j^{ex}(E_0) B_j^{da} \tag{34}$$

has to be added, where $Q_j^{ex}(E_0)$ is the cross section for excitation of an inner-shell electron to state j, and B_j^{da} is the appropriate branching ratio. This process will contribute below the IA threshold, where it will augment the double ionization.

Since net multiple ionization can result from many possible combinations of primary ionization (single or multiple) and decay pathways, the general expression for the cross section equivalent to Eq. (32) will be correspondingly more complicated if it is to include all contributions.

The notation $\sigma_{q,q+n}$ is also used to define the cross section for n-fold ionization of ionization stage q of a species.

In order to assist in deciding which processes should be taken into account in a particular case, we next give a brief discussion of the way in which the relative sizes of the different contributions to ionization vary with charge on the ion and with the ionic structure.

In hydrogenic ions, only direct ionization is important, and although indirect effects have been observed in high precision experiments in He-like ions, they only make a very small contribution to the total ionization cross section. In Li-like ions, however, EA has a significant effect, its contribution increasing in size with ionic charge from a few percent in Be^+ to much larger fractions for more highly charged systems. REDA is also important in these ions. In order to obtain accurate cross sections, both effects should be taken into account.

Indirect ionization is not important in Be-like ions or in ions with ground configuration $2p^q$ and may safely be ignored in these ions unless they are very highly stripped. It does, however, become important in Na-like and Mg-like ions because of the possibility of exciting the inner $2p^6$ shell: both EA and REDA increase in magnitude as the charge on the ion increases. The effect is small but observable in Mg^+ and must be taken into account in all ions with these configurations.

For more complex ions, EA and REDA become increasingly important, even for singly charged systems, because of the increased ease with which subvalent shells may be excited. The relative magnitudes of direct and indirect ionization vary in an irregular manner both with ionic charge and with nuclear charge, with the indirect contribution dominant in some ions, comparable with direct ionization in others, and small in yet others. Radiative decay of autoionizing states becomes important for more highly charged systems, but although it reduces the indirect effect, it never entirely destroys it.

Thus, when performing theoretical estimates, each case must be considered separately. When all factors are taken into account, it must be concluded that total ionization cross sections are not amenable to the simplistic procedure of scaling along isoelectronic or isonuclear sequences. At best, scaling may be done, with great caution, for the individual contributions, over restricted ranges of z.

3. EXPERIMENTAL METHODS

Electron–ion experiments have been reviewed in detail by many authors. The reader may consult, for instance, the books edited by Märk and Dunn[13] and by Brouillard.[14]

Beam–beam experiments are performed to obtain reliable information on the ionization process. The static target, which is widely used for neutral species, atoms, or molecules, is generally not appropriate to studies of electron impact ionization mechanisms for ions. This is due to the unstable character of ions, which interact strongly with residual gas molecules. One exception is the application of the electron beam ion source or trap (EBIS or EBIT) in some particular cases of highly charged systems.

3.1. Crossed Electron–Ion Beam Experiments

The crossed electron–ion beam method was developed more than 30 years ago and has produced single- and multiple-ionization cross sections for a very large number of ionic species, both negative and positive, including ions with charge states as high as 15 in the case of iron. In this method, ions are extracted from sources of various kinds, depending on the element and on the selected charge state. They are accelerated under a voltage of a few kilovolts to form a beam. This beam is mass and charge analyzed and crosses an electron beam, for which it constitutes a well-defined target (in terms of species as well as charge state). After beam interaction, product ions are magnetically separated from the primary beam, which is collected by means of a Faraday cup. Product ions are detected individually. For these experiments, the ribbon electron beam is considered as monoenergetic because the width of the energy distribution is of order of 1–2 eV, which is small compared to the covered energy range (from a few electron volts to several kilo-electron-volts).

The absolute determination of ionization cross sections requires the knowledge of the relevant parameters: fluxes of incident particles (ions and electrons) and the flux of product ions. In addition, owing to unavoidable beam density inhomogeneities, the overlap of intersecting beams must be taken into account.

3.1.1. Crossed-Beams Method

The initial method of investigation makes use of beams that interact steadily. In this situation, the cross section is given by

$$\sigma = \frac{v_e v_i N F}{v \sin(\theta) \left(\dfrac{I_i}{qe}\right)\left(\dfrac{I_e}{e}\right)} \tag{35}$$

In this expression, v_e and v_i, I_e and I_i, and e and qe are the velocities, currents, and charges of the electrons and ions, respectively. The relative velocity v between two colliding particles is

$$v = (v_e^2 + v_i^2 - 2v_e v_i \cos \theta)^{1/2} \tag{36}$$

with θ being the angle between the colliding beams (usually 90°). N is the count rate, and F is the form factor, which is expressed as a function of the unidimensional electron and ion current densities, $J_e(z)$ and $J_i(z)$, respectively:

$$F = \frac{\int_{-\infty}^{+\infty} J_e(z)\, dz \int_{-\infty}^{+\infty} J_i(z)\, dz}{\int_{-\infty}^{+\infty} J_e(z)\, J_i(z)\, dz} \tag{37}$$

This factor is the effective height over which the beams interact. Its estimation requires the measurement of the density profiles of the beams along the z-direction. Several methods have been developed for this measurement. In the most popular one, two slits of width Δz are moved simultaneously through the beams. The transmitted currents are recorded, and the F value is obtained from a numerical integration of the data. The form factor measurement is performed before or after the experiment itself. The background of the experiment is determined by chopping the electron beam.

3.1.2. Animated Beam Method

The animated beam method was proposed and applied to ionization studies, as an alternative to the crossed-beams method. In order to discard the form factor determination. In this method, the electron beam is swept across the ion beam in a linear seesaw motion at a constant speed u. This implies that each slice of one beam interacts with each slice of the other beam during the same period; it is easily seen that the integral over the beam density profiles is thus not needed anymore. The calculation shows that the ionization cross section is related to the measured quantities in the following way[15]:

$$\sigma = \frac{v_e v_i K u}{v \sin \theta \left(\dfrac{I_i}{qe}\right)\left(\dfrac{I_e}{e}\right)} \tag{38}$$

Here, K is the total number of events produced during one passage of the electron beam across the ion beam; that is,

$$K = \int_{-\infty}^{+\infty} dk(z') \tag{39}$$

where $dk(z')$ is the number of events produced when the beam axes are separated from each other by the distance z'. Assuming only that beam density profiles are stable during the measurement, possible fluctuations of the sweeping speed and of the particle currents are taken into account in the following way:

$$\frac{Ku}{I_e I_i} = \Delta z \int_{-\infty}^{+\infty} \frac{dk(z')}{\Delta t(z') I_e(z') I_i(z')} \tag{40}$$

For this purpose, the detector signal as well as the time interval and the current intensities are recorded separately at each position (z') of the electron gun during the motion; Δz is the width of the length interval of the motion.

A comparison of Eqs. (35) and (38) shows that the static beam method and the animated beam method are related through

$$Ku = NF \tag{41}$$

In order to achieve good precision, careful measurements are needed for all the parameters appearing in Eq. (35) or (38).

First of all, the kinematic parameters are given by

$$v_e = \sqrt{\frac{2eV_e}{m_e}} \quad \text{and} \quad v_i = \sqrt{\frac{2qeV_i}{m_i}} \tag{42}$$

where V_e and V_i are the electron and ion acceleration voltages, respectively. The true interaction energy E (eV) is obtained from Eq. (36) for beams interacting at right angles—which is the most frequent situation—and assuming $m_i \gg m_e$:

$$E = V_e + \frac{m_e}{m_i}(qV_i - V_e) \tag{43}$$

Corrections for contact and space-charge potentials are also taken into account and are on the order of a few electron volts. The width of the electron energy distribution, expressed as the full width at half maximum (FWHM), depends on the temperature of the emitter, that is, on the cathode material and on space-charge effects. This width is generally on the order of a few electron volts, which is low

Ionization of Atomic Ions by Electron Impact

enough for the electron beam to be considered as monoenergetic. However, this figure is too large for investigation of resonant processes. The secondary ion trapping effect is used to efficiently reduce the width to close to the width of a Maxwellian distribution[16]:

$$\Delta V \text{ (eV)} = 0.21 \, 10^{-3} \, T \text{ (K)} \tag{44}$$

where T, the cathode temperature, is on the order of 1000 K.

The intensities of incident particle currents are readily measured by means of good-precision electrometers by taking care that secondary emission of ions or electrons does not affect the measurement. The absolute determination of the number of product ions requires a lot of precautions. Among these, it must be ensured that the transmission between the collision region and the detector is 100%. This is achieved by a proper design of diaphragms along the beam transport line. Next, a careful determination of the detector efficiency is needed.

Improvements in the methods and in various applied techniques have allowed the cross section to be determined with absolute uncertainties that lie usually between ±5% and ±10% for 90% confidence limits.

Among the five experimental groups that have produced most of the data summarized in this chapter (groups at Oak Ridge National Laboratory, Culham Laboratory, Tokyo, Giessen, and Louvain-la-Neuve), two are using the animated beam method, which has been seen to be fully consistent with the steady beam method.

3.2. Electron Beam Ion Source and Trap

3.2.1. The Electron Beam Ion Source

The *electron beam ion source* (EBIS)[17] consists of a long solenoid (on the order of 1 m) producing a magnetic field on the order of a few teslas. The electron gun is located outside the field. It produces an intense electron beam at energies on the order of a few kilo-electron-volts. The electron beam moves along the field axis. Owing to magnetic compression, very high current densities can be achieved (up to 10^3 A/cm^2). After gas introduction, ions are created by electron impact. These ions are trapped inside the potential well formed by the electron beam space-charge effect. In addition, appropriate voltages applied to a set of electrodes surrounding the electron beam prevent low-energy ions from escaping along the trap axis. During the trapping period, ions are successively ionized. The pressure is kept as low as possible in order to reduce ion loss due to recombination. Ions are extracted, and the charge-state spectrum is obtained from a time-of-flight analysis. Spectra are taken for longer and longer confinement time—up to 10 seconds—showing the appearance of the highest charge states allowed by the electron energy. Modeling the evolution of the charge-state distribution produces single-ionization cross section values. Compared to the crossed-beams method, this method has a lower level of precision (the uncertainty in the data obtained by this method probably being on the order of ±30%). For this reason, this method gives only the general shape of the cross section and is not able to produce any detailed information on the ionization process.

3.2.2. The Electron Beam Ion Trap

Based essentially on a design similar to that of the EBIS, the electron beam ion trap (EBIT)[18] provides facilities for light detection and analysis when ions are stored indefinitely. Ionization cross sections are obtained from comparison with cross sections for the radiative recombination process that dominates light emission. This recent development will certainly produce in the future a lot of results for the very high charge states, which are—up to now—not produced by the present standard ion sources.

4. CROSS SECTIONS

In the past decade, a very large amount of data, both experimental and theoretical, has been produced. Many of the studies concern ionic species on which fusion plasma requirements have shed light. In accordance with the scope of this book, moderately charged systems ($z \leq 10$) only are reviewed here. The total number of concerned ions is 151, most of which may be treated in isoelectronic sequences.

Available data were reviewed and recommended cross sections and rates were given in three reports published by the Culham Laboratory.[19–21] Since the publication of these reports, as pointed out in the preceding sections, a lot of reliable theoretical and experimental data have appeared, in particular for multiply charged ions, including ions of the rare gases as well as metallic ions. For this reason, more precise information is presently available, and earlier recommended data may be slightly modified.

4.1. Hydrogen Isoelectronic Sequence

He^+. Accurate crossed-beams measurements for He^+ have been reported by Peart et al.[22] The measured data blend, to within the experimental uncertainty, with the Born calculations of Omidvar[23] and Peach[24] in the kilo-electron-volt region, where the Born approximation should be excellent. At lower energies, agreement between experiment and the distorted-wave-exchange calculations of Younger[25] is also good. As the best available data, we recommend the curve of Bell et al.,[19] which follows the experimental data very closely, merged with the calculations of Peach[24] above 8.6 keV. These data should be accurate to 5%.

Li^{2+}. The measurements of Tinschert et al.[26] are in very good agreement with the Lotz formula and with Younger's calculations[25] up to 1 keV. Our recommendation follows the experimental data.

Be^{3+} and B^{4+}. Accurate assessments for these two ions have already been made by Moores,[27] and these data have been adopted unmodified for the present compilation.

C^{5+}, O^{7+}, and Ne^{9+}. For these ions, cross sections have been calculated by the relativistic distorted-wave method and the results merged with the Born calculations[24] above about 8 keV.

4.2. Helium Isoelectronic Sequence

4.2.1. Ground State Ions

Li^+. Cross sections for Li^+ have been measured by Lineberger et al.,[28] Wareing and Dolder,[29] and Peart and Dolder.[30] Peart et al.[22] extended the measurements from 3 to 25 keV. They found excellent agreement with the Born approximation above 20 times threshold. The Bethe approximation was not found to agree with experiment until the energy reached 50 times threshold, at which point relativistic effects are just beginning to become significant. Measurements by Müller et al.[31] were mainly concerned with resolution of structure due to indirect ionization at low energy, but this amounts to no more than 0.3% of the background, and we will not consider it here. Distorted-wave calculations[25] are in reasonable agreement with experiment below the peak in the cross section. The curve we recommend is the same as that given by Bell et al.,[19] which is based on the experimental data.

Be^{2+}, B^{3+}, C^{4+}, and O^{6+}. For B^{3+} Moores[27] recommends a cross section based on a fit to the experimental data of Crandall et al.,[32] which agrees well with the distorted-wave calculation of Younger[33] at low energy, merged with the Born calculation of Peach[24] at very high energy. For Be^{2+} it was recommended that the B^{3+} curve be scaled using Eq. (11). For He-like C^{4+} and O^{6+} the Coulomb–Born calculations of Attaourti et al.[34] are in good agreement with the experimental data of Rachafi et al.,[35] and our recommended cross sections are based on these data merged with the Burgess and Chidichimo formula at high energies.

Ne^{8+}. Recently, experimental results[36] have been obtained for Ne^{8+} over a wide energy range, from below threshold up to 6 keV. They do not show the presence of metastable states. These results are in good agreement with the distorted-wave calculations of Younger[37] as well as with the Lotz formula. They are recommended.

4.2.2. Metastable States

For ionization out of the metastable $1s2s\ ^{1,3}S$ states of He-like C and O ions, we also recommend the Coulomb–Born calculations of Attaourti et al.[34] merged with the formula of Burgess and Chidichimo at very high energy. For Be and B we recommend the cross sections obtained by Moores[27] by scaling the C^{4+} calculations. For Ne^{8+} the same procedure is followed starting from the O^{6+} calculated data.

4.3. Lithium Isoelectronic Sequence

Although much attention has been paid to the lithium isoelectronic sequence, there remain discrepancies among and between experimental and theoretical results. For Be and B ions Moores[27] recommends data based on theory. For the more highly charged members of the Li sequence, excitation–autoionization contributions become significant. For C^{3+} we recommend the experimental data of Crandall et al.,[38] which agree well with the energy scan results of Hofmann et al.[39] For O^{5+} we recommend the experimental data of Rinn et al.[40] The experimental data of Defrance et al.[41] for Ne^{7+}, though of low precision, agree well with the calculations of Sampson et al.[42] below about 1 keV. The three measured points of Donets and Ovsyannikov[43] above 2 keV also agree with the results of the same calculation. We recommend therefore the Sampson et al. calculated cross section, which includes autoionization.

4.4. Be sequence: B^+, C^{2+}, O^{4+}, and Ne^{6+}

Experimental data are available for all the members of the Be sequence.[36,44] The ion beam consists of a mixture of ions in the $2s^2$ ground state and in the $2s2p$ metastable states, and it was found that the metastable content was very high (50–90%) except in the case of Ne^{6+}. In the latter case, large signal fluctuations were observed below the ground state ionization threshold, depending on source conditions. A set of data was obtained with negligible metastable content. The cross section enhancement due to excitation–autoionization does not exceed 10%. On the basis of these remarks, experimental data are recommended. Pure ground state cross sections may be obtained from the theoretical results of Younger.[45]

4.5. B sequence: C^+ and O^{3+}

The C^+ ion has been the subject of many experimental investigations.[46,47] The recommended data follow the data of Yu et al.[46] which extend up to 2500 eV. For O^{3+}, the ground state cross section is extracted from the experimental data of Crandall et al.[32]

4.6. O^+ and O^{2+}

Experimental and theoretical results for O^+ and O^{2+} are in good agreement with each other. Recommendations are those previously given.[19]

4.7. Rare Gases

4.7.1. Neon

Ne^+. This ion has been the subject of many experimental investigations.[20] The results of Man et al.[48] seem to have reached the highest possible precision (4.1%, 90% confidence limit) in this type of measurement. They are recommended.

Ne^{2+}. Crossed-beams results of Matsumuto et al.[49] are recommended.

Ne^{3+}. The experimental results of Gregory et al.[50] show a significant contribution of metastable states. The direct ionization distorted-wave calculation of Younger[51] as well as the plane-wave Born approximation of McGuire[52] are in very good agreement with the measurements. Experimental results are recommended.

Ne^{4+} and Ne^{5+}. Up to now, experimental results have been obtained from EBIS ion spectra[43] at high energy only. They generally overestimate the cross sections by more then 30%. For this reason, recommended cross sections are derived from scaling of oxygen ion results.

Ne^{6+} to Ne^{9+}. These ions are included in the isoelectronic sequences (see above).

4.7.2. Argon

Cross sections for all the concerned charge states have been measured in crossed beams.

Ar^+. There is some disagreement between the various experimental results of Woodruff et al.,[53] Müller et al.,[54] and Man et al.[48] The distorted-wave calculation of Younger[55] tends to underestimate these results over the whole energy range. This seems to be due to the absence of excitation–autoionization contributions in the calculation. This effect, clearly observed in the low-energy region, is probably due to excitation of $3s3p^5nl$ ($n > 3$) states. Data of Man et al.[48] are recommended.

Ar^{2+}. There is also some disagreement among the various experimental results obtained for Ar^{2+} by four groups: Man et al.,[56] Müller et al.,[57] Matsumuto et al.,[49] and Mueller et al.[58] This disagreement may probably be partially attributed to differences in the fraction of metastable ions. The results show no evident structure. Results of Man et al.[56] should be recommended as the best data presently available.

Ar^{3+}, Ar^{4+}, and Ar^{5+}. Available experimental data for these ions come from Gregory et al.,[50] Crandall et al.,[32] and Gregory et al.[59] for charge states 3, 4, and 5, respectively. They are mutually consistent and are, therefore, recommended.

Ar^{6+}. The measurements of Howald et al.[60] show the presence of a significant fraction of metastable states and a very large contribution of EA processes corresponding to $2p \rightarrow 3p$ transitions. These experimental results are recommended.

Ar^{7+}. Two sets of measurements by Rachafi et al.[61] and by Zhang et al.[62] are available. In these experiments, the presence of long-lived metastable autoionizing states in the sodium-like ion beam produces important signals below the ground state ionization threshold. For this reason, the precision of the data is very poor—on the order of 20% only. The 45-state close-coupling calculation of Tayal[63] includes the dominant indirect excitation–autoionization and resonant processes. There is satisfactory agreement between experiments and theory below 400 eV only. In the

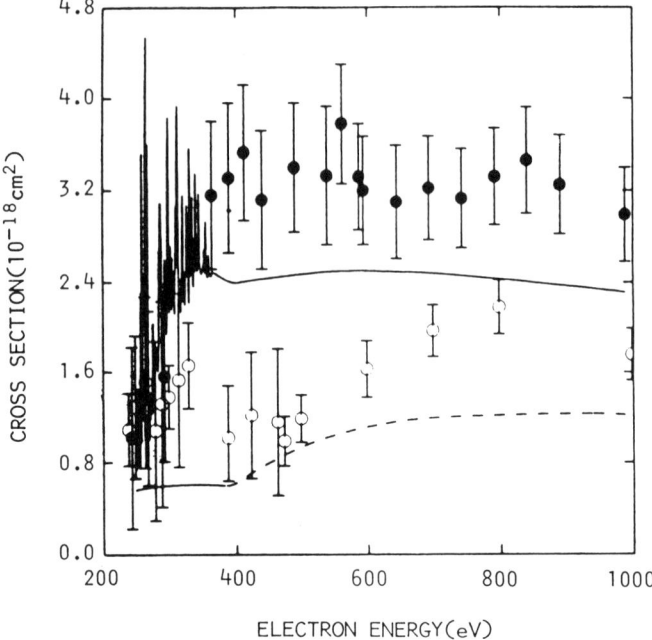

Figure 2. Electron impact ionization of Ar^{7+}. —) **R**-matrix 45-state close-coupling approximation (Ref. 63); - - -) direct ionization cross sections; ○ and ●) experimental values, Refs. 61 and 62, respectively. (Adapted from Ref. 63.)

40 to 1000-eV energy range, there is large disagreement between the aforementioned results, but they tend to merge at higher energies (Fig. 2). Both sets of experimental data are fitted to produce the recommended data.

Ar^{8+}. Two sets of experimental data are also available for this ion, one obtained by Defrance et al.[64] and the other by Zhang et al.[65] Both of them exhibit the presence of spurious signal below the ground state ionization threshold. In addition, the results of Zhang et al. are affected by a 3% metastable ion beam component. For these reasons, uncertainties are much larger then 10%. The results of Defrance et al. agree with the distorted-wave calculation of Younger.[66] They are recommended for the ground state ionization cross section (Fig. 3).

Ar^{9+}. The crossed-beams experimental results of Rachafi et al.[67] are in good agreement with the theoretical results of Laghdas et al.[68] The experimental results are recommended.

Ar^{10+}. The crossed-beams experimental results of Zhang et al.[69] are the only available results for this ion. They are recommended.

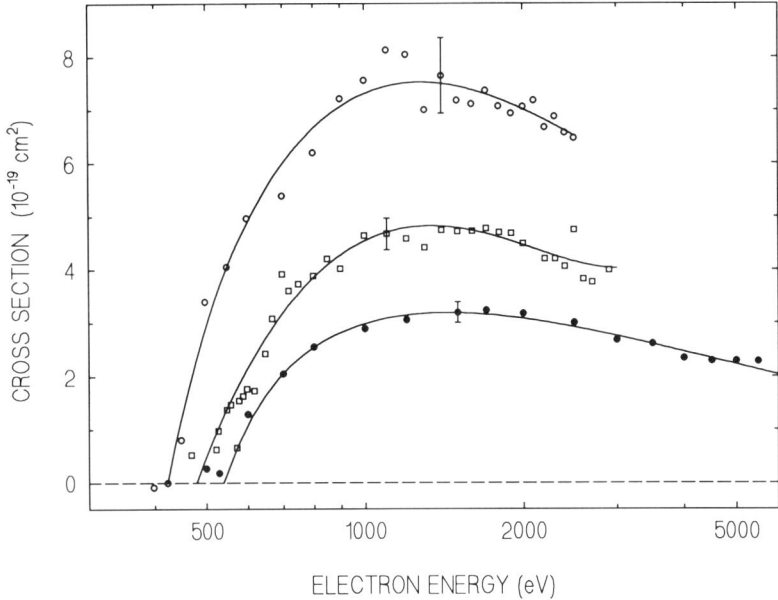

Figure 3. Electron impact ionization of Ar^{8+} (○), Ar^{9+} (□), and Ar^{10+} (●), from Refs. 65, 67, and 69, respectively. Fitting curves are also reproduced.

4.7.3. Krypton

Kr^+. The experimental results of Tinschert et al.[70] systematically exceed those of Man et al.[48] by about 10%. The absolute error on the latter is lower, so that they are recommended.

Kr^{2+}. The differences among the various sets of experimental data published by Gregory,[71] Tinschert et al.,[70] Danjo et al.[72] and Man et al.[56] exceed the size of the absolute error of the measurements. This discrepancy can only be partially explained by large differences in the metastable population of the beams, which results from the use of different types of ion sources. The structure observed only by Danjo et al.[72] may also be explained by the presence of high-lying metastable states.

The results of Man et al.[56] should be considered as a best estimation of the ionization cross section for the ground state ions.

Kr^{3+} ($4p^3$ $^4S_{3/2}$). The measurement of Gregory et al.[50] shows an important metastable component, assumed to be statistically determined, that is, up to 80%. Again, excitation–autoionization takes place close to the threshold. The results of Tinschert et al.[70] exceed those of Gregory et al. by more than 20% around the maximum. They are recommended, although they do not precisely correspond to the ground state cross section.

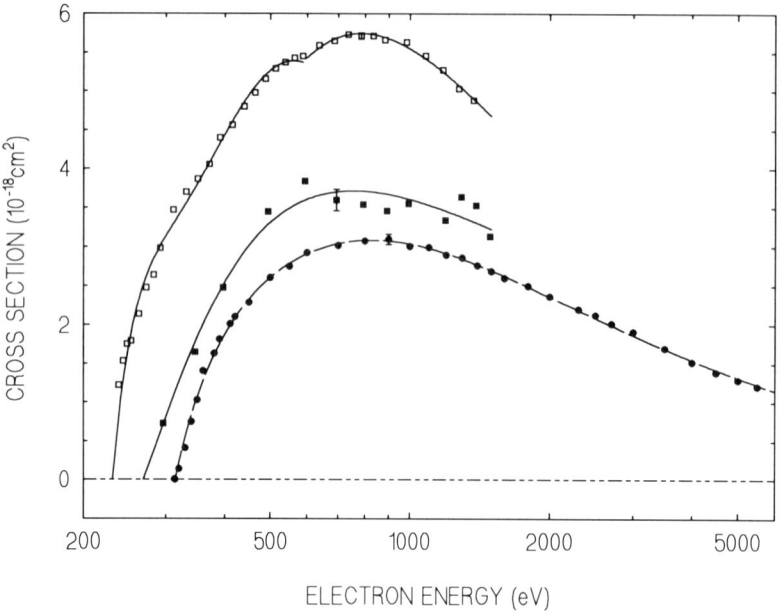

Figure 4. Electron impact ionization of Kr^{8+} (□), Kr^{9+} (■), and Kr^{10+} (●), from Refs. 75, 76, and 77, respectively. Fitting curves are also reproduced.

Kr^{4+} to Kr^{7+}. Experimental data[73] for charge states 4, 5, and 7 and theoretical data[74] for charge states 4 to 7 have been recently published. An important discrepancy between the experimental results and the configuration-average distorted-wave approximation results is frequently observed in cases in which excitation–autoionization dominates the process. It is seen that the proper inclusion of term dependence resolves this discrepancy. For this reason, the theoretical results may be recommended for the Kr^{6+} ion while the experimental ones are recommended for charge states 4, 5, and 7.

Kr^{8+}. Crossed-beams experimental results[75] (Figure 4) are recommended.

Kr^{9+}. The unpublished experimental data[76] are not complete. They follow the Lotz formula prediction. They may be recommended.

Kr^{10+}. Crossed-beams experimental results[77] are recommended.

4.7.4. Xenon

Xe^{1+}. The recommended curve is based on the crossed-beams data of Man et al.[48]

Xe^{2+}. Contrary to the results of Danjo et al.[72] and Achenbach et al.[78] the experimental results of Man et al.[56] more or less coincide with those of Griffin et

$al.^{79}$ and they show no presence of high-lying metastable states. The calculation[79] produces the position of the excitation–autoionization peak for $4d$ electrons, but not the magnitude of the process. In addition, other indirect processes that are seen to play a role in the low-energy region are not included in the calculation. Results of Man et al.[56] are recommended as the best values.

Xe^{3+}. The experimental results[50] show a large bump around 90 eV. It is attributed to excitation of $4d$ electrons. There is no agreement with the plane-wave Born-approximation (PWBA)-calculations.[52] Experimental results are recommended.

Xe^{4+}, Xe^{5+}, and Xe^{6+}. For these ions, recommended cross section curves follow the experimental measurements.[21,79,80]

Xe^{8+}. Crossed-beams experimental results[75] are recommended.

For Xe^{7+}, Xe^{9+}, and Xe^{10+}, no data are available.

4.8. Metallic Ions

4.8.1. Titanium

Four members (z = 1, 2, 3, and 5) of the titanium homonuclear sequence have received attention from experimentalists (Diserens et al.,[81] Mueller et al.,[82] Falk et al.,[83] and Chantrenne et al.[84]). The ground state configurations of Ti^+, Ti^{2+}, and Ti^{3+} are $3p^63d^24s$ ^4F, $3p^63d^2$ ^3F, and $3p^63d$ ^2D, respectively. For these three ions, the metastable population is of minor importance. Very large contributions from excitation–autoionization are superimposed on direct ionization close to the ionization threshold. This effect results from $\Delta n = 0$ transition, the promotion of an electron from the $3p$ to the $3d$ subshell. As a consequence, a simple estimate of the direct ionization cross section through the Lotz formula is almost an order of magnitude too low. For Ti^{3+}, the experimental results of Falk et al.[83] are largely overestimated by the distorted-wave dipole approximation of Griffin et al.,[85] and many-state **R**-matrix calculation[86,87] is required to obtain agreement. This is illustrated in Fig. 5.

The case of Ti^{5+} is very similar to that of Fe^{9+}, which also belongs to the chlorine isoelectronic sequence ($3s^23p^5$). A large number of ions are seen to be in metastable excited states $3p^43d$, a situation which produces a large increase of the cross section in the low-energy region.

The experimental data concerning z = 1, 2, and 3 were included in the second Culham Laboratory report,[20] so that recommended cross sections and rates may be extracted from this reference. For Ti^{5+}, the situation is more complex, owing to the presence of a large number of excited states in the experiment. For this reason, the ground state cross section should be based on calculated values by Younger.[88] For other charge states, approximate cross sections are provided by scaling from other ions belonging to the same isoelectronic sequences: Cr^{6+}, Cr^{8+}, Cr^{10+}, Ar^{3+}, Ar^{5+}, and Ar^{6+} for charge states 4 and 6 to 10, respectively.

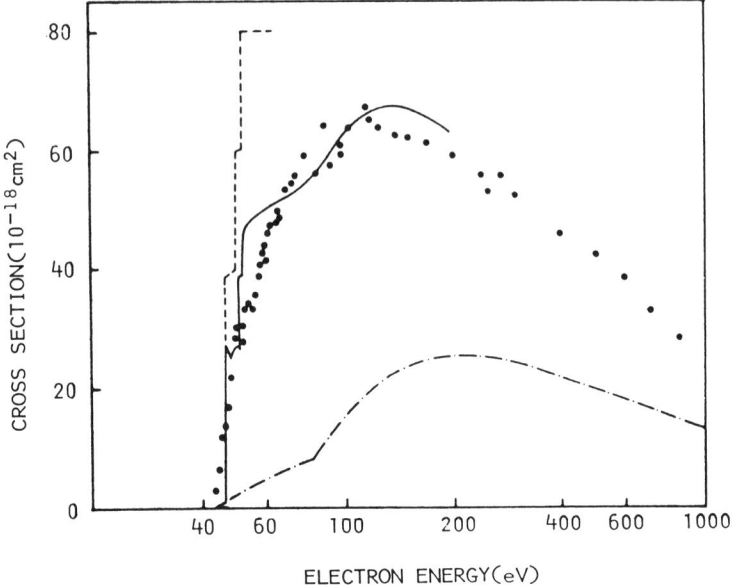

Figure 5. Electron impact ionization of Ti^{3+}. — · —) Lotz formula for direct ionization; — and – – –) R-matrix calculation (Ref. 86) and distorted-wave calculation (Ref. 85), respectively, for excitation–autoionization plus Lotz formula for direct ionization, •) experimental results (Ref. 83). (Adapted from Ref. 86.)

4.8.2. Vanadium

There is no experimental result—and only one theoretical result,[89] for V^{5+}—available in this sequence. Cross sections are obtained by scaling from data for Fe^{4+}, Ti^{1+}, Ti^{2+}, Ti^{3+}, Cr^{6+}, Cr^{7+}, Cr^{8+}, Fe^{11+}, Cr^{10+}, and Fe^{13+} for charge states 1 to 10, respectively.

4.8.3. Chromium

Five charge states have been the subject of experimental investigations: $z = 1$, 6, 7, 8, and 10. No detailed calculation has been performed for any of the members of the sequence.

The experimental results of Man et al.,[90] for Cr^+ show that the beam consists of a mixing of the ground state ($3p^6 3d^5\ ^7S$) and the first metastable states ($3p^6 3d^4 4s$, ^6D, ^4D, or ^4G). All these states contribute to the cross section through direct and indirect processes. In particular, similarly to the case of the low charge states of titanium, the excitation of a 3p electron to the $3p^5 3d^5 4s$, $3p^5 3d^4 4s^2$, and $3p^5 3d^6$ configurations is responsible for an important onset around the corresponding threshold (38 eV).

For Cr^{6+}, Cr^{7+}, and Cr^{8+}, a large signal is observed[91] below the ground state ($3s^2 3p^n$, $n = 6,5,4$) threshold. The onset corresponds precisely to the metastable components of the first excited configurations: $3s^2 3p^{n-1} 3d$. The comparison with

Ionization of Atomic Ions by Electron Impact

the prediction of the Lotz formula shows that these ion beams should be almost purely metastable. However, for Cr^{10+}, the metastable fraction is negligible, the observed threshold being consistent with the ground state ionization potential.

In the four cases, an enhancement of the cross section is observed at energies higher than 500 eV. This enhancement is due to excitation of $2p$ electrons to the $3d$ shell, leading to autoionizing excited configurations. For Cr^{10+}, the contribution of the EA process to the cross section is nearly 50%.

Recommended cross sections for charge states 2–4 and 9 are obtained from scaling Ti or Fe data.[20,92]

4.8.4. Iron

In the second Culham Laboratory report,[20] available information was almost complete for these fundamental ions, detailed reports having already been published.[93] Theoretical results have been obtained in the Average Configuration Statistical

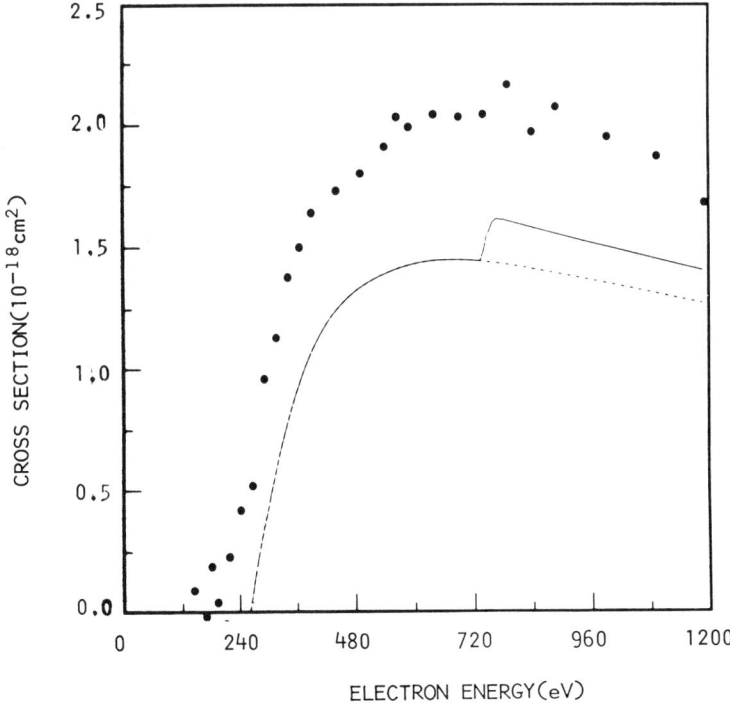

Figure 6. Electron impact ionization of Fe^{9+}. ——) Total cross section from the ground state configuration in the average configuration statistical model (Ref. 93); ---) direct ionization cross section only; •) experimental results (Ref. 95). (Adapted from Ref. 93.)

Model (ACSM) for the direct ionization cross section and for the excitation cross section from the ground state configuration as well as from some excited configurations. Experimental results are available for ion charges 1,[94] 2,[82] and 5, 6, and 9.[95] Experimental data unquestionably show the presence of ions in some of the metastable excited states. Although it is not clear what percentages of the ion population are in the ground state and in the long-lived excited states, respectively, it is seen that in some particular cases (Fe^{9+}, for instance; Fig. 6) the ground state may not be dominant. For this reason, calculations have been performed for the observed excited configurations. The comparison of experimental and theoretical cross sections also clearly demonstrates that, as expected, direct ionization calculations largely underestimate the results and, consequently, that excitation–autoionization may contribute very significantly to the observed signal. However, some of the excited levels or sublevels are located above the double-ionization threshold, so that they are able to partially contribute to double ionization, according to the corresponding branching ratio.

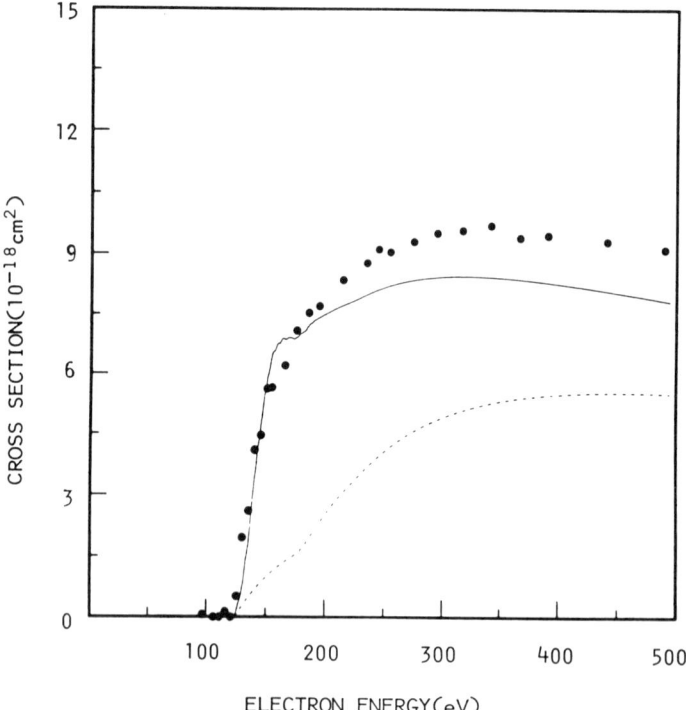

Figure 7. Electron impact ionization of Fe^{5+}. —) Total cross section from the $3p^63d^24s$ excited configuration in the average configuration statistical model (Ref. 93); – – –) direct ionization cross section only; •) experimental results (Ref. 95). (Adapted from Ref. 93.)

Ionization of Atomic Ions by Electron Impact

Direct ionization calculations are known to underestimate experimental results, whereas inclusion of excitation–autoionization processes tends to overestimate them (Fig. 7). For this reason, it is recommended that reliable experimental data be used when available, that is, for iron ions of charge 1, 2, 5, 6, and 9, but one must be aware that the metastable state population in the plasma may differ appreciably from that in crossed-beams experiments. For other charge states, recommended values come from calculations that include excitation–autoionization for the ground state configuration only, as the presence of metastable states has not been demonstrated.

4.8.5. Nickel

The situation for nickel is very similar to that for iron. A detailed report has been published.[96] Experimental results have been obtained at the Culham Laboratory[97] for the singly charged ion and at the Oak Ridge National Laboratory for charge states 3[98] and 5–8.[99]

4.8.6. Copper

Data are very scarce for copper: Cu^{2+} and Cu^{3+} have been experimentally investigated by Gregory and Howald,[98] and, up to now, theoretical data have been obtained for only the first three charge states by Pindzola et al.[100] It is seen that the metastable component of the beam is of low importance. The results are well described by the distorted-wave calculations for direct ionization of $3d$ shell electrons. As expected, the semiempirical Lotz formula considerably overestimates the cross section.

4.8.7. Gallium

Only the first charge state has received attention from experimentalists[101] and theoreticians.[102] Although the metastable population is very low, a few percent only, the spectrum is complex. Results show that EA processes play a dominant role very close to the ground state ionization threshold through $3d^{10}4s^2$–$3d^94s^24p$ transitions. However, the details of the increase of the cross section in the low-energy region have not been explained. A search for resonances and indirect ionization thresholds by means of an energy-resolved electron beam was conducted by Peart and Underwood.[103] The most important conclusion of the latter experiment is that the optically allowed transition to the $^1P_1$ level has a much smaller role than predicted by the calculations.

4.8.8. Molybdenum

The Mo^+ beam in the experiment of Man et al.[104] is supposed to contain a mixing of ground ($4d^5\ ^6S$) and metastable ($4d^45s\ ^6D$) state ions. These states are separated by only 1.6 eV, so that it was not possible to analyze precisely the composition of the beam. The scaled Born approximation of McGuire[105] seriously underestimates the experimental cross section, particularly at low energies. This discrepancy has not been explained, whereas above 35 eV, autoionization following excitation to $4p^54d^45snl$ should be responsible for the structure observed around this energy.

4.8.9. Tungsten

The structure of $W^+(5d^46s\ ^6D)$ makes this case very complicated, because the threshold is not known with good precision. The experimental results of Montague and Harrison[106] were compared to distorted-wave calculations by Pindzola and Griffin.[107] As is frequently the case for low-charge-state complex ions, the distorted-wave results[108] overestimate the cross section. Incidentally, the experimental results allow a more precise determination of the threshold, yielding a value of 16.1 ± 1.0 eV.

5. PARAMETRIC REPRESENTATION OF THE CROSS SECTIONS

The recommended cross sections are fitted in the form adopted in the Culham Laboratory reports[19]:

$$\sigma(E) = \frac{1}{IE}\left[A \ln\left(\frac{E}{I}\right) + \sum_{i=0}^{N} B_i \left(1 - \frac{I}{E}\right)^i\right]$$

where E is the incident electron energy (eV) and I is the ionization potential (eV). All the ionization potentials (in eV) and the fitting coefficients B_i (in units of 10^{-13} cm^2 eV^2) for the various subshells are given in Table I. In addition, excitation–autoionization cross sections have also been included in the fitting procedure for both the ground state and the excited configurations, when data are available. In the "Remarks" column, the cases in which data for metastable ions are available and the energy range of validity of the given fitting coefficients in units of the ionization potential are indicated. The type of data is indicated in the next column according to the following notation: E and T stand for experimental and theoretical data, respectively, and S means that the cross section is scaled classically by comparison with an ion of the same isoelectronic sequence, which is indicated in the adjacent column. In this case, the fitting coefficients are identical for both ions. The reference numbers 19, 20, and 21 refer to the three Culham Laboratory reports, from which the fitting coefficients have been extracted when these references are cited in the table. The column noted "basis" gives the ion on which the scaling is based.

Table I. Parametric Representation of the Cross Sections for Electron Impact Ionization of Atomic Ions

Atom	Ion	I (eV)	A	B_1	B_2	B_3	B_4	B_5	Remarks	Type of data[a]	Basis for scaled data	Reference
2 Helium	He(1+)	54.4	0.1845	0.089	0.131	0.388	−1.901	1.354		E		19
3 Lithium	Li(1+)	75.6	0.722	−0.149	−1.301	1.944				E		19
	Li(2+)	122.4	0.514	−0.198	−0.016	0.082				E		26
4 Beryllium	Be(1+)	18.2	0.1095	0.3004	−0.293	1.138	−1.388	0.8915		T		27
	Be(2+)	153.9	0.4603	−0.655	0.2742	3.857	−5.677	3.3451		SE	B(3+)	27
		32.2	0.0905	0.343	0.1088	0.1175			Singlet	ST	C(4+)	27
	[b]	35.3	0.4673	−0.221	−0.042	0.1064	−0.224		Triplet	ST	C(4+)	27
	Be(3+)	217.7	0.1755	0.2323	−0.502	2.321	−2.957	1.725		T		27
5 Boron	B(1+)	25.1	1.595	−0.542	−0.861	0.184				E		44
	B(2+)	37.9	0.1222	0.1391	0.2238	−0.146	−0.309	0.7722		T		27
	B(3+)	259.4	0.4603	−0.655	0.2742	3.857	−5.677	3.3451		E		27
		56.6	0.0905	0.343	0.1088	0.1175			Singlet	ST	C(4+)	27
	[b]	60.8	0.4673	−0.221	−0.042	0.1064	−0.224		Triplet	ST	C(4+)	27
	B(4+)	340.2	0.1768	0.2509	−0.569	2.563	−3.335	1.902		T		27
6 Carbon	C(1+)	24.4	0.9306	0.298	−1.972	2.134				E		46
	C(2+)	47.9	1.295	−0.096	−1.235	−0.006				E		44
	C(3+)	64.5	0.45	−0.318	1.026	−2.859	1.995			E		20
	C(4+)	392.1	0.7758	−0.194	−0.046	0.4718				T		34
		87.7	0.0905	0.343	0.1088	0.1175			Singlet	T		34
	[b]	93.1	0.4673	−0.221	−0.042	0.1064	−0.224		Triplet	T		34
	C(5+)	490	0.3553	0.0505	−0.15	0.3667				T		24
8 Oxygen	O(1+)	35.12	1.526	−0.593	−0.399	−0.583	3.235			E		19
	O(2+)	54.9	1.066	0.442	0.475	−2.961	4.47			E		19
	O(3+)	77.4	1.045	−0.652	1.299	0	0			E		19
		68.6	0.561	−0.61	4.652	−8.94	6.735		Metastable	E		19

(*continued*)

Table I. (continued)

Atom	Ion	I (eV)	A	B_1	B_2	B_3	B_4	B_5	Remarks	Type of data[a]	Basis for scaled data	Reference
	O(4+)	113.9	0.4528	1.305	−2.757	3.374				E		44
	O(5+)	138.1	0.6408	−0.287	−0.119	−0.178				E		40
		138.1	0.6408	−275.7	723.69	−475.6			$x > 3.98$			
	O(6+)	739.3	0.8049	−0.092	−0.275	0.8899	−0.255			T		34
	[b]	170.3	0.4428	−0.009	−0.04	0.0682	−0.335		Singlet	T		34
	[b]	178.2	0.3191	0.0031	−0.324	0.4462			Triplet	T		34
	O(7+)	871.4	0.408	−0.012	−0.058	0.1975				T		24
10 Neon	Ne(1+)	41.1	3.333	−2.944	−1.225	1.226				E		48
	Ne(2+)	63.5	5.1559	−3.836	−1.314	−1.976				E		49
	Ne(3+)	92.5	0.785	1.709	−10.85	41.5				E		20
	Ne(4+)	126.2	1.066	0.442	0.475	−2.961	−58.05	30.72		SE	O(2+)	20
	Ne(5+)	157.9	1.045	−0.652	1.299	0	4.47			SE	O(3+)	20
	Ne(6+)	207.5	0.8747	−0.282	0.7244	−0.645	0			E		36
	Ne(7+)	239.1	0.355	0.047	0.077	−0.211	0.112			E		20
		239.1	0.677	−0.345	2.769	−4.059	1.52		$x > 3.75$			
	Ne(8+)	1196	1.2821	−0.237	−0.928	0.8597				E		36
	[b]	281.1	0.4428	−0.009	−0.04	0.0682	−0.335		Singlet	ST	O(6+)	
	[b]	291.2	0.3191	0.0031	−0.324	0.4462			Triplet	ST	O(6+)	
	Ne(9+)	1362.2	0.4603	−0.028	−0.138	0.165				T		24
18 Argon	Ar(1+)	27.4	25.14	−24.31	92.63	−30.87				E		48
		27.4	2.206	3.727	−11.4	9.076			$x > 2.336$			
	Ar(2+)	40.9	1.148	2.665	−4.148	3.237				E		56
	Ar(3+)	52.3	1.186	−1.18	11.05	−30.79	36.62	−15.42		E		20
	Ar(4+)	75	1.574	0.722	−2.687	1.856				E		20
		126	2.798	4.114	−3.103	0.438			$x > 2.93$			
	Ar(5+)	91	1.17	0.843	−2.877	1.958				E		20
		200	3.771	16.163	−34.95	20.853			$x > 2.53$			

Ion								Range	Code	Ref ion	Note
Ar(6+)	124.3	−2.813	4.254	−1.844	7.023			$x > 1.68$	E		60
Ar(7+)	124.3	4.604	−13.82	30.843	−25.47				E		62
Ar(8+)	143.5	1.938	−1.609	1.866					E		64
Ar(9+)	422.5	4.216	−0.25	−2.596	2.4099				E		67
Ar(10+)	478.7	4.1496	−1.844	1.0784	0.5331	−1.81			E		69
	539	3.4068	0.2527	−3.486	1.3729						
22 Titanium											
Ti(1+)	13.6	0.618	0.048	0.185	−0.74	0.522	5.945		E		20
	28	4.026	8.92	−30.07	31.73	−17.97	7.738				
Ti(2+)	27.5	2.876	−2.321	5.91	−10.31	−1.333	−6.092	$x > 2.5$	E		20
Ti(3+)	43.1	3.57	0.531	−1.323	−7.778	14.92			E		20
Ti(4+)	99.2	0.672	−1.153	0.416	−0.876				SE	Cr(6+)	
Ti(5+)	119.4	5.334	−0.493	−6.746	1.995				E		84
	119.4	1.739	−1.451	1.564	4.739						
Ti(6+)	140.3	−2.399	1.651	5.785	7.827			$x > 3.35$	SE	Cr(8+)	
	140.3	10.409	−4.54	−15.88	7.009						
Ti(7+)	168.5	1.186	−1.18	11.05	−30.79	36.62	−15.42	$x > 2.59$	SE	Ar(3+)	
Ti(8+)	193.2	−0.39	4.114	−9.045	13.466				SE	Cr(10+)	
	193.2	3.164	−18.03	51.616	−39.28			$x > 2.016$			
Ti(9+)	215.9	1.17	0.843	−2.877	1.958				SE	Ar(5+)	
	546	3.771	16.163	−34.95	20.853			$x > 2.53$	E		
Ti(10+)	265.2	−2.813	4.254	−1.844	7.023			$x > 1.68$	SE	Ar(6+)	
	265.2	4.604	−13.82	30.843	−25.47						
23 Vanadium											
V(1+)	14.6	2.659	−1.068	−0.396	1.036				ST	Fe(4+)	20
	19.5	6.503	−0.278	−5.663	3.86			$x > 1.79$			
V(2+)	29.3	0.618	0.048	0.185	−0.74	0.522	5.945		SE	Ti(1+)	
	29.3	4.026	8.92	−30.07	31.73	−17.97	7.738	$x > 2.5$			
V(3+)	46.7	2.876	−2.321	5.91	−10.31	−1.333	−6.092		SE	Ti(2+)	
V(4+)	65.2	3.57	0.531	−1.323	−7.778	14.92			SE	Ti(3+)	
V(5+)	128.1	0.672	−1.153	0.416	−0.876				SE	Cr(6+)	
V(6+)	150.2	−15.13	23.925	−16.25	39.531			$x > 3.23$	SE	Cr(7+)	
	150.2	3.6784	−48.38	125.1	−80.45						

(continued)

Table I. (continued)

Atom	Ion	I (eV)	A	B_1	B_2	B_3	B_4	B_5	Remarks	Type of data[a]	Basis for scaled data	Reference
	V(7+)	173.7	−2.399	1.651	5.785	7.827				SE	Cr(8+)	
		173.7	10.409	−4.54	−15.88	7.009			$x > 2.59$			
	V(8+)	205.8	1.665	−0.575	5.343	1.145				SE	Fe(11+)	
	V(9+)	230.5	−0.39	4.114	−9.045	13.466				SE	Cr(10+)	
		230.5	3.164	−18.03	51.616	−39.28			$x > 2.016$			
	V(10+)	255	2.4909	−1.685	2.0889	2.0426				SE	Fe(13+)	
24 Chromium	Cr(1+)	16.5	3.329	−1.97	−5.967	8.701				E		90
	Cr(2+)	31	2.659	−1.068	−0.396	1.036				ST	Fe(4+)	20
		41.2	6.503	−0.278	−5.663	3.86			$x > 1.79$			
	Cr(3+)	49.1	2.888	0.753	−21.24	66.29	−75.23	29.08		SE	Fe(5+)	20
	Cr(4+)	69.3	2.876	−2.321	5.91	−10.31	−1.333	7.738		SE	Ti(2+)	20
	Cr(5+)	90.6	3.57	0.531	−1.323	−7.778	14.92	−6.092		SE	Ti(3+)	20
	Cr(6+)	114	4.5746	−2.398	−2.015					E		91
		114	2.6839	−41.77	94.504	−52.96			$x > 4.43$			
	Cr(7+)	185.9	−15.13	23.925	−16.25	39.531				E		91
		185.9	3.6784	−48.38	125.1	−80.45			$x > 3.23$			
	Cr(8+)	212.2	−2.399	1.651	5.785	7.827				E		91
		212.2	10.409	−4.54	−15.88	7.009			$x > 2.59$			
	Cr(9+)	244.4	1.665	−0.575	5.343	1.145				SE	Fe(11+)	92
	Cr(10+)	267.8	−0.39	4.114	−9.045	13.466				E		91
		267.8	3.164	−18.03	51.616	−39.28			$x > 2.016$			
26 Iron	Fe(1+)	16.2	2.124	−0.53	−9.617	29.17	−40.81	19.29		E		20
	Fe(2+)	30.6	4.42	4.81	5.692	12.977	6.698			E		20
	Fe(3+)	54.8	2.258	−0.843	0.622	0.01				T		20
		80	6.113	−1.148	1.583	−0.034			$x > 2.10$			
	Fe(4+)	75	2.659	−1.068	−0.396	1.036				T		20
		100	6.503	−0.278	−5.663	3.86			$x > 1.79$			

Ionization of Atomic Ions by Electron Impact

	Ion									
	Fe(5+)	99	4.397	0.098	−2.681	3.048			E	20
	Fe(6+)	125	4.612	0.819	−5.126	4.029			E	20
	Fe(7+)	151.1	26.99	−14.7	−47.28	26.44		$x > 1.55$	T	93
	Fe(8+)	151.1	2.554	3.313	−4.898	3.609			T	20
		233.6	233.6	2.856	0.209	0.488	−0.379		T	20
	Fe(9+)	262.1	2.457	0.289	0.484	−0.413			E	20
		214	2.432	−0.676	0.598			Metastable	E	20
		262.1	1.95	−0.946	11.125	−7.958		$x > 2.1$	T	20
	Fe(10+)	290.3	2.034	0.41	0.253	−0.321			T	20
28 Nickel	Ni(1+)	18.2	2.679	−2.865	6.289	−15.95			E	20
	Ni(2+)	36.2	3.453	−2.008	−1.621	0.889			T	20
	Ni(3+)	54.9	3.497	−1.698	−1.723	2.329			E	20
		70	6.044	2.131	−18.37	15.459		$x > 2.27$		
	Ni(4+)	76.1	2.89	5.458	−17.98	19.37			T	96
	Ni(5+)	103.8	10.413	−5.168	−6.03	−1.477			E	99
	Ni(6+)	131.3	9.304	−3.378	−6.299	0.151			E	99
	Ni(7+)	160	9.784	−2.243	−9.426	1.362			E	99
	Ni(8+)	191.2	9.225	−1.77	−10.37	2.257			E	99
	Ni(9+)	224.6	5.072	2.061	5.281	9.87			E	96
		224.6	3.03	2.57	−1.035	0		$x > 4.007$	T	
	Ni(10+)	320.4	6.621	−4.101	2.691	−7.598			T	96
		320.4	0.0741	1	7.256	0		$x > 2.75$		
29 Copper	Cu(1+)	20.3	2.554	−0.522	−3.671	1.805	0		T	22
	Cu(2+)	36.8	1.705	−0.936	0.741	−0.068		−0.06	E	21
		36.8	5.114	−3.331	−3.778	−0.256		$x > 2.44$		
	Cu(3+)	55.2	2.459	−1.63	0.705	0.288	0.028		E	21
		65	5.604	−1.883	−8.43	7.358		$x > 2.17$		
	Cu(4+)	103	3.497	−1.698	−1.723	2.329			SE	Ni(3+)
		131	6.044	2.131	−18.37	15.459		$x > 2.27$		
	Cu(5+)	130	2.89	5.458	−17.98	19.37			ST	Ni(4+)
	Cu(6+)	157	10.413	−5.168	−6.03	−1.477			SE	Ni(5+)

(continued)

Table I. (continued)

Atom	Ion	I (eV)	A	B_1	B_2	B_3	B_4	B_5	Remarks	Type of data[a]	Basis for scaled data	Reference
	Cu(7+)	185	9.304	-3.378	-6.299	0.151				SE	Ni(6+)	
	Cu(8+)	212	9.784	-2.243	-9.426	1.362				SE	Ni(7+)	
	Cu(9+)	240	9.225	-1.77	-10.37	2.257				SE	Ni(8+)	
	Cu(10+)	267	5.072	2.061	5.281	9.87				ST	Ni(9+)	
		267	3.03	2.57	-1.035	0			$x > 4.007$			
31 Gallium	Ga(1+)	20.5	2.63	-0.551	-3.69	1.28				E		21
	Ga(2+)	30.7	3.109	-2.278	-0.503	-3.262	-0.159	-0.382		SE	Zn(1+)	21
		30.7	0.815	-0.36	0.327	-0.081	0.002	0.002	$x > 5.57$			
	Ga(3+)	64.2	2.554	-0.522	-3.671	1.805	0	-0.06		ST	Cu(1+)	21
	Ga(4+)	96	1.705	-0.936	0.741	-0.068				SE	Cu(2+)	21
		96	5.114	-3.331	-3.778	-0.256			$x > 2.44$			
	Ga(5+)	130	2.459	-1.63	0.705	0.288	0.028	0.025		SE	Cu(3+)	21
		153	5.604	-1.883	-8.43	7.358			$x > 2.17$			
	Ga(6+)	164	3.497	-1.698	-1.723	2.329				SE	Ni(3+)	
		209	6.044	2.131	-18.37	15.459			$x > 2.27$			
	Ga(7+)	199	2.89	5.458	-17.98	19.37				ST	Ni(4+)	
	Ga(8+)	233	10.413	-5.168	-6.03	-1.477				SE	Ni(5+)	
	Ga(9+)	268	9.304	-3.378	-6.299	0.151				SE	Ni(6+)	
	Ga(10+)	302	9.784	-2.243	-9.426	1.362				SE	Ni(7+)	
36 Krypton	Kr(1+)	24.4	2.861	0.093	0.462	-0.57	-0.029	-0.023		E		21
	Kr(2+)	36.95	2.623	2.881	-4.673	5.059	-3.477			E		56
	Kr(3+)	52.5	2.185	5.497	-9.118	3.626				E		70
	Kr(4+)	64.7	4.013	1.215	-5.747	-1.71				E		73
	Kr(5+)	78.5	-21.05	26.31	-1.011	24.87				E		73
		78.5	5.621	-3.305	-0.792	-3.484			$x > 1.656$			
	Kr(6+)	111	-0.144	0.2812	20.51	-22.88				T		74
		111	2.682	3.892	-13.39	13.46			$x > 1.58$			

Ionization of Atomic Ions by Electron Impact

Element	Ion											Ref
	Kr(7+)	125.8	6.655	−1.175	−9.093	4.64				E		73
	Kr(8+)	230.8	−85.64	102.5	−11.54	131.12				E		75
	Kr(9+)	230.8	0.185	1.4067	8.153	14.55			$x > 2.60$	E		76
	Kr(10+)	268.2	11.345	−6.397	6.065	−9.753				E		77
		308.2	4.3215	7.851	−11.22	13.639				E		21
42 Molybdenum	Mo(1+)	13.6	1.348	−1.19	0.89	−0.092	0.004			E		21
		25	7.318	−1.803	−1.547	−5.614	−1.181		$x > 2.43$			
	Mo(2+)	29										
	Mo(3+)	43										
	Mo(4+)	58										
	Mo(5+)	72	5.008	2.219	6.545	−45.53	58.321	−22.88		SE	Zr(3+)	21
	Mo(6+)	124										
	Mo(7+)	145										
	Mo(8+)	167										
	Mo(9+)	189										
	Mo(10+)	217										
54 Xenon	Xe(1+)	21.2	3.587	−0.274	0.0346	−2.651				E		48
	Xe(2+)	32.1	4.0224	2.0222	−10.47	7.6249				E		56
		32.1	3.1468	−6.766	24.169	−20.84			$x > 2.18$			
	Xe(3+)	43	3.273	−0.909	4.322	−5.617	−0.362			E		21
		43	0.745	1.073	13.398	−11.72	0.006		$x > 1.28$			
	Xe(4+)	55	2.635	9.371	−12.84	5.474	0.087	0.105		E		21
		55	7.307	4.564	−24.72	15.743	0.006		$x > 3.64$			
	Xe(5+)	68	3.722	12.033	−16.21	5.557	0.475	0.9		E		21
		68	9.909	19.051	−68.05	43.142	0.633	1.377	$x > 3.53$			
	Xe(6+)	90.7	8.225	3.066	74.384	−348.7	493.1	−218.6		E		21
	Xe(7+)	112										
	Xe(8+)	182.2	3.912	4.771	12.29	−5.147				E		75
	Xe(9+)	202										
	Xe(10+)	233										

(continued)

Table I. (continued)

Atom	Ion	I (eV)	A	B_1	B_2	B_3	B_4	B_5	Remarks	Type of data[a]	Basis for scaled data	Reference
74 Tungsten	W(1+)	16.1	2.103	−0.955	0.043	0.542				E		21
		16.1	3.564	3.776	−15.13	7.5			$x > 5.28$			
	W(2+)	25										
	W(3+)	39										
	W(4+)	53	4.543	1.048	−2.921	1.615	0.364			SE	Ta(3+)	21
	W(5+)	67	4.084	1.981	−7.957	5.579				SE	Hf(3+)	21
		80.7	9.633	2.623	−22.94	12.39			$x > 1.66$			
	W(6+)	120										
	W(7+)	141										
	W(8+)	162										
	W(9+)	183										
	W(10+)	204										

[a] Abbreviations: E, Experimental; T, theoretical; S, scaled on the data for the ion identified in the adjacent column.
[b] Metastable state.

6. CONCLUSIONS

Among collisional atomic processes playing a role in magnetic fusion edge plasmas, electron impact ionization of ions is probably the most extensively studied: single-ionization data, experimental or theoretical, are presently available for 104 of the 151 ions falling within the scope of this book, and scaling provides information for over 29 others. The corresponding numbers from the three Culham Laboratory reports are 52 (data) and 54 (scaling). The significant progress that has been made, as evidenced by the doubling of the number of results of good accuracy since the publication of the Culham Laboratory reports, is due to a combined effort of theoreticians and experimentalists. Missing results, experimental or theoretical, essentially involve hydrogen-like ions ($z > 3$) and metallic ions, Ti, V, Cu, Ga, Mo, and W. Owing to the hazardous character of scaling, future efforts should be concentrated on moderately charged metallic ions so that these may be efficiently incorporated in plasma modeling.

REFERENCES

1. C. Joachain, *Quantum Collision Theory*, Third Ed., North-Holland, Amsterdam (1983).
2. A. Burgess, Proceedings of the Symposium on Atomic Collision Processes in Plasmas, UKAEA Report 4818, 63 Culham Laboratory, 1964.
3. A. Burgess, H. P. Summers, D. M. Cochran, and R. W. P. McWhirter, *Mon. Not. R. Astron. Soc.* **179**, 275 (1977).
4. G. H. Wannier, *Phys. Rev.* **90**, 817 (1953).
5. W. Lotz, *Astron. J. Suppl.* **14**, 207 (1967).
6. W. Lotz, *Z. Phys.* **216**, 241 (1968).
7. M. J. Seaton, *Planet. Space Sci.* **12**, 55 (1964).
8. A. Burgess and M. C. Chidichimo, *Mon. Not. R. Astron. Soc.* **203**, 1269 (1983).
9. M. Gryzinski, *Phys. Rev. A* **138**, 336 (1965).
10. L. B. Golden and D. H. Sampson, *J. Phys. B* **10**, 2229 (1977).
11. K. J. Reed, M. H. Chen, and D. L. Moores, *Phys. Rev. A* **44**, 4336 (1991).
12. M. S. Pindzola, D. C. Griffin, and C. Bottcher, in *Atomic Processes in Electron–Ion and Ion–Ion Collisions* (F. Brouillard, ed.), NATO ASI Series B, No. 145, Plenum, New York (1986).
13. T. D. Märk and G. H. Dunn (eds.), *Electron Impact Ionization*, Springer-Verlag, Berlin (1985).
14. F. Brouillard (ed.), *Atomic Processes in Electron–Ion and Ion–Ion Collisions*, NATO ASI Series B, No. 145, Plenum, New York (1986).
15. P. Defrance, F. Brouillard, W. Claeys, and G. Van Wassenhove, *J. Phys. B* **14**, 103 (1981).
16. A. Müller, K. Tinschert, G. Hofmann, E. Salzborn, and G. H. Dunn, *Phys. Rev. Lett.* **61**, 70 (1988).
17. E. D. Donets and P. Ovsyannikov, Joint Institute for Nuclear Research, R10780, Dubna, 1977.
18. K. L. Wong, P. Beiersdorfer, M. H. Chen, R. E. Marrs, K. J. Reed, J. H. Scofield, D. A. Vogel, and R. Zasadzinski, *Phys. Rev. A* **48**, 2850 (1993).
19. K. L. Bell, H. B. Gilbody, J. G. Hughes, A. E. Kingston, and F. J. Smith, Atomic and Molecular Data for Fusion, Part I, Recommended Cross Sections and Rates for Electron Ionization of Light Atoms and Ions, Culham Laboratory, Report CLM-R216, 1982; *J. Phys. Chem. Ref. Data* **12**, 891 (1983).
20. M. A. Lennon, K. L. Bell, H. B. Gilbody, J. G. Hughes, A. E. Kingston, M. J. Murray, and F. J. Smith, Atomic and Molecular Data for Fusion, Part 2, Recommended Cross Sections and Rates

for Electron Ionization of Atoms and Ions: Fluorine to Nickel, Culham Laboratory, Report CLM-R270, 1986; *J. Phys. Chem. Ref. Data* **17**, 1285 (1988).
21. M. J. Higgins, M. A. Lennon, J. G. Hughes, K. L. Bell, H. B. Gilbody, E. A. Kingston, and F. J. Smith, Atomic and Molecular Data for Fusion, Part 3, Recommended Cross Sections and Rates for Electron Ionization of Atoms and Ions: Copper to Uranium, Culham Laboratory, Report CLM-R294, 1989.
22. B. Peart, D. S. Walton, and K. T. Dolder, *J. Phys. B* **2**, 1347 (1969).
23. K. Omidvar, *Phys. Rev. A* **17**, 212 (1969).
24. G. Peach, International Atomic Energy Agency Consultants Meeting on Atomic Database for Be and B, Vienna, June 1991, Report INDC(NDS)-254, prepared by R. Janev.
25. S. Younger, *J. Quant. Spectrosc. Radiat. Transfer* **26**, 329 (1981).
26. K. Tinschert, A. Müller G. Hoffman, K. Huber, R. Becker, D. C. Gregory, and E. Salzborn, *J. Phys. B.* **22**, 531 (1989).
27. D. L. Moores, *Atomic and Plasma-Material Interaction Data for Fusion* (Nucl. Fusion, Supplement) **3**, 97 (1992).
28. W. C. Lineberger, J. W. Hooper, and E. W. McDaniel, *Phys. Rev.* **141**, 151 (1966).
29. J. B. Wareing and K. T. Dolder, *Proc. Phys. Soc.* **91**, 887 (1967).
30. B. Peart and K. T. Dolder, *J. Phys, B* 1872 (1968).
31. A. Müller, G. Hofmann, B. Weissbecker, M. Stenke, K. Tinschert, M. Wagner, and E. Salzborn, *Phys. Rev. Lett.* **63**, 758 (1989).
32. D. H. Crandall, R. A. Phaneuf, and D. C. Gregory, Electron Impact Ionization of Multicharged Ions, Oak Ridge National Laboratory, Report ORNL/TM-7020, 1979.
33. S. M. Younger, *Phys. Rev. A* **22**, 111 (1980).
34. Y. Attaourti, P. Defrance, A. Makhoute, and C. J. Joachain, *Phys. Scr.* **T43**, 578 (1991).
35. S. Rachafi, M. Zambra, Zhang Hui, M. Duponchelle, J. Jureta, and P. Defrance, *Phys. Scr.* **T28**, 12 (1989).
36. M. Duponchelle, Thesis, Université Catholique de Louvain, 1993; M. Duponchelle, E. M. Oualim, Zhang Hui, and P. Defrance, to be published.
37. S. M. Younger, *Phys. Rev. A* **22**, 1425 (1980).
38. D. H. Crandall, R. A. Phaneuf, B. E. Hasselquist, and D. C. Gregory, *J. Phys. B* **12**, L249 (1979).
39. G. Hofmann, A. Müller, K. Tinschert, and E. Salzborn, *Z. Phys. D* **16**, 113 (1990).
40. K. Rinn, D. C. Gregory, L. J. Wang, R. A. Phaneuf, and A. Müller, *Phys. Rev. A* **36**, 595 (1987).
41. P. Defrance, S. Chantrenne, S. Rachafi, D. S. Belic, J. Jureta, D. C. Gregory, and F. Brouillard, *J. Phys. B* **23**, 2333 (1990).
42. D. H. Sampson, R. E. H. Clark, and A. D. Parks, *J. Phys. B* **12**, 3257 (1979).
43. E. D. Donets and V. P. Ovsyannikov, *Sov. Phys. JETP* **53**, 466 (1981).
44. R. A. Falk, G. Stefani, R. Camilloni, G. H. Dunn, R. A. Phaneuf, D. C. Gregory, and D. H. Crandall, *Phys. Rev. A* **28**, 91 (1983).
45. S. M. Younger, *Phys. Rev. A* **24**, 1278 (1981).
46. D. Yu, Thesis, Université Catholique de Louvain, 1990; D. Yu, H. Zhang, M. Zambra, and P. Defrance, unpublished results (1990).
47. K. L. Aitken, M. F. A. Harrison, and R. D. Rundel, *J. Phys. B* **4**, 1189 (1971).
48. K. F. Man, A. C. H. Smith, and M. F. A. Harrison, *J. Phys. B* **20**, 5865 (1987).
49. A. Matsumuto, A. Danjo, S. Ohtani, H. Suzuki, H. Tawara, T. Takayanagi, K. Wakiya, I. Yamada, M. Yoshino, and T. Hirayama, *J. Phys. Soc. Jpn.* **59**, 902 (1990).
50. D. C. Gregory, P. F. Dittner, and D. H. Crandall, *Phys. Rev. A* **27**, 724 (1983).
51. S. M. Younger, private communication (1982), cited in Ref. 50.
52. E. J. McGuire, *Phys. Rev. A* **28**, 2091 (1983).
53. P. R. Woodruff, M. C. Hublet, and M. F. A. Harrison, *J. Phys. B* **9**, L305 (1978).
54. A. Müller A, K. Huber, K. Tinschert, R. Becker, and E. Salzborn, *J. Phys. B* **18**, 2993 (1985).

55. S. M. Younger, *Phys. Rev. A* **25**, 3396 (1982).
56. K. F. Man, A. C. H. Smith, and M. F. A. Harrison, *J. Phys. B* **26**, 1365 (1993).
57. A. Müller, E. Salzborn, R. Frodl, R. Becker, H. Klein, and H. Winter, *J. Phys. B* **13**, 1877 (1980).
58. D. W. Mueller, T. J. Morgan, G. H. Dunn, D. C. Gregory, and D. H. Crandall, *Phys. Rev A* **31**, 2905 (1985).
59. D. C. Gregory, D. H. Crandall, R. A. Phaneuf, A. M. Howald, G. H. Dunn, R. A. Falk, D. W. Mueller, and T. J. Morgan, Electron Impact Ionization of Multicharged Ions at ORNL: 1980–1984, Oak Ridge National Laboratory, Report ORNL/TM-9501, 1985.
60. A. M. Howald, D. C. Gregory, F. W. Meyer, R. A. Phaneuf, A. Müller, N. Djuric, and G. H. Dunn, *Phys. Rev. A* **33**, 3779 (1986).
61. S. Rachafi, D. S. Belic, M. Duponchelle, J. Jureta, M. Zambra, H. Zhang, and P. Defrane, *J. Phys. B* **24**, 1037 (1991).
62. Y. Zhang, C. B. Reddy, R. S. Smith, D. E. Golden, D. W. Mueller, and D. C. Gregory, *Phys. Rev. A* **45**, 2929 (1992).
63. S. S. Tayal, *Phys. Rev. A* **49**, 2561 (1994).
64. P. Defrance, S. Rachafi, J. Jureta, F. W. Meyer, and S. Chantrenne, *Nucl. Instrum. Methods B* **23**, 265 (1987).
65. Y. Zhang, C. B. Reddy, R. S. Smith, D. E. Golden, D. W. Mueller, and D. C. Gregory, *Phys. Rev. A* **44**, 4368 (1991).
66. S. M. Younger, *Phys. Rev. A* **23**, 1138 (1981).
67. S. Rachafi, J. Jureta, and P. Defrance, in *Proceedings of the 15th International Conference on Physics of Electronic and Atomic Collisions* (Brighton) (J. Geddes, H. B. Gilbody, A. E. Kingston, C. J. Latimer, and H. J. R. Walters, eds.), Book of Contributed Papers, p. 378, Queens University of Belfast (1987).
68. K. Laghdas, R. H. G. Reid, C. J. Joachain, and P. G. Burke, in *Proceedings of the 17th International Conference on Physics of Electronic and Atomic Collisions* (Aarhus) (T. Andersen, B. Fastrup, F. Folkmann, and H. Knudsen, eds.), Book of Contributed Papers, p. 368, Aarhus University (1993).
69. H. Zhang, Thesis, Université Catholique de Louvain, H. Zhang, M. Duponchelle, E. M. Oualim, and P. Defrance, unpublished results.
70. K. Tinschert, A. Muller, G. Hofmann, C. Achenbach, R. Becker, and E. Salzborn, *J. Phys. B* **20**, 1121 (1987).
71. D. C. Gregory, *Nucl. Instrum. Methods B* **10–11**, 87 (1985).
72. A. Danjo, A. Matsumuto, S. Ohtani, H. Suzuki, H. Tawara, K. Wakiya and M. Yoshino, *J. Phys. Soc. Jpn.* **53**, 4091 (1984).
73. M. E. Bannister, X. Q. Guo, and T. M. Kojima, *Phys. Rev. A* **49**, 4676 (1994).
74. T. W. Gorczyca, M. S. Pindzola, N. R. Badnell, and D. C. Griffin, *Phys. Rev. A* **49**, 4682 (1994).
75. M. E. Bannister, D. W. Mueller, L. J. Wang, M. S. Pindzola, D. C. Griffin, and D. C. Gregory, *Phys. Rev. A* **38**, 38 (1988).
76. D. C. Gregory, K. Rinn, and L. J. Wang, unpublished results (1986), listed in D. C. Gregory and M. E. Bannister, Electron Impact Ionization of Multicharged Ions at ORNL: 1985–1992, Oak Ridge National Laboratory, Report ORNL/TM-12729, 1994.
77. E. M. Oualim, Thesis, Université Catholique de Louvain, 1994; E. M. Oualim, M. Duponchelle, and P. Defrance, *Proceedings of the 7th International Conference on the Physics of Highly Charged Ions* (Vienna); (F. Aumayr, G. Betz, and H. P. Winter, eds.) *Nuclear Instruments and Methods B*, 1995 (in press).
78. C. Achenbach, A. Müller, E. Salzborn, and R. Becker, *J. Phys. B* **17**, 1405 (1984).
79. D. C. Griffin, C. Bottcher, M. S. Pindzola, S. M. Younger, D. C. Gregory, and D. H. Crandall, *Phys Rev. A* **29**, 1729 (1984).
80. D. C. Gregory and D. H. Crandall, *Phys. Rev. A* **27**, 2338 (1983).

81. M. J. Diserens, A. C. H. Smith, and M. F. A. Harrison, *J. Phys. B* **21**, 2129 (1988).
82. D. W. Mueller, T. J. Morgan, G. H. Dunn, D. C. Gregory, and D. H. Crandall, *Phys. Rev. A* **31**, 2905 (1985).
83. R. A. Falk, G. H. Dunn, D. C. Gregory, and D. H. Crandall, *Phys. Rev. A* **27**, 762 (1983).
84. S. J. Chantrenne, D. C. Gregory, M. J. Buie, and M. S. Pindzola, *Phys. Rev. A* **41**, 140 (1990).
85. D. C. Griffin, C. Bottcher, and M. S. Pindzola, *Phys. Rev. A* **25**, 1374 (1982).
86. P. G. Burke, W. C. Fon, and A. E. Kingston, *J. Phys. B* **17**, L733 (1984).
87. T. W. Gorczyca, M. S. Pindzola, D. C. Griffin, and N. R. Badnell, *J. Phys. B* **27**, 2399 (1994).
88. S. M. Younger, *Phys. Rev A* **25**, 3396 (1982).
89. S. M. Younger, *Phys. Rev. A* **26**, 3177 (1982).
90. K. F. Man, A. C. H. Smith, and M. F. A. Harrison, *J. Phys. B* **20**, 2571 (1987).
91. M. Sataka, S. Ohtani, D. Swenson, and D. C. Gregory, *Phys. Rev. A* **39**, 2397 (1989).
92. D. C. Gregory, L. J. Wang, F. Meyer, and K. Rinn, *Phys. Rev. A* **35**, 3256 (1987).
93. M. S. Pindzola, D. C. Griffin, C. Bottcher, S. M. Younger, and H. T. Hunter, *Nucl. Fusion, Special Supplement* **21** (1987).
94. R. G. Montague, M. J. Diserens, and M. F. A. Harrison, *J. Phys. B* **17**, 2085 (1984).
95. D. C. Gregory, F. W. Meyer, A. Müller, and P. Defrance, *Phys. Rev. A* **34**, 3657 (1986).
96. M. S. Pindzola, D. C. Griffln, C. Bottcher, M. J. Buie, and D. C. Gregory, Electron Impact Ionization Data for the Nickel Isonuclear Sequence, Oak Ridge National Laboratory, Report ORNL/ TM-11202, 1990.
97. R. G. Montague and M. F. A. Harrison, *J. Phys. B* **18**, 1419 (1985).
98. D. C. Gregory and A. M. Howald, *Phys. Rev. A* **34**, 97 (1986).
99. L. J. Wang, K. Rinn, and D. C. Gregory, *J. Phys. B* **21**, 2117 (1988).
100. M. S. Pindzola, D. C. Griffin, C. Bottcher, D. C. Gregory, A. M. Howald, R. A. Phaneuf, D. H. Crandall, G. H. Dunn, D. W. Mueller, and T. J. Morgan, Survey of Experimental and Theoretical Electron-Impact Ionization Cross Sections for Transition Metal Ions in Low Stages of Ionization, Oak Ridge National Laboratory, Report ORNL/TM-9436, 1985.
101. W. T. Rogers, G. Stefani, R. Camilloni, G. H. Dunn, A. Z. Msezane, and R. J. W. Henry, *Phys. Rev. A* **25**, 737 (1982).
102. M. S. Pindzola, D. C. Griffin, and C. Bottcher, *Phys. Rev. A* **25**, 211 (1982).
103. B. Peart and J. R. A. Underwood, *J. Phys. B* **23**, 2343 (1990).
104. K. F. Man, A. C. H. Smith, and M. F. A. Harrison, *J. Phys. B* **20**, 1351 (1987).
105. E. J. McGuire, *Phys. Rev. A* **16**, 73 (1977).
106. R. G. Montague and M. F. A. Harrison, *J. Phys. B* **17**, 2707 (1984).
107. M. S. Pindzola and D. C. Griffin, *Phys. Rev. A* **46**, 2486 (1992).
108. S. M. Younger, *Phys. Rev. A* **35**, 2841 (1987).

Chapter 8

The Dependence of Electron Impact Excitation and Ionization Cross Sections of H_2 and D_2 Molecules on Vibrational Quantum Number

M. Capitelli and R. Celiberto

1. INTRODUCTION

Modeling of H_2/D_2 plasmas requires the knowledge of a large number of cross sections involving electron–molecule interactions (electronic excitation, dissociation, and ionization). A lot of information—both experimental and theoretical—has been obtained to this end in these last years. On the experimental side, many cross sections have been obtained from investigations of electron–molecule interactions in crossed-beams experiments[1] as well as by deconvolution of transport properties by a Boltzmann analysis of swarm data.[2] On the theoretical side, many quantum-mechanical methods including resonant theories, close-coupling methods, distorted-wave approximations with and without exchange processes, and variational methods, have been applied.[3]

The information that has been obtained from these studies has been used by various authors to prepare consistent data bases for H_2 and D_2 to be used in plasma modeling. However, these data refer to the various processes involving the ground

---

M. CAPITELLI and R. CELIBERTO • Dipartimento di Chimica, Universitá di Bari and Centro di Studio per la Chimica dei Plasmi del Consiglio Nazionale delle Ricerche, Bari, Italy.

Atomic and Molecular Processes in Fusion Edge Plasmas, edited by R. K. Janev. Plenum Press, New York, 1995.

vibrational state of H_2 and D_2 molecules so that their use is limited to plasmas in which the presence of vibrationally excited molecules can be neglected.

Under many circumstances (multicusp magnetic plasmas, fusion edge plasmas), however, the vibrational excitation of H_2 and D_2 cannot be neglected, so that complete sets of cross sections involving the whole vibrational manifold of the diatom should be known. The data on these cross sections is very sparse compared with the corresponding data for the ground state.

From the experimental point of view, it is, in fact, very difficult to prepare H_2/D_2 molecules in vibrationally excited states so that information about cross sections involving vibrationally excited states essentially comes from theoretical considerations.

Of course, the calculation of complete sets (state-to-state) of cross sections is time-consuming so that semiclassical methods have been used in the past to obtain them. We refer in particular to the semiclassical Gryzinski[4] approximation as extended by Bauer and Bartky[5] to molecular systems. Use of this approximation has been and continues to be open to question because of *ad hoc* assumptions contained in the original treatment. In the case of H_2 and D_2 systems, this approximation seems to give satisfactory results, as can be appreciated by comparison with experimental and more suitable theoretical values. Validation of this theory was restricted for a long time to collision cross sections involving the ground vibrational level of the diatom. Only recently have more accurate quantum-mechanical approaches[6,7] been used for calculating complete sets of excitation cross sections for the following processes:

$$H_2(v_i) + e^- \rightarrow H_2(b^3\Sigma_u^+) + e^- \rightarrow H(n=1) + H(n=1) + e^- \quad \text{(a)}$$

$$H_2(v_i)/D_2(v_i) + e^- \rightarrow H_2(B^1\Sigma_u^+; C^1\Pi_u)/D_2(B^1\Sigma_u^+; C^1\Pi_u) + e^- \quad \text{(b)}$$

where $H_2(v_i)/D_2(v_i)$ represents the molecules in the *v*th vibrational level of ground electronic state $X^1\Sigma_g^+$. Comparison of these data with Gryzinski's results seems to validate these last cross sections, so we are now able to discuss the accuracy of the data base for H_2 and D_2 collected in the past few years and used by our group to study the influence of vibrational excited states on the bulk properties of H_2/D_2 plasmas under different experimental conditions. In particular, we will compare our cross sections, when possible, with complete sets of cross sections derived by Buckman and Phelps[8] from the Boltzmann analysis of swarm data for the interaction of electrons with $v_i = 0$ molecules and with the values reported by Rescigno and Schneider[6] and by Celiberto and Rescigno[7] for $v_i > 0$ states.

Of course, for $v_i = 0$, other cross section data[9] exist. Some of these cross sections are probably better than the data of Buckman and Phelps; however, their data can be considered as one of the best data bases in the literature, especially for plasma modeling applications. The data base, however, does not cover important processes

2. RESONANT VIBRATIONAL EXCITATION

Complete sets of data for state-to-state cross sections for vibrational excitation of H_2/D_2 by low-energy electrons, that is, for the process

$$H_2(v_i)/D_2(v_i) + e^- \rightarrow H_2^-/D_2^- \rightarrow H_2(v_f)/D_2(v_f) + e^- \qquad (c)$$

have been obtained by Atems and Wadehra[10] through application of the so-called resonant theory.[11] The corresponding cross section can be written as

$$\sigma^{res}_{v_i \rightarrow v_f} = \frac{k_f}{k_i} \int d\hat{k}_f |T^{res}_{i \rightarrow f}|^2 \qquad (1)$$

where k_i and k_f are the incident and the final electron moment, respectively, and $\hat{k}_f$ is d unit vector. $T^{res}_{i \rightarrow f}$ is the usual transition matrix, written as

$$T^{res}_{i \rightarrow f} = B \int dR \int dq \, \Psi_f^* V_{eT} \, \phi(q) \xi(R) \qquad (2)$$

where Ψ_f is the wave function describing the incident electron and the target in the final asymptotic state, V_{eT} is the interaction potential between the impinging electron and the molecule, ϕ and ξ are the electronic and the nuclear wave function, respectively, for the resonant molecular state, q is the totality of all electronic coordinates, and R is the internuclear distance. The constant B is given by

$$B = -\frac{m}{2\pi\hbar^2} \frac{1}{A(k_i)A(k_f)} \qquad (3)$$

where m is the electron mass, $\hbar$ is Planck's constant, and $A(k)$ is the amplitude of the free-electron wave function. Details of the theory can be found elsewhere.[11]

A sample of cross sections are presented in Figs. 1 and 2 as a function of energy for different v_i/v_f states. In particular, we show in Fig. 1 the cross sections for the 0/1, 0/2, 0/3, and 0/6 vibrational transitions for D_2. A dramatic decrease in the multiquantum energy exchange cross section compared with the one-quantum cross section is evident (compare 0/1 and 0/6 transitions). As well, Fig. 2 shows an increase in the one-quantum cross section with increasing initial vibrational quantum number, the 6/7 cross section being up to two orders of magnitude larger than the 0/1 cross section.

The accuracy of $0/v_f$ cross sections can be tested by comparing theoretical and experimental results. This has been done by different authors, and, in general, satisfactory agreement has been obtained. Here, we compare the cross sections obtained by Atems and Wadehra[10] with those derived by Buckman and Phelps[8] from a Boltzmann analysis of swarm data. Figure 1 shows this comparison for the

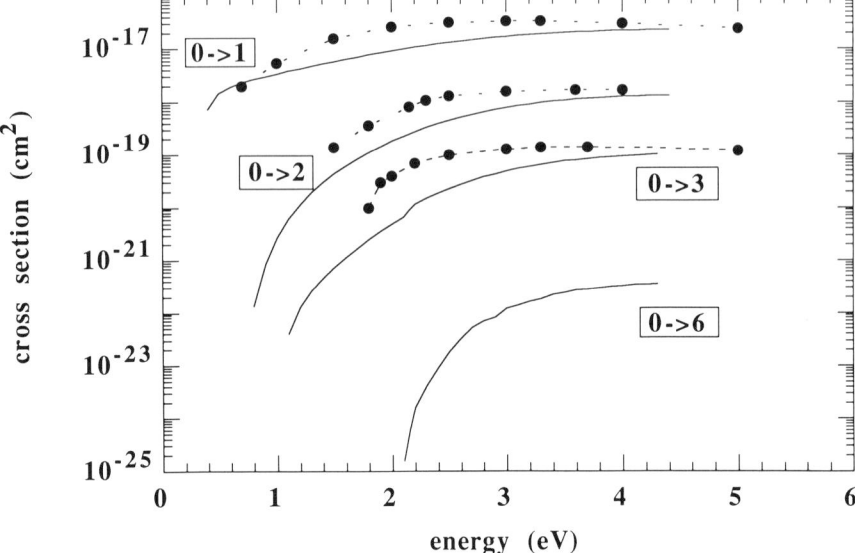

Figure 1. Cross sections for the process $D_2(X^1\Sigma_g^+, v_i=0) + e^- \to D_2^- \to D_2(X^1\Sigma_g^+, v_f) + e^-$ as a function of incident energy for different transitions. —) Ref. 10; ●) Ref. 8.

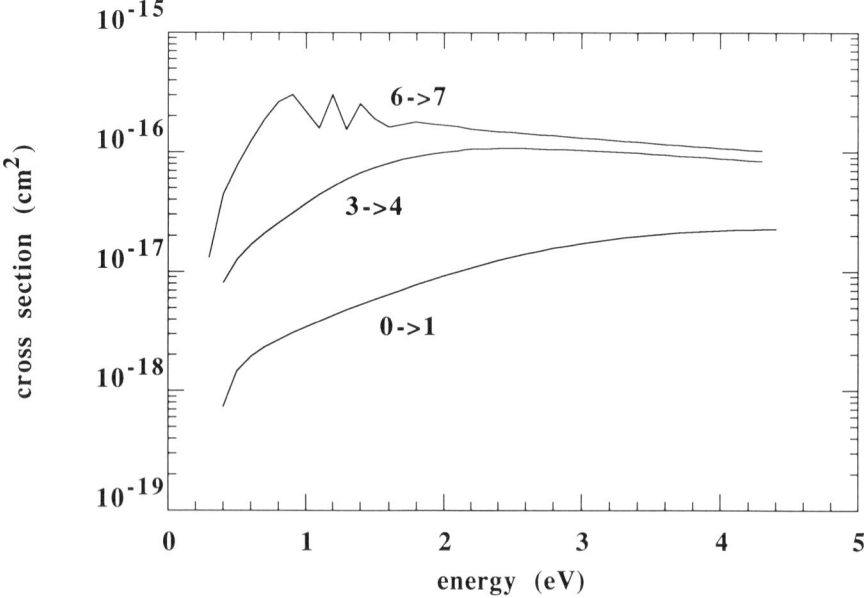

Figure 2. Cross sections for the process $D_2(X^1\Sigma_g^+, v_i) + e^- \to D_2^- \to D_2(X^1\Sigma_g^+, v_f) + e^-$ as a function of incident energy for different transitions. (From Ref. 10.)

0/1, 0/2, and 0/3 transitions. It can be seen that the agreement becomes worse and worse with increasing v_f. The same occurs for $v_i > 0$ as can be appreciated by comparing Atems and Wadehra's cross sections with other quantum-mechanical calculations.[12]

3. DISSOCIATIVE ATTACHMENT CROSS SECTION

Complete sets of cross sections for the dissociative attachment process, that is, for the process

$$H_2(v_i)/D_2(v_i) + e^- \rightarrow H_2^-/D_2^- \rightarrow H/D + H^-/D^- \qquad (d)$$

have been obtained by Atems and Wadehra,[11] again in the framework of resonant theory. These data, which also include the dependence of the cross section on the initial rotational quantum number, are very important for the codes describing negative-ion production.

The working equation can be written as

$$\sigma_{DA} = \frac{1}{A^2(k_i)} \frac{m}{k_i} \frac{1}{4\pi} \frac{K}{M} \lim_{R \to \infty} |\xi_f(R)|^2 \qquad (4)$$

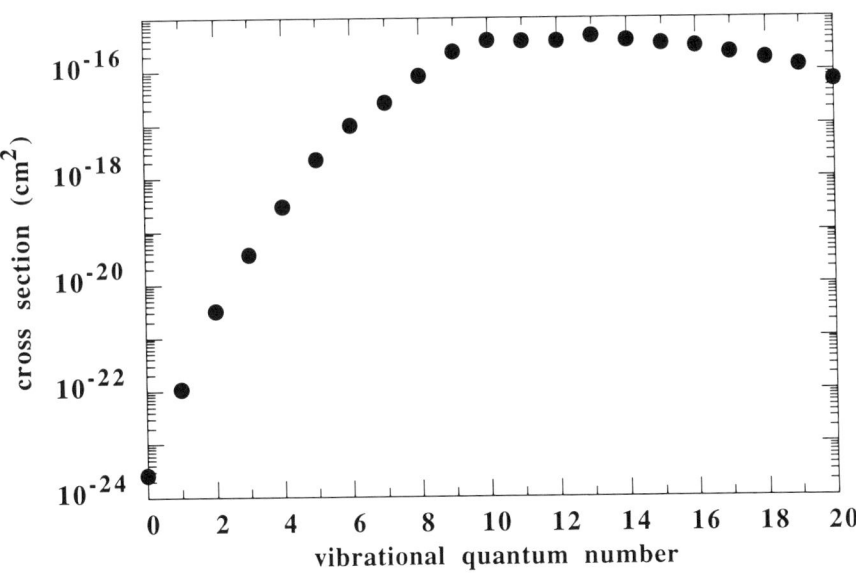

Figure 3. Cross sections (maximum values) for the process $D_2(X^1\Sigma_g^+, v_i) + e^- \rightarrow D_2^- \rightarrow D^- + D$ as a function of vibrational quantum number. (From Ref. 10.)

where $\hbar^2 K^2/2M = E_{tot} - V^-(\infty)$, E_{tot} is the total energy (incident electron energy + vibrational target energy), $V^-(R)$ is the potential energy for the resonant state, $\zeta_J(R)$ is the nuclear wave function for the resonant state, depending on the internuclear distance R and on the rotational quantum number J, and M is the nuclear reduced mass.

These cross sections are characterized by their dramatic dependence on the initial quantum number, as can be appreciated by inspection of Fig. 3. In this figure we have plotted the maximum value of the cross section as a function of vibrational quantum number for D_2. It can be seen that the cross section increases by orders of magnitude as v_i increases from 0 to 10. Note also that dissociative attachment from $v_i = 0$ is a very ineffective process: this means that the activation of the attachment process requires a preliminary vibrational excitation of D_2 molecules, this consideration being the basis of multicusp magnetic plasmas for the production of negative ions.

The accuracy of Atems and Wadehra's cross sections can be tested by comparing them with other calculated values. Good agreement with the cross sections obtained by Gauyacq[13] is observed, whereas some discrepancy is found in the comparison with the quantum-mechanical values of Mundel et al.[14]

4. ELECTRONIC EXCITATION

Our data base essentially consists of cross sections calculated by using the classical Gryzinski theory as well as the quantum-mechanical impact parameter method. Before presenting the results, we will briefly describe the basic equations characterizing the two methods.

The total cross section $\sigma_{v_i}^{\alpha_i \to \alpha_f}(E)$ for a state-to state electronic excitation can be expressed as

$$\sigma_{v_i}^{\alpha_i \to \alpha_f}(E) = \sum_{v_f} \sigma_{v_i,v_f}^{\alpha_i \to \alpha_f}(E) + \int d\varepsilon \frac{d\sigma_{v_i,\varepsilon}^{\alpha_i \to \alpha_f}(E)}{d\varepsilon} \quad (5)$$

where α_i and α_f indicate the initial and the final electronic state, respectively, v_i and v_f are the quantum numbers of the vibrational levels belonging to the α_i and the α_f electronic state, respectively, ε is the energy of the continuum vibrational levels of the α_f state, and E is the incident electron energy. The two terms on the right-hand side of Eq. (5) represent the contributions to the total cross section coming from, respectively, bound-to-bound and bound-to-continuum (dissociative) excitations.

Let us now discuss briefly the evaluation of these cross sections in the Gryzinski and impact parameter approximations. In the Gryzinski method, the bound-to-bound cross sections take the very simple form

$$\sigma_{v_i,v_f}^{\alpha_i \to \alpha_f}(E) = q_{v_i,v_f} N_e \sigma_G(E) \quad (6)$$

where q_{v_i,v_f} is the Franck–Condon factor, defined as

$$q_{v_i,v_f} = \left| \int_0^\infty dR \, \chi_{v_f}^{\alpha_f}(R) \chi_{v_i}^{\alpha_i}(R) \right|^2 \tag{7}$$

where χ_v^α is the wave function for the vth bound vibrational level belonging to the α electronic state, and R is the internuclear distance. In Eq. (6) $\sigma_G(E)$ is the cross section expressed in the Gryzinski approximation. For the explicit form of $\sigma_G(E)$, the reader is referred to Refs. 5 and 15. In this context, we just want to stress that this cross section is obtained in a completely classical scheme leading to two different expressions for optically allowed and forbidden transitions. N_e is the so-called "effective number of equivalent electrons," that is, the number of target electrons that can undergo excitation. Bauer and Bartky[5] give some empirical rules for determining its value. In our calculations, however, N_e has been selected in some cases by normalization of the calculated cross sections to the available $v_i = 0$ experimental data.[16]

The cross section for bound-to-continuum excitations is given by

$$\frac{d\sigma_{v_i,\varepsilon}^{\alpha_i \to \alpha_f}(E)}{d\varepsilon} = q_{v_i,\varepsilon} N_e \sigma_G(E) \tag{8}$$

where $q_{v_i,\varepsilon}$ is the Franck–Condon density, defined as

$$q_{v_i,\varepsilon} = \left| \int_0^\infty dR \, \chi_\varepsilon^{\alpha_f}(R) \chi_{v_i}^{\alpha_i}(R) \right|^2 \tag{9}$$

where $\chi_\varepsilon^{\alpha_f}$ is the wave function for the continuum vibrational levels of the α_f electronic state. Throughout our calculations with the Gryzinski method, the Franck–Condon density has been expressed in the δ-function approximation,[17] which assumes that the wave function $\chi_\varepsilon^{\alpha_f}$ behaves as a Dirac δ-function peaked at the classical turning point R_{ctp} of the molecular potential $V(R)$ of the α_f electronic state. The explicit expression of $q_{v_i,\varepsilon}$ is given by[17]

$$q_{v_i,\varepsilon} = \left(\frac{dV(R)}{dR}\right)_{R=R_{ctp}}^{-1} \left| \chi_{v_i}^{\alpha_i}(R_{ctp}) \right|^2 \tag{10}$$

In the impact parameter approximation,[7,18,19] the bound-to-bound cross sections are given by

$$\sigma_{v_i,v_f}^{\alpha_i \to \alpha_f}(E) = S_{v_i,v_f}^{\alpha_i \alpha_f} \cdot D_{v_i,v_f}^{\alpha_i \alpha_f}(E) \tag{11}$$

where the "structural factor" is written as

$$S_{v_i,v_f}^{\alpha_i\alpha_f} = \frac{m^2 e^2}{3g_i\hbar^4}(2-\delta_{\Lambda_f,0})(2-\delta_{\Lambda_i,0}) \cdot \left|\int_0^\infty dR\, \chi_{v_f}^{\alpha_f}(R)\, M_{\Lambda_f,\Lambda_i}(R)\, \chi_{v_i}^{\alpha_i}(R)\right|^2 \quad (12)$$

with m and e the electron mass and charge, respectively, and g_i a degeneracy factor. M_{Λ_i,Λ_f} is the usual electronic dipole transition moment, depending on the initial and final quantum numbers of the projection of the electronic angular momentum on the internuclear axis. The structural factor is analogous to the Franck–Condon factor appearing in the Gryzinski method. The "dynamical factor" $D_{v_i,v_f}^{\alpha_i\alpha_f}(E)$ describes in a semiclassical fashion the behavior of the incident electron. It depends on the transition energy, defined as

$$\Delta E_{v_i,v_f}^{\alpha_i\alpha_f} = E_{v_f}^{\alpha_f} - E_{v_i}^{\alpha_i} \quad (13)$$

where E_v^α is the vth vibrational eigenvalue of the α electronic state. The explicit expression of $D_{v_i,v_f}^{\alpha_i\alpha_f}(E)$ is quite complicated, so that once again the reader is referred to the literature[7,18,19] for more details.

The impact parameter method as formulated by Hazi[18] and Redmon et al.[19] takes into account only the electric dipole transitions and neglects the electron exchange, so that the method can be applied only to optically allowed excitations.

The dissociative cross sections is given by

$$\frac{d\sigma_{v_i,\varepsilon}^{\alpha_i\to\alpha_f}(E)}{d\varepsilon} = S_{v_i,\varepsilon}^{\alpha_i\alpha_f} D_{v_i,\varepsilon}^{\alpha_i\alpha_f}(E) \quad (14)$$

where the structural and dynamical factors can be obtained from the corresponding definitions for the bound-to-bound case by formally substituting v_f with ε, the energy in the continuum.

Allowed Transitions

Figures 4 and 5 show the cross sections calculated by various authors for electronic excitation to the first singlet states of D_2:

$$D_2(v_i = 0) + e^- \to D_2(B^1\Sigma_u^+;\, C^1\Pi_u) + e^- \quad (e)$$

The cross sections compared in these figures are those calculated by the Gryzinski method,[20] by Buckman and Phelps[8] (B-P), and by Celiberto and Rescigno[7] (C-R) according to the impact parameter method. We note a good agreement between the B-P cross sections and the C-R ones, both of them exceeding those obtained from the Gryzinski calculations by approximately a factor of 2. However, both the B-P and C-R cross sections are a factor of 2 larger than the experimental ones (see, e.g., Ref. 7). This last consideration was used in Refs. 16 and 20 to reduce the Gryzinski cross section by setting equal to 1 the number of equivalent electrons (N_e) entering into the theory.

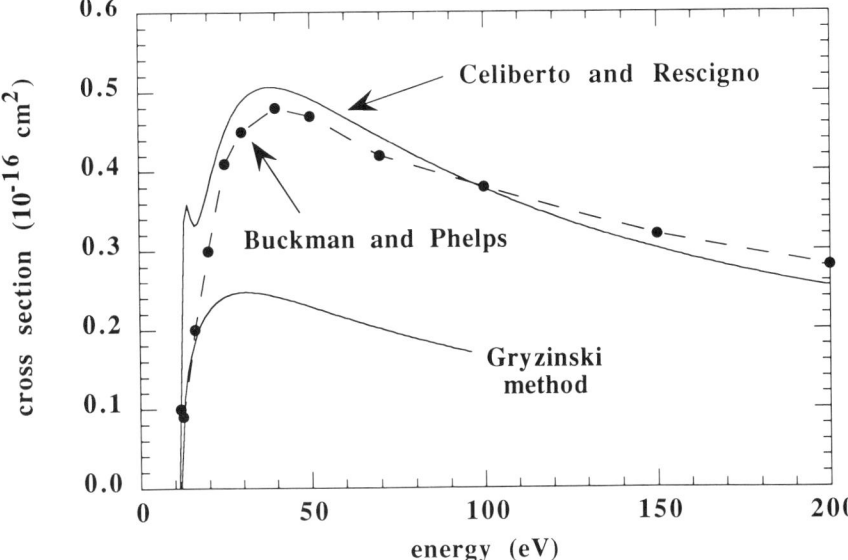

Figure 4. Cross sections for the process $D_2(X^1\Sigma_g^+, v_i = 0) + e^- \to D_2(B^1\Sigma_u^+) + e^-$ as a function of energy calculated by different authors.

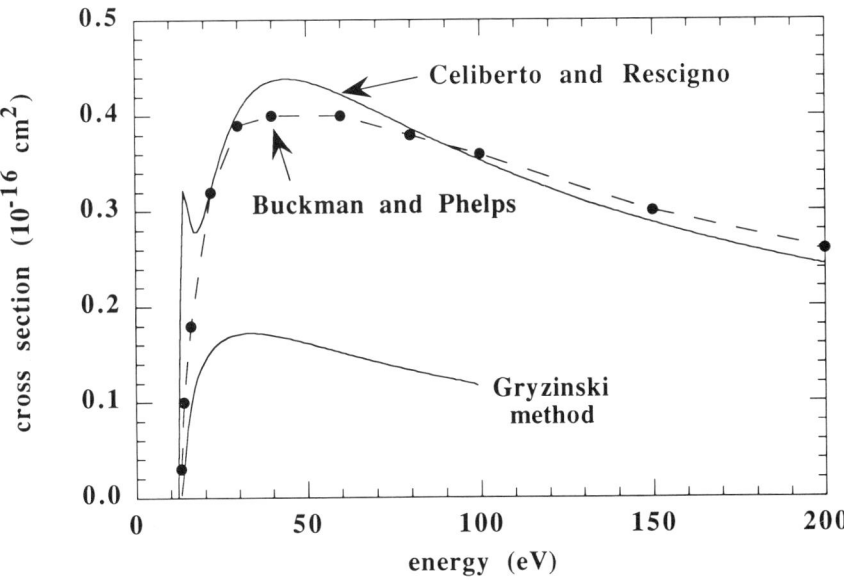

Figure 5. Cross sections for the process $D_2(X^1\Sigma_g^+, v_i = 0) + e^- \to D_2(C^1\Pi_u) + e^-$ as a function of energy calculated by different authors.

In Figs. 6 and 7, the excitation cross sections at fixed energy ($E = 35$ eV and $E = 40$ eV, respectively) for the same process calculated by the impact parameter method as a function of v_i are compared with the corresponding results obtained by using the modified Gryzinski method. In both cases, the impact parameter cross sections exceed by approximately a factor of 2 (the same factor as observed in Figs. 4 and 5) the corresponding Gryzinski cross sections. This is true for excitation to all vibrational levels of the $C^1\Pi_u$ state and up to $v = 10$ for the excitation of the $B^1\Sigma_u^+$ state. This is an indirect confirmation of the satisfactory accuracy of the complete sets of electron–molecule cross sections used in our code and calculated with the modified Gryzinski approximation, especially taking into account the fact that the present quantum-mechanical calculations for $v_i = 0$ are a factor of 2 larger than the experimental ones.

The same factor is present in the so-called E–V cross sections,[20,21] that is, the cross sections for the process

$$D_2(v_i = 0) + e^- \rightarrow D_2(B^1\Sigma_u^+; C^1\Pi_u) + e^- \rightarrow D_2(v_f) + e^- + h\nu \tag{f}$$

The cross section can be calculated according to the following equation:

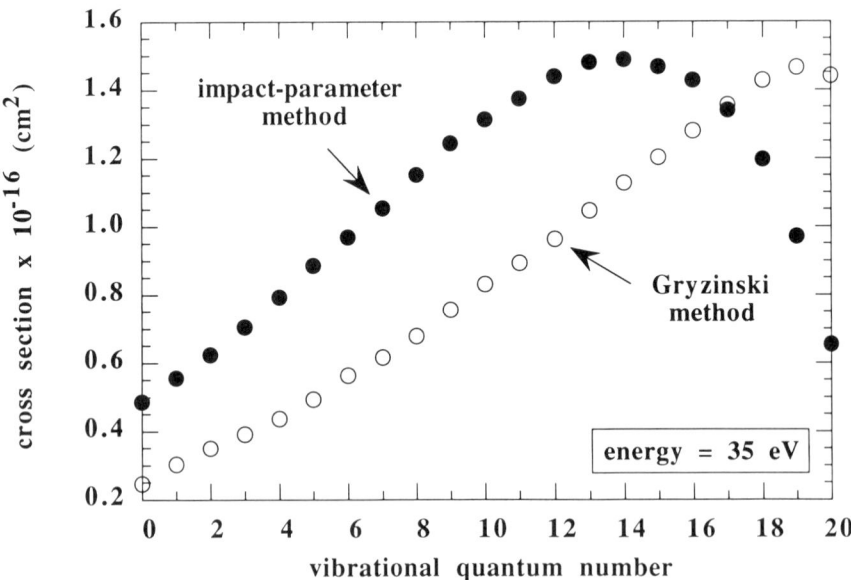

Figure 6. Cross sections for the process $D_2(X^1\Sigma_g^+, v_i) + e^- \rightarrow D_2(B^1\Sigma_u^+) + e^-$ at fixed incident energy as a function of vibrational quantum number, calculated according to the impact parameter and Gryzinski methods.

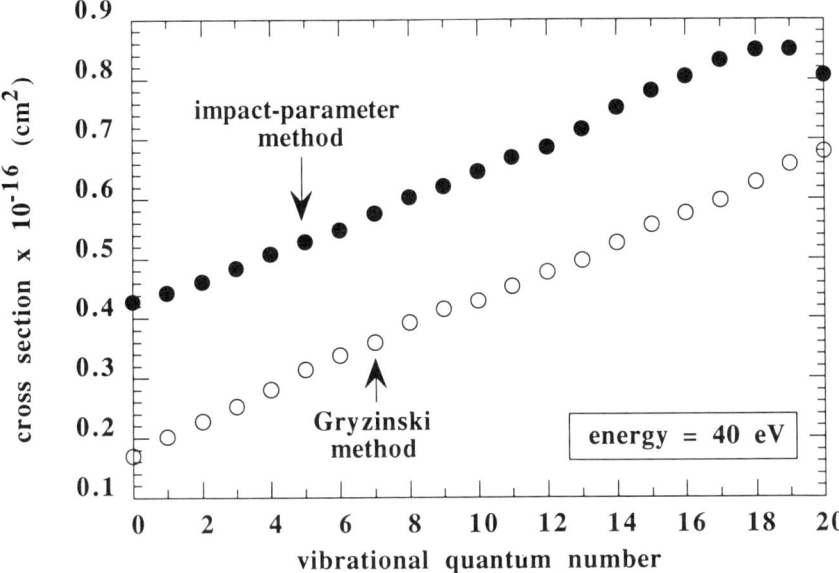

Figure 7. Cross sections for the process $D_2(X^1\Sigma_g^+, v_i) + e^- \rightarrow D_2(C^1\Pi_u) + e^-$ at fixed incident energy as a function of vibrational quantum number, calculated according to the impact parameter and Gryzinski methods.

$$\sigma_{v_i,v_f}^{\alpha_i \rightarrow \alpha_f \rightarrow \alpha_i}(E) = \sum_{v'} \sigma_{v_i,v'}^{\alpha_i \rightarrow \alpha_f}(E) \frac{A_{\alpha_i \rightarrow \alpha_f}(v',v_f)}{E_{v'}} \quad (15)$$
$$\sum_{v_f} A_{\alpha_i \rightarrow \alpha_f}(v',v_f) + \int_{\varepsilon_{th}} d\varepsilon_f A_{\alpha_i \rightarrow \alpha_f}(v',\varepsilon_f)$$

where v_i and v_f represent the initial and the final vibrational quantum number of the α_i electronic state of the molecule, v' is the vibrational quantum number of the excited α_f state, $A_{\alpha_i \rightarrow \alpha_f}$ are the Einstein probabilities linking v' levels with the bound(v_f)/continuum(ε_f) vibrational levels of the α_i state, $E_{v'}$ is the eigenvalue of the v'-th vibrational level, ε_{th} is the dissociative threshold energy of the ground electronic state, and $\sigma_{v_i,v'}^{\alpha_i \rightarrow \alpha_f}(E)$ is the excitation cross section [see eq. (11)].

In Fig. 8 the cross sections obtained by the Gryzinski method are compared with the corresponding ones obtained by the impact parameter method[22] for different $v_i = 0 / v_f$ transitions. Once again, the factor of 2 in the excitation cross sections propagates in the E–V cross sections.

Figure 9 shows the cross sections for process (f) as a function of v_f at fixed energy for both the H_2 and the D_2 system.[22] Inspection of this figure shows an isotopic effect. Note also that vibrational excitation through this mechanism should

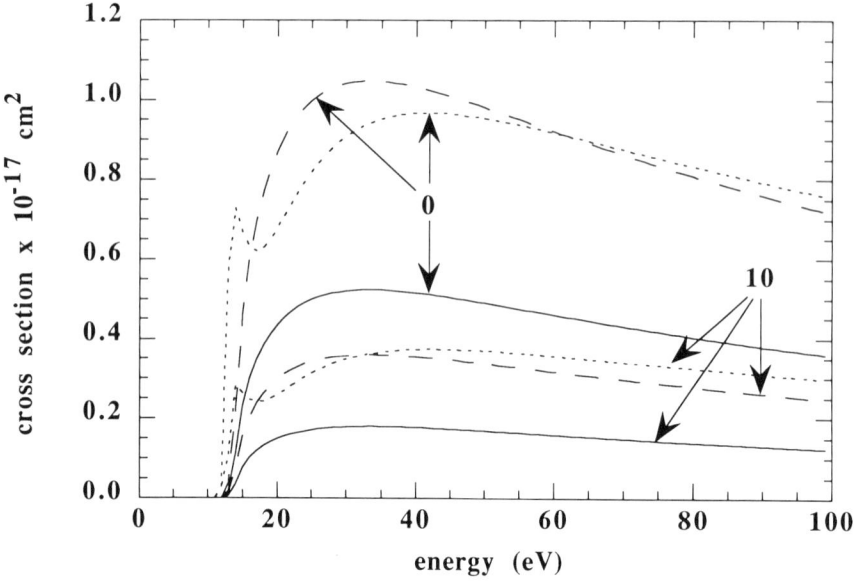

Figure 8. Cross sections for the process $e^- + D_2(X^1\Sigma_g^+, v_i = 0) \to e^- + D_2(B^1\Sigma_u^+; C^1\Pi_u) \to e^- + D_2(X^1\Sigma_g^+, v_f) + hv$ for different v_f values, calculated according to different methods. —) Ref. 20; ···) Ref. 22; ---) results from Ref. 20 multiplied by 2.)

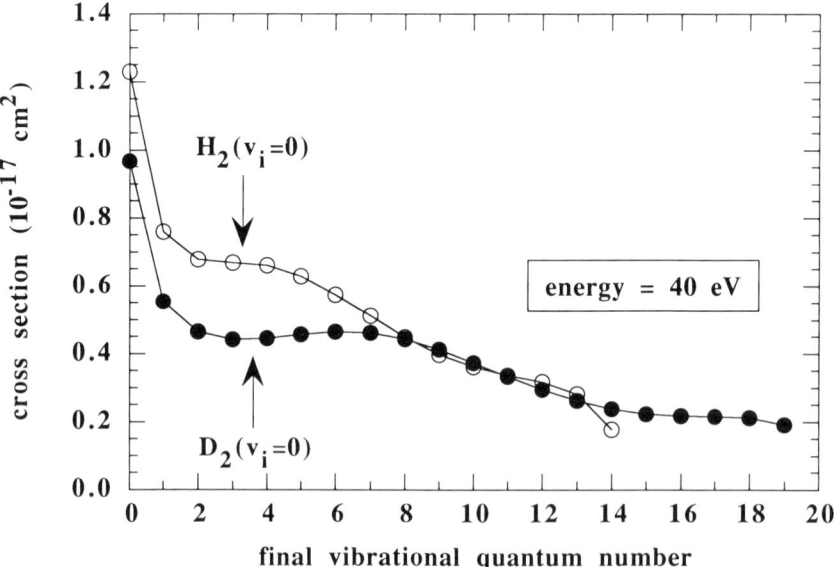

Figure 9. Cross sections for the process $D_2(X^1\Sigma_g^+, v_i = 0) + e^- \to D_2(B^1\Sigma_u^+; C^1\Pi_u) + e^- \to D_2(X^1\Sigma_g^+, v_f) + e^- + hv$ and for the corresponding process for H_2 at fixed energy as a function of final vibrational quantum number. (From Ref. 22.)

Electron Impact Excitation and Ionization Cross Sections

Table I. E–V Cross Sections as a Function of Incident Energy for the Process:

$$D_2(X^1\Sigma_g^+, v_i=0) + e^- \rightarrow D_2(B^1\Sigma_u^+, C^1\Pi_u) + e^- \rightarrow D_2(X^1\Sigma_g^+, v_f=0\text{–}20) + e^- + h\nu$$

Cross section (10^{-18} cm^2) for v_f equal to:

E (eV)	0	1	2	3	4	5	6	7	8	9	10	11	12	13	14	15	16	17	18	19	20
12	0.3	0.7	0.4	0.2	0.4	0.2	0.3	0.3	0.2	0.3	0.3	0.3	0.4	0.4	0.5	0.5	0.8	0.2	0.0	0.0	0.0
13	5.8	3.5	2.9	2.9	2.7	3.2	2.9	3.1	2.8	1.9	1.1	1.0	1.0	1.0	1.1	1.2	1.2	1.4	1.5	1.4	1.0
14	7.3	4.0	3.4	3.3	3.3	3.4	3.5	3.5	3.4	3.1	2.9	2.6	2.2	2.0	1.8	1.6	1.3	1.3	1.4	1.3	0.9
15	6.7	3.8	3.2	3.0	3.1	3.1	3.2	3.2	3.1	2.9	2.6	2.4	2.1	1.9	1.7	1.6	1.5	1.5	1.4	1.3	0.9
16	6.3	3.6	3.0	2.9	2.9	3.0	3.0	3.0	2.9	2.7	2.5	2.2	2.0	1.8	1.6	1.5	1.5	1.4	1.4	1.3	0.9
17	6.2	3.6	3.0	2.9	2.9	2.9	3.0	3.0	2.8	2.6	2.4	2.2	1.9	1.7	1.6	1.5	1.5	1.4	1.4	1.3	0.9
18	6.3	3.7	3.0	2.9	2.9	3.0	3.0	3.0	2.9	2.7	2.4	2.2	1.9	1.7	1.6	1.5	1.5	1.5	1.4	1.3	0.9
19	6.5	3.8	3.2	3.0	3.0	3.1	3.1	3.1	3.0	2.8	2.5	2.2	2.0	1.8	1.7	1.6	1.5	1.5	1.5	1.4	0.9
20	6.8	4.0	3.3	3.1	3.2	3.2	3.3	3.2	3.1	2.9	2.6	2.3	2.1	1.9	1.7	1.6	1.6	1.6	1.6	1.4	1.0
25	8.2	4.8	4.0	3.8	3.8	3.9	4.0	3.9	3.7	3.5	3.1	2.8	2.5	2.2	2.0	1.9	1.9	1.9	1.9	1.7	1.2
30	9.1	5.2	4.4	4.2	4.2	4.3	4.4	4.3	4.1	3.8	3.5	3.1	2.8	2.5	2.3	2.1	2.1	2.1	2.0	1.9	1.3
35	9.5	5.5	4.6	4.4	4.4	4.5	4.6	4.5	4.4	4.0	3.7	3.3	2.9	2.6	2.4	2.2	2.2	2.1	2.1	1.9	1.3
40	9.7	5.5	4.6	4.4	4.5	4.6	4.7	4.6	4.4	4.1	3.7	3.3	2.9	2.6	2.4	2.2	2.1	2.1	2.1	1.9	1.3
50	9.5	5.5	4.6	4.4	4.4	4.5	4.6	4.6	4.4	4.1	3.7	3.3	2.9	2.6	2.4	2.2	2.2	2.2	2.1	1.9	1.3
60	9.2	5.2	4.4	4.2	4.2	4.3	4.4	4.4	4.2	3.9	3.6	3.2	2.8	2.5	2.3	2.1	2.0	2.0	2.0	1.8	1.2
80	8.3	4.8	4.0	3.8	3.8	3.9	4.0	4.0	3.8	3.6	3.3	2.9	2.6	2.3	2.1	1.9	1.8	1.8	1.8	1.6	1.1
100	7.6	4.3	3.6	3.5	3.5	3.6	3.7	3.6	3.5	3.3	3.0	2.6	2.3	2.1	1.9	1.7	1.7	1.6	1.6	1.5	1.0

Table II. E–V Cross Sections as a Function of Incident Energy for the Process:

$$D_2(X^1\Sigma_g^+, v_i = 5) + e^- \rightarrow D_2(B^1\Sigma_u^+; C^1\Pi_u) + e^- \rightarrow D_2(X^1\Sigma_g^+, v_f = 0\text{--}20) + e^- + h\nu$$

Cross section (10^{-18} cm^2) for v_f equal to:

E (eV)	0	1	2	3	4	5	6	7	8	9	10	11	12	13	14	15	16	17	18	19	20
11	3.7	5.9	4.2	4.3	7.6	9.3	6.8	4.9	4.0	3.1	2.2	1.5	1.3	1.7	1.9	1.5	0.7	0.5	1.2	1.0	0.4
12	6.0	6.1	4.7	5.2	7.4	9.9	7.0	6.3	5.7	4.3	2.8	2.3	2.4	2.5	2.1	1.5	0.7	0.6	1.2	0.9	0.5
13	5.6	6.0	4.8	5.1	7.2	9.7	6.9	6.2	5.7	4.3	2.9	2.3	2.4	2.5	2.3	1.6	1.0	1.1	1.8	1.5	0.8
14	5.3	5.8	4.7	5.1	7.2	9.7	7.1	6.3	5.7	4.3	2.9	2.3	2.3	2.5	2.3	1.6	0.9	1.1	1.7	1.4	0.7
15	5.2	5.8	4.8	5.2	7.4	10.	7.4	6.5	5.9	4.4	3.0	2.3	2.3	2.5	2.3	1.7	0.9	1.0	1.7	1.4	0.7
16	5.4	6.0	5.0	5.5	7.8	10.	7.7	6.8	6.2	4.6	3.1	2.4	2.4	2.6	2.4	1.8	1.0	1.0	1.8	1.4	0.7
17	5.6	6.3	5.3	5.7	8.2	11.	8.1	7.2	6.5	4.9	3.2	2.5	2.5	2.7	2.6	1.8	1.0	1.1	1.8	1.4	0.7
18	5.9	6.6	5.5	6.0	8.6	11.	8.5	7.5	6.8	5.1	3.4	2.6	2.6	2.8	2.7	1.9	1.1	1.1	1.9	1.5	0.7
20	6.5	7.2	6.0	6.5	9.3	12.	9.1	8.0	7.3	5.5	3.6	2.8	2.8	3.0	2.9	2.1	1.1	1.2	2.0	1.6	0.8
25	7.5	8.3	6.7	7.2	10.	14.	10.	8.8	8.0	6.0	4.0	3.1	3.2	3.4	3.2	2.3	1.3	1.4	2.3	1.8	0.9
30	7.9	8.7	7.0	7.4	11.	14.	10.	9.1	8.2	6.2	4.1	3.3	3.4	3.6	3.3	2.4	1.4	1.5	2.5	1.9	1.0
35	8.1	8.8	7.0	7.4	11.	14.	10.	9.0	8.2	6.2	4.1	3.3	3.4	3.6	3.3	2.4	1.4	1.5	2.5	2.0	1.0
40	8.0	8.7	6.9	7.3	10.	14.	9.9	8.8	8.1	6.1	4.0	3.3	3.4	3.5	3.2	2.3	1.4	1.5	2.5	2.0	1.0
50	7.7	8.4	6.6	6.9	9.8	13.	9.3	8.3	7.6	5.7	3.8	3.1	3.2	3.4	3.1	2.2	1.3	1.5	2.4	1.9	1.0
60	7.3	7.9	6.2	6.5	9.2	12.	8.7	7.8	7.1	5.3	3.6	2.9	3.1	3.2	2.9	2.0	1.2	1.4	2.3	1.8	0.9
80	6.6	7.0	5.5	5.7	8.1	11.	7.7	6.8	6.3	4.7	3.1	2.6	2.7	2.8	2.5	1.8	1.1	1.3	2.0	1.6	0.8
100	5.9	6.3	4.9	5.1	7.2	9.7	6.8	6.1	5.6	4.2	2.8	2.3	2.4	2.5	2.3	1.6	1.0	1.1	1.8	1.5	0.8

Electron Impact Excitation and Ionization Cross Sections

Table III. E–V Cross Sections as a Function of Incident Energy for the Process:

$$D_2(X^1\Sigma_g^+, v_i = 10) + e^- \rightarrow D_2(B^1\Sigma_u^+; C^1\Pi_u) + e^- \rightarrow D_2(X^1\Sigma_g^+, v_f = 0\text{--}20) + e^- + h\nu$$

E (eV)	\multicolumn{21}{c}{Cross section (10^{-18} cm^2) for v_f equal to:}
	0	1	2	3	4	5	6	7	8	9	10	11	12	13	14	15	16	17	18	19	20
10	3.7	4.1	5.9	6.6	5.9	4.4	3.2	4.4	6.0	5.7	6.6	4.9	4.5	2.9	2.3	2.9	1.7	1.5	2.1	0.6	0.7
11	8.2	5.3	8.1	7.4	8.1	5.2	5.5	5.7	8.0	9.2	12.	9.1	6.3	4.3	4.0	3.3	1.9	1.5	1.9	0.6	0.6
12	7.6	5.5	8.3	7.7	8.2	5.3	5.5	5.8	8.0	9.1	12.	8.9	6.3	4.4	3.9	3.5	2.1	2.4	2.8	0.8	0.7
13	7.4	5.7	8.5	8.1	8.5	5.5	5.5	6.0	8.2	9.3	12.	9.0	6.4	4.4	3.9	3.6	2.1	2.3	2.7	0.8	0.7
14	7.5	5.9	8.9	8.6	8.9	5.8	5.7	6.3	8.5	9.7	12.	9.3	6.7	4.6	4.0	3.7	2.2	2.4	2.8	0.8	0.8
15	7.9	6.2	9.3	9.1	9.4	6.1	6.0	6.6	9.0	10.	13.	9.7	7.1	4.8	4.2	3.9	2.3	2.4	2.9	0.9	0.8
16	8.3	6.5	9.8	9.5	9.8	6.4	6.3	6.9	9.4	11.	14.	10.	7.4	5.0	4.4	4.1	2.4	2.5	3.1	0.9	0.8
17	8.7	6.8	10.	9.8	10.	6.7	6.5	7.2	9.8	11.	14.	11.	7.7	5.2	4.5	4.3	2.5	2.7	3.2	0.9	0.9
18	9.1	7.0	11.	10.	11.	6.9	6.8	7.4	10.	11.	15.	11.	8.0	5.4	4.7	4.4	2.6	2.8	3.3	1.0	0.9
20	9.8	7.4	11.	11.	11.	7.2	7.2	7.8	11.	12.	15.	12.	8.4	5.7	5.0	4.7	2.8	3.0	3.5	1.0	1.0
25	11.	7.8	12.	11.	12.	7.5	7.7	8.2	11.	13.	17.	12.	8.9	6.1	5.4	5.0	2.9	3.2	3.8	1.1	1.1
30	11.	7.9	12.	11.	12.	7.5	7.8	8.2	11.	13.	17.	13.	8.9	6.1	5.5	5.0	3.0	3.3	3.9	1.2	1.1
40	11.	7.5	11.	10.	11.	7.1	7.4	7.8	11.	12.	16.	12.	8.5	5.9	5.4	4.7	2.8	3.2	3.8	1.1	1.0
50	10.	6.9	10.	9.3	10.	6.6	7.0	7.2	9.9	11.	15.	11.	7.9	5.5	5.0	4.4	2.6	3.1	3.6	1.1	1.0
60	9.7	6.4	9.6	8.6	9.4	6.1	6.5	6.7	9.2	11.	14.	10.	7.3	5.1	4.7	4.1	2.4	2.9	3.3	1.0	0.9
80	8.5	5.6	8.3	7.4	8.1	5.2	5.7	5.8	8.0	9.3	12.	9.1	6.4	4.5	4.1	3.5	2.1	2.5	2.9	0.9	0.8
100	7.6	4.9	7.4	6.5	7.2	4.6	5.0	5.1	7.1	8.2	11.	8.1	5.6	3.9	3.7	3.1	1.9	2.2	2.6	0.8	0.7

Table IV. E–V Cross Sections as a Function of Incident Energy for the Process:

$$D_2(X^1\Sigma_g^+, v_i = 15) + e^- \rightarrow D_2(B^1\Sigma_u^+; C^1\Pi_u) + e^- \rightarrow D_2(X^1\Sigma_g^+, v_f = 0\text{–}20) + e^- + h\nu$$

Cross section (10^{-18} cm^2) for v_f equal to:

E (eV)	0	1	2	3	4	5	6	7	8	9	10	11	12	13	14	15	16	17	18	19	20
9	6.1	8.5	5.3	2.1	4.6	4.3	2.2	4.3	3.6	2.7	4.4	3.2	3.9	5.0	5.6	11.	5.7	3.9	1.7	2.3	1.0
10	7.8	12.	5.9	3.1	6.7	4.3	4.4	4.8	5.0	4.3	5.2	5.6	5.0	8.3	11.	25.	10.	3.9	1.6	2.1	0.9
11	7.7	12.	6.4	3.1	6.7	4.7	4.2	5.1	5.0	4.3	5.5	5.5	5.2	8.2	10.	24.	10.	6.3	2.6	2.6	1.6
12	7.9	13.	6.7	3.1	6.8	4.9	4.1	5.3	5.1	4.3	5.7	5.5	5.4	8.2	10.	24.	10.	6.2	2.6	2.6	1.5
13	8.2	13.	7.0	3.2	7.1	5.3	4.2	5.6	5.3	4.4	6.0	5.6	5.6	8.4	11.	24.	10.	6.4	2.6	2.8	1.5
14	8.7	14.	7.4	3.3	7.4	5.6	4.4	5.9	5.6	4.6	6.3	5.9	5.9	8.8	11.	25.	11.	6.6	2.7	2.9	1.6
15	9.1	15.	7.8	3.5	7.8	5.8	4.6	6.2	5.9	4.9	6.6	6.1	6.2	9.2	11.	26.	11.	6.9	2.8	3.1	1.6
16	9.5	15.	8.1	3.7	8.1	6.1	4.8	6.5	6.1	5.1	6.9	6.4	6.5	9.6	12.	27.	12.	7.2	3.0	3.2	1.7
17	9.9	16.	8.3	3.8	8.4	6.3	5.0	6.7	6.4	5.3	7.1	6.7	6.7	10.	12.	29.	12.	7.5	3.1	3.3	1.8
18	10.	16.	8.5	4.0	8.6	6.4	5.2	6.8	6.5	5.4	7.3	6.9	6.9	10.	13.	30.	13.	7.8	3.2	3.4	1.8
20	11.	17.	8.8	4.1	9.0	6.6	5.5	7.1	6.8	5.7	7.6	7.2	7.1	11.	14.	31.	13.	8.2	3.4	3.6	2.0
30	11.	17.	8.9	4.4	9.3	6.6	5.9	7.1	7.0	6.0	7.7	7.6	7.3	11.	14.	33.	14.	8.7	3.6	3.7	2.2
40	10.	16.	8.3	4.1	8.8	6.1	5.6	6.6	6.6	5.7	7.2	7.2	6.8	11.	14.	32.	13.	8.3	3.5	3.5	2.1
50	9.6	15.	7.6	3.8	8.1	5.6	5.2	6.1	6.1	5.3	6.6	6.7	6.3	10.	13.	30.	13.	7.7	3.2	3.3	2.0
60	8.9	14.	7.0	3.6	7.5	5.1	4.9	5.6	5.6	4.9	6.1	6.2	5.8	9.3	12.	28.	12.	7.2	3.0	3.0	1.8
80	7.7	12.	6.0	3.1	6.5	4.4	4.3	4.8	4.9	4.2	5.2	5.4	5.0	8.0	10.	25.	10.	6.3	2.6	2.6	1.6
100	6.8	11.	5.3	2.7	5.8	3.8	3.8	4.3	4.3	3.7	4.6	4.8	4.4	7.1	8.9	22.	9.0	5.6	2.3	2.3	1.4

produce a plateau in the vibrational distribution of H_2/D_2 owing to the form of the E–V cross sections, this plateau being at the basis of the formation of negative ions from vibrationally excited molecules through the dissociative attachment process.

It should be also noted that the recent calculations performed by Celiberto et al.[22] for H_2 are in good agreement with Hiskes's phenomenological results[21] for different v_i / v_f transitions. E–V cross sections for D_2 involving $v_i = 0, 5, 10$, and 15 are presented in Tables I–IV, respectively.

5. DISSOCIATION PROCESSES

Electron impact dissociation processes are important channels for the production of atomic H/D in a discharge. We can distinguish essentially two mechanisms whereby dissociation is induced by electrons. In the first one, dissociation is induced through direct electronic excitation of the repulsive part of a bound electronically excited state, and in the second one, through excitation of completely repulsive electronically excited states. For the first case, the most important processes involve the excitation of optically allowed states, whereas the second one involves the excitation of (spin) forbidden electronically excited states. Let us examine the two cases separately.

5.1. Allowed Transitions

The impact parameter method has been recently used to calculate the dissociative excitation cross section for the process

$$D_2(v_i) + e^- \rightarrow D_2(B^1\Sigma_u^+; C^1\Pi_u) + e^- \rightarrow D(n = 1) + D(n = 2) + e^- \qquad (g)$$

Complete sets of cross sections for both H_2 and D_2 are now available.[23]

In Figs. 10 and 11 the cross sections for the above process are plotted as a function of v_i at fixed energy ($E = 40$ eV). The nonmonotonic trend of both cross sections as a function of v_i is a result of the intricate form of the Franck–Condon density connecting the vibrational states of the ground electronic state with the vibrational continuum states of the electronically excited state. Apart from the isotopic effect, we can also note that these cross sections are strongly activated by the vibrational energy. In fact, in the plateau region of the plots, corresponding to $4 < v_i < 10$, the cross sections are approximately a factor of 10 higher than the corresponding cross sections near $v_i = 0$. Note, however, that the $v_i = 0$ cross section is very small, especially in comparison to the dissociation cross section involving triplet states (see below). Thus, this conclusion should be taken with caution since the dissociation cross section over the repulsive part of bound states strongly depends on the number of electronically excited states considered. As an example, Fig. 12 shows the total dissociation cross sections yielding atomic D in $n = 2$ and $n = 3$, that is, for the processes

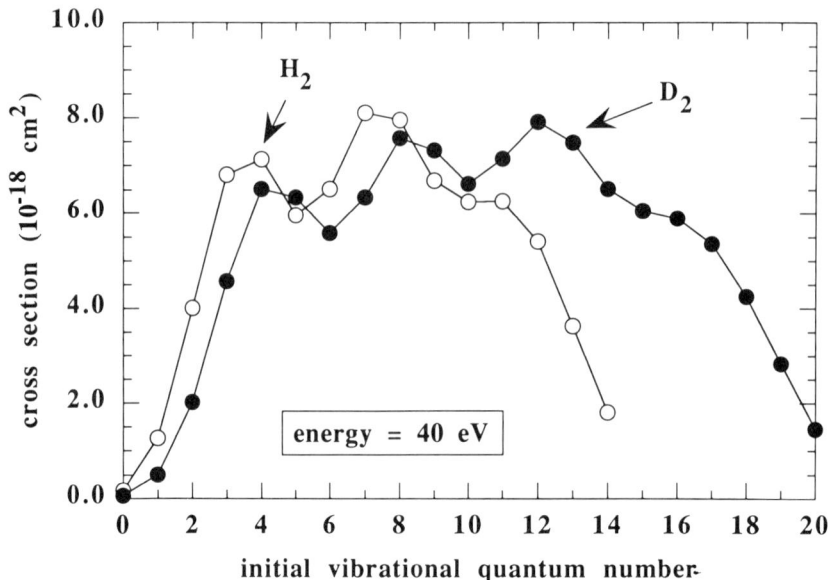

Figure 10. Cross sections for the process $D_2(X^1\Sigma_g^+, v_i) + e^- \to D_2(B^1\Sigma_u^+) + e^- \to D(n=1) + D(n=2) + e^-$ and for the corresponding process for H_2 at fixed energy as a function of initial vibrational quantum number. (From Ref. 23.)

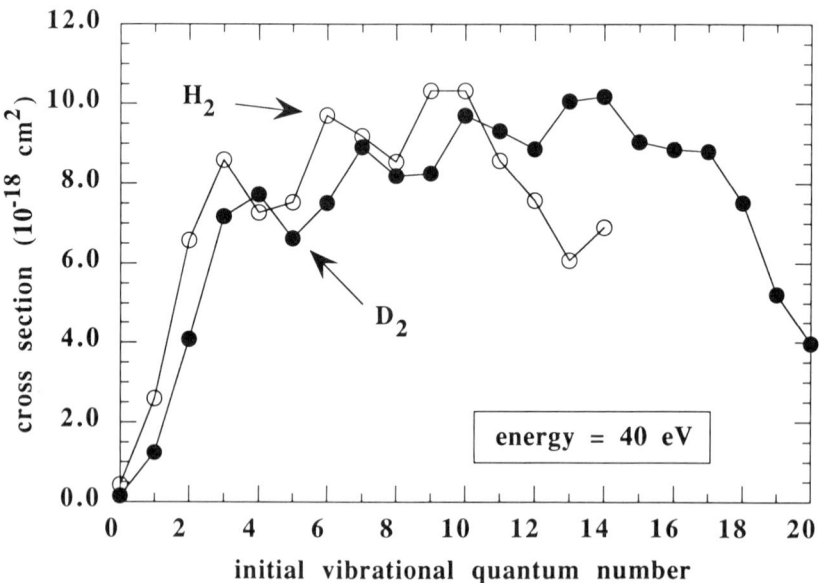

Figure 11. Cross sections for the process $D_2(X^1\Sigma_g^+, v_i) + e^- \to D_2(C^1\Pi_u) + e^- \to D(n=1) + D(n=2) + e^-$ and for the corresponding process for H_2 at fixed energy as a function of initial vibrational quantum number. (From Ref. 23.)

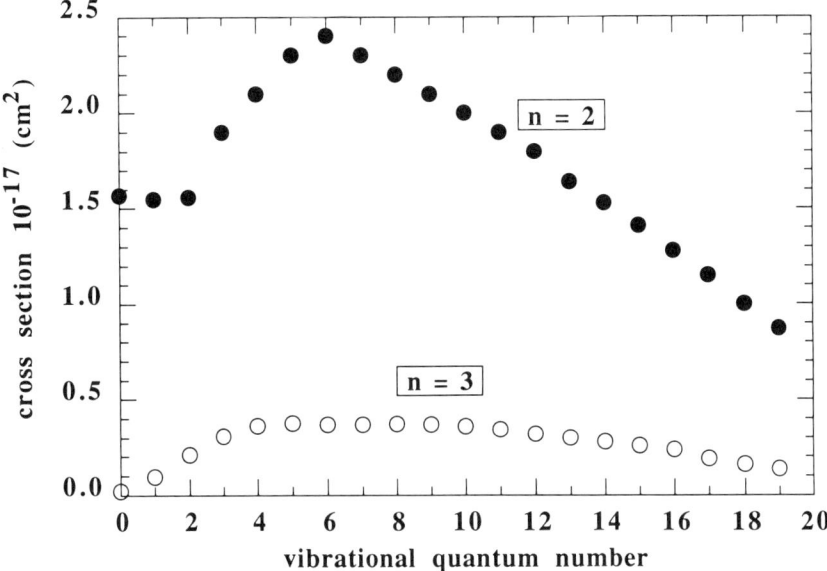

Figure 12. Cross sections (maximum value) for the process $D_2(X^1\Sigma_g^+, v_i) + e^- \rightarrow D + D(n = 2, 3) + e^-$ as a function of initial vibrational quantum number.

$$D_2(v_i) + e^- \rightarrow D_2^* + e^- \rightarrow D(n = 1) + D(n = 2, 3) + e^- \quad (h)$$

as a function of v_i. In this case, we have considered the maximum value of the cross section for each v_i so that the incident energy of the electrons changes a little as a function of v_i. These calculations have been performed with the Gryzinski approximation.[24] In this case, we see that the cross sections for $v_i = 0$ are of the order of 1.5×10^{-17} cm^2 and of 2.0×10^{-19} cm^2 for the processes yielding $D(n = 2)$ and $D(n = 3)$, respectively. These values should be compared with the corresponding B-P values, which are 1.3×10^{-17} cm^2 and 9.4×10^{-19} cm^2, respectively.

5.2. Spin-Forbidden Transitions

The most important channel for H_2/D_2 dissociation is the excitation of the completely repulsive state $b^3\Sigma_u^+$, that is, the reaction

$$D_2(v) + e^- \rightarrow D_2(b^3\Sigma_u^+) + e^- \rightarrow D(n = 1) + D(n = 1) + e^- \quad (i)$$

A lot of work—both experimental and theoretical—has been done on this reaction for H_2 in $v_i = 0$. The most recent results seem to show good agreement between rigorous quantum-mechanical approaches and Gryzinski's approach.[16] A satisfactory agreement has also been obtained for the cross sections involving $v_i > 0$.[16] In

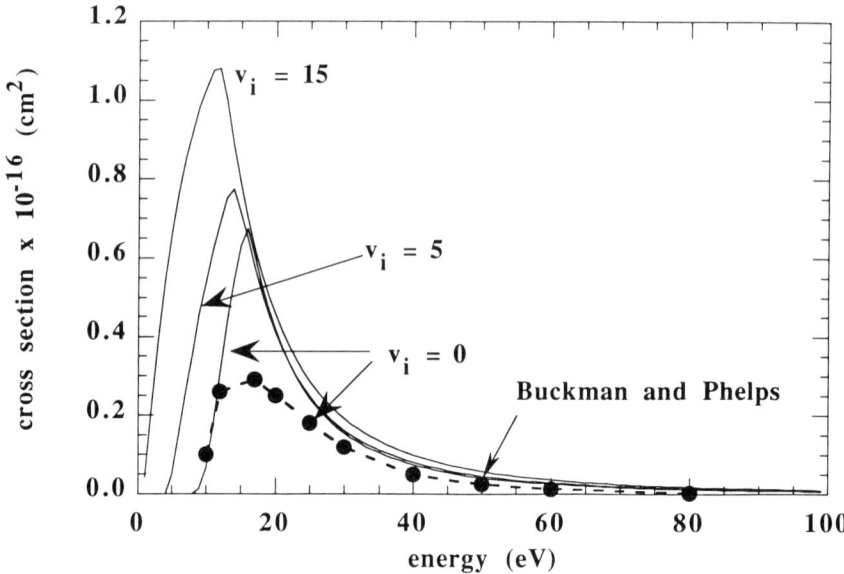

Figure 13. Cross sections for the process $D_2(X^1\Sigma_g^+, v_i) + e^- \to D_2(b^3\Sigma_u^+) + e^- \to 2D(n=1) + e^-$ as a function of electron energy for different vibrational quantum numbers. —) Adapted from Ref. 20; ●) from Ref. 8.

the latter case, no experimental data exist so that the validation of the Gryzinski's approach cross sections that we obtained has been performed by comparing them with the results of Rescigno and Schneider's[6] quantum-mechanical calculations as well as with those from Hiskes's[25] phenomenological approach.

Figure 13 (see also Table V) shows cross sections for D_2 dissociation as a function of incident energy for different vibrational levels, together with the $v_i = 0$ results of Buckmann and Phelps.[8] It may be noted that as v_i increases, the maximum value of the cross section increases and the relative threshold energy decreases. The agreement with the B-P results is not so good; in fact, the latter can differ from our results by as much as a factor of 3. We note, however, that the method used by Buckman and Phelps, namely, the deconvolution of transport data through a Boltzmann analysis of cross sections, does not yield a unique set of inelastic cross sections. In this case, this method underestimates the cross sections for process (i) not only in comparison to our results but also in comparison to the numerous other experimental and theoretical values reported in the literature.[16]

The other two excitations to triplet states are usually considered as dissociative processes. We refer to

$$D_2(v) + e^- \to D_2(a^3\Sigma_g) + e^- \to D(n=1) + D(n=1) + e^- \tag{j}$$

Table V. Dissociative Excitation Cross Sections as a Function of ΔE^a for the Process:
$$D_2(X^1\Sigma_g^+, v_i) + e^- \to D_2(b^3\Sigma_u^+) + e^- \to 2D + e^-$$

	Cross section (cm²) for v_i equal to:			
ΔE (eV)	0 (E_{th} = 4.7 eV)	5 (E_{th} = 3.0 eV)	10 (E_{th} = 1.6 eV)	15 (E_{th} = 0.70 eV)
4	2.5(−19)^b	2.4(−17)	4.4(−17)	6.1(−17)
6	8.0(−18)	4.2(−17)	6.3(−17)	8.1(−17)
8	3.0(−17)	5.8(−17)	7.7(−17)	9.4(−17)
10	5.2(−17)	7.0(−17)	8.8(−17)	1.0(−16)
11	6.0(−17)	7.0(−17)	8.9(−17)	1.1(−16)
12	5.7(−17)	6.3(−17)	7.9(−17)	1.0(−16)
13	5.0(−17)	5.8(−17)	7.1(−17)	8.8(−17)
14	4.4(−17)	5.2(−17)	6.3(−17)	7.8(−17)
15	3.9(−17)	4.6(−17)	5.7(−17)	6.9(−17)
18	2.8(−17)	3.4(−17)	4.2(−17)	5.1(−17)
20	2.3(−17)	2.7(−17)	3.4(−17)	4.2(−17)
22	1.9(−17)	2.3(−17)	2.8(−17)	3.5(−17)
25	1.4(−17)	1.7(−17)	2.2(−17)	2.7(−17)
28	1.1(−17)	1.4(−17)	1.7(−17)	2.1(−17)
30	9.6(−18)	1.2(−17)	1.5(−17)	1.8(−17)
32	8.3(−18)	1.0(−17)	1.3(−17)	1.6(−17)
35	6.8(−18)	8.4(−18)	1.0(−17)	1.3(−17)
40	4.9(−18)	6.1(−18)	7.7(−17)	9.4(−18)
50	2.9(−18)	3.6(−18)	4.5(−17)	5.5(−18)
60	1.8(−18)	2.3(−18)	2.9(−17)	3.5(−18)
80	8.7(−19)	1.1(−18)	1.4(−17)	1.5(−18)
90	6.3(−19)	8.0(−19)	1.0(−18)	1.0(−18)

$^a \Delta E = E - E_{th}$, where E is the incident electron energy, and E_{th} is the threshold energy of the process.
$^b (-x) = \times 10^{-x}$.

$$D_2(v) + e^- \to D_2(c^3\Pi_u) + e^- \to D(n=1) + D(n=1) + e^- \quad (k)$$

Both processes raise D_2 to a bound excited state. However, the $a^3\Sigma_g$ state is radiatively connected with the repulsive $b^3\Sigma_u^+$ state so that excitation is followed by the radiative transition

$$D_2(a^3\Sigma_g) \to D_2(b^3\Sigma_u^+) + h\nu \to D(n=1) + D(n=1) + h\nu \quad (l)$$

More uncertain is the fate of the $D_2(c^3\Pi_u)$ bound state. This state cannot radiatively decay to the repulsive $b^3\Sigma_u^+$ state so that it can be considered a metastable state.[26] Its energy, however, is only 0.5 eV less than that of the $a^3\Sigma_g$ state so that heavy-particle collisions can transform $c^3\Pi_u$ into $a^3\Sigma_g$, which can radiatively decay to the repulsive $b^3\Sigma_u^+$.

The cross sections for these processes have not been yet calculated. However, the corresponding results for H_2 have been reported in Ref. 15 for different vibrational levels.

6. IONIZATION

The electron impact ionization process mainly involves the excitation of the bound $^2\Sigma_g^+$ state of H_2^+, the excitation of the repulsive part of the bound $^2\Sigma_g^+$ state, and the excitation of the completely repulsive $^2\Sigma_u^+$ state, leading to:

$$D_2(v) + e^- \rightarrow D_2^+(^2\Sigma_g^+) + 2e^- \qquad (m)$$

$$D_2(v) + e^- \rightarrow D_2^+(^2\Sigma_g^+) + 2e^- \rightarrow D(n=1) + D^+ + 2e^- \qquad (n)$$

$$D_2(v) + e^- \rightarrow D_2^+(^2\Sigma_u^+) + 2e^- \rightarrow D(n=1) + D^+ + 2e^- \qquad (o)$$

Despite the apparent simplicity of the processes, no quantum-mechanical calculations have been reported. Simple formulas for calculating the cross sections for these processes are again provided by the Gryzinski method.[27,28]

The bound-to-bound ionization cross section [process (m)] is given by

$$\sigma_{v_i}^I(E) = N_e \pi e^4 \sum_{v_f} \frac{q_{v_i,v_f} g_{v_i,v_f}^G(x)}{\Delta E_{v_i,v_f}} \qquad (16)$$

where q_{v_i,v_f} is the Franck–Condon factor and

$$g_{v_i,v_f}^G(x) = \frac{1}{x}\left(\frac{x-1}{x+1}\right)^{3/2} \left\{1 + \frac{2}{3}\left(1 - \frac{1}{2x}\right) \ln\left[2.7 + (x-1)^{1/2}\right]\right\} \qquad (17)$$

with

$$x = \frac{E}{\Delta E_{v_i,v_f}} \qquad (18)$$

where $\Delta E_{v_i,v_f}$ is the transition energy linking the v_i and v_f vibrational levels belonging to the electronic states of the neutral and the ionized molecule, respectively.

The dissociative ionization cross section [processes (n) and (o)] is written as

$$\sigma_{v_i}^I(E) = N_e \pi e^4 \int d\varepsilon \frac{g_{v_i,\varepsilon}^G(x) q_{v_i,\varepsilon}}{\Delta E_{v_i,\varepsilon}} \qquad (19)$$

where $q_{v_i,\varepsilon}$ represents the Franck–Condon density.

Complete sets of cross sections calculated according to this method have been recently obtained by Celiberto et al. for H_2 and D_2.[20,28] These results have been validated by comparison with experimental data[16] for $v_i = 0$.

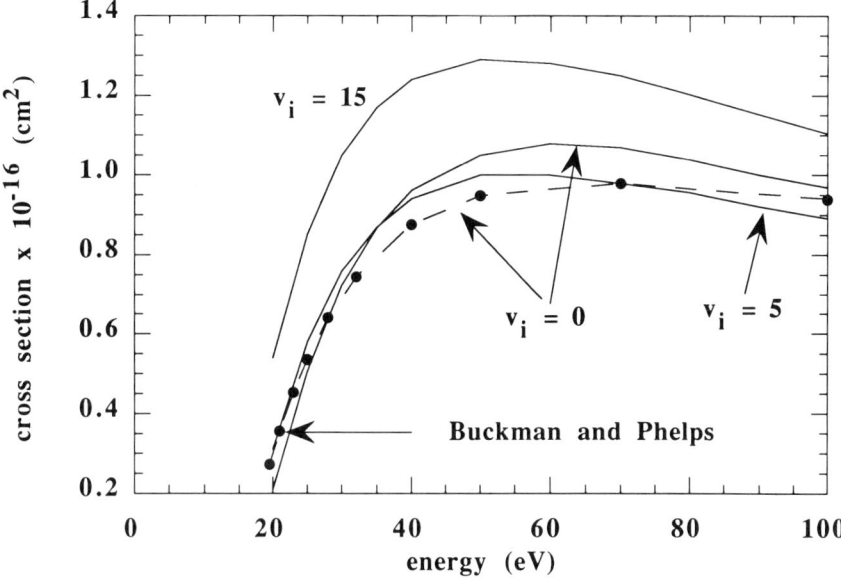

Figure 14. Cross sections for the process $D_2(X^1\Sigma_g^+, v_i) + e^- \to D_2^+(^2\Sigma_g^+) + 2e^-$ as a function of electron energy for different vibrational quantum numbers. —) Adapted from Ref. 20; ●) from Ref. 8.

Figure 14 shows the ionization cross sections for process (m) for different v_i values. The corresponding cross sections for $v_i = 0$ obtained by Buckman and Phelps[8] are also shown and are in excellent agreement with the calculated values. Note also the nonmonotonic trend of the cross sections with increasing vibrational quantum number. This behavior reflects the Franck–Condon density linking the vibrational states of the ground electronic state of D_2 with the corresponding vibrational states of the $^2\Sigma_g^+$ state of D_2^+.

A monotonic behavior is, on the other hand, exhibited by the dissociative ionization cross sections [process (o)] as a function of vibrational quantum number as can be seen in Fig. 15. This is due to the fact that an increase in the vibrational quantum number of the ground state is accompanied by an increase in the Franck–Condon region connecting D_2 and D_2^+ ($^2\Sigma_u^+$ state). This is also reflected in the monotonic increase of the maximum value of the cross section with increasing vibrational quantum number. A nonmonotonic behavior is also observed in the dissociative ionization process over the repulsive part of the bound state (see Ref. 28).

A comparison with the experimental $v_i = 0$ cross section suffers from the fact that Gryzinski's theory completely neglects the so-called indirect dissociative ionization. Keeping this point in mind, we have compared in Fig. 16 the theoretical

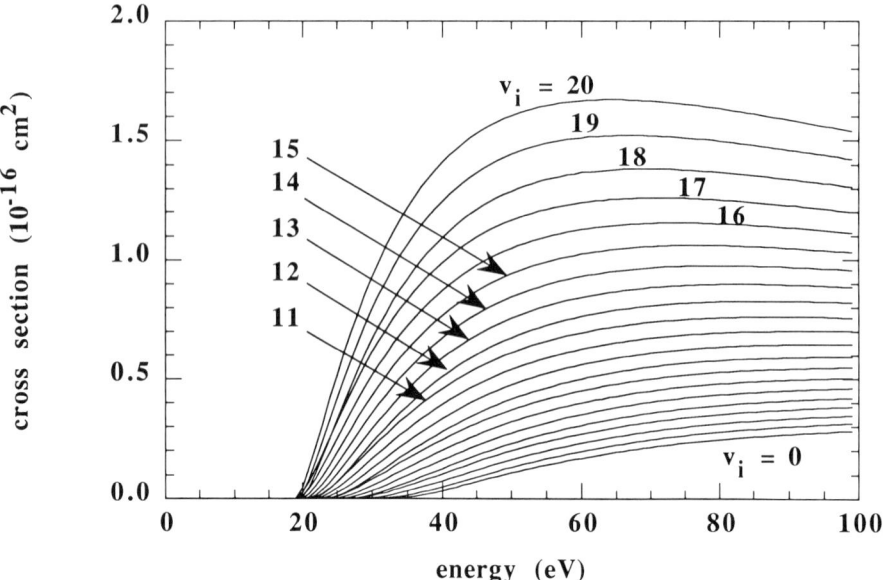

Figure 15. Cross sections for the process $D_2(X^1\Sigma_g^+, v_i) + e^- \to D_2^+(^2\Sigma_u^+) + 2e^- \to D(n=1) + D^+ + 2e^-$ as a function of electron energy for different vibrational quantum numbers. (From Ref. 20.)

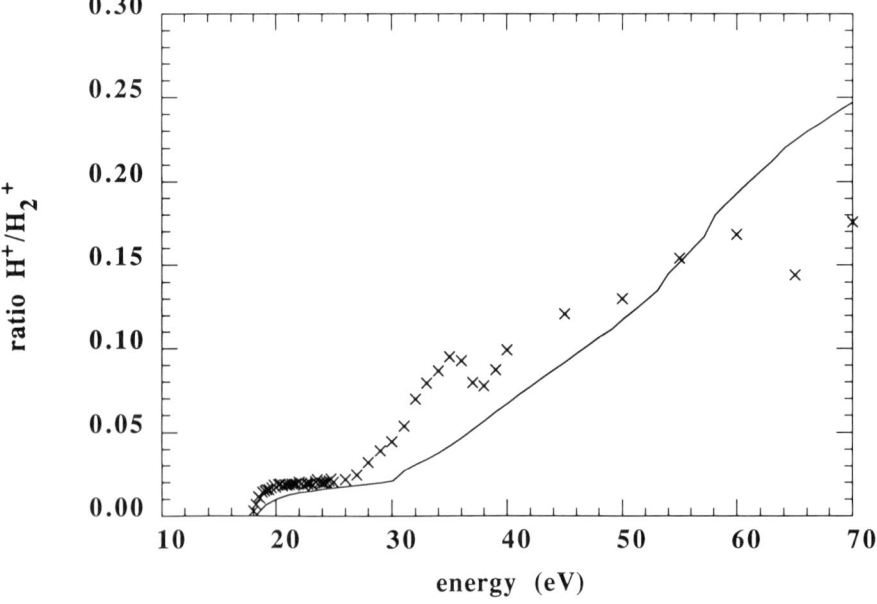

Figure 16. Experimental and theoretical H^+/H_2^+ yield as a function of incident electron energy. (From Ref. 28.)

ratio of dissociative ionization (the sum of the cross sections for the three processes) to ionization (the cross section leading to H_2^+) with the experimental H^+/H_2^+ yield. It can be seen that there is a qualitative agreement between theory and experiment, the theory overestimating the experimental values, especially at high electron energy. This means that we are probably overestimating the dissociative ionization cross section even though we are neglecting the indirect dissociative processes. These processes manifest themselves in Fig. 16, especially in the energy range of 30–40 eV.

Tables VI–VIII contain some numerical values for the different ionization channels.

Table VI. Ionization Cross Sections as a Function of Incident Energy for the Process:

$$D_2(X^1\Sigma_g^+, v_i) + e^- \rightarrow D_2^+ (^2\Sigma_g^+) + 2e^-$$

	Cross section (cm^2) for v_i equal to:			
E (eV)	0	5	10	15
16	2.4(–19)[a]	7.1(–18)	1.5(–17)	2.1(–17)
17	2.9(–18)	1.3(–17)	2.2(–17)	3.0(–17)
18	8.1(–18)	1.9(–17)	2.9(–17)	3.8(–17)
19	1.4(–17)	2.6(–17)	3.6(–17)	4.6(–17)
20	2.1(–17)	3.2(–17)	4.3(–17)	5.4(–17)
21	2.7(–17)	3.8(–17)	4.9(–17)	6.1(–17)
22	3.3(–17)	4.3(–17)	5.5(–17)	6.8(–17)
23	3.9(–17)	4.9(–17)	6.0(–17)	7.4(–17)
24	4.5(–17)	5.4(–17)	6.5(–17)	8.0(–17)
25	5.1(–17)	5.8(–17)	6.9(–17)	8.5(–17)
30	7.2(–17)	7.6(–17)	8.8(–17)	1.0(–16)
31	7.6(–17)	7.9(–17)	9.0(–17)	1.1(–16)
32	7.9(–17)	8.1(–17)	9.3(–17)	1.1(–16)
33	8.2(–17)	8.3(–17)	9.5(–17)	1.1(–16)
34	8.4(–17)	8.5(–17)	9.7(–17)	1.2(–16)
35	8.7(–17)	8.7(–17)	9.9(–17)	1.2(–16)
36	8.9(–17)	8.9(–17)	1.0(–16)	1.2(–16)
37	9.1(–17)	9.0(–17)	1.0(–16)	1.2(–16)
38	9.3(–17)	9.2(–17)	1.0(–16)	1.2(–16)
39	9.5(–17)	9.3(–17)	1.0(–16)	1.2(–16)
40	9.6(–17)	9.4(–17)	1.1(–16)	1.2(–16)
60	1.1(–16)	1.0(–16)	1.1(–16)	1.3(–16)
80	1.0(–17)	9.6(–17)	1.0(–16)	1.2(–16)
100	9.7(–17)	8.9(–17)	9.5(–17)	1.1(–16)

[a] $(-x) = \times 10^{-x}$.

Table VII. Dissociative Ionization Cross Sections as a Function of ΔE^a for the Process:

$$D_2(X^1\Sigma_g^+, v_i) + e^- \rightarrow D_2^+(^2\Sigma_g^+) + 2e^- \rightarrow D + D^+ + 2e^-$$

	Cross section (cm^2) for v_i equal to:			
ΔE (eV)	0 (E_{th} = 18.1 eV)	5 (E_{th} = 16.4 ev)	10 (E_{th} = 15.1 ev)	15 (E_{th} = 14.1 eV)
2	1.4(−19)^b	7.7(−19)	4.9(−19)	2.9(−19)
5	3.7(−19)	2.9(−18)	2.2(−18)	1.3(−18)
6	4.5(−19)	3.6(−18)	2.8(−18)	1.7(−18)
7	5.3(−19)	4.3(−18)	3.5(−18)	2.1(−18)
8	6.0(−19)	5.0(−18)	4.1(−18)	2.5(−18)
9	6.8(−19)	5.7(−18)	4.7(−18)	2.9(−18)
10	7.6(−19)	6.4(−18)	5.4(−18)	3.5(−18)
15	1.5(−18)	1.3(−17)	1.0(−17)	6.5(−18)
20	1.8(−18)	1.6(−17)	1.4(−17)	8.7(−18)
21	1.9(−18)	1.6(−17)	1.4(−17)	9.1(−18)
22	1.9(−18)	1.7(−17)	1.4(−17)	9.3(−18)
23	2.0(−18)	1.7(−17)	1.5(−17)	9.6(−18)
24	2.0(−18)	1.8(−17)	1.5(−17)	9.8(−18)
25	2.0(−18)	1.8(−17)	1.6(−17)	1.0(−17)
30	2.2(−18)	2.0(−17)	1.7(−17)	1.1(−17)
35	2.3(−18)	2.0(−17)	1.8(−17)	1.1(−17)
40	2.4(−18)	2.1(−17)	1.8(−17)	1.2(−17)
50	2.5(−18)	2.2(−17)	1.9(−17)	1.2(−17)
60	2.5(−18)	2.2(−17)	1.9(−17)	1.2(−17)
80	2.4(−18)	2.1(−17)	1.8(−17)	1.2(−17)

$^a \Delta E = E - E_{th}$, where E is the incident electron energy, and E_{th} is the threshold energy of the process.
$^b (-x) = \times 10^{-x}$.

7. ELECTRONIC EXCITATION FROM ELECTRONICALLY EXCITED STATES

When the electron density is very high, as in the divertor edge plasmas, not only the vibrational states of the ground electronic state of H_2/D_2 can be populated, but also the electronically excited states can have large concentrations. Thus, we are obliged to calculate sets of cross sections involving electronically excited states.

Our group, in particular, started calculating the cross sections in the framework of the impact parameter approximation for the following processes[29]:

$$H_2(B^1\Sigma_u^+, v_i = 0-14) + e^- \rightarrow H_2(I^1\Pi_g) + e^- \tag{p}$$

$$H_2(B^1\Sigma_u^+, v_i = 0-14) + e^- \rightarrow H_2(I^1\Pi_g) + e^- \rightarrow 2H + e^- \tag{q}$$

Table VIII. Dissociative Ionization Cross Sections as a Function of ΔE^a for the Process:

$$D_2(X^1\Sigma_g^+, v_i) + e^- \to D_2^+(^2\Sigma_u^+) + 2e^- \to D + D^+ + 2e^-$$

ΔE (eV)	0 (E_{th} = 18.1 eV)	5 (E_{th} = 16.4 ev)	10 (E_{th} = 15.1 ev)	15 (E_{th} = 14.1 eV)
3	1.7(–39)[b]	5.2(–28)	1.4(–20)	3.4(–18)
12	1.3(–20)	3.5(–18)	1.3(–17)	3.4(–17)
15	3.0(–19)	6.0(–18)	1.9(–17)	5.4(–17)
20	2.3(–18)	1.1(–17)	3.5(–17)	7.2(–17)
21	2.8(–18)	1.2(–17)	3.7(–17)	7.4(–17)
22	3.3(–18)	1.3(–17)	3.9(–17)	7.7(–17)
23	3.9(–18)	1.5(–17)	4.1(–17)	7.9(–17)
24	4.4(–18)	1.7(–17)	4.3(–17)	8.2(–17)
25	4.9(–18)	1.9(–17)	4.5(–17)	8.4(–17)
26	5.4(–18)	2.0(–17)	4.6(–17)	8.6(–17)
27	5.9(–18)	2.1(–17)	4.8(–17)	8.7(–17)
28	6.5(–18)	2.3(–17)	4.9(–17)	8.9(–17)
29	7.0(–18)	2.4(–17)	5.1(–17)	9.1(–17)
30	7.5(–18)	2.5(–17)	5.2(–17)	9.2(–17)
31	8.0(–18)	2.6(–17)	5.3(–17)	9.3(–17)
32	8.6(–18)	2.7(–17)	5.5(–17)	9.5(–17)
33	9.2(–18)	2.8(–17)	5.6(–17)	9.6(–17)
34	9.8(–18)	2.9(–17)	5.7(–17)	9.7(–17)
35	1.1(–18)	3.0(–17)	5.8(–17)	9.8(–17)
40	1.5(–17)	3.4(–17)	6.2(–17)	1.0(–16)
50	2.2(–17)	3.9(–17)	6.7(–17)	1.1(–16)
60	2.5(–17)	4.3(–17)	6.9(–17)	1.1(–16)
80	2.9(–17)	4.6(–17)	7.1(–17)	1.1(–16)

[a] $\Delta E = E - E_{th}$, where E is the incident electron energy, and E_{th} is the threshold energy of the process.
[b] $(-x) = \times 10^{-x}$.

Figure 17 shows the corresponding cross sections as a function of v_i at fixed energy. These cross sections exhibit a completely different trend from that observed for the cross sections involving the excitation of electronic states starting from the ground electronic state (compare Fig. 17 with Figs. 6 and 7). It should be also noted that the contribution of the dissociative channel starts to be important from $v_i = 3$ on, becoming essential at $v_i = 10$. At the same time, the contribution of excitation to the bound part of the potential strongly decreases with increasing v_i. Once again, this behavior is related to the Franck–Condon matrix (bound–bound, bound–continuum) linking the two electronic excited states.

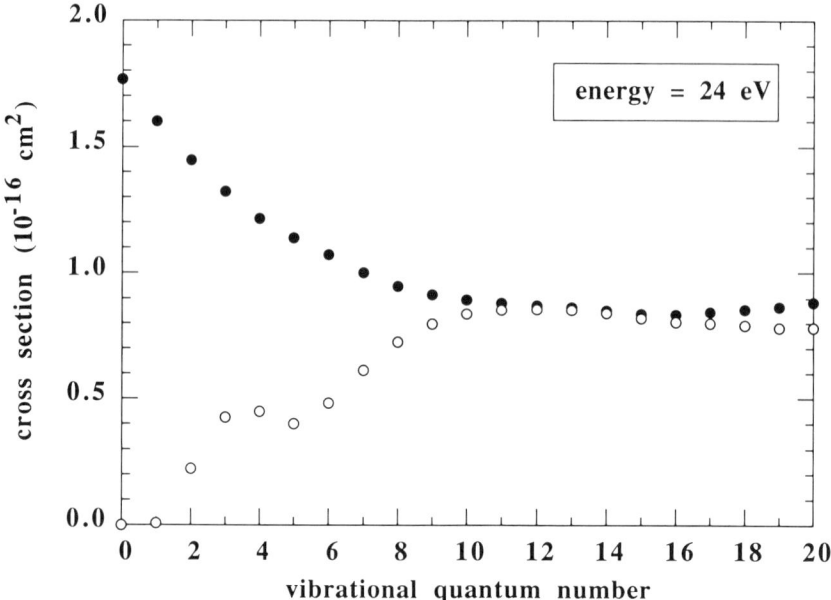

Figure 17. Cross sections for the processes $H_2(B^1\Sigma_u^+, v_i = 0\text{–}14) + e^- \rightarrow H_2(I^1\Pi_g) + e^-$ (●) and $H_2(B^1\Sigma_u^+, v_i = 0\text{–}14) + e^- \rightarrow H_2(I^1\Pi_g) + e^- \rightarrow 2H(n=2) + e^-$ (○) at a fixed energy as a function of initial vibrational quantum number.

8. CONCLUSION

The results presented in this chapter have shown the dependence of cross sections for H_2 and D_2 on the initial vibrational quantum number of the target. This dependence exhibits a nonmonotonic trend so that is very difficult to give general rules for an analytical behavior.

In general, we can see that a large dependence on v_i is exhibited by the so-called resonant processes (i.e., vibrational excitation and dissociative attachment) whereas a weaker dependence is shown by excitation and dissociation processes. In the latter case, a monotonic increase of the cross sections with increasing v_i is shown by processes leading to electronically repulsive states while an intricate behavior is exhibited by excitation to the repulsive part of bound states.

The results presented here constitute part of a data base, which is continuously being updated by our group, to be used for modeling situations in which vibrationally excited $H_2(v)/D_2(v)$ molecules play a non-negligible role in affecting the macroscopic properties of H_2/D_2 plasmas. Of course, other important processes such as deactivation/activation energy exchange of $H_2(v)/D_2(v)$ in the gas phase with molecules and with parent atoms as well as with the metallic walls of the container play an important role in forming the nonequilibrium vibrational distributions of the two molecules.[16,30] Moreover, we cannot forget the electron impact processes with the $(H_2^+/H_3^+)/(D_2^+/D_3^+)$ ions as well as their reactions with neutrals, which should be considered for any actual plasma conditions.

ACKNOWLEDGMENTS. This work was partially supported by Ministero Università e Ricerca Scientifica e Tecnologica 40% 93 and by European Economic Community Contract No. EBRCHRXCT920003, "Negative ion production by volume and surface processes in plasmas." The authors thank U. T. Lamanna and M. Cacciatore for several discussions and J. M. Wadehra for the complete tables of E–V and dissociative attachment cross sections.

REFERENCES

1. H. Erhardt, in *Nonequilibrium Processes in Partially Ionized Gases* (M. Capitelli and J. N. Bardsley, eds.), Plenum Press, New York (1990), Vol. B220, p. 19.
2. M. Hayashi, in *Nonequilibrium Processes in Partially Ionized Gases* (M. Capitelli and J. N. Bardsley, eds.), Plenum Press, New York (1990), Vol. B220, p. 333.
3. J. N. Bardsley, in *Nonequilibrium Processes in Partially Ionized Gases* (M. Capitelli and J. N. Bardsley, eds.), Plenum Press, New York (1990), Vol. B220, p. 1; P. G. Burke, *The Physics of Electronic and Atomic Collisions*, AIP Conf. Proc., No. 295, American Institute of Physics, New York (1993), p. 26.
4. M. Gryzinski, *Phys. Rev. A* **138**, 336 (1965); *Phys. Rev. A* **138**, 322 (1965).
5. E. Bauer and C. D. Bartky, *J. Chem. Phys.* **43**, 2466 (1965).
6. T. N. Rescigno and B. I. Schneider, *J. Phys. B.* **21**, L691 (1988).
7. R. Celiberto and T. N. Rescigno, *Phys. Rev. A* **43**, 1939 (1993).
8. S. J. Buckman and A. V. Phelps, Joint Institute for Laboratory Astrophysics Information Center Report 27, May 1985, p. 1.
9. H. Tawara, Y. Itikawa, N. Nishimura, and M. Yoshino, *J. Phys. Chem. Ref. Data* **19**, 617 (1990); K. Smith and A. H. Glasser, *Comput. Phys. Comm.* **54**, 391 (1989); R. K Janev, W. D. Langher, K. Evans, and T. E. Post, in *Elementary Processes in Hydrogen and Helium Plasmas*, Springer-Verlag, Berlin (1987).
10. D. E. Atems and J. M. Wadehra, *Phys. Rev. A* **42**, 5201 (1990); see also D. E. Atems and J. M. Wadehra, *Production and Neutralization of Negative Ions and Beams*, AIP Conf. Proc., No. 210, American Institute of Physics, New York (1990), p. 121.
11. J. M. Wadehra, in *Nonequilibrium Vibrational Kinetics* (M. Capitelli, ed.), *Topics in Current Physics*, Vol. 39, Springer-Verlag, Berlin (1986), p. 191.
12. A. Klonover and U. Kaldor, *J. Phys. B* **12**, 3797 (1979).
13. J. P. Gauyacq, *J. Phys. B* **18**, 1859 (1985).
14. C. Mundel, C. Berman, and W. Domcke, *Phys. Rev. A* **32**, 181 (1985).
15. M. Cacciatore and M. Capitelli, *Chem. Phys.* **55**, 67 (1981).
16. C. Gorse, R. Celiberto, M. Cacciatore, A. Laganà, and M. Capitelli, *Chem. Phys.* **161**, 211 (1992).
17. M. Cacciatore, M. Capitelli, and M. Dilonardo, *Chem. Phys.* **34**, 193 (1978).
18. A. U. Hazi, *Phys. Rev. A* **23**, 2232 (1981).
19. M. J. Redmon, B. C. Garrett, L. T. Redmon, and C. W. McCurdy, *Phys. Rev. A* **32**, 3354 (1985).
20. R. Celiberto, P. Cives, M. Cacciatore, M. Capitelli, and U. T. Lamanna, *Chem. Phys. Lett.* **169**, 69 (1990).

21. J. R. Hiskes, *J. Appl. Phys.* **70**, 3409 (1991).
22. R. Celiberto, M. Capitelli, and U. T. Lamanna, *Chem. Phys.* **183**, 101 (1994).
23. R. Celiberto, U. T. Lamanna, and M. Capitelli, *Phys. Rev. A* **50**, 4778 (1994).
24. R. Celiberto, M. Capitelli, M. Cacciatore, and C. Gorse, *Chem. Phys.* **133**, 355 (1989).
25. J. R. Hiskes, *Production and Neutralization of Negative Ions and Beams*, AIP Conf. Proc., No. 287, American Institute of Physics, New York (1994), p. 155.
26. C. Ottinger and T. Rox, *Phys. Lett. A* **161**, 135 (1991).
27. M. Cacciatore, C. Gorse, and M. Capitelli, *J. Phys. D.* **13**, 575 (1980).
28. R. Celiberto, M. Capitelli, and M. Cacciatore, *Chem. Phys.* **140**, 209 (1990).
29. R. Celiberto, U. T. Lamanna, N. Durante, and M. Capitelli, work in progress.
30. M. Capitelli, M. Cacciatore, and R. Celiberto, in *Advances in Atomic, Molecular and Optical Physics*, Vol. 33, (M. Inokuti, ed.), Academic Press, New York (1994), p. 322; M. A. Cacciatore, M. Capitelli, and R. Celiberto, *Nucl. Fusion, Supplement* **2**, 65 (1992).

Chapter 9

Electron–Molecular Ion Collisions

J. B. A. Mitchell

1. INTRODUCTION

Collisions between electrons and molecular ions are among the most complex processes encountered in atomic and molecular physics since the range of initial, final, and intermediate states can be enormous. This leads to great difficulties for the theoretician who tries to model these processes and for the experimentalist who tries to conduct a well-defined measurement of them. For this reason, electron–molecular ion collisions are more poorly understood than other collision processes, and, indeed, for complex molecular ions, only the most rudimentary information is starting to become available.[1] Even hydrogenic ion collisions are the subject of controversy and, despite several decades of study, still remain an enigma in many cases. This chapter will present a discussion of the state of our knowledge of these processes.

In discussing electron–molecular ion processes of relevance to thermonuclear fusion plasmas, we will confine ourselves in this chapter to collisions involving the

---

J. B. A. MITCHELL • Department of Physics, University of Western Ontario, London, Ontario, Canada N6A 3K7.

Atomic and Molecular Processes in Fusion Edge Plasmas, edited by R. K. Janev. Plenum Press, New York, 1995.

hydrogenic species H_2^+, D_2^+, HD^+, H_3^+, H_2D^+, HD_2^+, and D_3^+ and to nonhydrocarbon impurity species such as CO^+, CO_2^+, and O_2^+. No electron–ion collision work has been performed with tritiated species owing to the problems associated with handling the radioactive material, but from the following discussion it will be seen that generally the cross sections are not expected to be too different from those for the hydrogen- and deuterium-based molecules.

2. H_2^+

H_2^+ is the simplest molecular ion and so has received much attention over the years both from experimentalists and from theoreticians. Whereas it is tractable theoretically and some accurate calculations of collision processes involving H_2^+ have been performed, this ion presents a serious problem from an experimentalist's viewpoint. This problem can be understood by examining Fig. 1, which shows the potential energy curves of H_2^+ and its neutral precursor, H_2. It is seen that the minima of the potential wells of H_2 and H_2^+ are displaced from each other so that when H_2 is ionized, by, for example, electron or photon impact, the resulting ion is formed in a variety of vibrational states. Because H_2^+ does not have a dipole moment, the radiative lifetimes of these states are extremely long (10^6 s). This means that when H_2^+ ions are produced in a conventional ion source, using hydrogen as a precursor gas, the resulting beam of ions has many vibrational states populated. Most of the early collision work on H_2^+ used such beams. This problem was recognized of course, and Von Busch and Dunn[2] performed a pioneering study of the population of states found in a beam of H_2^+ molecules produced using an electron impact ion source. By fitting experimental data from a measurement of the photodissociation of H_2^+ ions to theoretical estimates of the cross sections for the photodissociation of individual vibrational levels, they found that the measured population was similar to the population that one would expect based upon a Franck–Condon (vertical ionization) analysis of the ionization process. The population factors are listed in Table I, and population factors for D_2^+ are listed in Table II. The population factors for these two ions are shown graphically in Figs. 2 and 3.

A number of other workers have examined the population distribution of hydrogen molecular ions[3–6] and have found results very similar to those of Von Busch and Dunn. Given the different ion sources used in these studies, it is clear that the excited vibrational levels of H_2^+ are remarkably resistant to collisional deexcitation in hydrogen gas. It is, however, possible to modify the vibrational population distribution chemically, via reaction with the rare gases helium and neon.[7–11] The reactions

$$H_2^+(v) + He \rightarrow HeH^+ + H$$

and

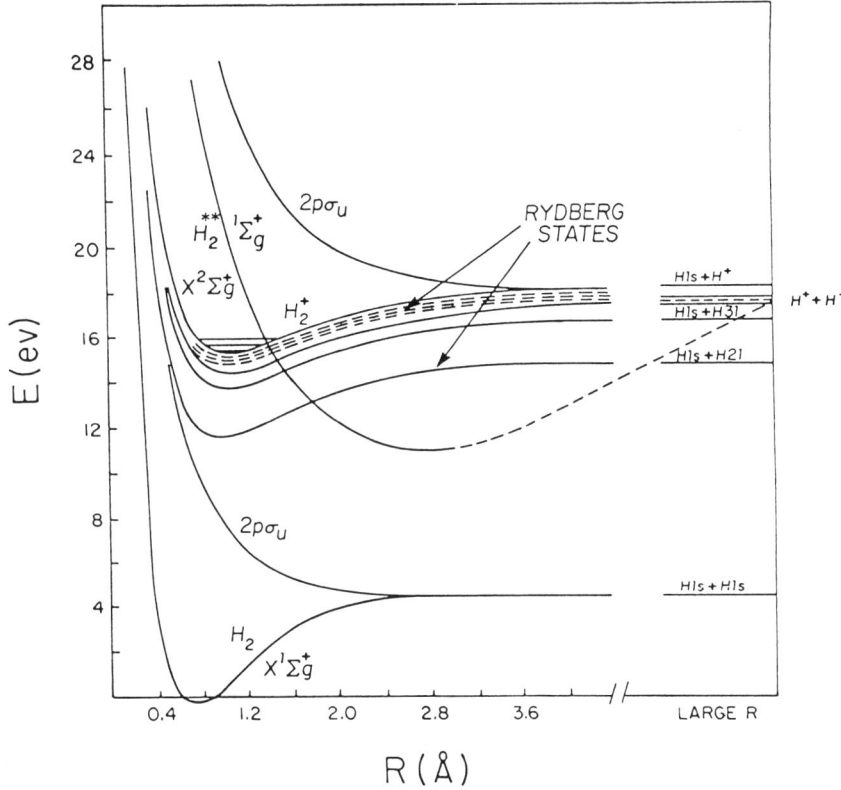

Figure 1. Potential energy curves for H_2 and H_2^+.

$$H_2^+(v) + Ne \rightarrow NeH^+ + H$$

are endothermic for $v \leq 2$ and $v \leq 1$, respectively. Because these reactions have large rate coefficients when exothermic, they can rapidly remove vibrationally excited H_2^+ molecules so that the resulting vibrational distribution of the remaining H_2^+ molecules is low. In practice, Van der Donk et al.[11] have found that a population of H_2^+ in the ground vibrational state only can be produced, even when helium is used. This is because of the reaction[7]

$$H_2^+(v) + He \rightarrow H_2^+(v = 0) + He$$

which can quench remaining excited states.

In the case of edge plasmas in thermonuclear fusion machines, it is possible that hydrogen molecules are formed via the recombination of hydrogen atoms on

Table I. Population Factors for Vibrational Levels of $H_2^{+\ a}$

| | Population factor | | Energy (eV) below |
v	Von Busch and Dunn	Franck–Condon	dissociation limit
0	0.119	0.092	2.645
1	0.190	0.162	2.374
2	0.188	0.176	2.118
3	0.152	0.155	1.877
4	0.125	0.121	1.651
5	0.075	0.089	1.44
6	0.052	0.063	1.243
7	0.037	0.044	1.059
8	0.024	0.030	0.890
9	0.016	0.021	0.734
10	0.0117	0.0147	0.593
11	0.0082	0.0103	0.465
12	0.0057	0.0072	0.351
13	0.00374	0.0051	0.252
14	0.00258	0.0036	0.168
15	0.00175	0.0024	0.100
16	0.00109	0.0016	0.0491
17	0.00056	0.0008	0.017
18	0.00012	0.0002	0.002

[a] From Ref. 2.

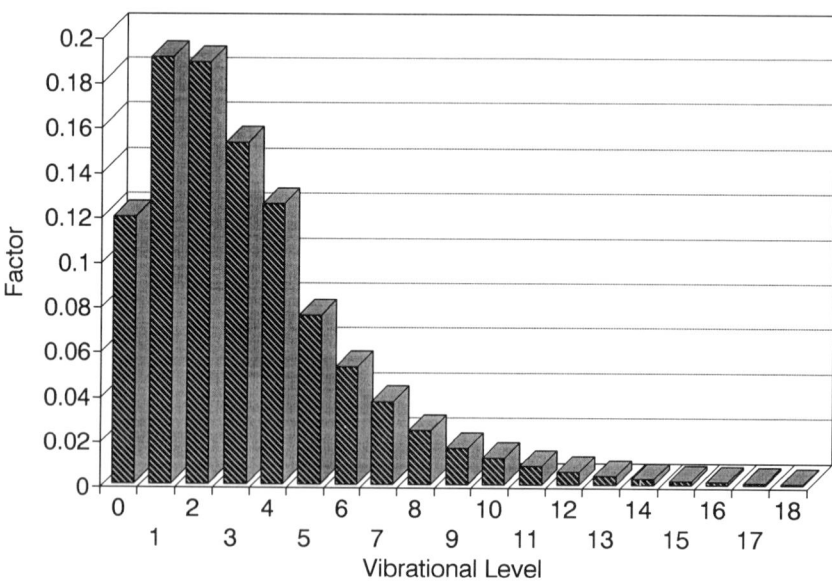

Figure 2. Population factors for the vibrational states of H_2^+ following electron impact ionization. The histogram shows Franck–Condon distributions calculated by Von Busch and Dunn.[2]

Table II. Population Factors for D_2^+ [a]

v	Population factor		Energy (ev) below dissociation limit
	Von Busch and Dunn	Franck–Condon	
0	0.0448	0.03426	2.685
1	0.1038	0.08547	2.490
2	0.1407	0.12412	2.303
3	0.1476	0.13839	2.123
4	0.1337	0.13205	1.951
5	0.1106	0.11416	1.787
6	0.085	0.09238	1.630
7	0.063	0.07156	1.480
8	0.042	0.05383	1.337
9	0.034	0.03973	1.201
10	0.024	0.02899	1.072
11	0.017	0.02104	0.950
12	0.0122	0.01524	0.835
13	0.0088	0.01106	0.727
14	0.0064	0.00805	0.625
15	0.0048	0.00589	0.531
16	0.0036	0.00433	0.443
17	0.0025	0.00320	0.363
18	0.00185	0.00238	0.290
19	0.00138	0.00178	0.225
20	0.00102	0.00132	0.167
21	0.00075	0.00098	0.118
22	0.00054	0.00072	0.076
23	0.00037	0.00050	0.044
24	0.00023	0.00032	0.020
25	0.00011	0.00016	0.006
26	0.00002	0.00004	0.001

[a]From Ref. 2.

surfaces, and these molecules can be in a vibrationally excited state. This would mean that H_2^+ ions formed from these excited molecular precursors would have a different vibrational population distribution than those obtained from the ionization of ground state H_2. Given that the Franck–Condon model for H_2 ($v = 0$) ionization appears to yield a good picture of the H_2^+ vibrational distribution, if the population of the H_2 molecules in the plasma can be measured, using, for example, VUV absorption techniques, then the vibrational population of H_2^+ ions can be estimated. This does mean, however, that collisions between electrons and molecular ions in specific vibrational states should be understood, and in many cases, as we shall see below, the required information is not currently available.

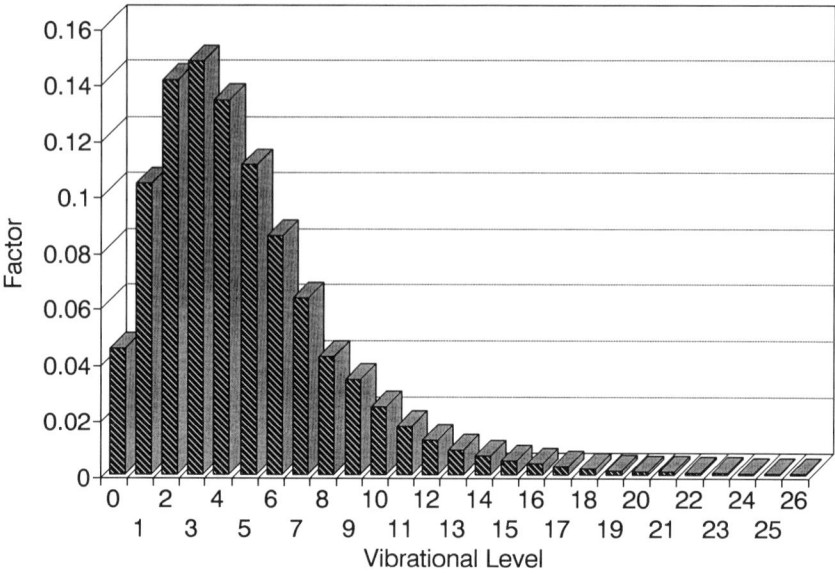

Figure 3. Population factors for the vibrational states of D_2^+ following electron impact ionization. The histogram shows Franck–Condon distributions calculated by Von Busch and Dunn.[2]

2.1. Dissociative Recombination

Dissociative recombination (DR) is a process in which a molecular ion captures an electron, forming a doubly excited intermediate state that subsequently breaks up into atomic or molecular fragments, thereby releasing the recombination energy in the form of kinetic energy of these dissociation products. Such a state is shown in Fig. 1. At low energies (<1 eV), DR is often very rapid. The cross section, however, falls off with increasing energy, and so processes such as dissociative excitation may be more important for the energy range encountered in edge plasmas.

Total cross sections for the dissociative recombination of H_2^+ with all vibrational levels populated have been measured by Peart and Dolder[12] and Auerbach et al.,[9] and the results are shown in Fig. 4. The cross section is seen to be large (few × 10^{-14} cm^2) at low collision energies (0.01 eV) but decreases as E^{-1}.

Cross sections for the dissociative recombination of H_2^+ ($v = 0$) have been measured by Van der Donk et al.,[11] and these are shown in Fig. 5. Also shown is a theoretical multichannel quantum defect theory (MQDT) calculation by Schneider et al.,[13] and it can be seen that this agrees very well with the experimental data. The resonances in the cross section are due to the influence of the neutral Rydberg states that lie just below the H_2^+ ion state. Nakashima et al.[14] have also used MQDT to

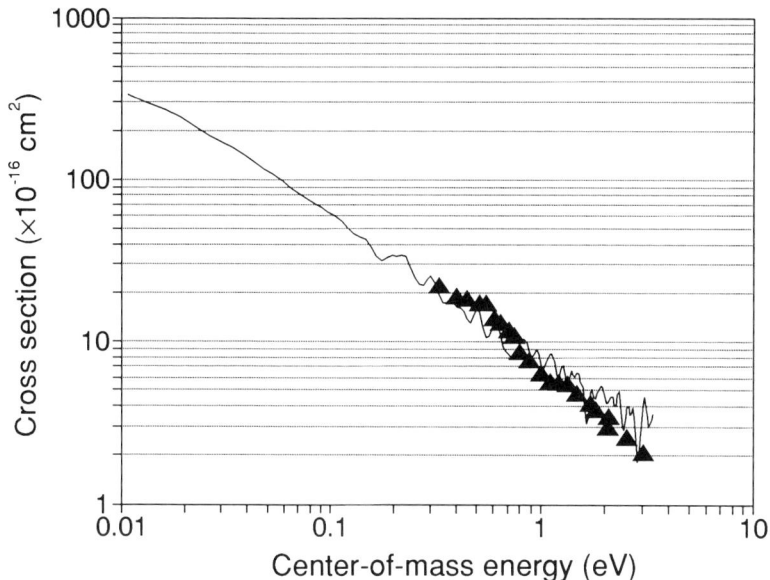

Figure 4. Dissociative recombination cross sections for H_2^+ (all v). —). Auerbach et al.[9]; ▲) Peart and Dolder.[12]

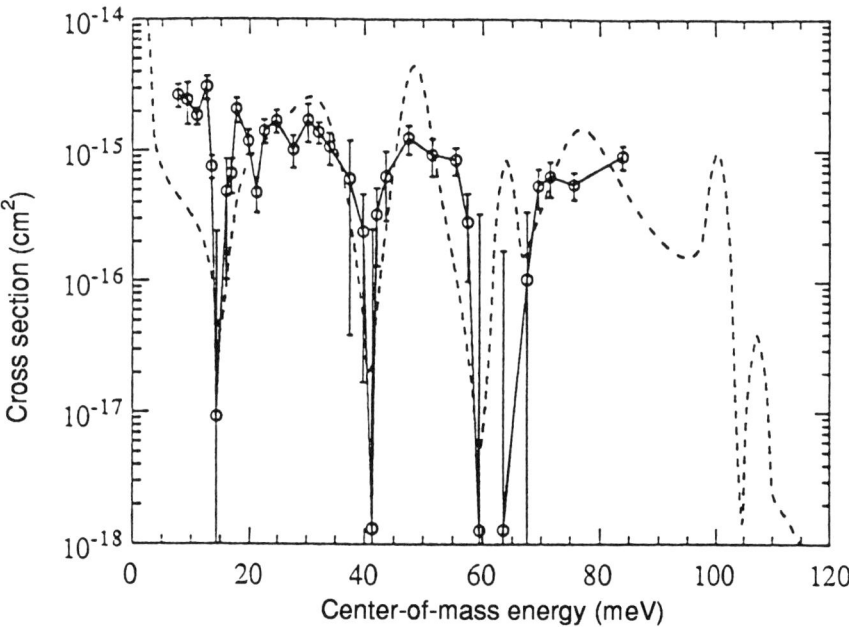

Figure 5. Experimental and theoretical cross sections for the dissociative recombination of H_2^+ ($v = 0$). ○) Experimental measurements of Van der Donk et al.[11]; ---) calculation by Schneider et al.[13] (From Schneider et al.[13])

calculate DR cross sections for H_2^+, HD^+, and D_2^+ ions in vibrational states $v = 0,1,2,3$, and 4, and, more recently, Takagi[15] has extended this work to include the effects of rotational excitation in the initial ions. Figures 6a–f show some of the results for the specific cases of $H_2^+(v = 0–5, J = 0)$.

It can be seen from Fig. 1 that for $H_2^+(v = 0)$ recombination, only products with $n = 2$ can be formed, assuming that coupling to the state with $n = 1$ is radiative and therefore very slow. Higher quantum state products are energetically forbidden, at least at low energies. For vibrationally excited state products, however, this is no longer the case. In Table III we have listed the possible product states available as the internal initial vibrational state and collision energy are increased. Dunn and co-workers[16,17] have investigated cross sections for the dissociative recombination of $H_2^+(v)$ leading to final products with $n = 2$ and $n = 4$ populated, i.e.,

$$e^- + H_2^+(v) \rightarrow H(1s) + H(2s)$$

$$e^- + H_2^+(v) \rightarrow H(1s) + H(n = 4)$$

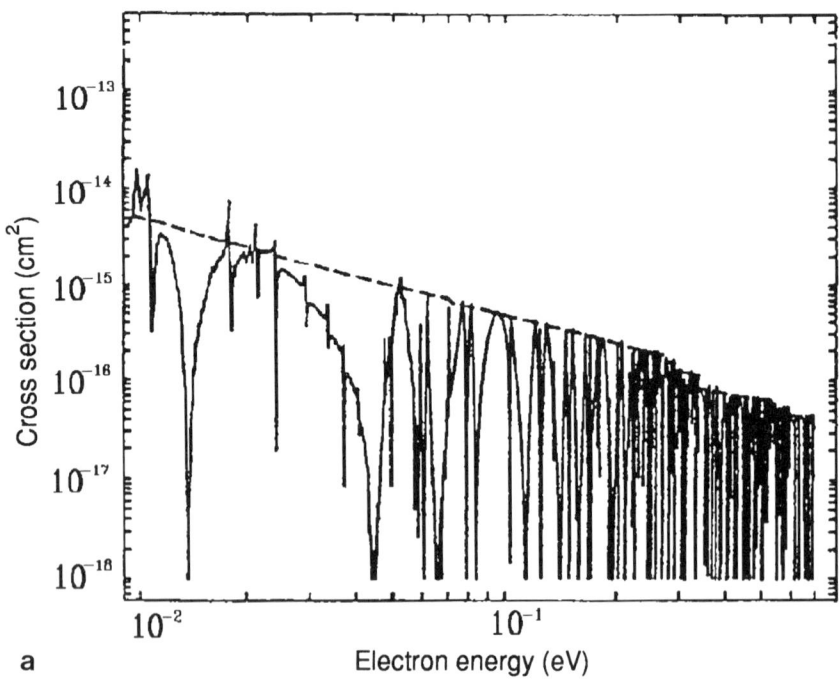

a

Figure 6. Theoretical cross sections for the dissociative recombination of H_2^+. (a) $v = 0, J = 0$; (b) $v = 1, J = 0$; (c) $v = 2, J = 0$; (d) $v = 3, J = 0$; (e) $v = 4, J = 0$; (f) $v = 5, J = 0$. (Cross sections provided by H. Takagi.)

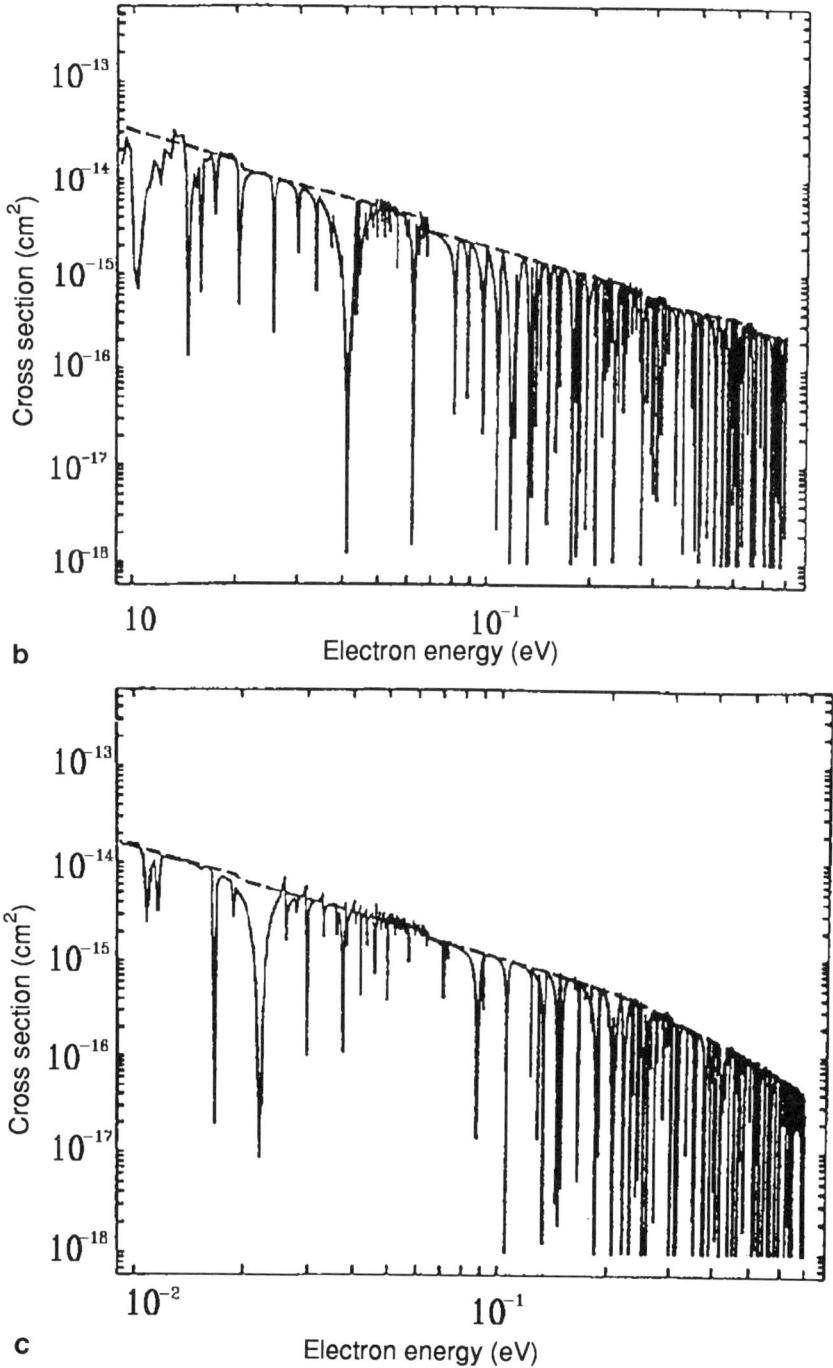

Figure 6. (continued)

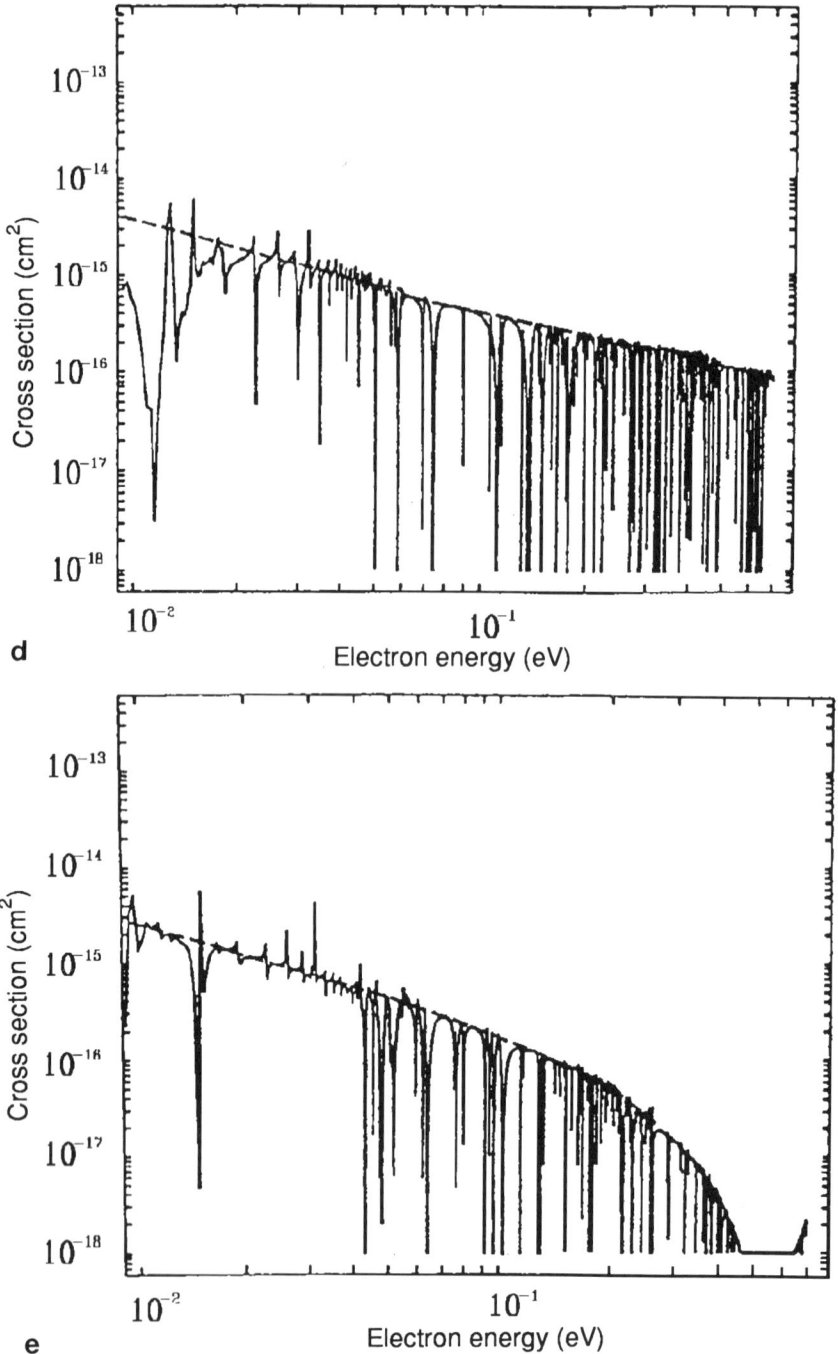

Figure 6. (continued)

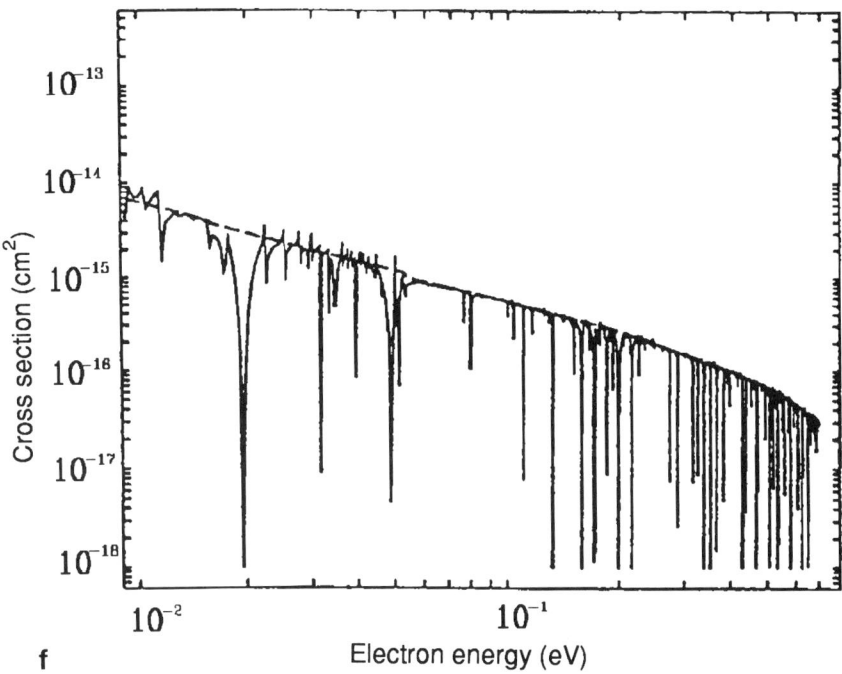

Figure 6. (continued)

They found in both cases that the cross sections accounted for about 10% of the total DR cross section. More recently, Mitchell et al.[18] studied the recombination of H_2^+ leading to the formation of hydrogen atoms in states with $n \geq 10$, and they found that this accounts for about 3% of the total cross section (Fig. 7). Both experimental and theoretical studies are currently in progress to determine the cross sections to form individual n states. These studies did indicate that highly vibrationally excited H_2^+ would exhibit very large individual collision cross sections, a phenomenon known as "super dissociative recombination."

Figure 8 shows dissociative recombination cross sections for D_2^+ measured by Peart and Dolder.[19] These ions had a vibrational distribution such as that listed in Table II and shown in Fig. 3. Also shown in the figure are Peart and Dolder's results for H_2^+, and it can be seen that there is a small difference between the two species. This reflects not only differences in the cross sections for individual vibrational levels[14] but also the differences in the vibrational level populations for the two ions.

Very recently, a new measurement of the recombination of HD^+ ions with electrons has been performed at the heavy-ion storage ring at Heidelberg in Germany.[20] This measurement covered the energy range from near zero up to 20

Table III. Kinetic Energy Released (eV) in Recombination of a 10-meV Electron with an H_2^+ Molecule in Vibrational Level v, Leading to the Dissociation Channel $H(1s) + H(nl)$

Dissociation channel	Kinetic energy released (eV) for v equal to:																		
	0	1	2	3	4	5	6	7	8	9	10	11	12	13	14	15	16	17	18
$H(1s)+H(2l)$	0.760	1.03	1.29	1.53	1.75	1.96	2.16	2.35	2.51	2.67	2.81	2.94	3.05	3.15	3.24	3.31	3.36	3.39	3.41
$H(1s)+H(3l)$	−1.13	−0.86	−0.60	−0.36	−0.14	0.07	0.27	0.460	0.62	0.78	0.92	1.05	1.16	1.26	1.35	1.42	1.47	1.50	1.52
$H(1s)+H(4l)$	−1.79	−1.52	−1.26	−1.02	−0.80	−0.59	−0.39	−0.20	−0.04	0.12	0.26	0.39	0.50	0.60	0.69	0.76	0.81	0.84	0.86
$H(1s)+H(5l)$	−2.10	−1.83	−1.57	−1.33	−1.11	−0.90	−0.70	−0.51	−0.35	−0.19	−0.05	0.08	0.19	0.29	0.38	0.45	0.50	0.53	0.55
$H(1s)+H(6l)$	−2.26	−1.99	−1.73	−1.49	−1.27	−1.06	−0.86	−0.67	−0.51	−0.35	−0.21	−0.08	0.03	0.13	0.22	0.29	0.34	0.37	0.39
$H(1s)+H(7l)$	−2.36	−2.09	−1.83	−1.59	−1.37	−1.16	−0.96	−0.77	−0.61	−0.45	−0.31	−0.18	−0.07	0.03	0.12	0.19	0.24	0.27	0.29
$H(1s)+H(8l)$	−2.43	−2.16	−1.90	−1.66	−1.44	−1.23	−1.03	−0.84	−0.68	−0.52	−0.38	−0.25	−0.14	−0.04	0.05	0.12	0.17	0.20	0.22
$H(1s)+H(9l)$	−2.47	−2.20	−1.94	−1.70	−1.48	−1.27	−1.07	−0.88	−0.72	−0.56	−0.42	−0.29	−0.18	−0.08	0.01	0.08	0.13	0.16	0.18
$H(1s)+H(10l)$	−2.50	−2.23	−1.97	−1.73	−1.51	−1.30	−1.10	−0.91	−0.75	−0.59	−0.45	−0.32	−0.21	−0.11	−0.02	0.05	0.10	0.13	0.15
$H(1s)+H(11l)$	−2.53	−2.26	−2.00	−1.76	−1.54	−1.33	−1.13	−0.94	−0.78	−0.62	−0.48	−0.35	−0.24	−0.14	−0.05	0.02	0.07	0.10	0.12
$H(1s)+H(12l)$	−2.55	−2.28	−2.02	−1.78	−1.56	−1.35	−1.15	−0.96	−0.80	−0.64	−0.50	−0.37	−0.26	−0.16	−0.07	0.00	0.05	0.08	0.10
$H(1s)+H(13l)$	−2.56	−2.29	−2.03	−1.79	−1.57	−1.36	−1.16	−0.97	−0.81	−0.65	−0.51	−0.38	−0.27	−0.17	−0.08	−0.01	0.04	0.07	0.09
$H(1s)+H(14l)$	−2.57	−2.30	−2.04	−1.80	−1.58	−1.37	−1.17	−0.98	−0.82	−0.66	−0.52	−0.39	−0.28	−0.18	−0.09	−0.02	0.03	0.06	0.08
$H(1s)+H(15l)$	−2.57	−2.30	−2.04	−1.80	−1.58	−1.37	−1.17	−0.98	−0.82	−0.66	−0.52	−0.39	−0.28	−0.18	−0.09	−0.02	0.03	0.06	0.08
$H(1s)+H(16l)$	−2.58	−2.31	−2.05	−1.81	−1.59	−1.38	−1.18	−0.99	−0.83	−0.67	−0.53	−0.40	−0.29	−0.19	−0.10	−0.03	0.02	0.05	0.07
$H(1s)+H(17l)$	−2.59	−2.32	−2.06	−1.82	−1.60	−1.39	−1.19	−1.00	−0.84	−0.68	−0.54	−0.41	−0.30	−0.20	−0.11	−0.04	0.01	0.04	0.06
$H(1s)+H(18l)$	−2.59	−2.32	−2.06	−1.82	−1.60	−1.39	−1.19	−1.00	−0.84	−0.68	−0.54	−0.41	−0.30	−0.20	−0.11	−0.04	0.01	0.04	0.06
$H(1s)+H(19l)$	−2.60	−2.33	−2.07	−1.83	−1.61	−1.40	−1.20	−1.01	−0.85	−0.69	−0.55	−0.42	−0.31	−0.21	−0.12	−0.05	0.00	0.03	0.05
$H(1s)+H(20l)$	−2.60	−2.33	−2.07	−1.83	−1.61	−1.40	−1.20	−1.01	−0.85	−0.69	−0.55	−0.42	−0.31	−0.21	−0.12	−0.05	0.00	0.03	0.05
$H(1s)+H(21l)$	−2.61	−2.34	−2.08	−1.84	−1.62	−1.41	−1.21	−1.02	−0.86	−0.70	−0.56	−0.43	−0.32	−0.22	−0.13	−0.06	−0.01	0.02	0.04
$H(1s)+H(22l)$	−2.61	−2.34	−2.08	−1.84	−1.62	−1.41	−1.21	−1.02	−0.86	−0.70	−0.56	−0.43	−0.32	−0.22	−0.13	−0.06	−0.01	0.02	0.04
$H(1s)+H(23l)$	−2.61	−2.34	−2.08	−1.84	−1.62	−1.41	−1.21	−1.02	−0.86	−0.70	−0.56	−0.43	−0.32	−0.22	−0.13	−0.06	−0.01	0.02	0.04
$H(1s)+H(24l)$	−2.62	−2.35	−2.09	−1.85	−1.63	−1.42	−1.22	−1.03	−0.87	−0.71	−0.57	−0.44	−0.33	−0.23	−0.14	−0.07	−0.02	0.01	0.03
$H(1s)+H(25l)$	−2.62	−2.35	−2.09	−1.85	−1.63	−1.42	−1.22	−1.03	−0.87	−0.71	−0.57	−0.44	−0.33	−0.23	−0.14	−0.07	−0.02	0.01	0.03

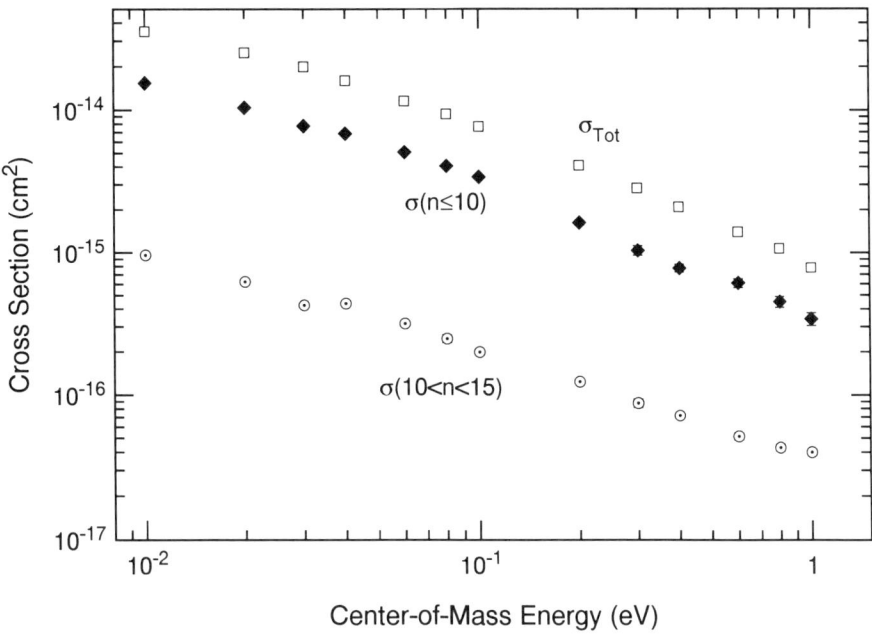

Figure 7. Measured cross sections for the dissociative recombination $e^- + H_2^+(v) \rightarrow H(n) + H(1s)$ into specified final channels. (From Mitchell et al.[18])

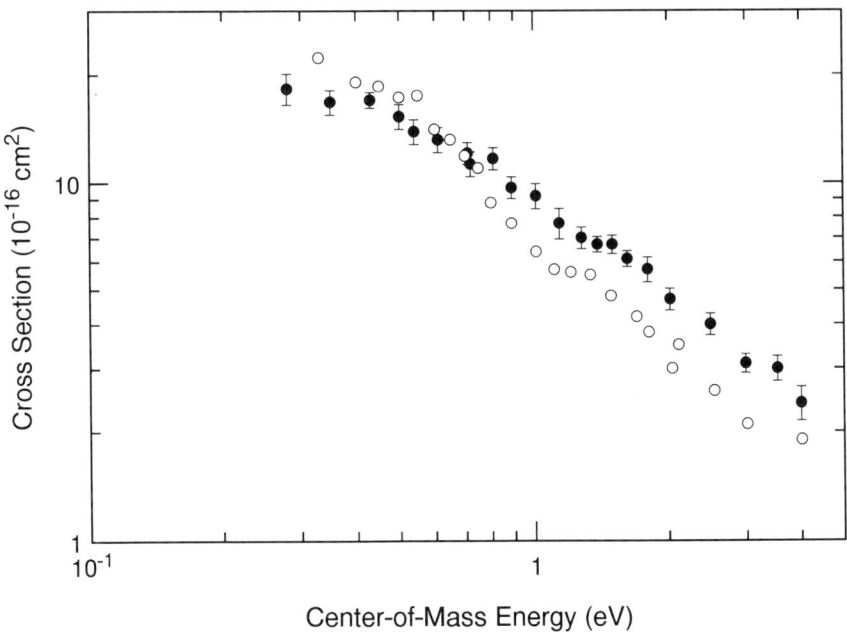

Figure 8. Experimental cross sections for the dissociative recombination of D_2^+ (●), measured by Peart and Dolder.[19] Also shown are the results obtained for H_2^+ (○), by Peart and Dolder.[12]

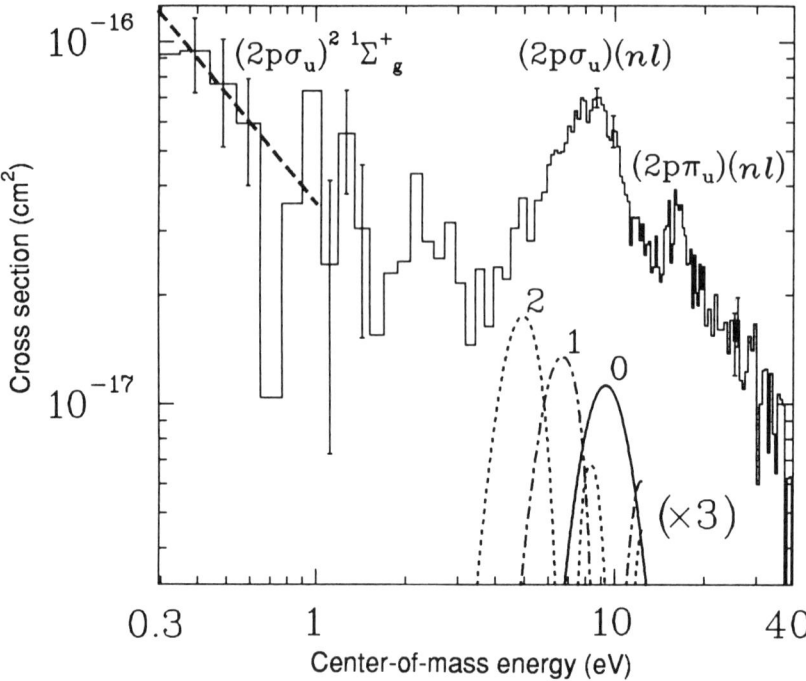

Figure 9. Experimental results for the dissociative recombination of $HD^+(v = 0)$ measured using an ion storage ring by Forck et al.[20]. The results are normalized to those of Hus et al.,[10] represented by the dashed line.

eV. In this particular study, the HD^+ ions were vibrationally cold, having been allowed to relax during several passes through the storage ring prior to recombination. The HD^+, being heteronuclear, has a dipole moment so that the excited vibrational states are short-lived. The results, normalized to those of Hus et al.,[10] are shown in Fig. 9. It can be seen that there is a strong peak in the cross section appearing at about 10 eV. Similar peaks have been seen in DR studies of HeH^+ and H_3^+,[21–23] and these peaks are due to recombination through repulsive Rydberg states converging to excited ion states.

2.2. Dissociative Excitation

Dissociative excitation is a process in which electron impact promotes a molecular ion to an excited state that either dissociates directly or is predissociated. Compared to atomic ion excitation, the excitation of molecular ions has been measured in very few instances, and this is unfortunate for it plays an important role in plasmas whose mean energies are in the tens of electron volts.

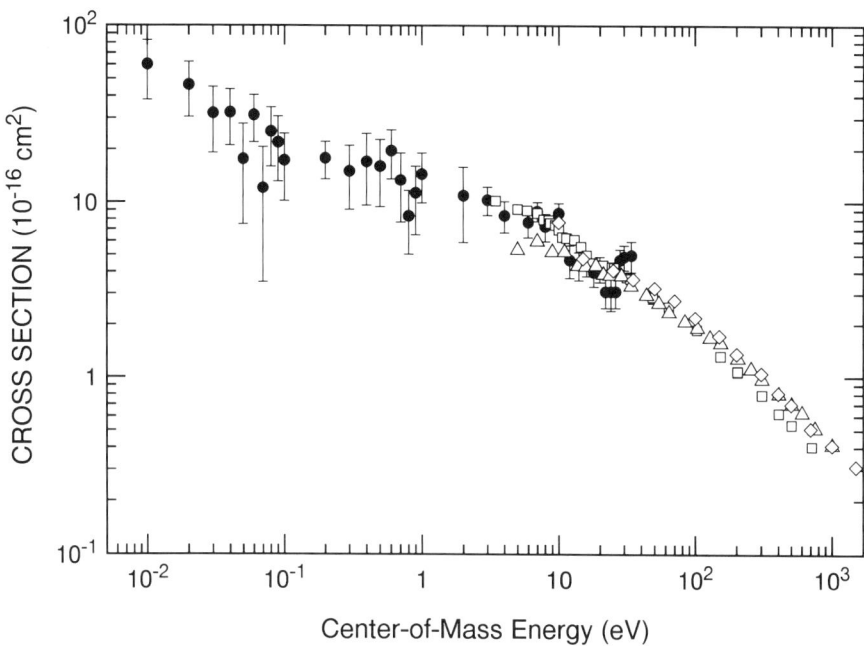

Figure 10. Measured cross sections for the dissociative excitation of H_2^+ (all v). (●) Yousif et al.[27]; (□) Peart and Dolder[24]; (△) Dunn and Van Zyl[25]; (◇) Dance et al.[26] (From Yousif et al.[27])

Peart and Dolder,[24] Dunn and Van Zyl,[25] and Dance et al.[26] have experimentally studied the dissociative excitation of H_2^+ formed with a distribution of vibrational states as discussed above:

$$e^- + H_2^+(1s\sigma_g, v) \to H_2^*(2p\sigma_u, 2p\pi_u) \to H(1s) + H^+ + e^-$$

The results of these studies are shown in Fig. 10. Also shown in this figure are some recent merged-beams results[27] that extend the energy range of the measurements down to 0.01 eV. It is seen that there is excellent agreement between all these measurements. Peek[28,29] has calculated dissociative excitation cross sections for individual vibrational states of H_2^+, and those for the $1s\sigma_g \to 2p\sigma_u$ transition are shown in Fig. 11. When Peek's results are combined with the vibrational level populations of Table I, it is found that there is excellent agreement between theory and experiment.

Hus et al.[10] examined the threshold region for H_2^+ (low v) and found cross sections several times higher than those calculated by Peek. In addition, the onset of the excitation process was seen at about 6.5 eV, which is 2.5 eV lower than the expected threshold for a $1s\sigma_g \to 2p\sigma_u$ transition. At the time of this measurement, it was believed that this was an indication of the presence of vibrationally excited

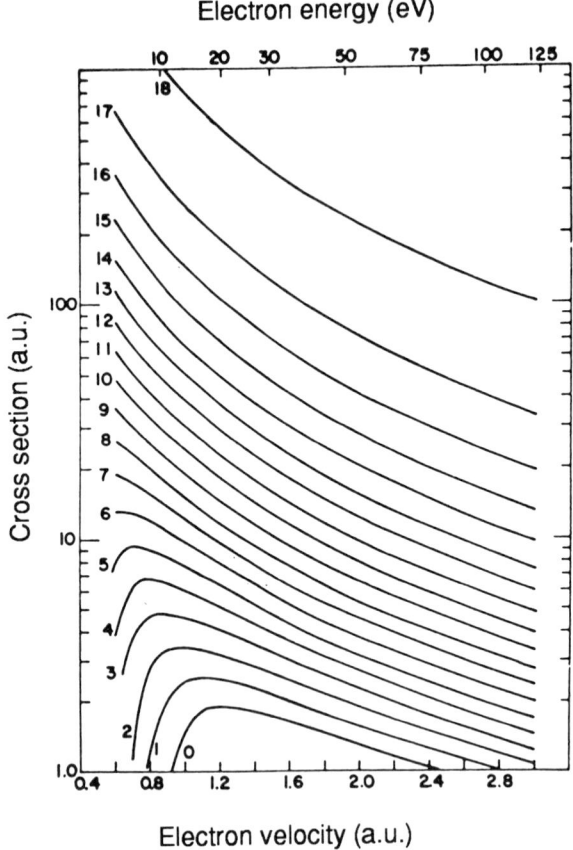

Figure 11. Theoretical cross sections for the dissociative excitation of H_2^+ in specified vibrational states, $v = 0–18$, as calculated by Peek.[28,29]

states in the ion beam although the observation of resonances in the recombination cross section argued against this. Very recently, storage ring measurements of this process have been made, and results very similar to those of Hus *et al.* have been found (D. Zajfman, private communication, 1993). It would seem that the early onset and large cross sections are a manifestation of an excitation to neutral Rydberg states lying below the $2p\sigma_u$ state that subsequently autodissociate. This is a topic that deserves further investigation.

2.3. Dissociative Ionization

Dissociative ionization is similar to dissociative excitation except that the second hydrogen atom is excited past its ionization limit:

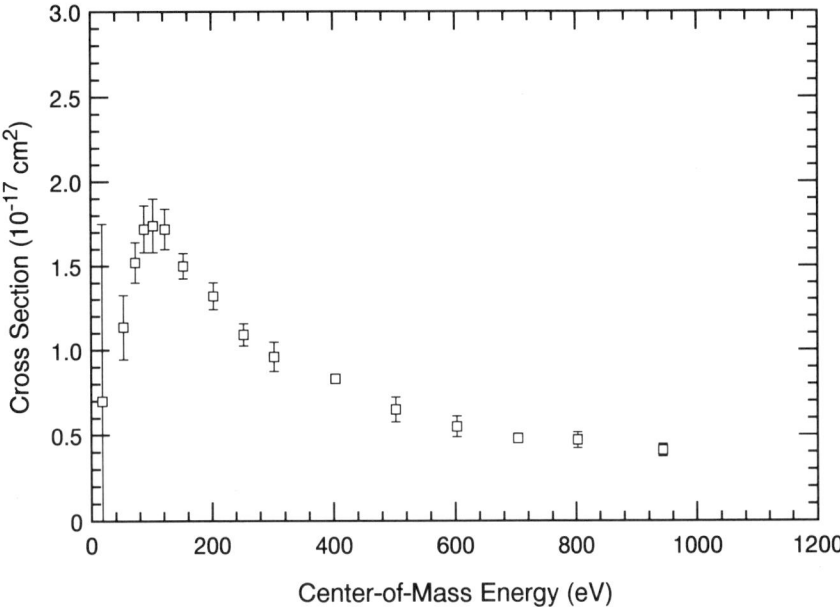

Figure 12. Experimental cross sections for the dissociative ionization of H_2^+ (all v). (Adapted from Peart and Dolder.[30])

$$e^- + H_2^+ \rightarrow H^+ + H^+ + 2e^-$$

This process has also been studied by Peart and Dolder,[30] and their results are shown in Fig. 12. Since the threshold is at 30 eV above the $v = 0$ level of H_2^+, Peart and Dolder have argued that there should be only a weak dependence on the state of excitation of the ion. It should be noted that even at the peak the cross sections are much smaller than for dissociative excitation.

2.4. Ion-Pair Formation

Peart and Dolder[31] have studied the recombination of H_2^+ ions leading to the formation of ion pairs, i.e.,

$$e^- + H_2^+ \rightarrow H^+ + H^-$$

As mentioned earlier, the recombination process arises via a transition from the ion ground state to an intermediate doubly excited state such as the $^1\Sigma_g^+$ state of H_2. At large internuclear distance R, this particular state has $H^+ + H^-$ as its dissociation limit, and so it might be expected that ion-pair formation would be a major decay channel for the recombination process. The cross section measurements are shown in Fig. 13 and refer to ions with a vibrational population distribution such as that listed in Table I. Since the $^1\Sigma_g^+$ state crosses the ion ground state in the vicinity of the $v = 1$ level, no threshold for ion-pair formation is seen. Had the ions been in their $v = 0$ level, then the onset would have occurred at about 4 eV. It can be seen from Fig. 13 that in fact the measured cross sections are much smaller than those for the neutral decay channel for H_2^+ recombination (Fig. 4). The reason for this is as follows. As the hydrogen atoms fly apart, gaining potential energy, the potential energy of the $^1\Sigma_g^+$ state decreases, eventually falling below the ionization potential of the ion. At this point, the recombination is stabilized against re-ionization. As the system proceeds toward large R, the $^1\Sigma_g^+$ state crosses a myriad of neutral Rydberg states, and at each crossing there is the possibility for the system to make a transition to a Rydberg state, thus dissociating to form neutral hydrogen atoms, $H(1s) + H(nl)$. Given the large number of Rydberg states, it is much more likely that such a crossing will occur than that the system will proceed to its final ion-pair limit. Hence, ion-pair formation is a very minor decay channel for the recombination process.

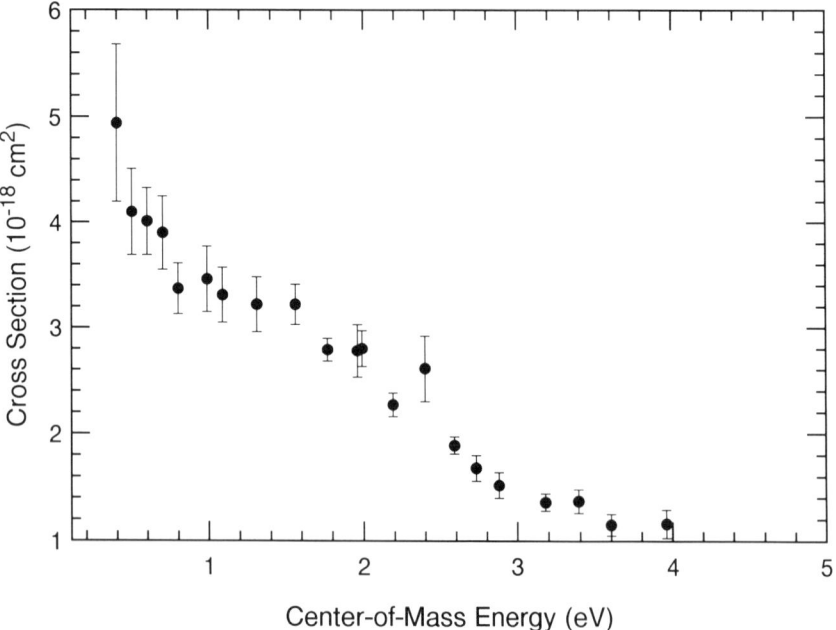

Figure 13. Experimental cross sections for ion-pair formation following the dissociative recombination of H_2^+ (all v). (Adapted from Peart et al.[31]).

3. H_3^+

H_3^+ is usually the dominant ion in a hydrogen plasma, as it is formed through the rapid reaction

$$H_2^+ + H_2 \rightarrow H_3^+ + H$$

Anicich and Futrell[32] have analyzed the vibrational populations of H_3^+ arising from this reaction, and their results are tabulated in Table IV and illustrated in Fig. 14. Also shown in Table IV and Fig. 15 are data for D_3^+ produced from the corresponding reaction between D_2^+ and D_2. The vibrational spacing of H_3^+ is 0.37 eV compared to a spacing of 0.26 eV for D_3^+. This means that when D_3^+ is formed, the vibrational population is strongly skewed toward higher excited states. H_3^+ has two vibrational modes, a symmetric stretch and an asymmetric stretch, and the energies and lifetimes of the vibrational states have been calculated by Carney and Porter[33] and, more recently, by Dinelli et al.[34] The asymmetric states can make dipole transitions to the symmetric ground vibrational state and so have millisecond lifetimes. The symmetric states are able to decay directly to the ground state only via electric quadrupole transitions, but they can make intermediate transitions to the asymmetric states so that they also generally have effective lifetimes in the millisecond range, although the first excited A_1 state has a lifetime for transition to the first E state of about 1 s.

Table IV. Relative Populations and Energies (above the $v = 0$ States) of the Vibrational Levels of H_3^+ and D_3^+ [a]

V	H_3^+		D_3^+	
	Population	Energy (eV)	Population	Energy (eV)
0	0.0341	0.0	0.0106	0.0
1	0.0919	0.372	0.0290	0.261
2	0.1601	0.744	0.0556	0.522
3	0.2197	1.116	0.0853	0.783
4	0.2303	1.488	0.1156	1.04
5	0.1387	1.86	0.1424	1.305
6	0.0796	2.232	0.1587	1.566
7	0.0299	2.604	0.1483	1.827
8	0.0109	2.976	0.1146	2.088
9	0.0037	3.348	0.0672	2.349
10	0.0010	3.720	0.0396	2.610
11	0.0001	4.092	0.0187	2.871
12			0.0083	3.132
13			0.0035	3.393
14			0.0014	3.654
15			0.0001	4.176

[a] From Ref. 32.

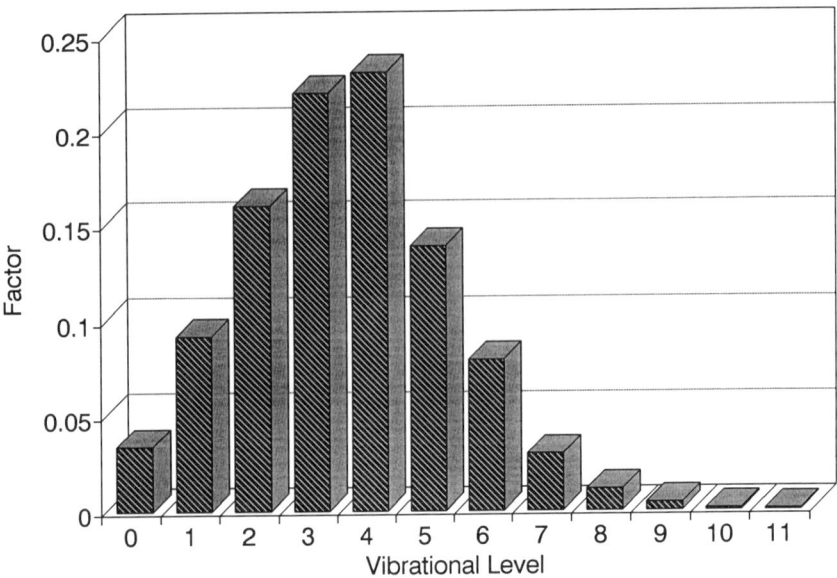

Figure 14. Population factors for the vibrational states of H_3^+ following production. The histogram has been plotted using the results of Anicich and Futrell.[32]

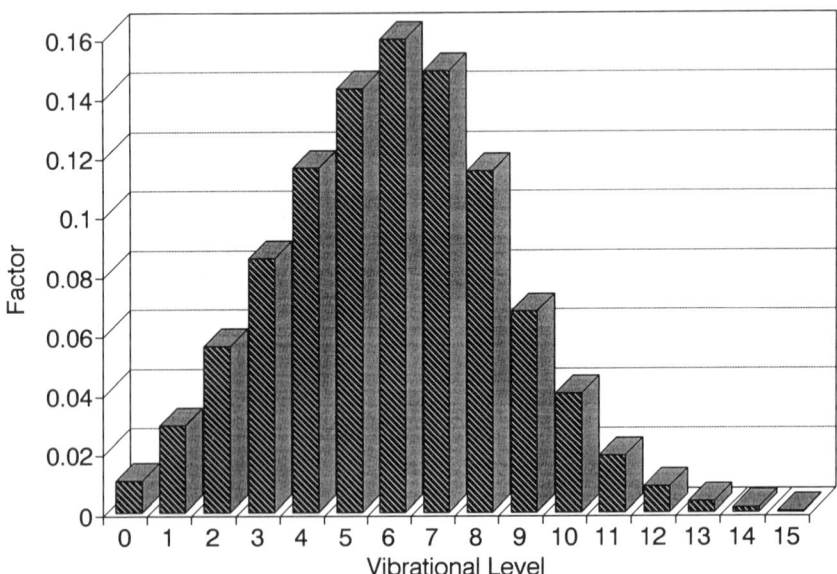

Figure 15. Population factors for the vibrational states of D_3^+ following production. The histogram has been plotted using the results of Anicich and Futrell.[32]

In addition to radiative deexcitation, H_3^+ ions can be vibrationally deexcited through the proton transfer reaction

$$H_3^+(v) + H_2(v' = 0) \rightarrow H_3^+(v = 0) + H_2(v')$$

This reaction was studied by Kim et al.,[35] who found a deactivation rate for the initial distribution of vibrationally excited states of 2.7×10^{-10} cm^3/s. A closer examination of this process by Blakley et al.,[36] however, revealed that while this figure is representative of deactivation rates for higher vibrational states, rates for the $v = 1$ and $v = 2$ levels were considerably smaller. Indeed, Blakley et al. found rates of 10^{-12} cm^3/s and 10^{-11} cm^3/s, respectively, for these states. This has very important consequences for the dissociative recombination of H_3^+, as will be discussed in the next section.

3.1. Dissociative Recombination

Dissociative recombination of H_3^+ has proved to be a very controversial subject, and many experimental studies have addressed this problem over the last few years. A number of reviews of the status of the problem have been given,[37,38] and only an overview will be presented here. Figure 16 shows the potential energy curves for H_3 and H_3^+,[39] and it can be seen that the neutral state through which the

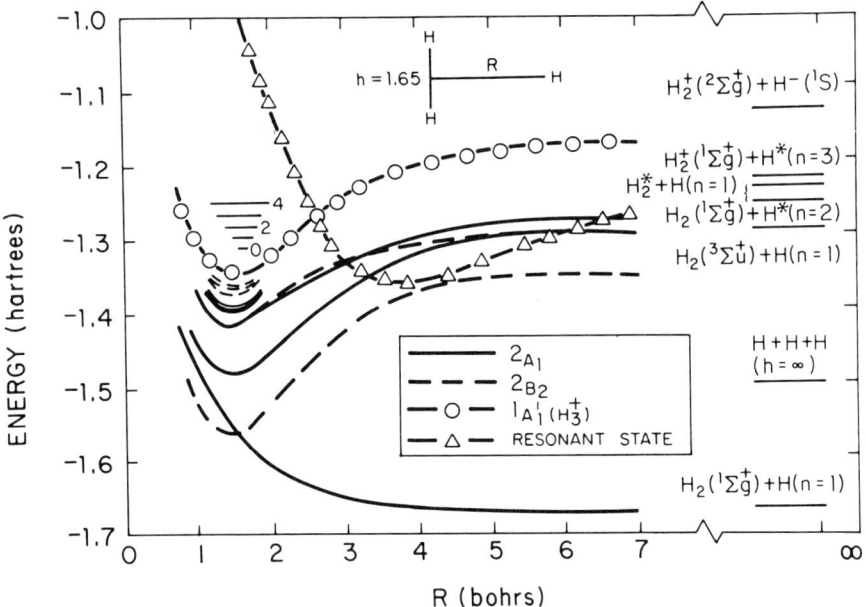

Figure 16. Potential energy curves for H_3 and H_3^+.[39]

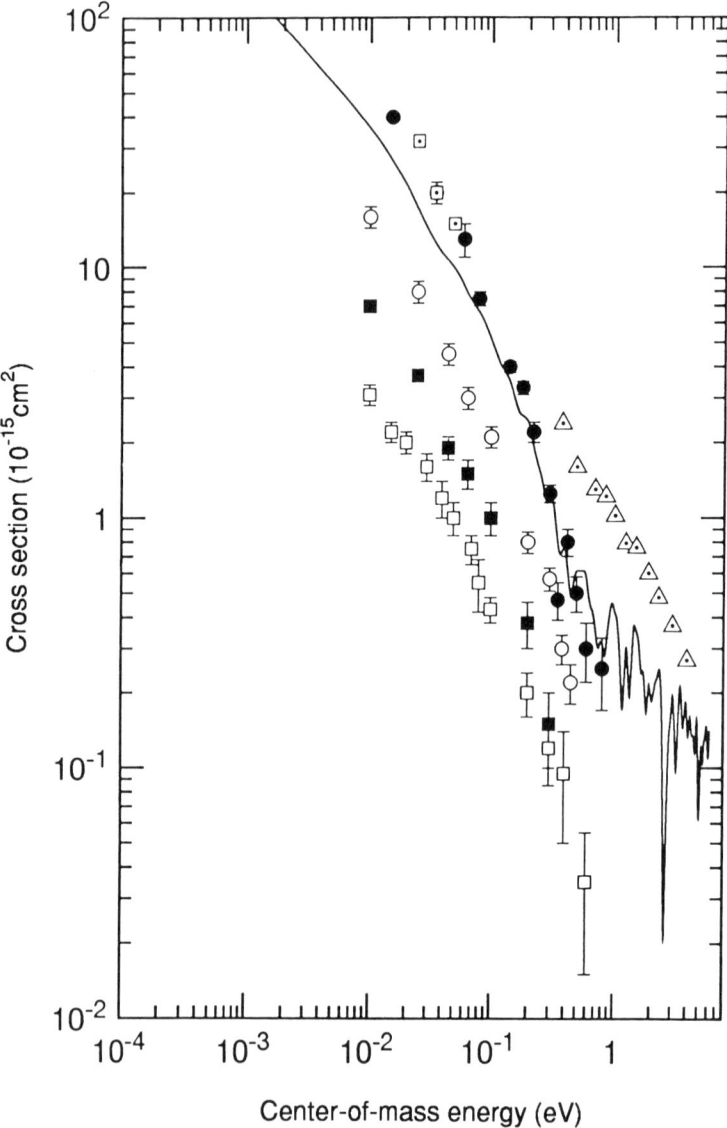

Figure 17. Dissociative recombination cross sections for H_3^+. —) Auerbach et al.[9]; △) Peart and Dolder[41]; ⊡) Leu et al.[42] (estimated from measured rate coefficients); ●, ○, and ■) Hus et al.[40] for successively lower vibrational state population; □) Hus et al.[40] for $v = 0$. (From Yousif et al.[48])

dissociative recombination process is expected to proceed intersects the ion state in the vicinity of the $v = 3$ level. This means that lower vibrational states might be expected to display lower recombination cross sections. Indeed, this has been observed in merged-beams experiments, where a small cross section corresponding to a rate coefficient of about 2×10^{-8} cm^3/s at room temperature was found for ground state ions[40] (Fig. 17).† A flowing afterglow Langmuir probe (FALP) experiment has also indicated a rate coefficient for ions in low vibrational states of about 2×10^{-8} cm^3/s at 300 K.[43,44] This finding is, however, contradicted by recent afterglow experiments by Canosa et al.[45] and by Amano,[46,47] who have found a rate coefficient of 1×10^7–2×10^{-7} cm^3/s. It has been proposed that the fact that the merged-beams results of Hus et al. are smaller than those from these afterglow experiments might possibly be due to electric field effects that could diminish the measured cross section.[48] Preliminary results from a low-field merged-beams experiment[49] have provided confirmation of this. When the field on the primary ion beam analyzer is reduced by more than an order of magnitude, the measured recombination cross section is found to increase to values close to those for excited ions. At the time of writing, this work is still in progress.

As in the case of HD$^+$, a storage ring measurement of H_3^+ recombination has been performed,[23] and these results are shown in Fig. 18. These ions should also be in the ground state, and it is interesting to note that the cross section obtained in this measurement is about a factor of 2.7 larger than that measured by Hus et al., but it still corresponds to a 300 K rate coefficient less than that obtained by Canosa et al.[45] and Amano[46,47] in their afterglow experiments. However, in a very recent paper from the same group,[50] the cross sections reported are a factor of 3 higher and now agree with the results of Canosa et al.

Insofar as the recombination of H_3^+ ions in a thermonuclear plasma is concerned, however, it is likely that the ions will still be vibrationally excited in this situation, and then the higher cross sections shown in Fig. 17 are appropriate.

Mitchell and co-workers[51,52] have studied the branching ratio for the dissociative recombination channels

$$e^- + H_3^+(v) \begin{cases} H + H + H & \text{(I)} \\ H_2 + H & \text{(II)} \\ H_3^* & \text{(III)} \end{cases}$$

using a modification of the merged-beams method. For vibrationally excited ions, they found that the ratio of channel I to channel II was about 2:1, whereas for ions in the $v = 0$ state only, the channels were about equally probable. Surprisingly, in the latter experiment it was found that channel III accounted for about 8% of the

---

†When vibrationally excited ions were studied, the measured cross sections correspond to a room temperature rate of 2×10^{-7} cm^3/s.

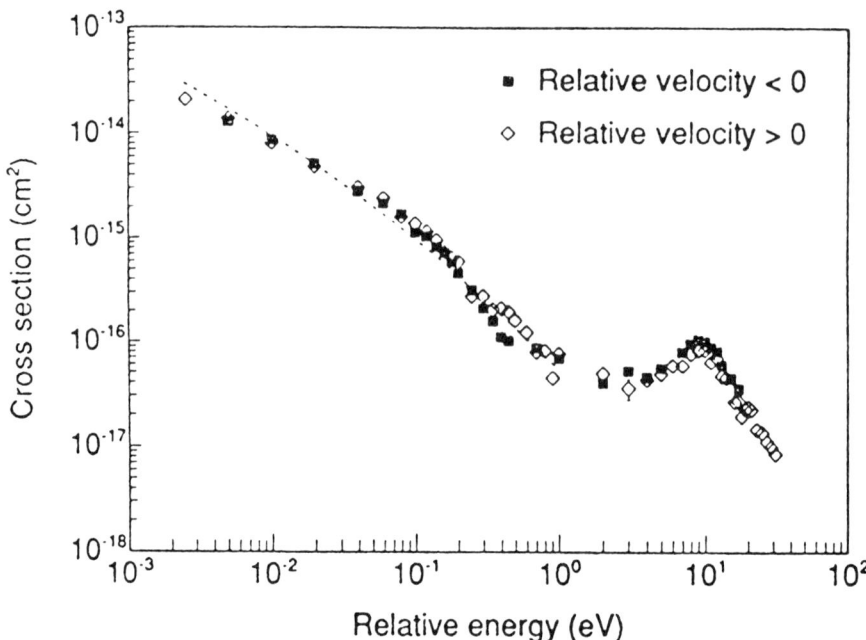

Figure 18. Dissociative recombination cross sections for H_3^+ ($v = 0$) measured using a storage ring. (From Larsson et al.[23])

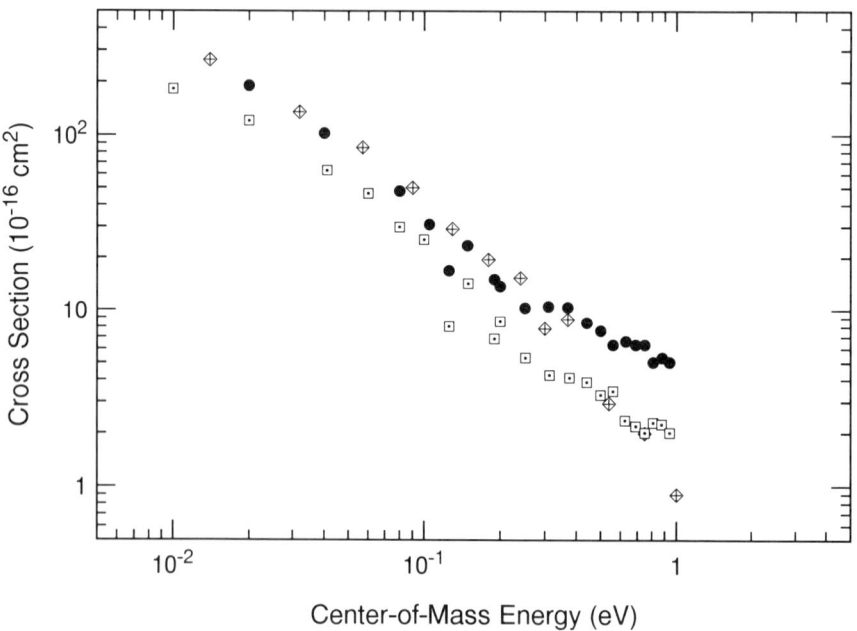

Figure 19. Experimental cross sections for the dissociative recombination of H_3^+ (⊕), HD_2^+ (●), and D_3^+ (□). Ions have excited vibrational states populated. (Adapted from Mitchell et al.[53])

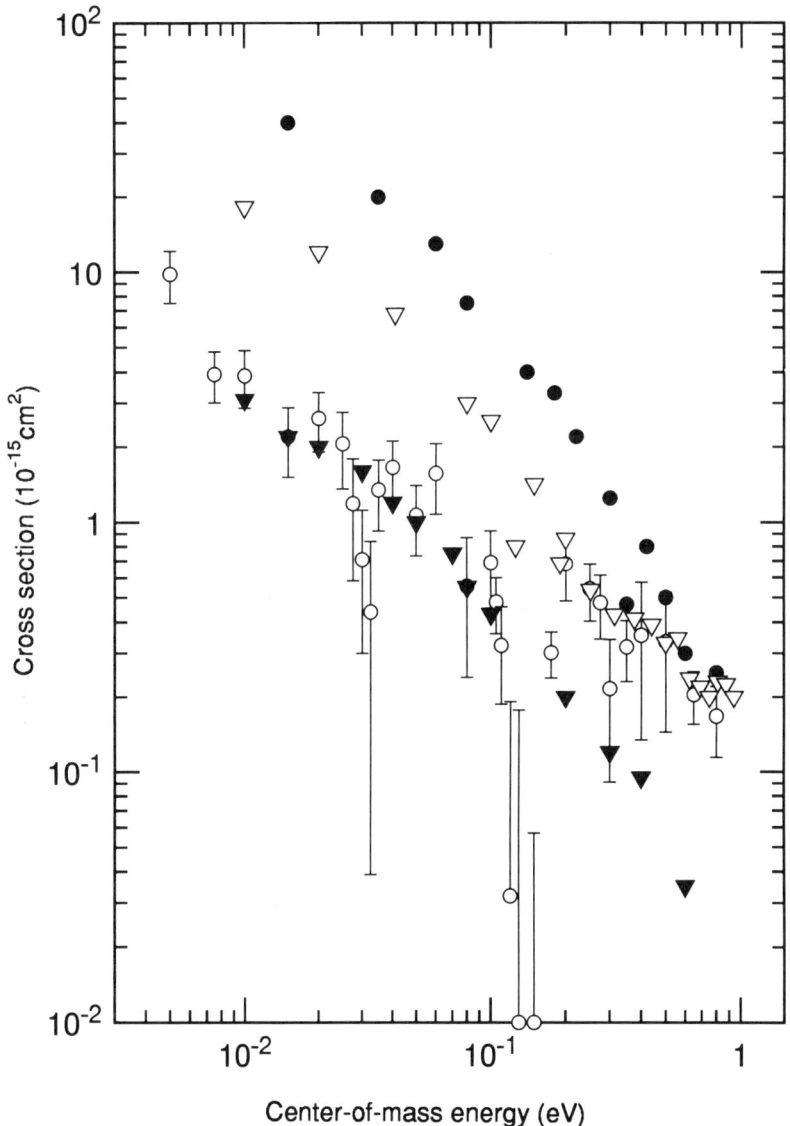

Figure 20. Cross sections for the dissociative recombination of D_3^+ and H_3^+. ○) D_3^+, $v = 0$; ▽) D_3^+, vibrationally excited ion; ▼) H_3^+, $v = 0$; ●) H_3^+ excited ions. (From Van der Donk et al.[54])

total measured cross section. (This channel had not been looked for in the excited ion measurement.) Since H_3 has an unbound ground state, this species must eventually dissociate via channel I or channel II. The measurement indicated, however, that the lifetime of the H_3^* molecules must have exceeded 10^{-7} s.

In 1984, Mitchell et al.[53] reported cross sections for the recombination of H_3^+, HD_2^+, and D_3^+ ions that had been prepared in a low-pressure ion source so that they were vibrationally excited. It was found that the D_3^+ ions exhibited a cross section that was about a factor of 2 lower than that for H_3^+. (The HD_2^+ cross sections were found to be closer to those for H_3^+ but exhibited a linear energy dependence instead of the disjointed dependence found for the other two isotopomers.) These results are shown in Fig. 19. More recently, Van der Donk et al.[54] have reexamined this problem but this time focusing on ground state ions. In this case, it was found that the H_3^+ and D_3^+ ions yielded similar recombination cross sections (Fig. 20). The reason for the difference noted in the earlier work with vibrationally excited ions can be explained by reference to Table IV and Figs. 14 and 15. It is seen that when D_3^+ ions are formed, a larger percentage of them are in vibrationally excited states, and, as noted above, the more excited vibrational states exhibit larger quenching rates than lower states. Hence, although the D_3^+ ions start out more excited, by the time they have exited the ion source and reached the interaction region of the merged-beams apparatus, they are less excited than the H_3^+ ions and therefore are found to have smaller recombination cross sections.

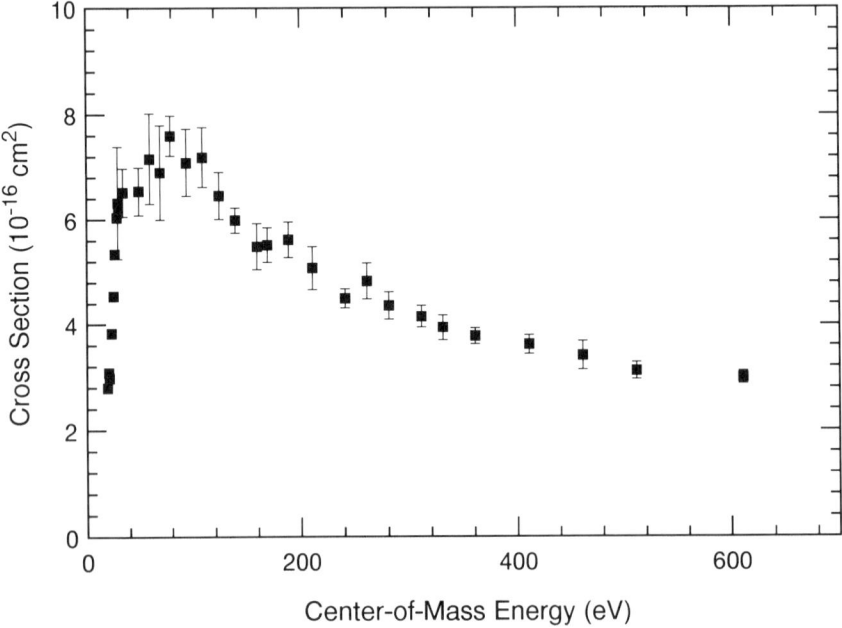

Figure 21. Cross sections for the dissociative excitation of H_3^+. (Adapted from Peart and Dolder.[55,56])

3.2. Dissociative Excitation

The dissociative excitation of H_3^+ ions has been measured by Peart and Dolder[55,56] and by Yousif et al.[57] The latter measurements refer to ions that were in their ground vibrational state.† In the case of Peart and Dolder's measurements, the ions were more excited, with at least the $v = 1$ state populated (Ref. 48). The results of Peart and Dolder cover the energy range from below threshold to 500 eV and are shown in Fig. 21. Comparison with Fig. 17 shows that dissociative excitation will dominate the H_3^+ electron–ion collisions in the energy region above 15 eV.

3.3. Ion-Pair Formation

An alternative decay channel for the dissociative recombination of H_3^+ is ion-pair formation, that is,

$$e^- + H_3^+ \rightarrow H_2^+ + H^-$$

This process involves a direct transition from the initial vibrational level of the ion ground state to the $^1A_1'$ state that has the ion pair as its dissociation limit. It can be seen from Fig. 16 that this state lies 5.4 eV above the ground vibrational level at its equilibrium internuclear distance, and so for cold ions, the process will exhibit a threshold at this energy.

Figure 22 shows the results of two separate experiments that have been carried out to study this process. The open circles in the figure are the results of Peart et al.,[58] obtained with an inclined-beams apparatus, while the closed circles are the merged-beams results of Yousif et al.[48] Both these measurements were made using ions created under high-pressure source conditions designed to produce cold ions. From the position of the threshold, it can be seen that the merged-beams experiment was more successful in this regard. It would seem that the inclined-beams experiment used ions that had some fraction in the $v = 1$ level. The agreement between these two measurements is, however, very good. The closed squares in the figure refer to measurements taken with ions created under lower pressure source conditions. The position of the threshold for these results shows that again $v = 1$ was populated but now occupied a larger fraction of the beam. As might be suspected, the cross section is seen to increase with increasing vibrational excitation, but without knowledge of the fractional population of the $v = 1$ level, this cannot be described quantitatively. The $^1A_1'$ state crosses the ion ground state in the vicinity of the $v = 4$ level, and presumably this state would exhibit the maximum cross section for ion-pair production without the presence of a threshold for onset. Studies of product branching ratios for H_2^+ have shown, however, that the neutral channel strongly dominates over ion-pair formation even when there is a favorable overlap between the ion ground state and the recombining state.[31] This is because the

---

†The merged-beams results covered only the threshold region and are not presented here.

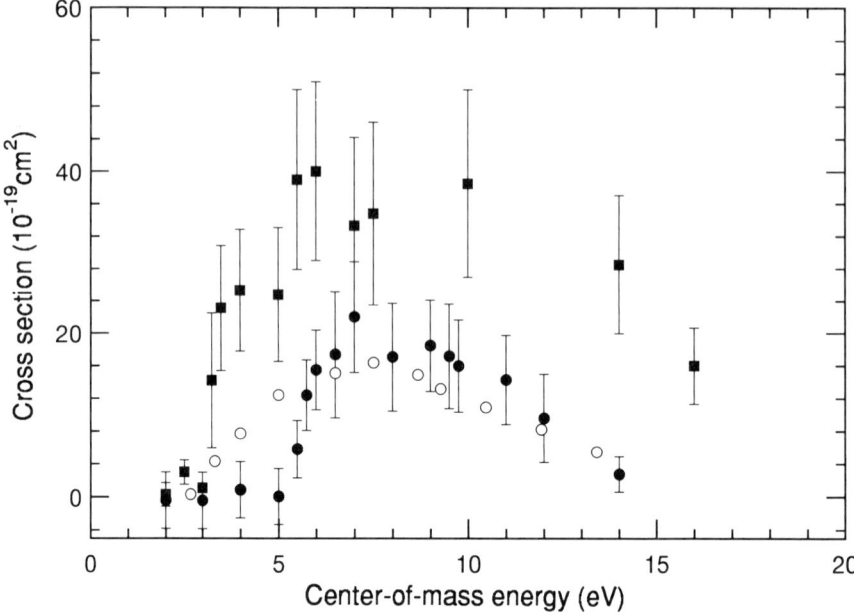

Figure 22. Cross sections for the ion-pair formation in the dissociative recombination of H_3^+. ●) H_3^+ ($v = 0$) (Yousif et al.[48]); ■) excited ions (Yousif et al.[48]; ○) cool ions (Peart et al.[58]). (From Yousif et al.[48])

dissociating system has many opportunities to undergo curve-crossing transitions with the neutral Rydberg states lying beneath the ion ground state. The likelihood of the recombining state surviving out to large R, and thus to the ion-pair limit, is small, and so it is not likely that the ion-pair channel will compete effectively with either the dissociative recombination or the dissociative excitation processes at any energy.

4. O_2^+

4.1. Dissociative Recombination

Cross sections for the dissociative recombination of O_2^+ ions have been measured over the energy range from 0.01 to 1 eV by Mul and McGowan[59] using a merged-beams apparatus and over the energy range from 0.1 to 8 eV by Walls and Dunn[60] using an ion trap. These results are shown in Fig. 23, and it is seen that there is good agreement between the two measurements. Franck–Condon factors for the ionization of ground state O_2 are listed in Table V. Figure 24 shows a histogram of this vibrational population.

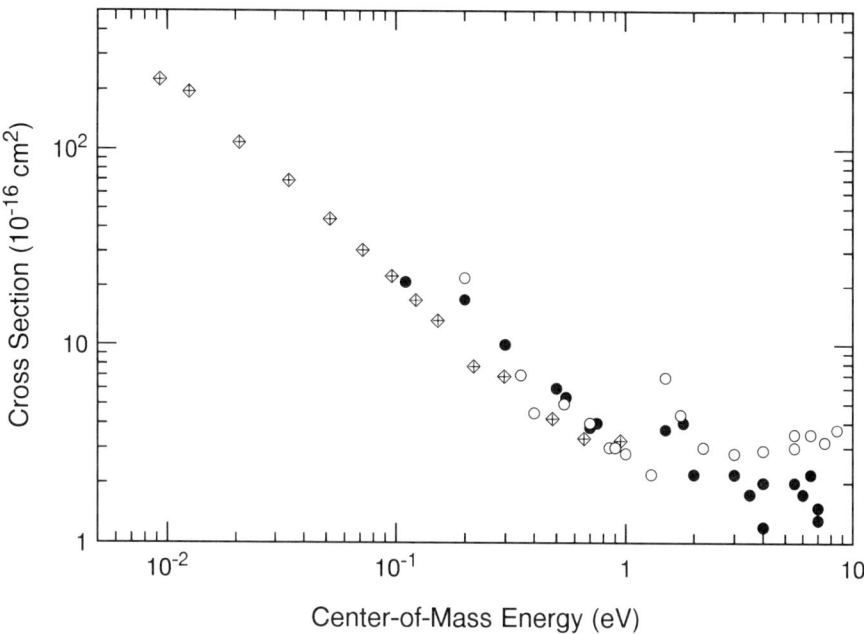

Figure 23. Dissociative recombination cross sections for O_2^+. ⊕) Mul and McGowan[59]; ○ and ●) Walls and Dunn.[60] (Adapted from Mul and McGowan.[59])

Figure 24 shows that the ions, when formed, will be predominantly in the $v = 0, 1, 2,$ and 3 levels. Since O_2^+ is homonuclear, excited vibrational levels are radiatively long-lived (10^5–10^7 s) and, because of the low pressures (10^{-10} Torr) used in ion trap experiments, would not have been collisionally quenched in the measurement of Walls and Dunn.[60] O_2^+ can, however, be rapidly deexcited via symmetric charge transfer collisions with O_2,[61] and so the Mul and McGowan results (where the ions were produced in an ion source operating at 0.1 Torr) probably refer to much colder ions. Deexcitation rates of O_2^+ with other gases such as H_2 and He are orders of magnitude smaller than with O_2, and so in an edge plasma environment the ions will probably be in an excited state.

Table V. Franck–Condon Factors for $O_2^+(v)$

v	Franck–Condon factor
0	0.188
1	0.365
2	0.290
3	0.123
4	0.030
5	0.004

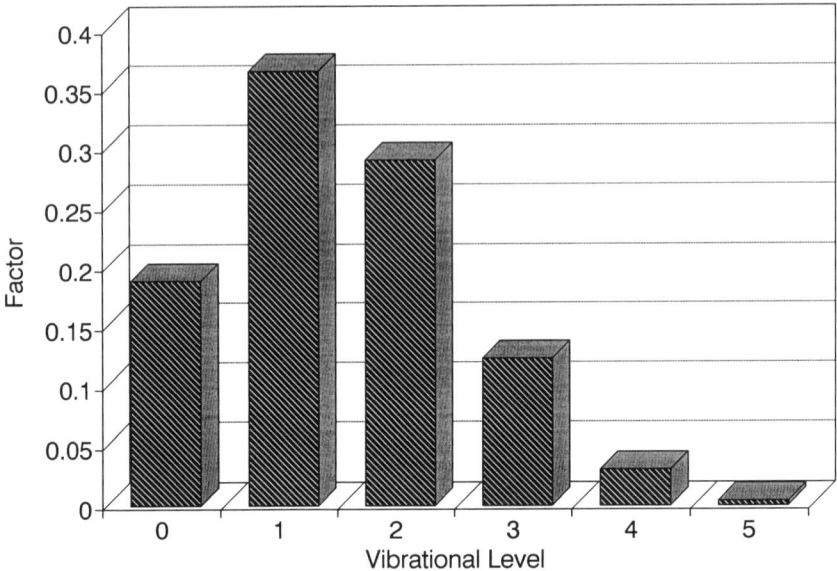

Figure 24. Population factors for O_2^+ following electron impact.

Guberman[62] has calculated the potential energy curves of valence states of O_2 that intersect the ion ground state and that participate in the dissociative recombination. There are many such states, and it would appear from experimental observations that the total recombination cross section is rather insensitive to the vibrational state of the ion. Guberman and Giusti-Suzor[63] have calculated the cross sections for the particular channel

$$e^- + O_2^+(v) \rightarrow O(^1S) + O(^1D)$$

including both direct and indirect recombination, and the results of these calculations are shown in Fig. 25. Only one state, the $^1\Sigma_u^+$ state that intersects the ion ground state at low vibrational levels, leads to the formation of excited $O(^1S)$ atoms, and so here it can be seen that there is a vibrational dependence.

In addition to cross section measurements, there have also been a number of rate coefficient measurements using afterglows and shock tubes (Ref. 64 and references therein). These have generally agreed well with each other, and the accepted value at 300 K is 2×10^{-7} cm^3/s.

Electron–Molecular Ion Collisions

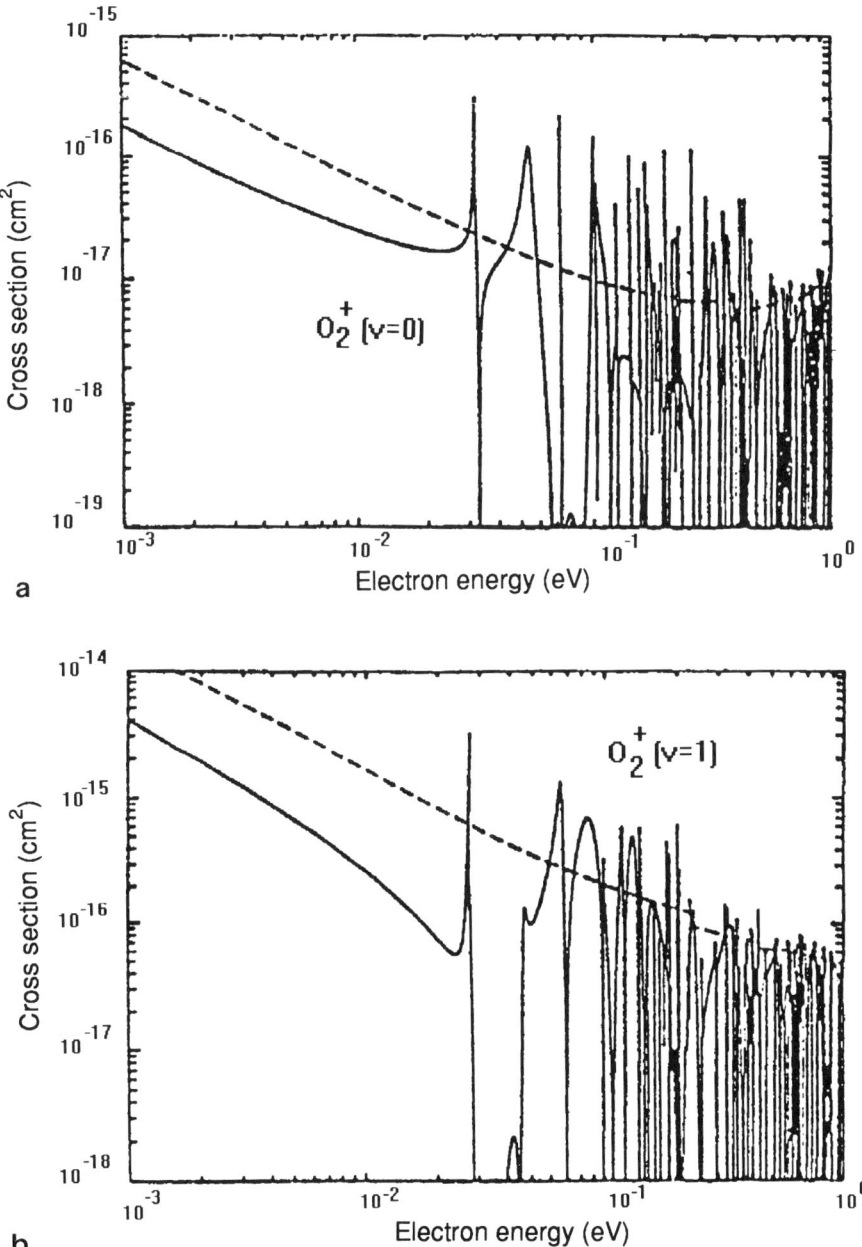

Figure 25. The full (—) and direct (---) cross sections for the $^1\Sigma_u^+$ dissociative route from the $v = 0$ (a), $v = 1$ (b), and $v = 2$ (c) states of O_2^+. (From Guberman and Giusti-Suzor.[63])

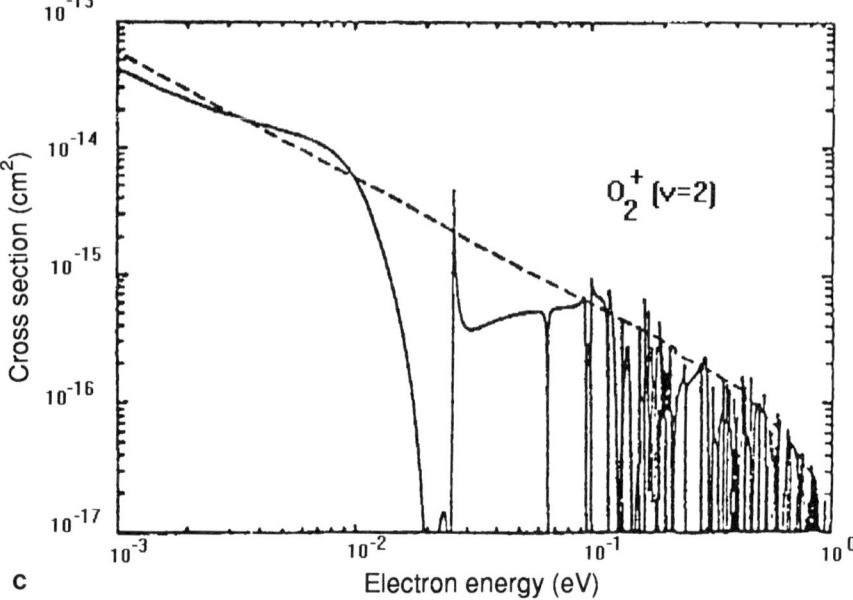

Figure 25. (continued)

4.2. Dissociative Excitation

The dissociative excitation of O_2^+:

$$e^- + O_2^+ \rightarrow O + O^+ + e^-$$

has been measured by Van Zyl and Dunn,[65] and the results are shown in Fig. 26. These measurements refer to an unidentified mixture of vibrational states. The cross section has a rather flat energy dependence above 100 eV, and in this region dissociative excitation will dominate over dissociative recombination.

5. CO^+

5.1. Dissociative Recombination

The dissociative recombination of CO^+ has been measured by Mitchell and Hus,[66] and the results are shown in Fig. 27. This measurement used ions with a range of vibrational states up to $v = 7$ populated. An analysis of Franck–Condon factors for CO ionization and for the radiative relaxation of excited states of CO^+ was used to predict the populations of the various levels, and the population factors obtained are listed in Table VI and illustrated in Fig. 28.

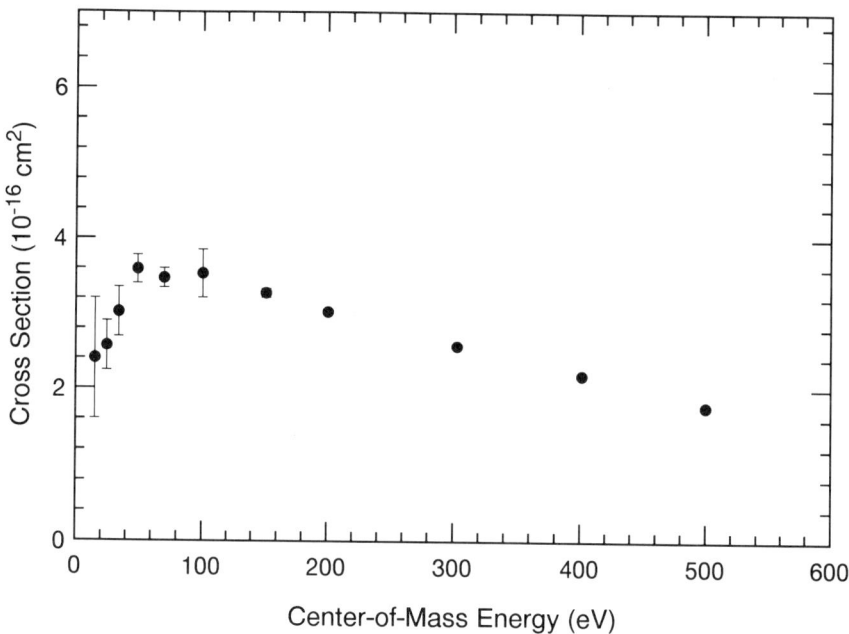

Figure 26. Cross sections for the dissociative excitation of O_2^+. (Adapted from Van Zyl and Dunn.[65])

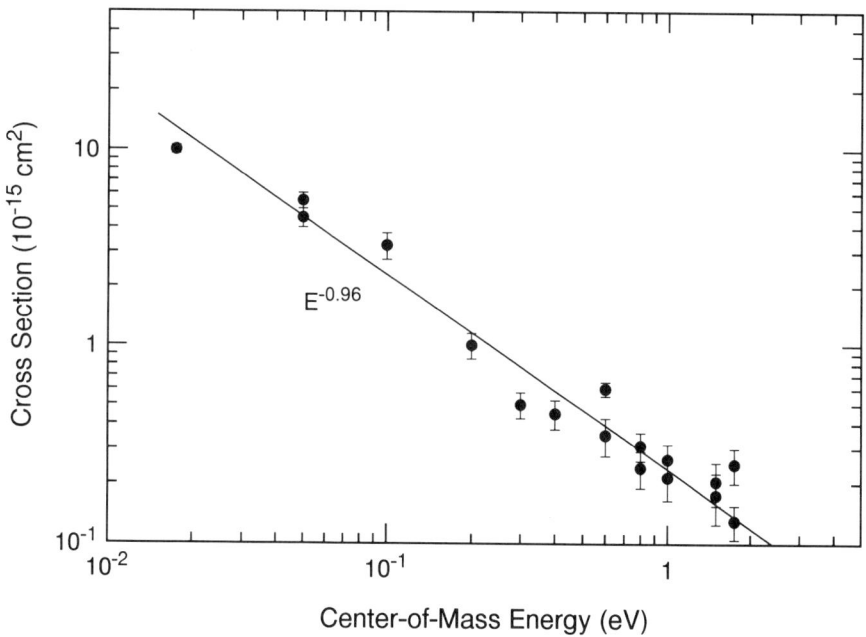

Figure 27. Dissociative recombination cross sections for CO^+. (Adapted from Mitchell and Hus.[66])

Table VI. Population Factors for the Vibrational Levels of Ground State CO^+ Following Production

v	Population
0	0.62
1	0.149
2	0.079
3	0.05
4	0.032
5	0.02
6	0.015
7	0.0076

5.2. Dissociative Excitation

The dissociative excitation of CO^+ has also been measured,[66] and the cross sections are shown in Fig. 29. Unfortunately, no information is available concerning the effects of initial vibrational excitation on either the recombination or the excitation cross sections. Again, dissociative excitation will dominate over recombination at energies above a few electron volts.

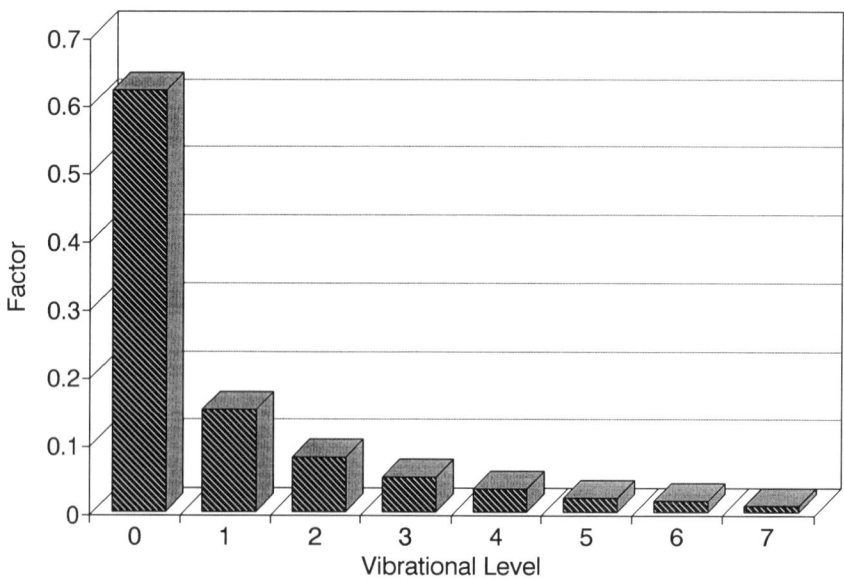

Figure 28. Population factors for CO^+ following electron impact and radiative relaxation.

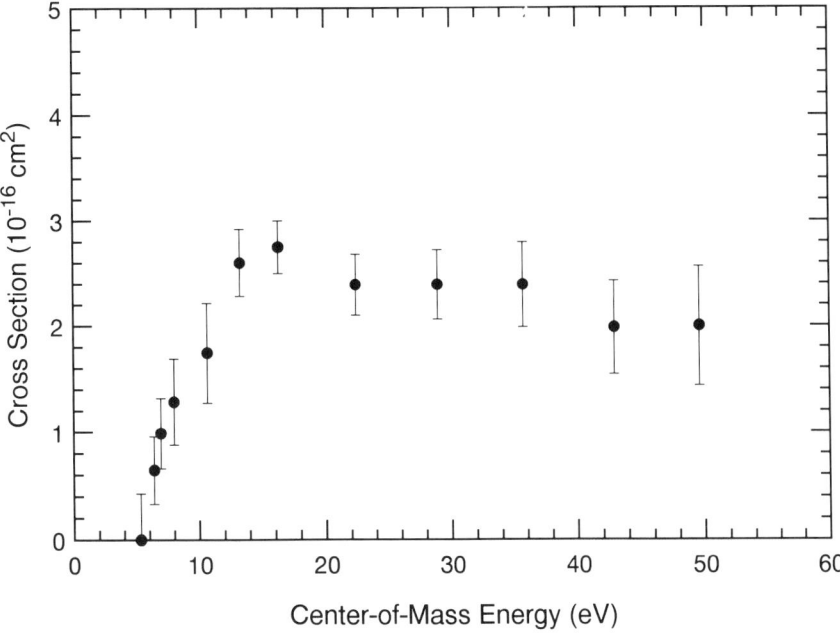

Figure 29. Cross sections for the dissociative excitation of CO^+. (Adapted from Mitchell and Hus.[66])

6. CO_2^+

Only the room temperature dissociative recombination rate coefficient[67] has been measured for CO_2^+. This has a value of 3.8×10^{-7} cm^3/s. Again, the vibrational state of the ions was not identified. No cross section data are available for either recombination or excitation.

7. SUMMARY

In reviewing the needed cross section data for edge plasma chemistry, it is obvious from the above that the situation is not very good for electron–molecular ion collisions. Our knowledge of the effects of initial vibrational excitation (not to speak of rotational excitation) is minimal, and a much clearer understanding of these effects will be needed before one can confidently build a model of the plasma chemistry, using these cross sections. Many processes have only been studied at thermal or near thermal energies. Much of the reason for this has been simple neglect. While the major focus of thermonuclear fusion-related atomic physics lay with the bulk plasma in fusion devices, the emphasis was heavily placed upon atomic collision processes, and our knowledge of this area is now very good. Now that more attention is being given to the cooler edge plasma, it is hoped that similar attention will be paid to molecular ion processes. With the advent of storage ring electron–ion experiments, the pendulum is already swinging in that direction.[68]

Relatively little attention has been paid to isotopic effects in these processes, and, as mentioned earlier, no experiments have been performed for tritiated species. Given that tritiated molecules (and deuterated and tritiated isotopomers) will be formed with quite different vibrational population distributions from those of the species discussed here, it is certainly worthwhile to consider examining these molecules and their collision processes.

REFERENCES

1. H. Abouelaziz, J. C. Gomet, D. Pasquerault, B. R. Rowe, and J. B. A. Mitchell, *J. Chem. Phys.* **99**, 237–243 (1993).
2. F. Von Busch and G. H. Dunn, *Phys. Rev. A* **5**, 1726–1743 (1972).
3. N. P. F. B. Van Asselt, J. G. Maas, and J. Los, *Chem. Phys.* **5**, 429–438 (1974).
4. J. B. Ozenne, J. Durup, R. W. Odom, C. Pernot, A. Tabche-Fouhaille, and M. Tadjeddine, *Chem. Phys.* **16**, 75–80 (1976).
5. A. G. Brenton, P. G. Fournier, B. L. Govers, E. G. Richard, and J. H. Beynon, *Proc. Roy. Soc. London, Ser. A* **395**, 111–125 (1984).
6. Y. Weijun, R. Alheit, and G. Werth, *Z. Phys. D.* **28**, 87–88 (1993).
7. L. P. Theard and W. T. Huntress, *J. Chem. Phys.* **60**, 2840–2848 (1974).
8. Z. Herman and V. Pacak, *Int. J. Mass Spectrom Ion Phys.* **24**, 355–358 (1977).
9. D. Auerbach, R. Cacak, R. Caudano, T. D. Gaily, C. J. Keyser, J. W. McGowan, J. B. A. Mitchell, and S. F. J. Wilk, *J. Phys. B* **10**, 3797–3820 (1977).
10. H. Hus, F. B. Yousif, C. Noren, A. Sen, and J. B. A. Mitchell, *Phys. Rev. Lett.* **60**, 1006 (1988).
11. P. Van der Donk, F. B. Yousif, J. B. A. Mitchell, and A. P. Hickman, *Phys. Rev. Lett.* **67**, 42 (1991).
12. B. Peart and K. T. Dolder, *J. Phys. B* **7**, 236 (1974).
13. I. F. Schneider, O. Dulieu, and A. Giusti-Suzor, *J. Phys. B* **24**, L289 (1991).
14. K. Nakashima, H. Takagi, and H. Nakamura, *J. Chem. Phys.* **86**, 726 (1987).
15. H. Takagi, *J. Phys. B* **26**, 4815 (1993).
16. R. A. Phaneuf, D. H. Crandall, and G. H. Dunn, *Phys. Rev. A* **11**, 528 (1975).
17. M. Vogler and G. H. Dunn, *Phys. Rev. A* **11**, 1983 (1975).
18. J. B. A. Mitchell, F. B. Yousif, P. J. T. Van der Donk, and T. J. Morgan, in *Dissociative Recombination: Theory, Experiment and Applications* (B. R. Rowe, J. B. A. Mitchell, and A. Canosa, eds.), Plenum Press, New York (1993), pp. 87–97.
19. B. Peart and K. T. Dolder, *J. Phys. B* **6**, L359 (1973).
20. P. Forck, M. Grieser, D. Habs, A. Lampert, R. Repnow, D. Schwalm, A. Wolf, and D. Zajfman, *Phys. Rev. Lett.* **70**, 426 (1993).
21. T. Tanabe, I. Katayama, N. Inoue, K. Chida, Y. Arakaki, T. Watanabe, M. Yoshizawa, S. Ohtani, and K. Noda, *Phys. Rev. Lett.* **70**, 422 (1993).
22. F. B. Yousif, J. B. A. Mitchell, A. Canosa, M. Rogelstad, A. Le Paddelec, and M. I. Chibisov, *Phys. Rev. A.* **49**, 4610 (1994).
23. M. Larsson, H. Danared, J. R. Mowat, P. Sigray, G. Sundstrom, L. Brostrom, A. Filevich, A. Kalberg, S. Mannervik, K. G. Rensfelt, and S. Datz, *Phys. Rev. Lett.* **70**, 430 (1993).

24. B. Peart and K. T. Dolder, *J. Phys. B* **5**, 860 (1972).
25. G. H. Dunn and B. Van Zyl, *Phys. Rev.* **154**, 40 (1967).
26. D. F. Dance, M. F. A. Harrison, R. D. Rundel, and A. C. H. Smith, *Proc Phys. Soc.* **92**, 577 (1967).
27. F. B. Yousif, A. and J. B. A. Mitchell, *Z. Physik D* in preparation.
28. J. Peek, *Phys. Rev.* **140**, A11 (1965).
29. J. Peek, *Phys. Rev.* **154**, 52 (1967).
30. B. Peart and K. T. Dolder, *J. Phys. B.* **6**, 2409 (1973).
31. B. Peart and K. T. Dolder, *J. Phys. B.* **8**, 1570 (1975).
32. V. G. Anicich and J. H. Futrell, *Int. J. Mass Spectrom Ion Processes* **55**, 189 (1983/1984).
33. G. D. Carney and R. N. Porter, *J. Chem. Phys.* **65**, 3547 (1976).
34. B. Dinelli, S. Miller, and J. Tennyson, *J. Mol. Spectrosc.* **153**, 718 (1992).
35. J. K. Kim, L. P. Theard, and W. T. Huntress, *Int. J. Mass Spectrom. Ion Phys.* **15**, 223 (1974).
36. C. R. Blakley, M. L. Vestal, and J. H. Futrell, *J. Chem. Phys.* **66**, 2392 (1977).
37. D. R. Bates, M. F. Guest, and R. A. Kendall, *Planet. Space Sci.* **41**, 9 (1993).
38. B. R. Rowe, J. B. A. Mitchell, and A. Canosa, *Dissociative Recombination: Theory, Experiment and Applications,* Plenum Press, New York (1993).
39. H. H. Michels and R. H. Hobbs, *Astrophys. J.* **286**, L27 (1984).
40. H. Hus, F. B. Yousif, A. Sen, and J. B. A. Mitchell, *Phys. Rev. A* **38**, 658 (1988).
41. B. Peart and K. T. Dolder, *J. Phys. B.* **7**, 1948 (1974).
42. M. T. Leu, M. A. Biondi, and R. Johnsen, *Phys. Rev. A* **8**, 413 (1973).
43. N. G. Adams, D. Smith, and E. Alge, *J. Chem. Phys.* **81**, 1778 (1984).
44. D. Smith and P. Spanel, *Int. J. Mass Spectrom. Ion Processes* **129**, 163 (1993).
45. A. Canosa, J. C. Gomet, B. R. Rowe, J. B. A. Mitchell, and J. L. Queffelec, *J. Chem. Phys.* **97**, 1028 (1992).
46. T. Amano, *Astrophys. J.* **329**, L121 (1988).
47. T. Amano, *J. Chem Phys.* **92**, 6492 (1990).
48. F. B. Yousif, P. J. T. Van der Donk, and J. B. A. Mitchell, *J. Phys. B* **26**, 4249 (1993).
49. J. B. A. Mitchell, M. Rogelstad, and F. B. Yousif, manuscript in preparation.
50. G. Sundstrom, J. R. Mowat, H. Danared, S. Datz, L. Brostrom, A. Filevich, A. Kalberg, K. G. Rensfelt, P. Sigray, M. af Ugglas, and M. Larsson, *Science* **263**, 785 (1994).
51. J. B. A. Mitchell, J. L. Forand, C. T. Ng, D. P. Levac, R. E. Mitchell, P. M. Mul, W. Claeys, A. Sen, and J. W. McGowan, *Phys. Lett.* **51**, 885 (1983).
52. J. B. A. Mitchell and F. B. Yousif, in *Microwave and Particle Beam Sources and Directed Energy Concepts* (H. E. Brandt, ed.), *Proc. SPIE* **1061**, 536 (1989).
53. J. B. A. Mitchell, C. T. Ng, L. Forand, R. Janssen, and J. W. McGowan, *J. Phys. B* **17**, L909 (1984).
54. P. Van der Donk, F. B. Yousif, and J. B. A. Mitchell, *Phys. Rev. A* **43**, 5971 (1991).
55. B. Peart and K. T. Dolder, *J. Phys. B* **7**, 1567 (1974).
56. B. Peart and K. T. Dolder, *J. Phys. B* **8**, L143 (1975).
57. F. B. Yousif, P. J. T. Van der Donk, M. Orakzai, and J. B. A. Mitchell, *Phys. Rev. A* **44**, 5653 (1991).
58. B. Peart, R. A. Forest, and K. T. Dolder, *J. Phys. B* **12**, 3441 (1979).
59. P. M. Mul and J. W. McGowan, *J. Phys. B* **12**, 1591 (1979).
60. F. L. Walls and G. H. Dunn, *J. Geophys. Res.* **79**, 1911 (1974).
61. H. Bohringer, M. Durup-Ferguson, D. W. Fahey, F. C. Fehsenfeld, and E. E. Ferguson, *J. Chem. Phys.* **79**, 4201 (1983).
62. S. L. Guberman, in *Physics of Ion–Ion and Electron–Ion Collisions* (F. Brouillard and J. W. McGowan, eds.), Plenum Press, New York (1983), pp. 167–200.

63. S. L. Guberman and A. Giusti-Suzor, *J. Chem. Phys.* **95**, 2602 (1991).
64. P. Spanel, L. Dittrichova, and D. Smith, *Int. J. Mass Spectrom. Ion Processes* **129**, 183 (1993).
65. B. Van Zyl and G. H. Dunn, *Phys. Rev.* **163**, 43 (1967).
66. J. B. A. Mitchell and H. Hus, *J. Phys. B.* **18**, 547 (1985).
67. C. S. Weller and M. A. Biondi, *Phys. Rev. Lett.* **19**, 59 (1967).
68. D. Zajfman, J. B. A. Mitchell, D. Schwalm, and B. R. Rowe, *Dissociative Recombination: Theory, Experiment and Applications III,* World Scientific, Singapore (1995).

Chapter 10

Energy and Angular Distributions of Secondary Electrons Produced by Electron Impact Ionization

Yong-Ki Kim

1. INTRODUCTION

Atomic and molecular cross sections for electron impact ionization are of fundamental importance in modeling the interaction of charged particles with matter. In order to follow the history of the ejected electrons and understand the subsequent events and products, not only the total ionization cross sections but often the cross sections differential in ejected electron energies and angles are also needed.

When the incident particle is an electron, the scattered and the ejected electrons are indistinguishable, and it is customary to call the faster of the two electrons the *primary* electron and the slower one the *secondary* electron.

The most detailed description of an ionizing collision includes angular and energy distributions of both the primary and the secondary electrons. This is the so-called triply differential cross section (TDCS), $d^3\sigma/dW\,d\Omega_s\,d\Omega_p$, where W is the energy of the secondary electron, Ω_s is the solid angle of the secondary electron, and Ω_p is the solid angle of the primary electron. Actually, one must also specify

---

YONG-KI KIM • National Institute of Standards and Technology, Gaithersburg, Maryland 20899.
Atomic and Molecular Processes in Fusion Edge Plasmas, edited by R. K. Janev. Plenum Press, New York, 1995.

the incident electron energy T and its energy loss E to describe the collision completely.

For a multishell target, T and W alone are insufficient to characterize the collision since an electron could be ejected from any one of the shells, provided $T \geq B + W$, where B is the binding energy of the ejected electron. When the energy loss E of the primary electron is measured instead of W, the appropriate TDCS is $d^3\sigma/dE\,d\Omega_s\,d\Omega_p$.

The doubly differential cross section (DDCS), $d^2\sigma/dW\,d\Omega_s$, is obtained when the TDCS is integrated over Ω_p. This is the angular distribution of the secondary electron. Occasionally, one needs to integrate the TDCS over Ω_s to obtain the angular distribution of the primary electron. In this case, it is more appropriate to refer to the energy loss E of the primary electron instead of W, yielding $d^2\sigma/dE\,d\Omega_p$.

When the DDCS of a secondary electron is integrated over Ω_s, the energy distribution of the secondary electron, $d\sigma/dW$, is obtained. Similarly, the energy-loss cross section of the primary electron, $d\sigma/dE$, results when $d^2\sigma/dE\,d\Omega_p$ is integrated over Ω_p.

Finally, the total ionization cross section (TICS), σ_i, is obtained by either integrating $d\sigma/dE$ over E or $d\sigma/dW$ over W. The relationships between these differential and integrated cross sections are listed below:

$$\frac{d^2\sigma}{dW\,d\Omega_s} = \int \frac{d^3\sigma}{dW\,d\Omega_s\,d\Omega_p}\,d\Omega_p \tag{1}$$

$$\frac{d^2\sigma}{dE\,d\Omega_p} = \int \frac{d^3\sigma}{dE\,d\Omega_s\,d\Omega_p}\,d\Omega_s \tag{2}$$

$$\frac{d\sigma}{dW} = \int \frac{d^2\sigma}{dW\,d\Omega_s}\,d\Omega_s \tag{3}$$

$$\frac{d\sigma}{dE} = \int \frac{d^2\sigma}{dE\,d\Omega_p}\,d\Omega_p \tag{4}$$

$$\sigma_i = \int \frac{d\sigma}{dW}\,dW \cong \int \frac{d\sigma}{dE}\,dE \tag{5}$$

The last two integrals in Eq. (5) may not lead to the same σ_i in targets with complicated shell structures because a large energy loss may lead to the ejection of more than one electron (e.g., initial ejection of an inner-shell electron), or a molecule may dissociate without ejecting an electron (e.g., when E is slightly above the molecular ionization potential).

There is no simple theory that provides reliable secondary electron cross sections for all T and W. Only very recently a promising theory for electron impact ionization based on a close-coupling scheme emerged.[1] This theory, however,

Energy and Angular Distributions of Secondary Electrons

requires a substantial computational effort even for the ionization of the hydrogen atom and is limited to the ionization of valence electrons. No comparable theory exists for molecules.

For medium to heavy atoms, the probability for ejection of more than one electron in a single collision may be significant. For instance, the ionization of an inner-shell electron will most likely lead to a series of Auger electrons, producing a multiply charged ion. The method described in this chapter will provide a reasonable estimate of the cross section for ejection of an electron from either a valence or an inner shell.

The inner-shell ionization cross section will indicate the relative importance of multiple ionization versus the dominant single ionization of valence electrons. In general, vacancies in the K shell of a medium to heavy atom (atomic number > 30) are likely to be filled by emitting an energetic photon, while vacancies near the valence shell are likely to be filled by the Auger process.

We will focus on SDCS ($d\sigma/dW$) and DDCS ($d^2\sigma/dW\,d\Omega_s$) for the emission of one secondary electron per collision in this chapter. Qualitative considerations are presented in Section 2, a general theoretical method to calculate SDCSs in Section 3, analytical formulas to derive DDCSs in Section 4, comparisons of theory with known experimental SDCSs and DDCSs in Section 5, and concluding remarks in Section 6. Discussions in this chapter are equally applicable to the ionization of atoms, ions, and molecules, although ionization of a molecule is more complicated because the remaining ion may further disintegrate into fragments. Cases in which ionization is accompanied by the excitation or fragmentation of the resulting ion are minor events compared to the cases in which the ion remains in its ground state. Only the latter process is described in this chapter.

2. QUALITATIVE CONSIDERATIONS

All collisions of an electron with an atom or molecule may be classified into two general categories—*hard* collisions with small impact parameters and *soft* collisions with large impact parameters. Hard collisions produce energetic secondary electrons with a sharply focused angular distribution around the binary peak angle θ_0. For the collision of a free electron with energy T and another at rest,

$$\cos\theta_0 = \sqrt{W/T} \tag{6}$$

where W is the ejected electron energy. This relationship is determined simply from the requirement of energy–momentum conservation.[2] The momentum p_0 of a bound electron is given by

$$p_0 a_0/\hbar \equiv k_0 a_0 = \sqrt{U/R} \tag{7}$$

where a_0 is the Bohr radius, $\hbar$ is Planck's constant divided by 2π, R is the Rydberg energy, and the average kinetic energy U is defined by

$$U \equiv \langle \mathbf{p}_0^2/2m \rangle \tag{8}$$

where m is the electron mass.

The width of the binary peak is qualitatively given by the sum of $\mathbf{p}_0$ and the momentum $\mathbf{p}_s$ of the ejected electron,

$$p_s a_0/\hbar \equiv k_s a_0 = \sqrt{W/R} \tag{9}$$

Hence, the width of the binary peak, $\Delta(\cos\theta_0)$, can be written as

$$|\Delta(\cos\theta_0)| = c \cdot \sin\theta_0 \arctan(k_0/k_s) \tag{10}$$

where c is a constant (≤ 1) representing the averaged direction of $\mathbf{p}_0$ with respect to $\mathbf{p}_s$.

In contrast, soft collisions are dominated by dipole interaction and generate mostly slow secondary electrons with a broad angular distribution. The angular distribution of photoionized electrons—representing a pure dipole interaction—has the general form

$$\frac{d\sigma(h\nu)}{d\Omega} = a + b\cos^2\theta \tag{11}$$

where a and b are functions of the photon energy $h\nu$ and depend on the properties of the target.

However, the correct angular distribution of secondary electrons produced by electron impact or ion impact will not be a simple superposition of these two qualitative angular distributions. Instead, amplitudes must be added before they are squared, often leading to angular distributions significantly altered by interference.

For electron impact collisions, the indistinguishability of the scattered and ejected electrons introduces further interference between the direct and exchange interaction terms. The net effect of the exchange interaction is to reduce integrated cross sections, but its effect on angular distribution is to create additional interference and a more complicated structure.

The angular distribution of *primary* electrons is very different from that of secondary electrons. The former peaks at $\theta = 0°$. Actually, the binary peak of secondary electrons moves toward the forward direction as W increases and occurs at $\theta_0 = 0°$ when $W = T$ according to Eq. (6).

In summary, SDCSs and TICSs can be understood qualitatively in terms of physical interactions, but DDCSs are subject to interference between amplitudes arising from different interactions, and the resulting angular distribution is likely to obscure the origins of the physical interactions.

3. ANALYTICAL MODEL FOR ENERGY DISTRIBUTIONS OF SECONDARY ELECTRONS

Kim and Rudd[3] have combined the binary-encounter theory[4] and the Bethe theory for electron impact ionization[5] and derived an analytical expression for the SDCS, or the energy distribution, of secondary electrons. This new theoretical model is called the binary-encounter-dipole (BED) model, and it provides formulas for calculation of the SDCS for electrons ejected from a given atomic or molecular orbital as a function of T and W. It requires knowledge of occupation number N, binding energy B, the average kinetic energy U, and the differential dipole oscillator strength df/dW of the orbital. It is convenient to write the energy variables in units of the binding energy B:

$$t = T/B \tag{12}$$

$$w = W/B \tag{13}$$

$$u = U/B \tag{14}$$

In the BED model, the SDCS is given by

$$\frac{d\sigma}{dW} = \frac{4\pi a_0^2 R^2 N}{B^3(t+u+1)} \left\{ \frac{(N_i/N) - 2}{t+1} \left(\frac{1}{w+1} + \frac{1}{t-w} \right) \right.$$

$$\left. + \left(2 - \frac{N_i}{N}\right) \left[\frac{1}{(w+1)^2} + \frac{1}{(t-w)^2} \right] + \frac{\ln t}{N(w+1)} \frac{df}{dw} \right\} \tag{15}$$

where N_i is the integral of df/dw:

$$N_i = \int_0^{(t-1)/2} \frac{df}{dw} dw \tag{16}$$

The constant N_i is the ionization part of the sum rule for f values commonly known as the Thomas–Kuhn–Reich sum rule. For the SDCS of an atom or molecule, Eq. (15) must be summed over all orbitals that contribute to the SDCS. Because of the large binding energies, inner orbitals do not contribute much to the total SDCS or TICS. Contributions from the valence orbital and a few more outer orbitals should account for most of the TICS.

The first term in the curly brackets of Eq. (15) arises from the interference between the direct and exchange interaction components in hard collisions, the second term from hard collisions based on the Mott cross section,[6] and the last term from the soft collisions (= dipole interaction) based on the Bethe theory.

The values of N, B, and U are easily obtained from the open literature or from computer codes for wave functions. These values are listed in Table I for small

Table I. Binding (B) and Kinetic (U) Energies (in eV) and Occupation Numbers (N) for Atoms

Orbital	H-like ions[a]			He			Ne			Ar		
	B	U	N	B	U	N	B	U	N	B	U	N
1s	Z^2R	Z^2R	1	24.59	39.51	2	866.9	1259.1	2	3202.9	4192.9	2
2s							48.47	141.88	2	326.0	683.1	2
2p							21.60	116.02	6	249.18	651.4	6
3s										29.24	103.5	2
3p										15.82	78.07	6

[a]Z = atomic number.

atoms and in Table II for small molecules.[2] The values of U for He and H_2 are from correlated wave functions, and those for Ne are from Hartree–Fock wave functions. Data on df/dw are available only for simple atoms and molecules, though they can be either calculated from wave functions or deduced from experimental photoionization cross sections. Oscillator strength data are usually in graphical or tabular form and are not necessarily convenient for modeling. Differential oscillator strengths for H, He, Ne, and H_2 have been fit to a power series,[3] and their coefficients are listed in Table III. The differential oscillator strengths have been converted from experimental photoionization cross sections, except for H. Analytical fits to the df/dw of Ar and N_2 in terms of Gaussian functions are given in Ref. 2, along with N, B, and U for a large number of molecules.

Note that the coefficients of the df/dw for the hydrogen atom in Table III can also be used for hydrogen-like ions when B and U are scaled by Z^2, as shown in Table I. The uncertainties of these fitted df/dw are negligible compared to the expected accuracy of the BED model.

Table II. Binding (B) and Kinetic (U) Energies (in eV) and Occupation Numbers (N) for Molecules

Orbital	H_2			N_2			O_2		
	B	U	N	B	U	N	B	U	N
$1\sigma_g$	15.43	15.98	2	409.9	601.78	2	543.5	794.84	2
$1\sigma_u$				409.9	602.68	2	543.5	795.06	2
$2\sigma_g$				37.3	69.53	2	40.3	78.19	2
$2\sigma_u$				18.78	62.45	2	25.69	90.40	2
$1\pi_u$				16.96	55.21	4	18.88	72.24	4
$3\sigma_g$				15.59	44.27	2	16.42	60.08	2
$1\pi_g$							12.07	82.14	2

Table III. Power Series Fit to df/dw [a]

Coefficient[b]	H 1s	He 1ss	H$_2$ 1σ_g	Ne 2p,I[c]	Ne 2p,II[c]	Ne 2s
a				4.8791		
b	−2.2473(−2)			−2.8820	−5.8514	1.7769
c	1.1775	1.2178(1)	1.1262	−7.4711(−1)	3.2930(2)	2.8135
d	−4.6264(−1)	−2.9585(1)	6.3982		−1.6788(3)	−3.1510(1)
e	8.9064(−2)	3.1251(1)	−7.8055		3.2985(3)	6.3469(1)
f		−1.2175(1)	2.1440		−2.3250(3)	−5.2528(1)
g						1.5982(1)
N_i	4.343(−1)	1.605	1.173		6.963[d]	7.056(−1)

[a]Numbers in parenthesis are powers of 10.
[b]$df/dw = ay + by^2 + cy^3 + dy^4 + ey^5 + fy^6 + gy^7$, where $y = (w + 1)^{-1}$.
[c]2p, I for photon energies 21.60–48.47 eV; 2p, II for photon energies ≥ 48.47 eV.
[d]Sum of 2p, I and 2p, II.

4. ANALYTICAL MODEL FOR ANGULAR DISTRIBUTIONS OF SECONDARY ELECTRONS

In contrast to the situation for total ionization cross sections, *reliable* experimental data on SDCSs and DDCSs are scarce. In most experiments, DDCSs have been measured directly and then integrated over Ω_s to deduce SDCSs. Unfortunately, experimental DDCSs in the extreme forward (≤30°) and backward (≥150°) angles are unreliable in many cases owing to experimental difficulties. As a result, DDCSs derived from the analytical formulas presented below are less reliable than the SDCSs discussed in Section 3.

As discussed in Section 2, the angular distribution consists of a binary peak centered around θ_0 and a rise in the backward direction for low W. This backward rise is clearly present in the one set of clean experimental data available for He.[7] For other targets, experimental data in the extreme angles and at low W are too uncertain to allow definitive conclusions to be drawn, but there are "hints" that the backward rise exists. The magnitude of the backward rise at low W is known for He but not for other targets.

Rudd[8] proposed that a Lorentzian function,

$$F_1(W, \theta) = \frac{G_1}{1 + \left(\dfrac{\cos\theta - G_2}{G_3}\right)^2} \qquad (17)$$

be used to describe the binary peak. In Eq. (17), G_1 is a normalization constant so that the integral of the DDCS matches the SDCS [see Eq. (3)], G_2 fixes the location of the binary peak (for a bound electron), and G_3 determines the width of the binary peak. Not only does a Lorentzian function describe the observed angular distributions well, but it is also easy to integrate over Ω_s.

Based on the discussions in Section 2, we have found that the following choice reproduces the shape of known experimental DDCSs rather well for $W \geq 10$ eV:

$$G_2 = \sqrt{\frac{w+1}{t+1}} \equiv \cos\theta_b \tag{18}$$

$$G_3 = (2/3) \sin\theta_b \arctan(k_0/k_s) \tag{19}$$

where θ_b is the binary-peak angle (for a bound electron), and the factor "2/3" in Eq. (19) represents the average over the orientation of p_0. This factor changes rapidly from unity to $\frac{2}{3}$ between $W = 0$ and 10 eV according to the experimental data for He.[7]

The value of G_1 is affected by how we represent the backward rise. The rise diminishes as W increases, and one quantity that decreases with increasing W is the dipole interaction contribution to the SDCS. Therefore, we tentatively propose the following function to describe the backward rise in DDCS:

$$F_2(W, \theta) = G_4 \cos^2\theta \quad (\theta \geq 90°) \tag{20}$$

where

$$G_4 = \frac{D(t, w)}{t+1} \tag{21}$$

and

$$D(t, w) = \frac{4\pi a_0^2 R^2}{B^3(t+u+1)} \frac{\ln t \cdot df/dw}{w+1} \tag{22}$$

Note that $D(t, w)$ defined above is simply the dipole contribution to the SDCS. *This choice of F_2 is strictly empirical* and will have to be modified when we have a better theoretical understanding of the origin of the backward rise and more reliable experimental data with which to verify the theory.

Finally, we tentatively propose the following as a "semiempirical" BED model for the DDCS:

$$d^2\sigma/dW\,d\Omega_s = F_1(W, \theta) + F_2(W, \theta), \tag{23}$$

where the value of G_1 is determined by equating the integral of $F_1 + F_2$ to the SDCS:

Energy and Angular Distributions of Secondary Electrons

$$I_1 \equiv 2\pi \int_0^\pi F_1(W, \theta) \sin \theta \, d\theta = 2\pi G_1 G_3 \left(\arctan \frac{1 - G_2}{G_3} + \arctan \frac{1 + G_2}{G_3} \right) \quad (24)$$

$$I_2 \equiv 2\pi \int_{\pi/2}^\pi F_2(W, \theta) \sin \theta \, d\theta = (2\pi/3) G_4 \quad (25)$$

$$\frac{d\sigma}{dW} = I_1 + I_2 \quad (26)$$

The SDCS on the left-hand side of Eq. (26) is obtained from Eq. (15). Note that the value of the arctan functions in Eqs. (19) and (24) should be between 0 and π radians.

5. COMPARISONS WITH EXPERIMENT

An important test of the BED model presented here is on the total ionization cross section σ_i, defined in Eq. (5), which is obtained by integrating Eq. (15):

$$\sigma_i(t) = \frac{4\pi a_0^2 R^2 N}{B^2(t + u + 1)} \left[D(t) \ln t + \left(2 - \frac{N_i}{N}\right) \left(\frac{t-1}{t} - \frac{\ln t}{t+1} \right) \right] \quad (27)$$

where

$$D(t) \equiv N^{-1} \int_0^{(t-1)/2} \frac{1}{w+1} \frac{df}{dw} dw \quad (28)$$

Equations (27) and (28), along with the data in Tables I–III, generate σ_i for H, He, H_2, and Ne from T = threshold to a few keV with remarkable accuracy (10% or better). Once the accuracy of σ_i is verified, then Eq. (15) can be used not only to generate $d\sigma/dW$, but also to determine the angular distribution for arbitrary values of W using the equations presented in Section 4.

In Fig. 1, we compare the total ionization cross section, σ_i, derived for He from the present BED model (Eq. 27) with that derived from another theory[9] and with experiments.[10,11]

In Fig. 2, we compare the singly differential cross section, $d\sigma/dW$, obtained for He from the present BED model (Eq. 15) with experiments.[7,12] Although the BED results and the experimental data reported by Müller-Fiedler et al.[7] are in excellent agreement at T = 200 eV, the experimental data for W = 40 eV at larger incident energies are too small. These experimental results are inconsistent with the σ_i calculated from the BED model and with experimental σ_i, contrary to the claim made by Müller-Fiedler et al. that the SDCSs (at W = 40 eV and larger T) published by others are too high.

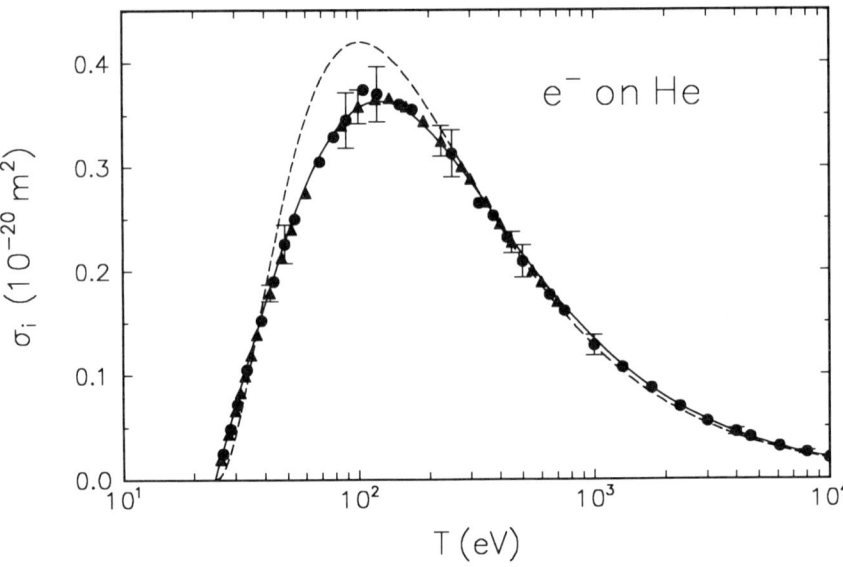

Figure 1. Comparison of σ_i of He calculated in the present work (—) with the distorted-wave Born cross section calculation by Younger[9] (- - -) and experimental data of Shah et al.[10] (●) and Montague et al.[11] (▲).

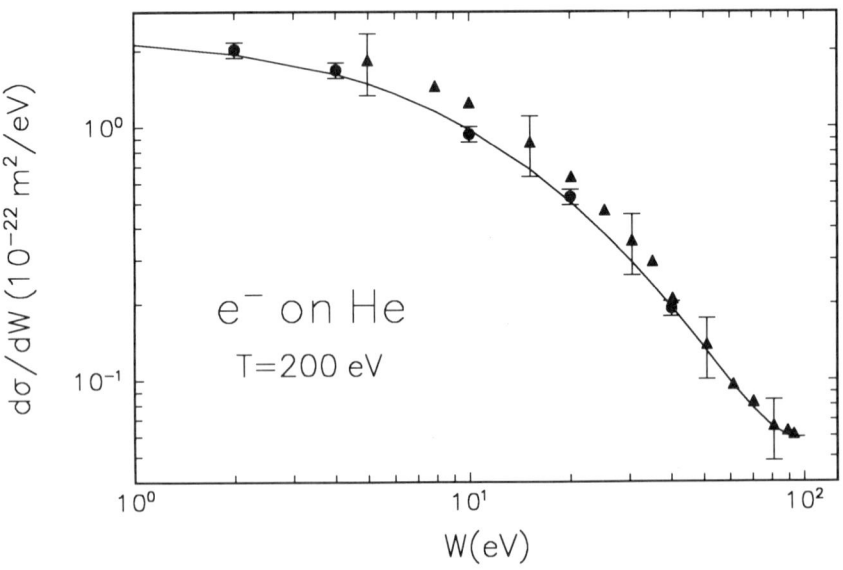

Figure 2. Comparison of $d\sigma/dW$ of He calculated in the present work (—) with the experimental data of Müller-Fiedler et al.[7] (●) and Opal et al.[12] (▲).

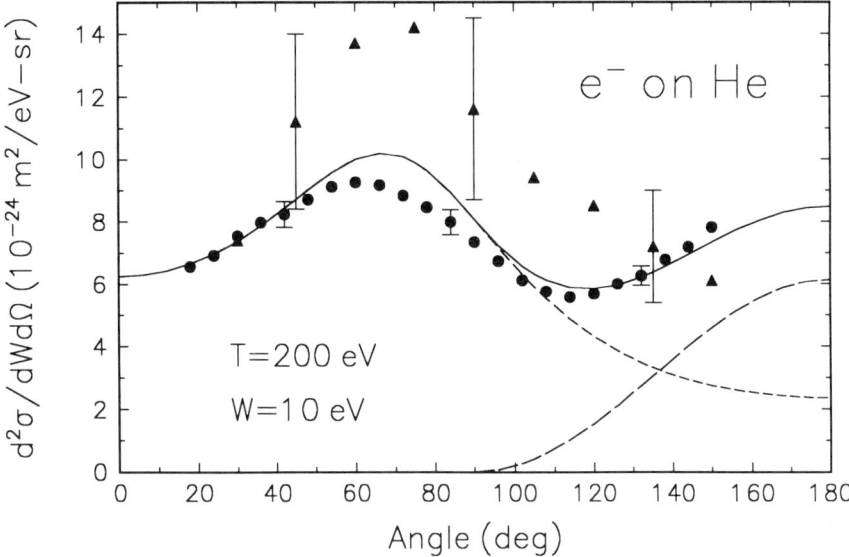

Figure 3. Comparison of $d^2\sigma/dW\,d\Omega_s$ of He calculated in the present work (—) with the experimental data of Müller-Fiedler et al.[7] (●) and Opal et al.[12] (▲). (– – –) Lorentzian function, $F_1(W, \theta)$ (Eq. 17); (— — —) backward rise represented by $F_2(W, \theta)$ (Eq. 20).

In Fig. 3, we compare the BED model's prediction for the doubly differential cross section, $d^2\sigma/dW\,d\Omega_s$, of He (Eq. 23) with experiments.[7,12] The DDCSs obtained by Opal et al.[12] for most targets tend to be too low at $\theta = 30°$ and $150°$. The fact that the normalization of the DDCS data by Opal et al. is too high is reflected in the high value of their SDCS in Fig. 2 at $W = 10$ eV.

In Fig. 4, we compare the BED model's results for the total ionization cross section of H_2 (Eq. 27) with the results from other theories[13,14] and experiment.[15] The classical trajectory Monte Carlo (CTMC) cross section calculated by Schultz et al.[14] falls below the experimental σ_i at $T \geq 100$ eV because the CTMC method does not account for the dipole contribution, which becomes significant at large T.

In Fig. 5, we compare our calculation of the singly differential cross section, $d\sigma/dW$, for H_2 (Eq. 15) with experiments.[12,16] Again, the normalization of the data by Opal et al. is too high and inconsistent with the known σ_i at $T = 500$ eV.

In Fig. 6, we compare the doubly differential cross section $d^2\sigma/dW\,d\Omega_s$ for H_2 derived from Eq. (23) with experiments.[12,16] The normalization of both sets of experimental data is too high, and this fact is reflected in the high values of their SDCSs in Fig. 5 at $W = 40$ eV. At this and larger values of W, the backward rise is insignificant.

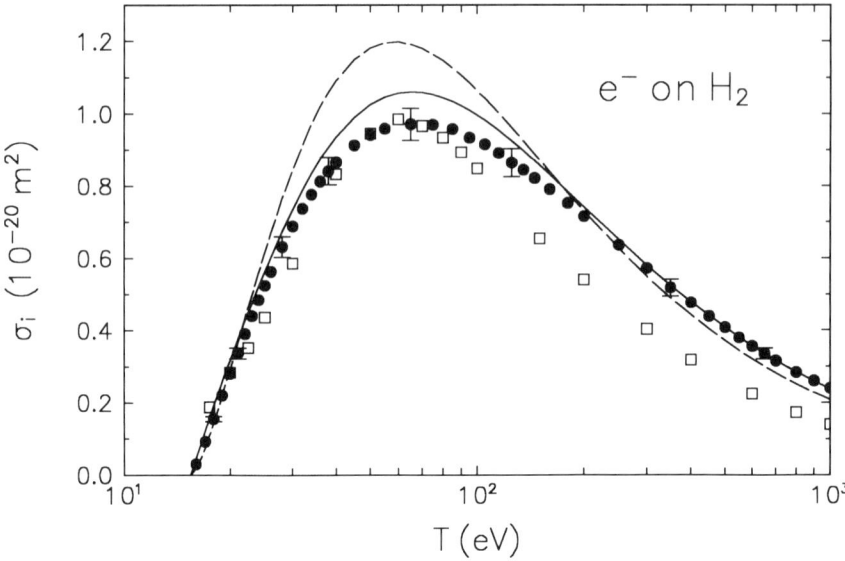

Figure 4. Comparison of σ_i of H_2 calculated in the present work (—) with the classical theory of Gryzinski[13] (— — —), the classical trajectory Monte Carlo (CTMC) theory of Schultz et al.[14] (□), and the experimental data of Rapp and Englander-Golden[15] (●).

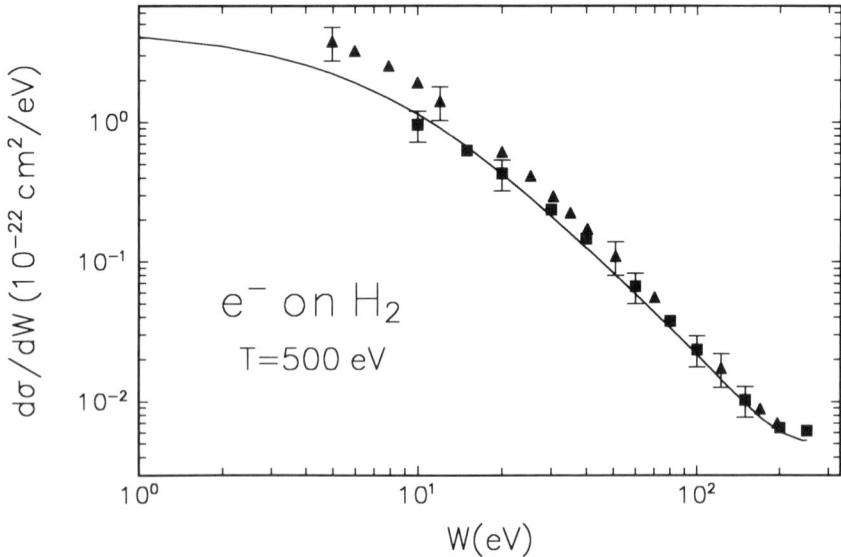

Figure 5. Comparison of $d\sigma/dW$ of H_2 calculated in the present work (—) with the experimental data of Rudd et al.[16] (■) and Opal et al.[12] (▲).

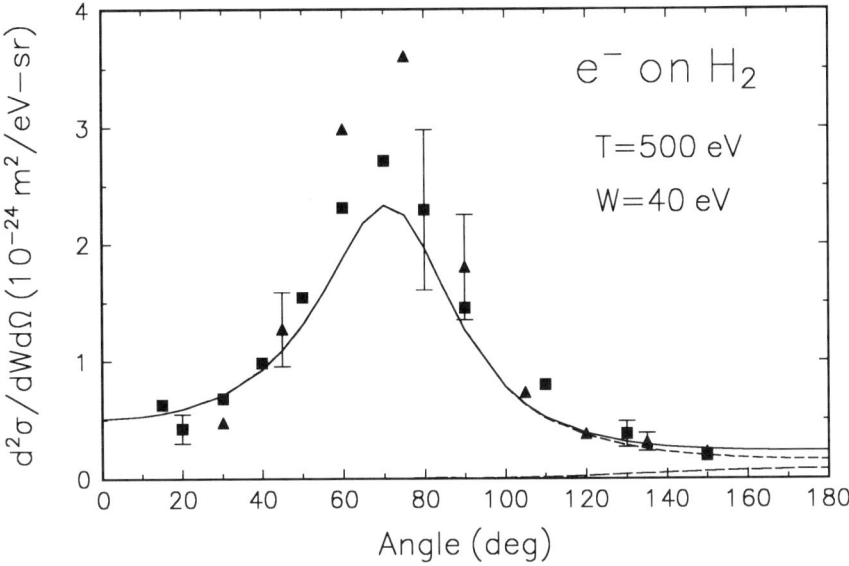

Figure 6. Comparison of $d^2\sigma/dW\,d\Omega_s$ of H_2 calculated in the present work (—) with the experimental data of Rudd et al.[16] (■) and Opal et al.[12] (▲). (- - -) Lorentzian function, $F_1(W, \theta)$ (Eq. (17); (— — —) backward rise represented by $F_2(W, \theta)$ (Eq. (20)).

6. CONCLUDING REMARKS

As can be seen in the examples presented here, the BED model provides analytical formulas for calculating σ_i, $d\sigma/dW$, and $d^2\sigma/dW\,d\Omega_s$, all of which are consistent with each other. The most important ingredient of the BED model is the differential dipole oscillator strength, df/dw. Although accurate df/dw are welcome, the BED model does not need "very" accurate df/dw; instead, the model needs df/dw expressed in terms of simple functions of the energy transfer $E = W + B$, preferably orbital by orbital. An effort by the present author to compile such data on df/dw is in progress, and the data will be distributed to the magnetic fusion research community as they become available. Meanwhile, the procedure described here is simple enough for anyone to apply to a wide variety of targets not addressed in this chapter.

For instance, Tables I and III include all the information needed to generate SDCSs and DDCSs of H-like ions. Comparisons of σ_i of He^+ and Li^{2+} based on the BED model with experiments suggest that the denominator $t + u + 1$ on the right-hand side of Eq. (15) should be replaced by $t + 1$ to obtain much better agreement.[3] Hence, this modified equation for $d\sigma/dW$, along with the constants in Tables I and III, should provide reliable SDCSs of H-like ions. Similarly, the $t + u$

+ 1 denominator on the right-hand side of Eq.(22) should also be replaced by $t + 1$, and the modified equation then used with other equations in Section 4 to generate DDCSs.

In addition to the references mentioned so far, there are more published experimental DDCSs on H,[17] He,[18] H_2,[19] N_2,[20] and H_2O.[21] Unfortunately, there are no direct measurements of SDCSs. Shyn's DDCSs of H,[17] however, do not exhibit binary peaks, suggesting serious experimental problems. Reference 7 includes angular distributions for both secondary and primary electrons, though the range of W values is limited to ≤ 40 eV. Reference 12 is the most comprehensive in that it contains DDCSs of He, N_2, and O_2 for wide ranges of $T \leq 2000$ eV and $W \leq 205$ eV, as well as DDCSs of Ne, Ar, Kr, Xe, H_2, CH_4, NH_3, H_2O, CO, C_2H_2, NO, and CO_2 for $T = 500$ eV and $W \leq 205$ eV. Although the DDCSs measured by Opal *et al.*[12] are too low in the extreme forward and backward angles, as mentioned earlier, their general shape is reliable. Their DDCSs were normalized either to the elastic scattering cross section at $\theta = 90°$ or to σ_i, which sometimes introduces normalization errors of 20–30%. Rudd and DuBois[22] reported DDCSs of He,[22] and of Ne, Ar, H_2, and N_2.[23] The normalization of DDCSs for $W \leq 30$ eV in Refs. 22 and 23 seems to be too low by as much as a factor of 2.

Finally, the reader is cautioned that the BED model is too simple to predict the ionization cross sections of more complex atoms and molecules with the high accuracy observed in the examples presented in this chapter. The accuracy of differential ionization cross sections based on the BED model may be judged by the accuracy of σ_i based on the same model, which can be checked against abundant experimental data. In the absence of such tests, it is safer to assume that $d\sigma/dw$ and $d^2\sigma/dW\,d\Omega_s$ based on the BED model will be reliable to within 20–30% in absolute magnitude, though the model is likely to produce more reliable shapes for these cross sections. Formulas presented in this chapter are based on nonrelativistic collision theory and therefore should not be used for $T \approx 10$ keV and above.

ACKNOWLEDGMENTS. This research was supported in part by the Office of Fusion Energy of the U.S. Department of Energy. The author would like to thank M. E. Rudd for long-standing collaboration on this subject.

REFERENCES

1. I. Bray, *Phys. Rev. A* **49**, 1066 (1994) and references therein.
2. M. E. Rudd, Y.-K. Kim, D. H. Madison, and T. J. Gay, *Rev. Mod. Phys.* **64**, 441 (1992), Appendix B.
3. Y.-K. Kim and M. E. Rudd, *Phys. Rev. A* **50**, 3954 (1994).
4. L. Vriens, in *Case Studies in Atomic Physics*, Vol. 1 (E. W. McDaniel and M. R. C. McDowell, eds.), North-Holland, Amsterdam (1969), p. 335.
5. H. Bethe, *Ann. Physik* **5**, 325 (1930).
6. N. F. Mott, *Proc. R. Soc. London, Ser. A* **126**, 259 (1930).

7. R. Müller-Fiedler, K. Jung, and H. Ehrhardt, *J. Phys. B* **19**, 1211 (1986).
8. M. E. Rudd, *Phys. Rev. A* **44**, 1644 (1991).
9. S. M. Younger, *J. Quant. Spectrosc. Radiat. Transfer* **26**, 329 (1981).
10. M. B. Shah, D. S. Elliot, P. McCallion, and H. B. Gilbody, *J. Phys. B* **21**, 2751 (1988).
11. R. G. Montague, M. F. A. Harrison, and A. C. H. Smith, *J. Phys. B* **17**, 3295 (1984).
12. C. B. Opal, E. C. Beaty, and W. K. Peterson, *At. Data* **4**, 209 (1972).
13. M. Gryzinski, *Phys. Rev.* **138**, A305, A322, A336 (1965).
14. D. R. Schultz, L. Meng, and R. E. Olson, *J. Phys. B* **25**, 4601 (1992).
15. D. Rapp and P. Englander-Golden, *J. Chem. Phys.* **43**, 1464 (1965).
16. M. E. Rudd, K. W. Hollman, J. K. Lewis, D. L. Johnson, R. R. Porter, and E. L. Fagerquist, *Phys. Rev. A* **47**, 1866 (1993).
17. T. W. Shyn, *Phys. Rev. A* **45**, 2951 (1992).
18. R. R. Goruganthu and R. A. Bonham, *Phys. Rev.* **34**, 103 (1986).
19. T. W. Shyn, W. E. Sharp, and Y.-K. Kim, *Phys. Rev. A* **24**, 79 (1981).
20. R. R. Goruganthu, W. G. Wilson, and R. A. Bonham, *Phys. Rev. A* **35**, 540 (1987).
21. M. A. Bolorizadeh and M. E. Rudd, *Phys. Rev. A* **33**, 882 (1986).
22. M. E. Rudd and R. D. DuBois, *Phys. Rev. A* **16**, 26 (1977).
23. R. D. DuBois and M. E. Rudd, *Phys. Rev. A* **17**, 843 (1978).

Chapter 11

Elastic and Related Cross Sections for Low-Energy Collisions among Hydrogen and Helium Ions, Neutrals, and Isotopes

D. R. Schultz, S. Yu. Ovchinnikov, and S. V. Passovets

1. INTRODUCTION

In the pursuit of ever more realistic, accurate models and diagnostic methods for fusion energy research, complete data bases of heavy-particle cross sections for excitation, ionization, charge transfer, and recombination have been sought for many years. Great progress has been made along these lines, but the type of reactions and the collision energy range investigated have been dictated primarily by the need to understand the physics of the central core plasma in such magnetically confined plasma devices as tokamaks. Contemporary interest in developing so-called "next-step" experimental reactors, such as ITER (International Thermonuclear Experimental Reactor) has, however, highlighted the need for the study of new atomic and molecular collision regimes. As described in great detail in the present volume and elsewhere,[1] engineering and physics issues are focused on (i) the edge plasma, which must be tailored to suppress the ingress of impurities into

---

D. R. Schultz • Physics Division, Oak Ridge National Laboratory, Oak Ridge, Tennessee 37831-6373.   S. Yu. Ovchinnikov and S. V. Passovets • Department of Physics and Astronomy, University of Tennessee, Knoxville, Tennessee 37996-1501. *Permanent address for S. Yu. Ovchinnikov and S. V. Passovets*: Ioffe Physical Technical Institute, St. Petersburg, Russia.

Atomic and Molecular Processes in Fusion Edge Plasmas, edited by R. K. Janev. Plenum Press, New York, 1995.

the core and to entrain them, and (ii) the divertor, which will be used for hydrogen recycling and heat (power) and particle (impurities, helium ash) exhaust. Because these plasma regimes are characterized by greatly lower temperatures and higher densities than the core, correspondingly different atomic, and even molecular, reactions play crucial roles.

Owing to these necessarily lower temperatures and higher densities in the edge and divertor regions, significant amounts of neutrals will be present. Since the cross sections for elastic ion–ion, ion–neutral, and neutral–neutral scattering can be large compared to the inelastic channels at very low collision energies, these processes play a dominant role in the momentum balance of these regions. In the production of atomic data relevant to fusion energy research, a data base for elastic and other cross sections related to transport properties has not been compiled in light of the heretofore heavy emphasis on inelastic, intermediate- to high-energy collisions. In particular, these slow, neutral particles are important because of their role in radiating and dissipating power and in providing a high-recycling region that shields plasma-facing components from the high heat and particle fluxes. (For a description of various aspects of modeling neutral gas transport in edge and divertor plasmas, see, e.g., Ref. 2.)

It is therefore our purpose here to tabulate in a useful way data that can be found in the literature and to provide new calculations of the relevant cross sections where the data are not available, or when they have not been presented in a way that can be readily utilized. A significant contribution to such efforts has been very recently made by Bachmann and Belitz,[3] who utilized a classical approximation to compute cross sections for proton impact of H, H_2, and He and for collisions of He^+ with He. We will refer to this work where appropriate in the present discussions, comparing the recommendations and methods of Bachmann and Belitz to those adopted here and recording their cross section fits for systems for which more detailed or comprehensive work has not yet been completed.

In the following sections we define the pertinent cross sections, describe the general theory and the approximations adopted for the present work, and tabulate the cross sections and fitting parameters used in order to present them in a form useful for incorporation into plasma modeling codes. Because the edge and divertor regions will be rich in neutral atomic hydrogen and molecular hydrogen recycled after recombination at the walls or introduced for fueling, these species will be of primary interest. In addition, since the ash of the fusion process must also be removed at the divertor, significant amounts of helium should also be present. Thus, we consider collisions among the various relevant ions and isotopes of H, H_2, and He. Since the edge plasma temperature is on the order of, say, 10 to 300 eV, and that of the divertor 1–50 eV, we consider center-of-mass collision energies in the range $0.001 \leq E_{cm} \leq 100$ eV. This range accommodates the fact that often the cross sections must be averaged over Maxwellian distributions so that knowledge of the cross section over a slightly larger range is useful. In addition, at the low end of this range comparisons can be made with data of astrophysical interest, thus providing an additional benchmark.

Indeed, much of the available data on the processes of interest comes from the astrophysical literature, as these processes play important roles in interstellar clouds and planetary atmospheres. The gaseous electronics literature is also an important source of information in this regard since these reactions are basic to the properties of mobility and diffusion. Finally, accurate interaction potentials, often derived from or validated by elastic scattering measurements, have been extensively discussed in the chemical physics literature and play a central role in much of quantum chemistry.

2. THE ELASTIC AND RELATED CROSS SECTIONS

When two particles collide without change of state, we term the event elastic scattering. A familiar example is the Rutherford scattering of two pointlike charges (e.g., proton–proton scattering), for which the differential cross section in atomic units is

$$\frac{d\sigma^{\text{Rutherford}}}{d\Omega_{\text{cm}}}(\theta_{\text{cm}}) = \left[\frac{Z_a Z_b}{4 E_{\text{cm}} \sin^2 \theta_{\text{cm}}}\right]^2 \tag{1}$$

Here θ_{cm}, Ω_{cm}, and E_{cm} are the center-of-mass (cm) scattering angle, solid angle, and collision energy, respectively, and Z_a and Z_b are the charges of the two fully stripped ions. We note that both classical and quantum-mechanical treatments of such a collision governed by an interaction potential $V(r) \sim r^{-1}$ yield the same result but that this is not true in general for other potentials [e.g., $V(r) \sim r^{-n}$, $n > 1$].

For simplicity, we will drop the subscript on θ and Ω because we will only consider their values in the center of mass in this work. We will, however, use both center-of-mass and laboratory (lab) collision energies so we will retain subscripts denoting this. For ease of comparison, we state the relationship between the center-of-mass and laboratory collisions energies:

$$E_{\text{cm}} = \frac{m_b}{m_a + m_b} E_{\text{lab}} = \frac{\mu}{m_a} E_{\text{lab}} \tag{2}$$

where $\mu = (m_a m_b)/(m_a + m_b)$ is the reduced mass given in terms of the individual particle masses, m_a and m_b. In addition, the transformation between the center-of-mass and laboratory frame differential cross sections may be found in any of a large number of elementary texts, such as Refs. 4–7, which are also excellent reference sources for atomic collision theory pertinent to the present considerations.

The total elastic cross section, namely,

$$Q_{el} = \int \frac{d\sigma_{el}}{d\Omega} d\Omega = 2\pi \int_0^\pi \frac{d\sigma_{el}}{d\Omega} \sin\theta \, d\theta \qquad (3)$$

may easily be shown to be the same in both the center-of-mass and laboratory frames. For Rutherford scattering, the term $\sin^{-4}\theta$ leads to a divergent differential cross section as $\theta \to 0$, and therefore an infinite total cross section. The origin of this behavior is the infinite range of the Coulomb potential, and an infinite cross section is obtained in the classical treatment of elastic scattering subject to any force of infinite range. That is, since the force is of infinite range, a nonzero deflection occurs even from arbitrarily large impact parameters, and therefore the cross section is infinite. Quantum-mechanically, however, accounting for the Heisenberg uncertainty principle, one finds that deflections through angles smaller than some minimum are unobservable, and thus the total cross section is finite (for potentials which approach zero faster than r^{-2} for large values of r). This minimum angle can be shown[7,8] to be given by

$$\theta_{min} \approx [(m_a + m_b)v_{cm}b_{max}]^{-1} \qquad (4)$$

where b_{max} approximates the range of the interaction, and v_{cm} is the velocity in the center of mass frame.

In a plasma, however, electrons tend to screen or shield the more slowly moving heavy ions, so that it may be more appropriate to consider their interaction through a "screened" or "shielded" Coulomb potential. Everhart et al.,[9] for example, have computed the elastic and momentum transfer cross sections for an exponentially screened Coulomb potential [i.e., $V(r) \sim r^{-1}e^{-r/\lambda}$, with λ a screening parameter such as the Debye length].

When either or both of the two particles have electronic structure (e.g., proton–atomic hydrogen scattering), there is no exact analytic classical or quantum-mechanical expression for the differential cross section. In fact, rather than being governed by a single interaction potential, the collision system may have many adiabatic potential energy curves, representing the open electronic transition channels. Even at very low energies where not even the first excitation threshold can be reached, charge transfer between two identical or nearly identical nuclei can take place with a probability almost as large as that for elastic scattering. Thus, above the first excitation threshold, elastic scattering is but one of the competing channels in a possibly complicated multichannel scattering problem, while even below this energy, it may be necessary to solve at least a two-channel problem.

2.1. Theoretical Approaches

Especially if the colliding particles, A and B, are dissimilar, so that resonant charge transfer is not possible, a classical approach may often be adequate to approximate well the elastic scattering since the ground state of the quasimolecule

(AB) may be well separated from the lowest-lying next excited level. In this case, the "classical deflection function" is determined from this potential numerically, and its derivative with respect to impact parameter is directly related to the elastic differential cross section. Detailed descriptions of this method may be found, for example, in the texts by Goldstein,[10] McDaniel,[7] and McDowell and Coleman[4] and the recent work by Bachmann and Belitz.[3] The advantage of this approach is its relative computational ease compared with quantum-mechanical treatments, but caution in its use must be taken. For example, the approximation of classical trajectories may not be sufficient, nor may the neglect of other channels such as resonant charge transfer. Also, as noted above, since the classical total cross section diverges as $\theta \to 0$, a cutoff angle must be used to prevent a divergent result for the total elastic cross section. As we will see below, the momentum and viscosity cross sections do not heavily weight small-angle scattering and thus may be computed potentially more accurately than the elastic cross section. Despite these disadvantages, the classical treatment is often remarkably accurate.

In order to compute the elastic differential cross section quantum-mechanically, for an arbitrary impact energy, one would have to solve a set of coupled differential equations (Schrödinger wave equations for each of the coupled channels that are energetically accessible). As we have indicated, at low collision energy, typically very few channels are important in addition to elastic scattering, but for collisions involving molecules such as H_2, thresholds for rovibronic excitation are much lower in energy than those for electronic excitation. To the level of detail appropriate in the present work, we will only consider at most two channels. That is, for systems in which resonant or nearly resonant charge transfer is significant (e.g., $H^+ + H$, $D^+ + H$), we will consider two channels (elastic and charge transfer). For other systems in which charge transfer can be neglected, we will therefore only consider elastic scattering. Even above the lowest excitation thresholds, we will neglect these other channels, since, up to 100 eV, cross sections for excitation and ionization are still considerably smaller than the elastic and charge transfer cross sections (see, e.g., Janev and Smith[11] for recommended total cross sections for excitation, ionization, and charge transfer in H^+ collisions with H in the energy range above about 100 eV).

In this approach, one requires the solution to the Schrödinger equation subject to each of the potential energy curves governing the interaction. For spherically symmetric potentials, a partial wave expansion is usually employed, where the wave function is expanded in terms of products of Legendre polynomials, $P_l(\cos \theta)$, and radial wave functions dependent on angular momentum, l. The wave function is typically determined by matching a numerically computed solution of the differential equation to an asymptotic form. This procedure results in the determination of a phase shift, δ_l, which, for a single channel, in turn yields scattering amplitudes,

$$f(\theta) = \frac{1}{2i\kappa} \sum_{l=0}^{\infty} (2l+1)(e^{2i\delta_l} - 1)P_l(\cos\theta) \quad (5)$$

where $\kappa = \mu v_a = m_b v_{cm}$. The differential cross section is given as the square of $f(\theta)$,

$$\frac{d\sigma}{d\Omega} = |f(\theta)|^2 \quad (6)$$

and integration over solid angle yields the elastic total cross section,

$$Q = \frac{4\pi}{\kappa^2} \sum_{l=0}^{\infty} (2l+1) \sin^2\delta_l \quad (7)$$

(As is well known, the phase shifts for scattering in a Coulomb potential are

$$\delta_l^{Coulomb} = \arg[\Gamma(l+1 + i\frac{-2Z_a Z_b}{v_{cm}})] \quad (8)$$

which may readily be shown to result in the Rutherford cross section.) Other forms of the amplitudes are appropriate if more than one channel is considered, and which take account of nuclear symmetry, particular cases of which will be given in the results sections below. Much more detailed descriptions of this method can be found in the scattering theory references mentioned above. McDaniel[7] also describes, in significant detail, experimental approaches to the determination of elastic total and differential cross sections.

Motivated by the desire to simplify the calculation of the phase shifts, various "semiclassical" approximations have been developed. In fact, this is the approach adopted in the present work because the need to compute a very large number of individual cross sections necessitates the reduction of overall computational effort. The semiclassical method is therefore described in more detail in Section 3.

2.2. Related Cross Sections

Knowledge of the elastic total and differential cross section, computed by theoretical means or measured experimentally, is critical to modeling the transport of ions and neutrals in cool dense gas or plasma. However, other related cross sections are actually of greater practical use. They may be measured through various parameters of the gas or plasma transport (see, e.g., Ref. 7) or calculated from the elastic differential cross section. Of these, we will be particularly concerned with the momentum transfer (Q_{mt}) and viscosity (Q_{vi}) cross sections, which we describe below and which we compute and survey in Section 4.

2.2.1. The Momentum Transfer Cross Section

One of the most important primitive quantities in the study of the mobility and diffusion of neutral or charged particles in a gas or plasma is the momentum transfer cross section, Q_{mt}, defined by

$$Q_{mt} = \int (1 - \cos\theta) \frac{d\sigma_{el}}{d\Omega} d\Omega \tag{9}$$

$$= 2\pi \int_0^\pi (1 - \cos\theta) \frac{d\sigma_{el}}{d\Omega} \sin\theta \, d\theta \tag{10}$$

To see how this quantity is related to momentum transfer, consider the elastic collision between two particles a and b. The linear momentum of particle a is simply μv_a. If θ is again the center-of-mass scattering angle, the change in the forward momentum of the particle is $\mu v_a(1 - \cos\theta)$. Thus, Q_{mt} is a measure of the average forward momentum lost in such collisions. Since backscattering retards the diffusion of particles in a gas or plasma, this loss of forward momentum thus determines the rate of diffusion, and therefore Q_{mt} is often referred to as the "diffusion cross section." Other important measures may be defined in terms of Q_{mt}, such as the momentum transfer mean free path, collision frequency, and fractional energy loss per collision.[7]

If the elastic differential cross section is dominated by forward scattering, then owing to the factor $(1 - \cos\theta)$ in the definition of the momentum transfer cross section, $Q_{el} > Q_{mt}$. Conversely, if backscattering dominates, $Q_{mt} > Q_{el}$. If the elastic cross section diverges as $\theta \to 0$, this factor may lead to a finite value of Q_{mt}.

2.2.2. The Viscosity Cross Section

Inversely related to the heat conductivity and viscosity of a gas or plasma is the "viscosity cross section." It too is defined as the integral over solid angle of the elastic differential cross section, weighted by a factor of $\sin^2\theta$ instead of $(1 - \cos\theta)$. Thus,

$$Q_{vi} = \int \sin^2\theta \frac{d\sigma_{el}}{d\Omega} d\Omega \tag{11}$$

$$= 2\pi \int_0^\pi \sin^3\theta \frac{d\sigma_{el}}{d\Omega} d\theta \tag{12}$$

Since the $\sin^2\theta$ factor is maximum at $\theta = \pi/2$ and goes to zero for $\theta \to 0$ or π, this factor emphasizes scattering near $\pi/2$ while deemphasizing either forward or backward scattering. Collisions resulting in scattering to center-of-mass angles near $\pi/2$ are more effective in inhibiting conductivity because such collisions tend to equalize the energy, and, therefore, the greater the rate of collisional equalization of energy, the smaller the viscosity and heat conduction. Clearly, if the elastic differential cross section is dominated by forward scattering, the viscosity cross section is much smaller than the elastic total cross section.

3. THE SEMICLASSICAL METHOD

In the present work we wish to compute quite accurately the elastic differential cross section for a large number of collision energies and systems and create efficient computer codes that may be applied to other systems as the appropriate potential energy curves are determined. However, we note that the computational effort required to do so using a fully quantum-mechanical approach is very large compared to that for either classical or semiclassical treatments. Thus, we have adopted a semiclassical method, which we find to be considerably faster than conventional methods of numerical integration of the Schrödinger equation yet which allows an accurate representation of the elastic cross section throughout the energy range of interest.

The semiclassical approximation has been described in detail by, for example, Ford and Wheeler,[12] Mott and Massey,[13] McDowell and Coleman,[4] Child,[14] and Nikitin and Umanskii.[15] It consists of approximating the phase shift by the leading term in an expansion in powers of $\hbar$. Usually, this amounts simply to the adoption of the Jeffreys–Wentzel–Kramers–Brillouin (JWKB) approximation. In addition, validity of the Born–Oppenheimer separation, in which the electronic and nuclear motions are decoupled, is required. Traditionally, this approach also relied on techniques for replacing the sum over angular momentum for large values of l with integration over impact parameter and the use of asymptotic expansions of the Legendre polynomials. Contemporary computational facilities allow us to directly perform these summations and to check the accuracy of computed Legendre functions. Below, we first describe the Massey–Mohr approximation and then the particular form of the semiclassical approximation that we utilize in the present work.

3.1. The Massey–Mohr Approximation

A very useful semiclassical method for estimating Q_{el} and Q_{mt} is provided by the Massey–Mohr theory of scattering from an attractive, inverse-power-law potential.[13,16] This occurs because the long-range force acting between particles typically plays the dominant role in determining the elastic cross section, and because these forces may often be represented by the form

$$V(r) \sim \frac{-C}{r^n} \tag{13}$$

where C is a positive constant and n is a positive integer. For example, the asymptotic form of the interaction between an ion of charge, say, q_a and an atom with dipole polarizability (pol) α_b may be approximated as

$$V_{pol}(r) = \frac{-q_a^2 \alpha_b}{2r^4} \tag{14}$$

Similarly, the asymptotic interaction of two atoms may be approximated by the van der Waals (W) form

$$V_W(r) \sim \frac{-C_W}{r^6} \tag{15}$$

We may derive the result of Massey and Mohr by beginning with the JWKB approximation to the phase shift, i.e.,

$$\delta_l^{JWKB} = \int_{r_o}^{\infty} \left[\kappa^2 - 2\mu V(r) - \frac{l(l+1)}{r^2} \right]^{1/2} dr - \int_{r_o'}^{\infty} \left[\kappa^2 - \frac{l(l+1)}{r^2} \right]^{1/2} dr \tag{16}$$

$$= \int_{r_o}^{\infty} I_1(r)\, dr - \int_{r_o'}^{\infty} I_2(r)\, dr \tag{17}$$

where, as before, $\kappa = \mu v_a = m_b v_{cm}$ is the wave number of the relative motion, and r_o and r_o' are the roots of the equations $I_1(r) = 0$ and $I_2(r) = 0$, respectively. Performing a binomial expansion on the square root, inserting the form $V(r) = -Cr^{-n}$, and making the Langer modification $l(l+1) \to (l+\tfrac{1}{2})^2$, we obtain the Massey–Mohr (MM) phase shift

$$\delta_l^{MM} = \frac{\mu C}{\kappa} \int_L^{\infty} [r^{n-1}(r^2 - L^2)^{1/2}]^{-1}\, dr = \frac{\mu C \kappa^{n-2}}{(l+\tfrac{1}{2})^{n-1}} f(n) \tag{18}$$

where we use the shorthand notation $L = (l+\tfrac{1}{2})/\kappa$ and

$$f(n) = \begin{cases} \dfrac{\pi}{2} & \text{if } n = 2 \\[4pt] 1 & \text{if } n = 3 \\[4pt] \dfrac{n-3}{n-2}\dfrac{n-5}{n-4}\cdots\dfrac{1}{2}\dfrac{\pi}{2} & \text{if } n \text{ even} \\[4pt] \dfrac{n-3}{n-2}\dfrac{n-5}{n-4}\cdots\dfrac{2}{3} & \text{if } n \text{ odd} \end{cases} \tag{19}$$

This approximation is reasonable for values of the phase shift such that $\delta_l < 0.5$, which occur for angular momenta larger than some value, say, l_o. We note that an equivalent result is obtained from the Born approximation in the limit of large l. For $l < l_o$, provided that there are no regions of stationary phase, δ_l oscillates rapidly, and we may replace the term $\sin^2 \delta_l$ in Eq. (7) with its mean value $\tfrac{1}{2}$. This is known as the "random phase approximation." The total cross section in this approximation is composed of two contributions, one from $l < l_o$, and another from $l \geq l_o$. The result is that

$$Q_{el}^{MM} = \pi \left(\frac{2n-3}{n-2} \right) \left(\frac{2\mu C}{\kappa} \right)^{2/(n-1)} \qquad (20)$$

We note that Mott and Massey[13] gave a slightly different expression, namely,

$$Q_{el}^{MM} = \pi \left(2 + \frac{4}{n-2} \right) \left(\frac{\mu C}{\kappa} \right)^{2/(n-1)} \qquad (21)$$

which is essentially equivalent (e.g., for $n = 4$, the values of the constant terms

$$\frac{2n-3}{n-2}(2)^{2/(n-1)} \quad \text{and} \quad 2 + \frac{4}{n-2} \qquad (22)$$

differ by less than 1%). Also, in order to display explicitly the dependence of this approximation on the center-of-mass energy, we may rewrite Eq. (20) as

$$Q_{el}^{MM} = \pi \left(\frac{2n-3}{n-2} \right) \left(\frac{2\mu C^2}{E_{cm}} \right)^{1/(n-1)} \qquad (23)$$

since it may easily be shown that $\kappa = \sqrt{(2\mu E_{cm})}$.

Dalgarno and McDowell[17] have extended this type of analysis to potentials of exponential form, that is $V(r) = ae^{-cr}$, and McDowell and Coleman[4] gave a Massey–Mohr approximation for any additive potential combining inverse-power-law and exponential potentials, i.e.,

$$V(r) = \sum_{i=1} a_i r^{b_i} e^{-c_i r} + \sum_{j=2} \frac{C_j}{r^{j+1}} \qquad (24)$$

These results are useful for various common forms used to represent interaction potentials, such as the well-known Lennard-Jones (12,6) potential. However, simple analytical approximations for the differential cross section, and hence the momentum and viscosity cross sections, are not often available. A better approach is to directly compute the diffusion cross section using an expression involving the phase shifts for an ion–atom scattering:

$$Q_{mt} = \frac{4\pi}{\kappa^2} \sum_{l=0}^{\infty} (l+1) \sin^2(\delta_l - \delta_{l-1}) \qquad (25)$$

(see, e.g., McDaniel[7] and Dalgarno et al.[18] who discuss in detail the mobility of ions in unlike and parent gases). If the ion and the atom are of the same species, then the gerade and ungerade potentials must be taken into account, and this expression must be modified. It must also be modified if the atom has nonzero nuclear spin (see, e.g., Hirschfelder et al.[19] and the example used below for H$^+$ + H and H + H scattering). Using the Massey–Mohr theory, the diffusion cross section can be obtained in this way for the polarization potential (Eq. 14) and is

$$Q_{mt}^{MM} = 2.210\pi \left(\frac{q_a^2 \alpha}{2E_{cm}}\right)^{1/2} \tag{26}$$

(see, e.g., Ref. 13). Since this result is essentially the same as that obtained in the well-known Langevin theory, this approximation is also called the Langevin diffusion cross section.

Clearly, these approximations are justified when no other more rigorous calculations are available or when the only information regarding the interaction potential is given, for example, by the atomic polarizability or the van der Waals constant. In particular, if the potential energy is known well enough to justify a fit using the form given by Eq. (24), then not much more effort would allow one to compute the full semiclassical or quantum-mechanical result. In order to illustrate the use of these approximations, in Fig. 1 we compare our semiclassical results with the fit made by Bachmann and Belitz[3] to their classical calculations and with the Massey–Mohr elastic and momentum transfer cross sections for collisions of a proton with atomic hydrogen. In computing the Massey–Mohr estimates, we have assumed that the interaction is dominated by the polarization interaction, and so $C = \alpha_{hydrogen}/2 = 2.25$ a.u. and $n = 4$. Thus, Eqs. (23) and (26) take the form

$$Q_{el,H^+ + H}^{MM} = 165 E_{cm}^{-1/3} \tag{27}$$

$$Q_{mt,H^+ + H}^{MM} = 10.4 E_{cm}^{-1/2} \tag{28}$$

As we will illustrate in more detail below, the elastic differential cross section is characterized at low energies by many oscillations, among them, effects of rainbow and glory scattering. The total cross section also has pronounced oscillations and resonance features. The origin of many of the velocity-dependent oscillations has been summarized, for example, by McDowell and Coleman,[4] who also illustrate the energy range in which the semiclassical approximation breaks down, in tunneling and orbiting regimes. Such a breakdown will be evident in the charge transfer cross sections computed within the present approach for energies less than about 0.01 eV for collisions of H$^+$, D$^+$, and T$^+$ with H, D, and T.

3.2. Practical Computational Schemes

If one were to try to compute the semiclassical phase shifts from Eq. (16) directly, one would quickly run into substantial numerical difficulties, since this expression is the difference of two very large numbers, computed as integrals of functions with upper bounds of infinity. Our approach is to use equivalent expressions that arrange for the extremely large numbers to cancel while the important

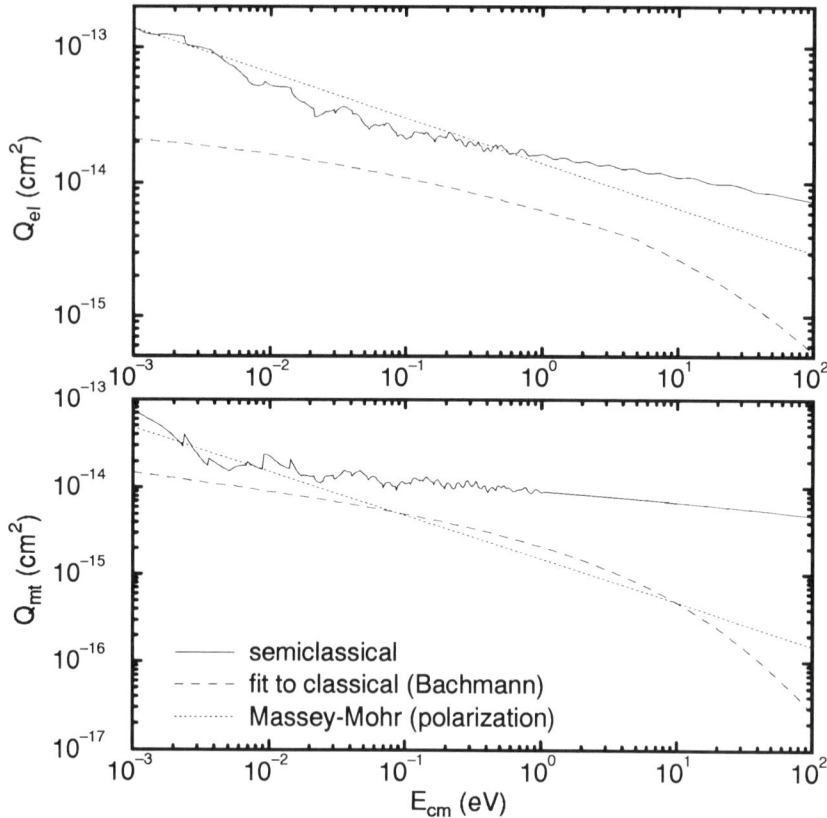

Figure 1. Comparison of the elastic (Q_{el}) and momentum transfer (Q_{mt}) cross sections for collisions of H^+ with H. Displayed are the present semiclassical method, the fit made by Bachmann and Belitz[3] to their classical approximation, and the Massey–Mohr approximation computed using the potential $V(r) = -\alpha_H/2r^4$.

difference is represented with high precision. Using the notation defined by Eqs. (16) and (17), after some rearrangement we find

$$\delta_l = \begin{cases} \int_{r_o}^{r_o'} I_1(r)\,dr + \int_{r_o'}^{\infty} \frac{2\mu V(r)\,dr}{I_1(r) + I_2(r)} & \text{if } r_o < r_o' \\ \int_{r_o'}^{r_o} I_2(r)\,dr + \int_{r_o}^{\infty} \frac{2\mu V(r)\,dr}{I_1(r) + I_2(r)} & \text{if } r_o > r_o' \end{cases} \quad (29)$$

In addition, one integral can be performed analytically, namely,

$$\int_{r_o'}^{r_o} I_2(r)\, dr = f(r_o) - f(r_o') \tag{30}$$

$$= f(r_o) \tag{31}$$

$$= r_o \left[\kappa^2 - \frac{(l+\tfrac{1}{2})^2}{r_o^2} \right]^{1/2} - (l+\tfrac{1}{2}) \cos^{-1}\left(\frac{l+\tfrac{1}{2}}{\kappa r_o} \right) \tag{32}$$

since $f(r_o') = 0$.

Given a potential function $V(r)$, the procedure begins by computing the roots ("turning points") r_o and r_o'. This is a very sensitive numerical task in that the argument includes the centripetal term $(l+\tfrac{1}{2})/2\mu r^2$ and thus will have a region for some small range of l in which the kinetic energy is almost tangential to it. Thus, great care must be taken to ensure that the outermost root is found, rather than skipping over it and finding the "easy" root far in the interior where the potential steeply rises, and that the "top-of-barrier" is itself not skipped by having too large a step size in the root-finding routine. Next, the integrals are computed very accurately by breaking the integration range into several (e.g., five) roughly logarithmically spaced zones. In each zone, the range is further divided into a possibly large number of subdomains, and Gauss–Legendre quadrature is performed in each. In each major zone, the number of subdomains is doubled until a predetermined convergence criterion is satisfied. Finally, the phase shifts are summed in such a manner that the cross sections appropriate for each system considered are produced, subject to the proper symmetry of the wave function and nuclear spin. These expressions are given below, along with a description of the procedure used to determine the potential energy curves, for each system considered. As described also below, a test case in which very accurate quantum-mechanically calculated phase shifts were available was utilized as a benchmark for the semiclassical computer code. As a general comment, we found that in order to obtain good agreement with these accurate quantum-mechanical results, we required potential energy curves accurate throughout the range extending from near the nuclei to the radii at which asymptotic expansions are very good. Also, any discontinuities introduced by matching potential energy curves computed with different methods in different regimes caused very large discrepancies. In brief, we had expected that potential energy curves good to a few percent might yield cross sections good to, say, ten percent. In fact, we found that the dependence was indeed much stronger than this naive expectation. The potential energy curves computed for $H^+ + H$ are probably good to a part in 10^6, while the phase shifts and cross sections are good to a few percent, in general.

3.3. Asymptotic Behavior

Before proceeding to describe the cross sections tabulated and calculated in this work, it is worth briefly mentioning the asymptotic behaviors expected for the cross sections since, in practice, one may have to extrapolate the present results to higher energies in order to accommodate a Maxwellian distribution of particle velocities and because these formulas suggest the form we should use for fitting the cross sections.

Considering only charged particle–atom collisions, classical arguments (see, e.g., Ref. 3) lead to the following predicted behaviors:

$$Q_{el}(E_{cm} \to \infty) \sim (a - b \ln E_{cm})^2 \tag{33}$$

$$Q_{mt}(E_{cm} \to \infty) \sim \frac{c}{E_{cm}^2} \tag{34}$$

$$Q_{vi}(E_{cm} \to \infty) \sim \frac{d}{E_{cm}^2} \tag{35}$$

From the semiclassical analysis summarized above, we find that

$$Q_{el}(E_{cm} \to \infty) \sim \frac{c}{E_{cm}^3} \tag{36}$$

$$Q_{mt}(E_{cm} \to \infty) \sim \frac{c}{E_{cm}^2} \tag{37}$$

Also, Dalgarno et al.[18] and others have shown that for ions in their parent gases, $Q_{mt} \approx 2Q_{ct}$ for sufficiently low energies, where Q_{ct} is the charge transfer cross section, which they also show to behave as $(a - b \ln E_{cm})^2$. Thus, this analysis yields

$$Q_{mt}(E_{cm} \to \infty) \sim (a - b \ln E_{cm})^2 \tag{38}$$

These forms and the expected behavior as $E_{cm} \to 0$ suggest that the cross sections could be fitted by the expression

$$\ln Q(E_{cm}) \approx \sum_{i=1}^{n} a_i (\ln E_{cm})^{(i-1)} \tag{39}$$

where n is typically less than 10. This also allows complete compatibility with the fitting procedure used by Bachmann and Belitz.[3] For each cross section given, we have used a least-squares procedure to determine the coefficients a_i. The number of coefficients used was determined to simultaneously both give a good fit and avoid spurious oscillations in the resulting fitted curve.

Of course, as energies much above the range considered here ($0.001 \leq E_{cm} \leq 100$ eV) are approached, we are well above the thresholds for many other channels (excitation, ionization) which will deplete the elastic channel to a larger and larger extent as energy is increased. Therefore, results of these formal approximations must be weighed against the results expected from experimental measurements or calculations taking into account the competition provided by these new channels.

4. SPECIFIC CROSS SECTIONS

In the following sections, we describe the calculation or source for the cross sections and fits given below.

4.1. $H^+ + H$

Noting the sensitivity of the computed phase shifts, and thus the desired cross sections, to the accuracy of the potential energy curves, we took great care in computing the two lowest curves for $H^+ + H$ ($1s\sigma_g$ and $2p\sigma_u$). Tabulated values which could be interpolated, parameterized fits, and the results of certain "standard" computer codes did not yield very good results. We used the computer code of Ovchinnikov and Solov'ev,[20] the development of which took into account the generally higher precision required in the study of mesic molecular potential energies (due to the greater mass of the muon). For very large radii, we used the asymptotic expansion of Damburg and Propin.[21] In addition, for the computation of the appropriate phase shifts and cross sections when the projectile ion and/or target atom was any of the isotopes of hydrogen (H, D, T), we utilized the procedure that Macek and co-workers[22] employed to account for the asymptotic energy differences and the shifts due to the mass differences.

A very accurate quantum-mechanical solution to the elastic scattering and charge transfer problem in the energy range of interest has been provided for $H^+ + H$[23,24] and for $H^+ + D$ and $D^+ + H$[25] by Hunter and Kuriyan. These works provided a benchmark for the development of our semiclassical approach, and we find very good agreement with the phase shifts that Hunter and Kuriyan tabulated extensively.[24] We also find good agreement with the elastic and charge transfer total and differential cross sections contained in these works. In addition, quite recently, Hodges and Breig[26] have performed numerical integrations of the radial Schrödinger equation for Hunter and Kuriyan's potential energy curves and confirmed their results independently. Our semiclassical results agree very well with their published elastic, momentum transfer, and charge transfer cross sections.

The equations needed to compute the elastic differential cross section for $H^+ + H$ which account for the electronic and nuclear symmetry can be found in Ref. 24. Using them, we have computed the cross sections for as many as 100 energies per decade in the range of $0.001 \leq E_{cm} \leq 100$ eV. Examples of the elastic and charge transfer differential cross sections are presented in Figs. 2 and 3. The rapid oscillation of these differential cross sections, as mentioned above, increases for

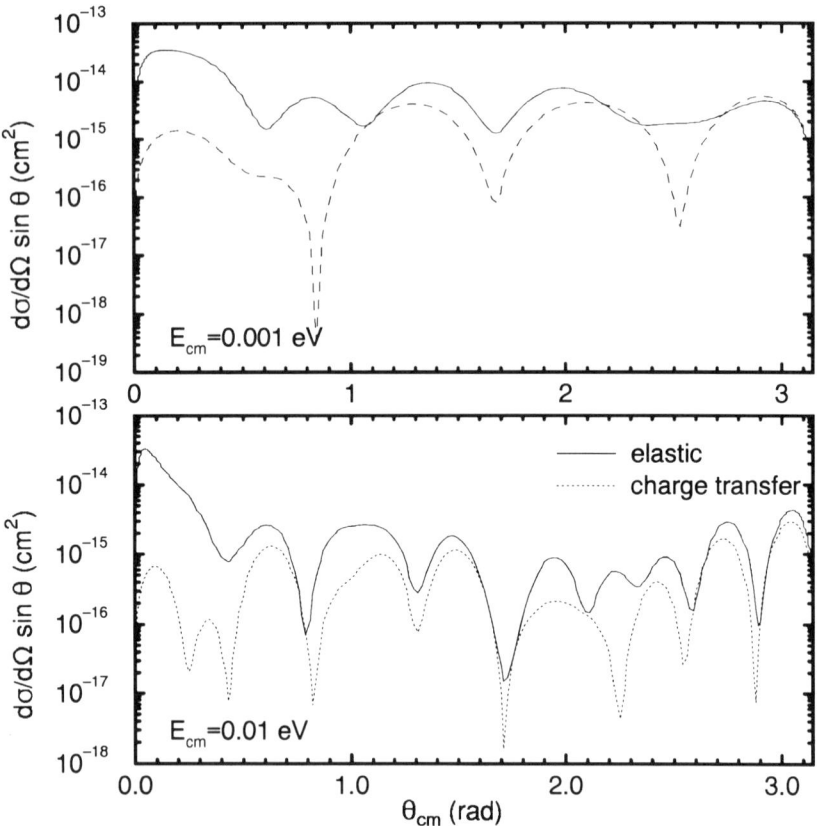

Figure 2. The elastic and charge transfer differential cross sections for collisions of $H^+ + H$ for E_{cm} = 0.001 and 0.01 eV computed using the present semiclassical approximation.

increasing impact energy because many more partial waves contribute to the scattering at higher impact energy.

In Fig. 4 we present our results for Q_{el}, Q_{mt}, Q_{vi}, and Q_{ct}. One may easily note the velocity-dependent oscillations described above. In particular, above 1 eV the elastic differential cross section becomes dominated by the forward and backward peaks, and thus the viscosity cross section can also be seen to dramatically decrease in magnitude. The breakdown of the semiclassical approach is also seen for energies below 0.01 eV for the charge transfer cross section, where actual resonances take on a sharp, discontinuous shape. McDowell and Coleman[4] described this effect in some detail. However, we note that the agreement with the quantum-mechanical

Elastic and Related Cross Sections

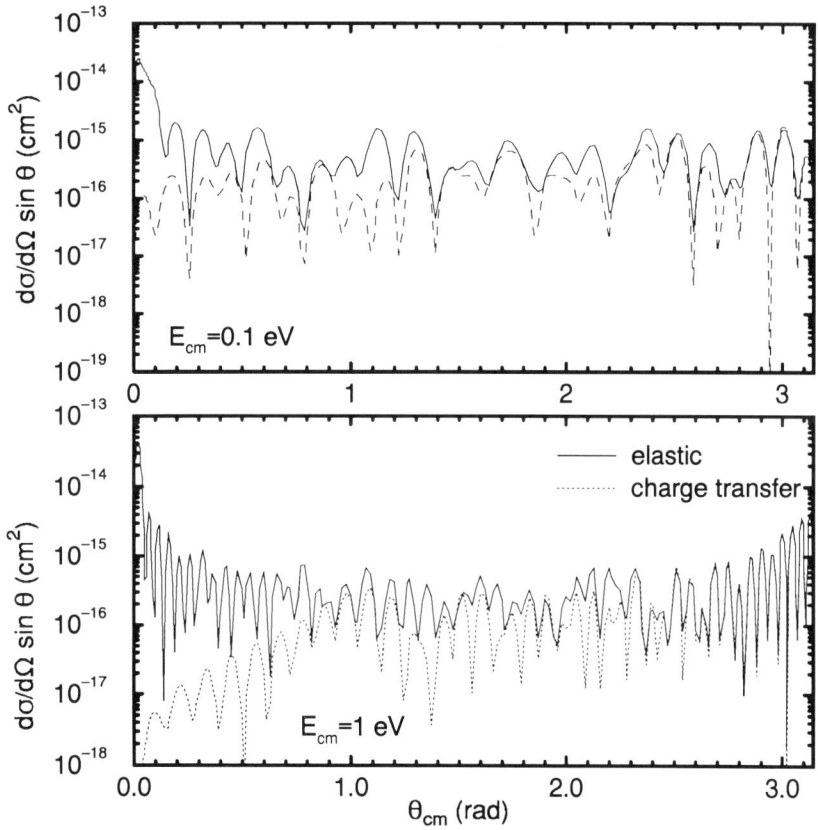

Figure 3. The same as Fig. 2, except for E_{cm} = 0.1 and 1 eV.

calculations of the elastic cross section remains good even below 0.01 eV. Below the range considered here, this may certainly not be the case.

The least-squares fits that we have performed are displayed in Fig. 5, and the coefficients are given in Table I. Since the cross sections decrease almost linearly on this log–log plot for energies above about 1 eV, in Table I we also give coefficients for fits including only terms up to the second power in $\ln E_{cm}$, since they will not oscillate as much. These restricted fits should be used above, say, 10 eV.

4.2. $D^+ + D$

Wadehra[27] computed the elastic and momentum transfer cross sections for $0.0005 < E_{cm} < 5$ eV for $D^+ + D$, using the Numerov method of integrating the radial

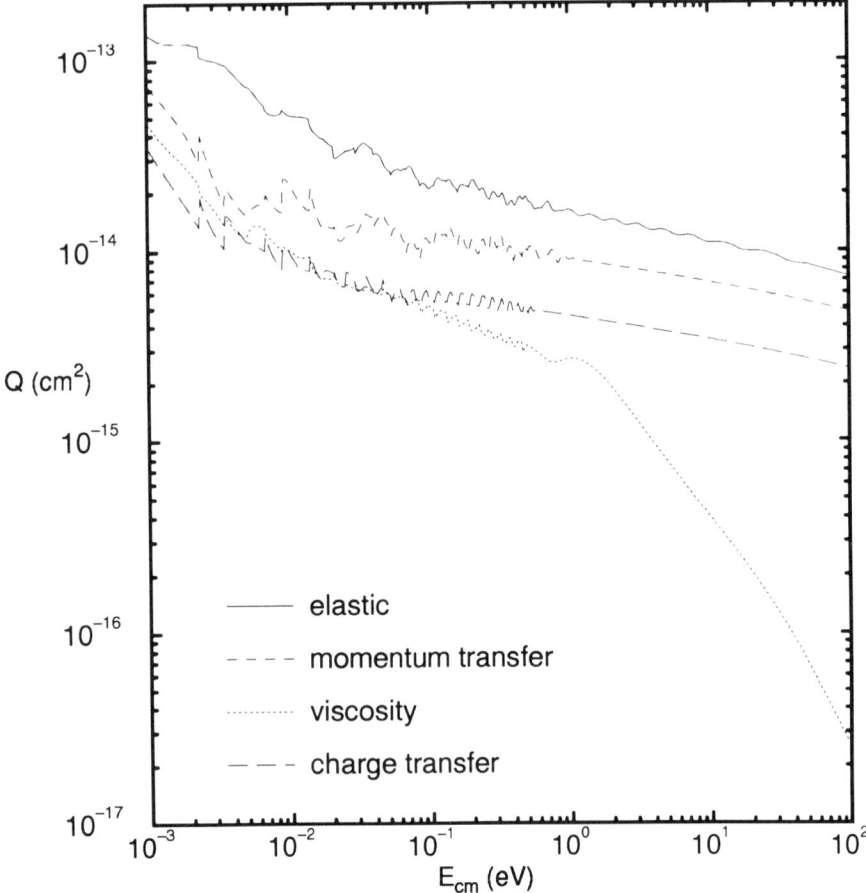

Figure 4. The elastic, momentum transfer, viscosity, and charge transfer cross sections for H$^+$ + H computed with the present semiclassical approximation.

Schrödinger equation and the numerically tabulated potential energy curves of Peek.[28] Beyond 30 a.u. a simple asymptotic form ($c_4 r^{-4} + c_6 r^{-6}$) was used, and for large values of angular momentum, the Born approximation to the phase shift was used. Various higher order transport cross sections were also computed, defined by

$$Q^{(k)} = \frac{2\pi}{C_k} \int_0^\pi (1 - \cos^k \theta) \frac{d\sigma}{d\Omega}(\theta) \sin \theta \, d\theta \qquad (40)$$

where

Elastic and Related Cross Sections

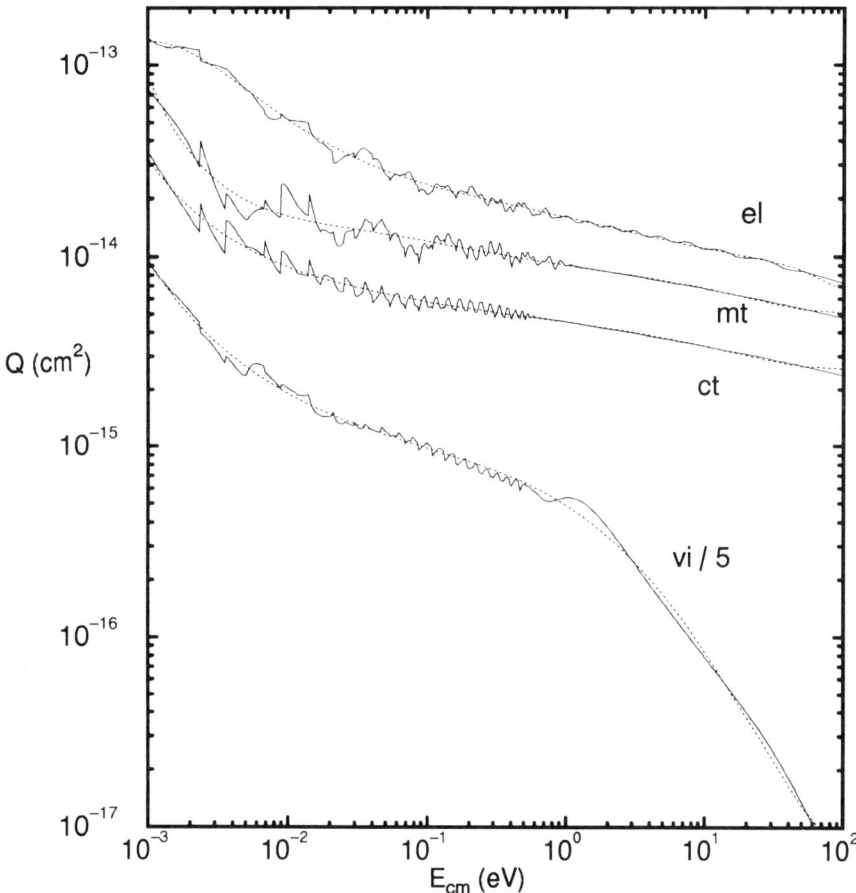

Figure 5. The same as Fig. 4 with the addition of the least-squares fits performed for these cross sections.

$$C_k = 1 - \frac{1 + (-1)^k}{2(l+1)} \qquad (41)$$

for $l = 1, 2, 3,$ and 4. Comparing this with our definition of the momentum transfer cross section, we see that for $k = 1$, Wadehra's $Q^{(1)} \equiv Q_{mt}$.

Wadehra's results agree reasonably well with ours, the differences being perhaps mainly attributable to the choice of potential energy curves. We expect that our procedure, in which these curves have been computed very accurately, should lead to a better result. Wadehra also pointed out a useful scaling between the results

Table I. Fitting Coefficients for the Elastic, Momentum Transfer, Viscosity, and Charge Transfer Cross Sections for H⁺ + H from the Present Semiclassical Approximation

	Elastic		Momentum transfer	
	$0.001 \leq E_{cm} \leq 10$ eV	$E_{cm} > 10$ eV	$0.001 \leq E_{cm} \leq 10$ eV	$E_{cm} > 10$ eV
a_1	–3.176304E+01	–3.175474E+01	–3.233966E+01	–3.231141E+01
a_2	–1.640510E–01	–1.650750E–01	–1.126918E–01	–1.386002E–01
a_3	–4.995055E–03		5.287706E–03	
a_4	6.936219E–04		–2.445017E–03	
a_5	1.423218E–03		–1.044156E–03	
a_6	–1.622733E–04		8.419691E–05	
a_7	–3.998204E–05		3.824773E–05	
	Viscosity		Charge transfer	
	$0.001 \leq E_{cm} \leq 10$ eV	$E_{cm} > 10$ eV	$0.001 \leq E_{cm} \leq 10$ eV	$E_{cm} > 10$ eV
a_1	–3.364341E+01	–3.224881E+01	–3.302872E+01	–3.301160E+01
a_2	–4.588546E–01	–1.277213E+00	–1.000134E–01	–1.358551E–01
a_3	–1.086684E–01		–4.543425E–03	
a_4	–1.629809E–02		–3.587916E–03	
a_5	1.191807E–03		–1.446320E–05	
a_6	4.366559E–04		9.322700E–05	
a_7	3.483356E–05		1.665865E–05	

for H⁺ + H and D⁺ + D. In particular, one finds that for any of the cross sections discussed,

$$Q^{H^+ + H}(E) \approx \frac{1}{2} Q^{D^+ + D}(\frac{1}{4}E) \tag{42}$$

In addition, he noted the well-known scaling of the momentum transfer cross section for energies above 5 eV of the form

$$Q_{mt} \sim (a - b \ln E_{cm})^2 \tag{43}$$

4.3. H⁺ + D, H⁺ + T, D⁺ + H, D⁺ + T, T⁺ + H, T⁺ + D, and T⁺ + T

Using Macek's method[22] and the present semiclassical approximation, we have computed the elastic, momentum transfer, viscosity, and charge transfer cross sections for the other seven combinations of the hydrogen ion–hydrogen atom collisions. For economy of space, we have not tabulated these results here, but we will distribute them to any interested reader. Also, certain scalings of these cross sections with regard to one another are apparent from the results and can be predicted from the fact that the phase shifts scale with the center-of-mass velocity in these collisions. However, that is not to say that the cross sections can all be accurately obtained from the $H^+ + H$ curves. Indeed, the asymptotic difference between the gerade and ungerade potential energy curves dramatically affects the charge transfer cross section at low energy and changes the shape of the elastic differential cross section, and therefore the momentum transfer and viscosity cross sections as well. For energies above, say, 1 eV, the cross sections are easily obtained in terms of scaling laws from the $H^+ + H$ cross sections.

4.4. H + H

The relevant cross sections for H + H have been considered recently by Jamieson et al.[29] but only for energies well below the range of interest. Similarly,

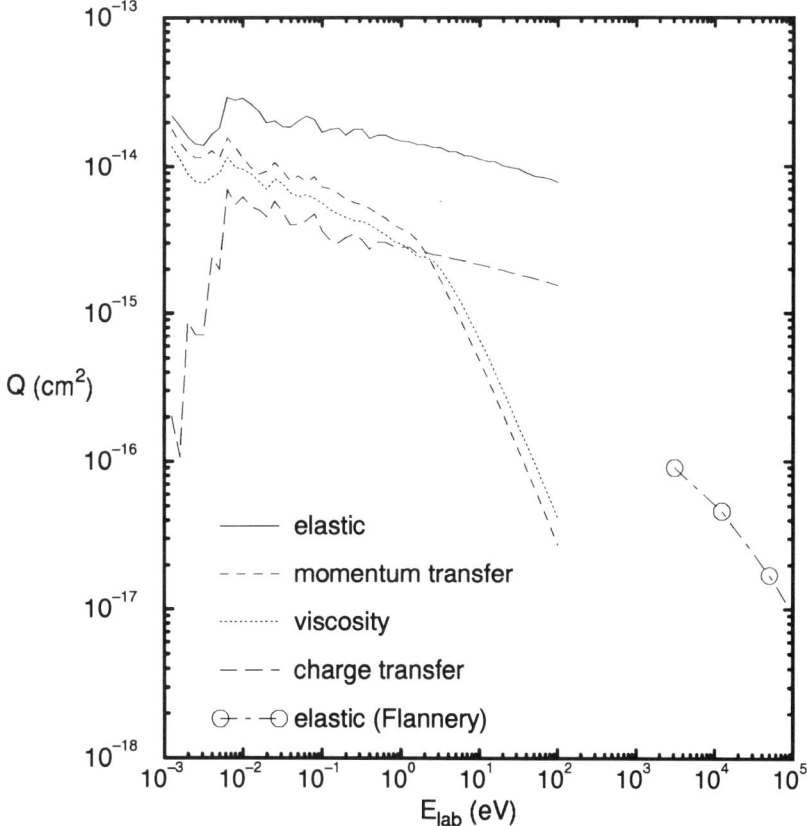

Figure 6. The elastic, momentum transfer, viscosity, and charge transfer cross sections for H + H computed with the present semiclassical approximation and the potential energy curves assembled by Jamieson et al.[29] Also shown is the four-state coupled-channels elastic cross section given by Flannery and McCann.[31]

Koyama and Baird[30] have considered the elastic cross sections for either singlet or triplet scattering for very low energies (below 0.001 eV). In addition, Flannery and McCann[31] performed Born approximation and four-state coupled-channels calculations, but at very high energies (3–112 keV). Using the potential energy curves assembled by Jamieson et al.[29] we have performed the semiclassical approximation and found that the singlet and triplet cross sections agree reasonably well with those of Koyama and Baird,[30] except that they overestimate by about a factor of 2 the resonance associated with the ($v = 14$, $l = 4$) level near 0.004 eV.

We display our results in Fig. 6, along with the elastic total cross sections at high energies given by Flannery and McCann. Apparently, the two-channel calculation that we perform, accounting for the elastic and charge transfer reactions, must overestimate the cross section in the limit of high energies, due to the depletion

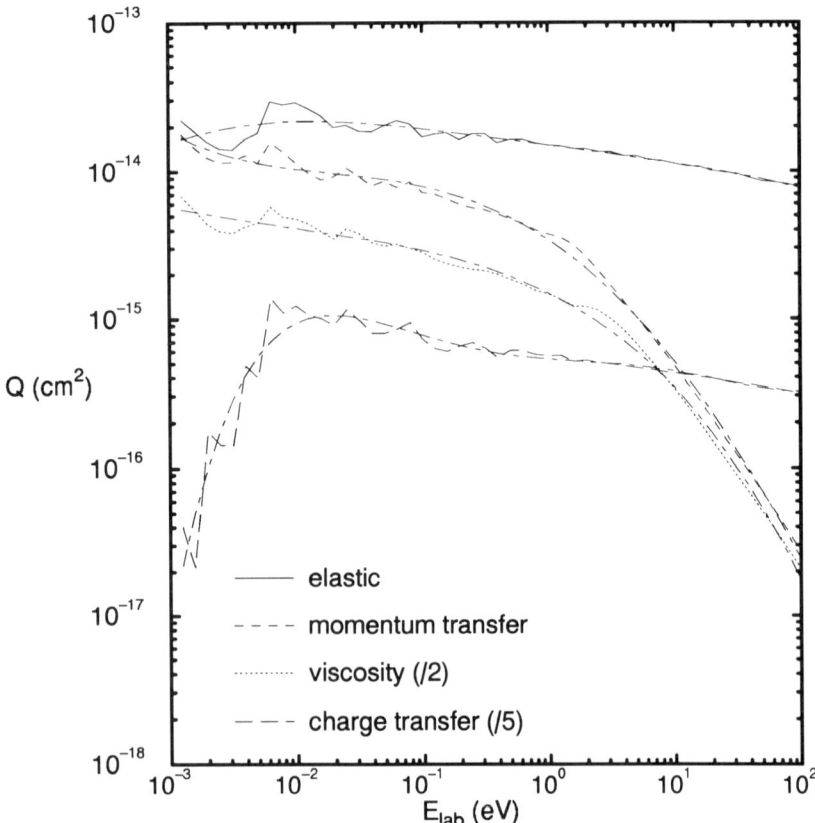

Figure 7. The same as Fig. 6 with the addition of the fits to the present semiclassical results.

Table II. Fitting Coefficients for the Elastic, Momentum Transfer, Viscosity, and Charge Transfer Cross Sections for H + H from the Present Semiclassical Approximation

	Elastic	Momentum transfer	Viscosity	Charge transfer
a_1	−3.183543E+01	−3.330843E+01	−3.344860E+01	−3.356106E+01
a_2	−1.188321E−01	−5.738374E−01	−4.238982E−01	−7.034667E−02
a_3	−5.605593E−04	−1.028610E−01	−7.477873E−02	2.287162E−02
a_4	2.665721E−04	−3.920980E−03	−7.915053E−03	−1.232475E−02
a_5	−2.999621E−04	5.964135E−04	−2.686129E−04	−1.011379E−03
a_6				4.768785E−04

caused by inclusion of excitation pathways. Since some uncertainty must still exist, and since the results are in reasonable agreement with the low-energy calculations of Koyama and Baird, we have still provided fits for the cross sections that we have computed, noting that the most caution should be used for energies above 1 eV. These fits are displayed in Fig. 7, and the coefficients are given in Table II.

4.5. H + He and D + He

The total cross section for elastic scattering has been measured for collisions of H and D with helium by Toennies et al.[32] for beam velocities of 350–3000 m/s. These authors have also computed the total cross section extending to lower velocities by fitting various published sections of the potential (i.e., matching a short-range exponential repulsion of the Born–Mayer form, a numerical calculation in the mid-range of internuclear separations, and a long-range multipolar attraction). In reasonable agreement with these measurements and covering an extended range of energies are quantal calculations performed by Gayet et al.[33] using a model potential approach. We present these data in Figs. 8 and 9. For H + He, the calculations of Gayet et al. differ by as much as about 40–50% from the calculations and experiment of Toennies et al. for the lower portion of the energy range covered by the experiment but converge to within about 5% for higher energies. Thus, without clear reason to do otherwise, we have taken the average of the two calculations for the lower portion of the range, approached the experiments in the middle, and followed the calculation of Gayet et al. to the high end of the range. In addition, since we are interested here only in center-of-mass energies greater than 10^{-3} eV, and because of the uncertainties in reading data from figures, we have ignored the fact that the calculations and experiment reflect an averaging over the 5 K thermal distribution of the helium target. Therefore, the uncertainty in the present tentative recommendation is approximately 25% for the lower energies and at least 2–5% for the higher energies.

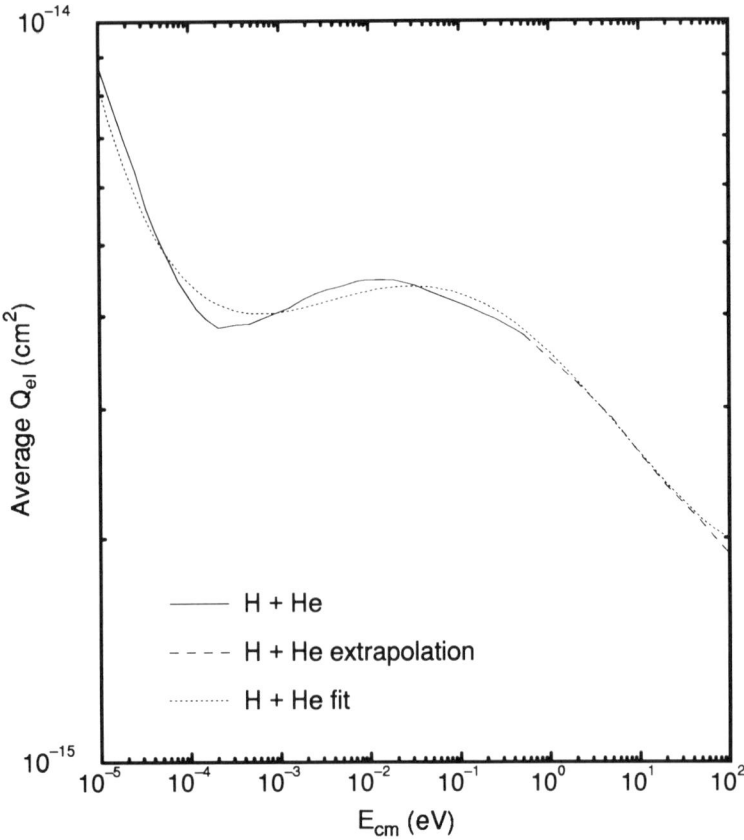

Figure 8. The average elastic cross section for H + He obtained as described in the text.

Additional calculations extending the range of these data, and subsequent study of the relevant potential energy curves, can benefit from the very recent quantum Monte Carlo calculations of Bhattacharya and Anderson.[34] They have also summarized and reviewed quite a few previous results of experiments and theories which have given information regarding the interaction potential governing the H–He system, and we therefore refer the interested reader to that work.

In order to extend the information found in these works to higher energies, we have extrapolated the cross section by matching in the region of 1 keV with the Born theory cross sections determined from Gillespie's[35] work. Also, in Fig. 10 we demonstrate the result of scaling the H + He data to approximate the D + He result

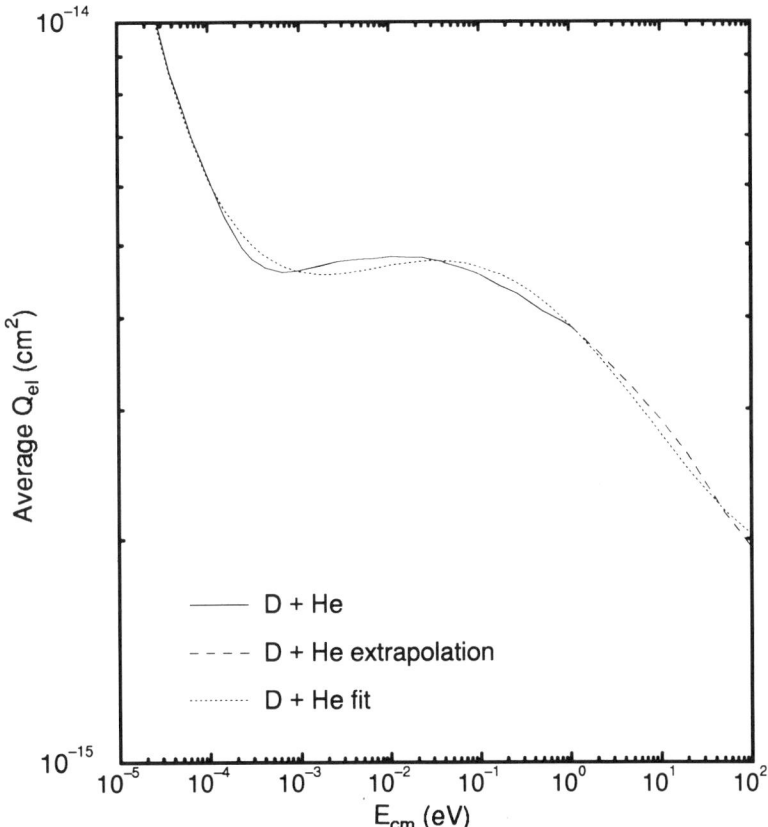

Figure 9. The average elastic cross section for D + He obtained as described in the text.

by transforming it to an equal center-of-mass velocity. Finally, in Table III we give the present fitting coefficients for these elastic total cross sections. Clearly, work is required to calculate the differential cross sections so that the transport cross sections may be tabulated as well.

4.6. H^+, H_3^+, H, H^-, and $H_2 + H_2$

Phelps[36] has recommended values for the momentum transfer cross section for $0.1 \leq E_{lab} \leq 10{,}000$ eV for a number of projectiles colliding with H_2, primarily based upon measurements of ionic mobility. In Fig. 11 we display these recommended curves, and we tabulate fitting coefficients for them in Table IV.

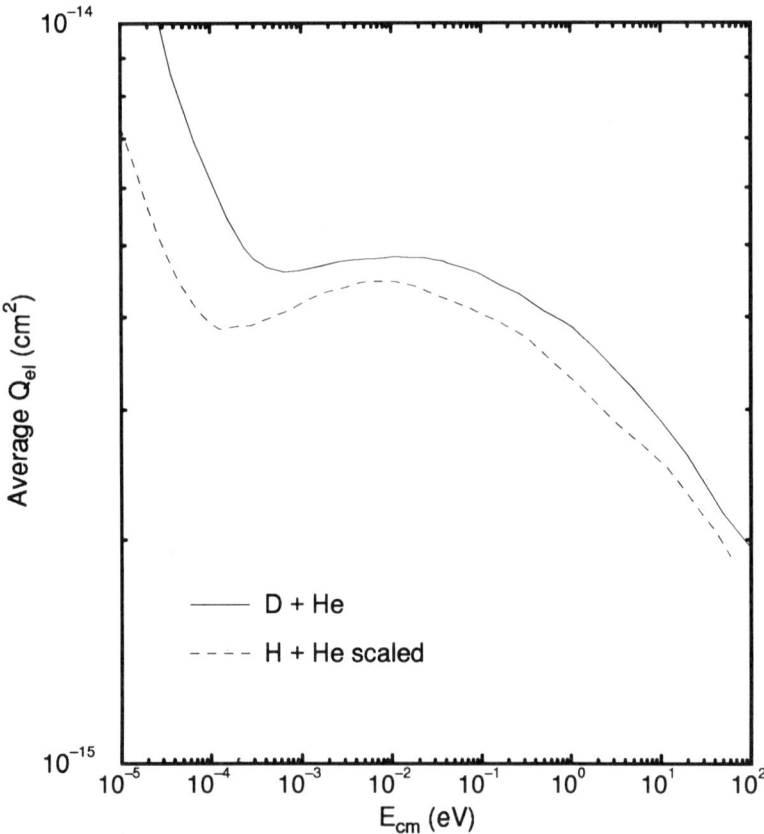

Figure 10. The average elastic cross sections for D + He and for H + He scaled for equal center-of-mass velocities.

Table III. Fitting Coefficients for the Elastic Cross Section in Collisions of H + He and D + He, Determined from the Experimental Measurements of Toennies et al.,[a] the Theory of Gayet et al.,[b] and an Extrapolation to the Born Cross Sections Given by Gillepsie[c]

	H + He	D + He
a_1	−3.326692E+01	−3.317723E+01
a_2	−1.155103E−01	−1.207214E−01
a_3	−1.350961E−02	−1.645774E−02
a_4	1.496878E−03	1.390332E−03
a_5	2.038736E−04	2.529289E−04

[a] Ref. 32.
[b] Ref. 33.
[c] Ref. 35.

Elastic and Related Cross Sections

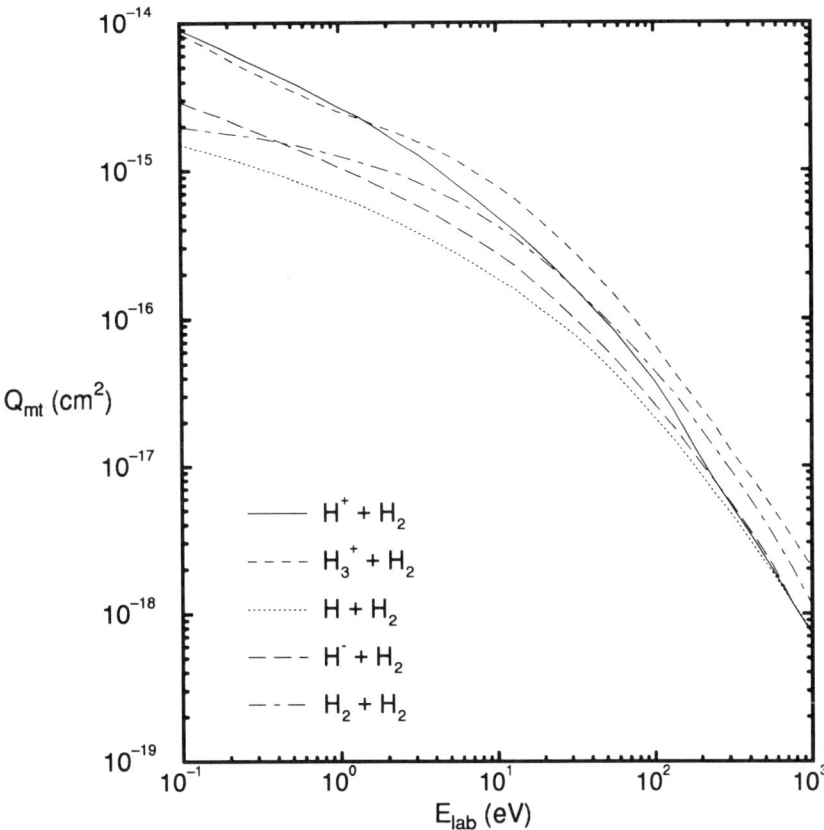

Figure 11. The momentum transfer cross sections recommended by Phelps[36] for H^+, H_3^+, H, H^-, and H_2 colliding with H_2 for $0.1 \leq E_{lab} \leq 1000$ eV.

Table IV. Fitting Coefficients Determined from the Recommended Momentum Transfer Cross Section Given by Phelps[a] for H^+, H_3^+, H, H^-, and $H_2 + H_2$, with $0.1 \leq E_{lab} \leq 1000$ eV

	$H^+ + H_2$	$H_3^+ + H_2$	$H + H_2$	$H^- + H_2$	$H_2 + H_2$
a_1	−3.355719E+01	−3.355777E+01	−3.495671E+01	−3.447844E+01	−3.430345E+01
a_2	−5.696568E−01	−3.943775E−01	−4.062257E−01	−4.683839E−01	−2.960406E−01
a_3	−4.089556E−02	−8.527676E−03	−3.820531E−02	−3.246911E−02	−6.382532E−02
a_4	−1.143513E−02	−2.651879E−02	−9.404486E−03	−1.007441E−02	−7.557519E−03
a_5	5.926596E−04	2.067432E−03	3.963723E−04	3.530742E−04	2.606259E−04

[a]Ref. 36

4.7. $H^+ + He$, $H^+ + H_2$, $He^+ + He$, and $He^{2+} + He$

Bachmann and Belitz have performed classical calculations for the elastic scattering for $H^+ + He$, $H^+ + H_2$, and $He^+ + He$, using a one-potential formalism. Fitting coefficients for these cross sections are given in their paper.[3] Also, elastic and momentum transfer cross sections for $He^{2+} + He$ have been calculated by Wadehra[27] as described above for $D^+ + D$, along with various other transport cross sections.

5. CONCLUSIONS

We have computed the elastic, momentum transfer, viscosity, and charge transfer cross sections in the energy range $0.001 \leq E_{cm} \leq 100$ eV as accurately as is feasible given the scope of the present considerations. These cross sections are important for modeling the energy and momentum balance and other properties relating to transport in the edge and divertor regions of next-step fusion reactors. As other accurate potential energy curves are tabulated, the present authors plan to perform additional semiclassical and new quantum-mechanical calculations. Because the calculation of accurate cross sections, even for systems in which the basic physics is understood, is a time-consuming and painstaking task, the present work must be continued to provide the required data for other systems. In addition, for the higher portion of the energy range of interest, it may be necessary to perform multichannel or optical potential calculations to properly account for the effects of open reaction channels other than elastic scattering and charge transfer.

ACKNOWLEDGMENTS. The authors wish to acknowledge helpful discussions with J. H. Macek and express their appreciation to him and A. Ovchinnikova for allowing us to use their procedure for scaling the $H^+ + H$ potential energy curves to other isotopes. In addition, we thank M. J. Jamieson for providing us with computer codes to produce the $H + H$ potential that he and his co-authors used in Ref. 29. Support of the theoretical work reported here has been provided by the U.S. Department of Energy, Office of Fusion Energy, through contract number DE-AC05-84OR21400 with Oak Ridge National Laboratory, managed by Martin Marietta Energy Systems, Inc. SYuO and SVP also acknowledge the support of the National Science Foundation under grant number PHY-9213953.

REFERENCES

1. R. K. Janev, in *Review of Fundamental Processes and Applications of Atoms and Ions* (C. D. Lin, ed.), World Scientific, Singapore (1993).
2. D. Reiter, in *Atomic and Plasma-Material Interaction Processes in Controlled Thermonuclear Fusion* (R. K. Janev and H. W. Drawin, eds.), Elsevier, Amsterdam (1993).

3. P. Bachmann and H. J. Belitz, Elastic Processes in Hydrogen–Helium Plasmas: Collision Data, Max Planck Institut für Plamsaphysik, Report IPP 8/2, 1993.
4. M. R. C. McDowell and J. P. Coleman, *Introduction to the Theory of Ion–Atom Collisions*, North-Holland, Amsterdam (1970).
5. C. J. Joachain, *Quantum Collision Theory*, North-Holland, Amsterdam (1983).
6. B. H. Bransden and C. J. Joachain, *Physics of Atoms and Molecules*, Longman, London (1983).
7. E. W. McDaniel, *Collision Phenomena in Ionized Gases*, John Wiley & Sons, New York (1964); *Atomic Collisions: Electron and Photon Projectiles*, John Wiley & Sons, New York (1989); *Atomic Collisions: Heavy Particle Projectiles*, John Wiley & Sons, New York (1993).
8. E. A. Mason, J. T. Vanderslice, and C. J. G. Raw, *J. Chem. Phys.* **40**, 2153 (1964).
9. E. Everhart, G. Stone, and R. J. Carbone, *Phys. Rev.* **99**, 1287 (1955); G. H. Lane and E. Everhart, *Phys. Rev.* **117**, 920 (1960).
10. H. Goldstein, *Classical Mechanics*, Addison-Wesley, Reading, Massachusetts (1980).
11. R. K. Janev and J. J. Smith, in *Atomic and Plasma-Material Interaction Data for Fusion (Nucl. Fusion, Supplement)* **4**, 1 (1993).
12. K. W. Ford and J. A. Wheeler, *Ann. Phys. (N.Y.)* **7**, 259 (1959).
13. N. F. Mott and H. S. W. Massey, *The Theory of Atomic Collisions*, Clarendon Press, Oxford (1965).
14. M. S. Child, *Molecular Collision Theory*, Academic Press, New York (1974).
15. E. E. Nikitin and S. Ya. Umanskii, *Theory of Slow Atomic Collisions*, Springer-Verlag, Berlin (1984).
16. H. S. W. Massey and C. B. O. Mohr, *Proc. R. Soc. London, Ser. A* **144**, 188 (1934).
17. A. Dalgarno and M. R. C. McDowell, *Proc. Phys. Soc. London Sect. A* **69**, 615 (1956).
18. A. Dalgarno, M. R. C. McDowell, and A. Williams, *Phil. Trans. R. Soc. London Ser. A* **250**, 411 (1958); A. Dalgarno, M. R. C. McDowell, and A. Williams, *Phil. Trans. R. Soc. London Ser. A* **250**, 426 (1958).
19. J. O. Hirschfelder, C. F. Curtiss, and R. B. Bird, *Molecular Theory of Gases and Liquids*, John Wiley & Sons, New York (1964).
20. S. Yu. Ovchinnikov and E. A. Solov'ev, *Comments At. Mol. Phys.* **22**, 69 (1988).
21. R. J. Damburg and R. Kh. Propin, *J. Phys. B* **1**, 681 (1968).
22. J. H. Macek and K. A. Jerjian, *Phys. Rev. A* **33**, 233 (1986); J. H. Macek and A. Ovchinnikova, private communication.
23. G. Hunter and M. Kuriyan, *Proc. R. Soc. London Ser. A* **353**, 575 (1977).
24. G. Hunter and M. Kuriyan, *At. Data Nucl. Data Tables* **25**, 287 (1980).
25. G. Hunter and M. Kuriyan, *Proc. R. Soc. London Ser. A* **358**, 321 (1977).
26. R. R. Hodges and E. L. Breig, *J. Geophys. Res.* **95**, 7697 (1991).
27. J. M. Wadehra, *Phys. Rev. A* **20**, 1859 (1979).
28. J. M. Peek, *J. Chem. Phys.* **43**, 3004 (1965).
29. M. J. Jamieson, A. Dalgarno, and J. N. Yukich, *Phys. Rev. A* **46**, 6956 (1992); see also the comment in E. Tiesinga, *Phys. Rev. A* **48**, 4801 (1993).
30. N. Koyama and J. C. Baird, *J. Phys. Soc. Jpn.* **55**, 801 (1986).
31. M. R. Flannery and K. J. McCann, *Phys. Rev. A* **9**, 1947 (1974).
32. J. P. Toennies, W. Welz, and G. Wolf, *Chem. Phys. Lett.* **44**, 5 (1976).
33. R. Gayet, R. McCarroll, and P. Valiron, *Chem. Phys. Lett.* **58**, 501 (1978).
34. A. Bhattacharya and J. B. Anderson, *Phys. Rev. A* **49**, 2441 (1994).
35. G. H. Gillespie, *Phys. Rev. A* **17**, 1284 (1978).
36. A. V. Phelps, *J. Phys. Chem. Ref. Data* **19**, 653 (1990); Erratum, *J. Phys. Chem. Ref. Data* **20**, 1339 (1991).

Chapter 12

Rearrangement Processes Involving Hydrogen and Helium Atoms and Ions

F. Brouillard and X. Urbain

1. INTRODUCTION

In this chapter, we deal with rearrangement processes occurring in binary collisions of hydrogen or helium atoms, molecules, or ions, that is, H, H_2, He, H^+, He^+, H_2^+, He^{2+}, and H^-. He^- is not taken into account as it is short-lived and unlikely, under thermonuclear plasma density conditions, to undergo a collision before decaying. For the same reason, we consider only collisions in which the partners are in their ground state or in a metastable state.

The eight collision partners mentioned above combine to form 36 collisional systems. Three of these, however, are trivial (H^+–H^+, He^{2+}–He^{2+}, H^+–He^{2+}), with no possibility of rearrangement processes.

A rearrangement process is understood here as one that modifies the grouping of the particles composing the system. Rearrangement processes thus include transfer or ejection of electrons (ionization), dissociations of molecules (molecular ions), associations of atoms (ions) into a molecular structure, and reactive colli-

---

F. BROUILLARD and X. URBAIN • Laboratoire de Physique Atomique et Moléculaire, Université Catholique de Louvain, Louvain-la-Neuve, Belgium.

Atomic and Molecular Processes in Fusion Edge Plasmas, edited by R. K. Janev. Plenum Press, New York, 1995.

sions, which are a combination of the last two processes. Reactive collisions, however, are not considered here as they are dealt with in Chapter 14.

The point of view adopted is that of a user of atomic data whose primary concern is not to understand the physical processes but to have confidence in the reliability of the data. We therefore made no attempt to discuss or even to present the theoretical models or fundamental mechanisms underlying the rearrangement processes. Instead, we have selected in the literature what we believe to be useful cross sections and gathered them in a set of 11 figures (Figs. 1–11). Comments are given in the text with respect to the origin and reliability of the data.

All data presented are experimental, except in one or two cases in which theoretical predictions are used to extrapolate experimental values, with which they are in good agreement, outside the experimental energy range. Theoretical predictions are suspect, as a rule, until they have been confirmed experimentally. On the other hand, the number of particular atomic processes is so large (infinite when the quantum states are specified) that theoretical modeling is a necessity not only from the standpoint of physical understanding but also for practical purposes. The strength of theoretical models is their capacity to produce an unlimited amount of data. Because of this, there is no sense to make systematic tables or graphs of theoretical data.

The compilation presented in this chapter is less elaborate and exhaustive than some others in the literature.[1-6] We think, however, that many readers may find some advantage in its compact form and will benefit from the updating that it provides. Complementary information can also be found in recent theoretical papers.[7-10]

2. EXPERIMENTAL METHODS

We will briefly review here the major difficulties that are encountered in experiments related to low-energy rearrangement processes and point out the most common sources of error in cross section measurements. Questions arise in relation to the interaction geometry, the control of the reactants, the specification of the processes, and the detection of the reaction products.

2.1. The Interaction of Two Beams

The measurement of the cross section of a collisional process between two particles (1 and 2) can always be regarded as involving the interaction of a beam of particle 1 intersecting a beam of particle 2. The case in which one of the beams is at rest is a special one but also a very common one because of its experimental simplicity. This case obtains when a beam interacts with a gaseous target. The relation between the cross section σ and the observed reaction rate dN/dt is then a simple one:

$$\frac{dN}{dt} = \sigma I_1 \int n_2(z)\, dz$$

I_1 is the beam intensity (number of particles per second), and n_2 is the density of gas, which varies in the direction (z) of the beam but can be taken as constant across it. The integral is the so-called "target thickness." It can usually be measured or calculated with good accuracy. The target thickness can be made large, up to the limit imposed by the condition of single collisions. As a result, the reaction rate is high even for very small cross sections.

This is not the case for beam–beam measurements because the achievable density in a beam is orders of magnitude lower than that in a gas target. For a beam of light atoms in the kilo-electron-volt energy range, the density is usually less than 10^6 atoms/cm^3, whereas a typical value for gas targets is 10^{12} atoms/cm^3.

A second drawback of beam–beam measurements is their sensitivity to the interaction geometry. This shows up in the relation linking the cross section with the reaction rate, which, for two beams of intensities I_1 and I_2 and particle velocities $\mathbf{v}_1$ and $\mathbf{v}_2$, intersecting at an angle θ, is written as:

$$\frac{dN}{dt} = \frac{\sigma v}{v_1 v_2 \sin\theta} I_1 I_2 F$$

where

$$v = |\mathbf{v}_1 - \mathbf{v}_2|$$

and

$$F = \frac{1}{I_1 I_2} \int J_1(y) J_2(y)\, dy$$

J_1 and J_2 are the one-dimensional current density profiles along the direction (y) perpendicular to both beams. They represent the number of particles passing per second through a 1-cm slit placed across the beam in the collision plane (plane of $\mathbf{v}_1$ and $\mathbf{v}_2$).

The quantity F (form factor) must be known in order to derive the cross section from beam–beam experiments. Its measurement is not easy and cannot be performed simultaneously with the main measurement. Changes in the form factor occurring during the data acquisition are thus not recorded and lead to fluctuations of the cross section. This uncertainty regarding the form factor is one of the main causes of inaccuracy in beam–beam measurements.

There is, however, a way to circumvent the problem. It is the so-called "animated" cross-beams method, proposed by Defrance et al.[11] This method consists in sweeping, repeatedly, one of the beams across the other with a constant velocity (u), from and to positions where the beams no longer overlap. It is easily shown that the number of reactions (K) occurring during one sweep no longer depends on the density profiles but solely on the intensities of the beams:

$$K = \frac{\sigma v}{v_1 v_2 \sin \theta} \frac{I_1 I_2}{u}$$

Use of the beam–beam method is unavoidable when both reactants are ions or, more generally, such particles that cannot be accumulated in a box. Excited atoms obviously belong to this category.

The beam–beam method also allows measurements to be conducted at low center-of-mass (CM) energies that cannot be reached in a beam–cell experiment. In the latter type of experiment, Coulomb repulsion impedes the formation of low-energy ion beams (below 100 eV) with useful intensity and thus also of atomic beams obtained by partially neutralizing an ion beam. In addition, the reaction products would have too little energy to be detectable. In experiments with intersecting beams, both beams can be fast while the CM energy, given by

$$E_{CM} = \frac{\mu}{m_1} E_1 + \frac{\mu}{m_2} E_2 - 2\sqrt{\frac{\mu}{m_1} \frac{\mu}{m_2} E_1 E_2} \cos \theta$$

where m_1, m_2, E_1, and E_2 are the masses and laboratory energies of the particles in the two beams, and μ is the reduced mass, remains small if θ is small. An extreme case is the merged-beams configuration, where $\theta = 0$, which allows measurements to be carried out down to below 0.01 eV.

2.2. Control of the Reactants

The control of the nature of the reactants is an essential part of a collision experiment. Ideally, the interacting partners should be single species in definite quantum states. This is not always feasible and is not necessary either. Measurements can be carried out in less resolved situations, where several states of the reactants are populated for instance, but the populations must then be controlled.

Accidental contamination of the reactants or uncontrolled excitation is often the reason for the discrepancies existing between the data published by different authors. The following examples illustrate this point.

The scattering among the data for ionization of helium by protons below 10 keV is probably due to a slight contamination of the helium target, as the ionization cross section of helium is exceptionally small. The different results that have been obtained for the ionization of helium in collisions with gases can almost certainly be ascribed to an uncontrolled population of the metastable state. The extent of formation of hydrogen in the 2p state via electron capture by protons in helium has recently been shown[12] to be several times smaller than found in earlier measurements. The presence of a minute amount of neutral hydrogen in the proton beam and its subsequent excitation in helium suffices to explain the higher results obtained previously.

The cross sections of H_2^+, especially for dissociative processes, are very sensitive to the vibrational excitation of the ion. Effects larger than 30% have been demonstrated by Williams and Dunbar.[13]

2.3. Specification of the Process

Measured cross sections refer to processes that are more or less completely specified, depending on the sophistication of the experiments. The process is defined by the information available on the reaction products. This information relates to (1) the nature of the products (species and charge state), and (2) their quantum state. In addition, the velocities of the products (energies and angles) are also to be specified for differential cross sections. However, most existing data are not differential cross sections, and the quantum states are often ignored. Even the nature of all the products is not always given.

In a rearrangement collision, for instance, between a fast He^{2+} ion and a stationary neutral target X, five processes can occur:

$$He^{2+} + X \begin{cases} \xrightarrow{\sigma_1} He^{2+} + X^+ + e^- & \text{(ionization)} \\ \xrightarrow{\sigma_2} He^+ + X^+ & \text{(electron transfer)} \\ \xrightarrow{\sigma_3} He^{2+} + X^{2+} + 2e^- & \text{(double ionization)} \\ \xrightarrow{\sigma_4} He^+ + X^{2+} + e^- & \text{(transfer ionization)} \\ \xrightarrow{\sigma_5} He + X^{2+} & \text{(double transfer)} \end{cases}$$

In standard experiments, one simply measures the currents of He and He^+ created in the He^{2+} beam, and these yield σ_5 and $(\sigma_2 + \sigma_4)$, respectively. In more sophisticated measurements, one extracts the slow X^+ and X^{2+} ions and the electrons out of the target and measures their separate currents, using, for example, time-of-flight techniques, to obtain $(\sigma_1 + \sigma_2)$, $(\sigma_3 + \sigma_4 + \sigma_5)$, and $(\sigma_1 + 2\sigma_3 + \sigma_5)$, respectively. Deriving the individual cross sections from the measurements is not as easy as it might seem. First, the set of equations one has for the five unknowns is underdetermined and can only be solved provided one of the cross sections can be regarded as negligible, and, second, the combined experimental errors can make the results meaningless.

Measurements that unambiguously identify the collisional processes are those using coincidence techniques, that is, measurements in which the simultaneous formation of two reaction products is detected. Coincidence techniques are especially well adapted to beam–beam experiments, in which they are almost obligatory. They are (often) the only way to overcome the major problem of beam–beam experiments, namely, the interaction of the beams with the residual gas that dominates, by many orders of magnitude, the beam-beam interaction and generates huge backgrounds.

2.4. Detection of Reaction Products

The rate of production of a reaction product is measured either by recording the corresponding (tiny) current or by counting individual particles on an appropriate detector. There is, in principle, no significant error in this part of the measurement when conventional, reliable instruments are used. However, this does not exclude the possibility that some discrepant data have suffered from a badly controlled detector efficiency.

A more important source of error is the incomplete collection of the reaction products. In some processes with a small cross section or involving a dissociation, the reaction products emerge with non-negligible transverse velocity. At low energy, the corresponding deflection angle can be large enough for the product to miss the detector. Many of the existing discrepancies in the low-energy measurements of the dissociation of H_2^+ in gases can be ascribed to this effect.[14]

3. CROSS SECTIONS

Experimental data for the cross sections of rearrangement processes in the binary collisions of hydrogen or helium atoms, molecules, or ions (H, H_2, He, H^+, He^+, H_2^+, He^{2+}, and H^-) are presented in Figs. 1–11. The sources of these data are identified below in Sections 3.1–3.11, respectively. The energy range considered is 0–10 keV. In all cases, the energy is the center-of-mass (CM) energy. When helium is involved, this energy refers to the mass 3 isotope. We have partitioned the data by considering separately the collisions of each of the eight species listed above. As the collisions of H and H^+ are divided into two groups that distinguish between the production of H^+ or H^- and H or H^-, respectively, this makes a total of 10 figures. An additional figure is provided for associative ionization and Penning ionization. The energy range in the latter case is 0–10 eV.

Clearly, some arbitrary choices have been made in partitioning the data among these 11 figures—for example, charge exchange between H and He^{2+} can be classified as a collision of H or as a collision of He^{2+}. In such cases, our choice was often made on the basis of practical considerations (cross section scale).

3.1. Collisions of H—Production of H^-

Figure 1 shows the cross sections for collisional processes involving H that result in the production of H^-. The data are taken from the measurements by Hill

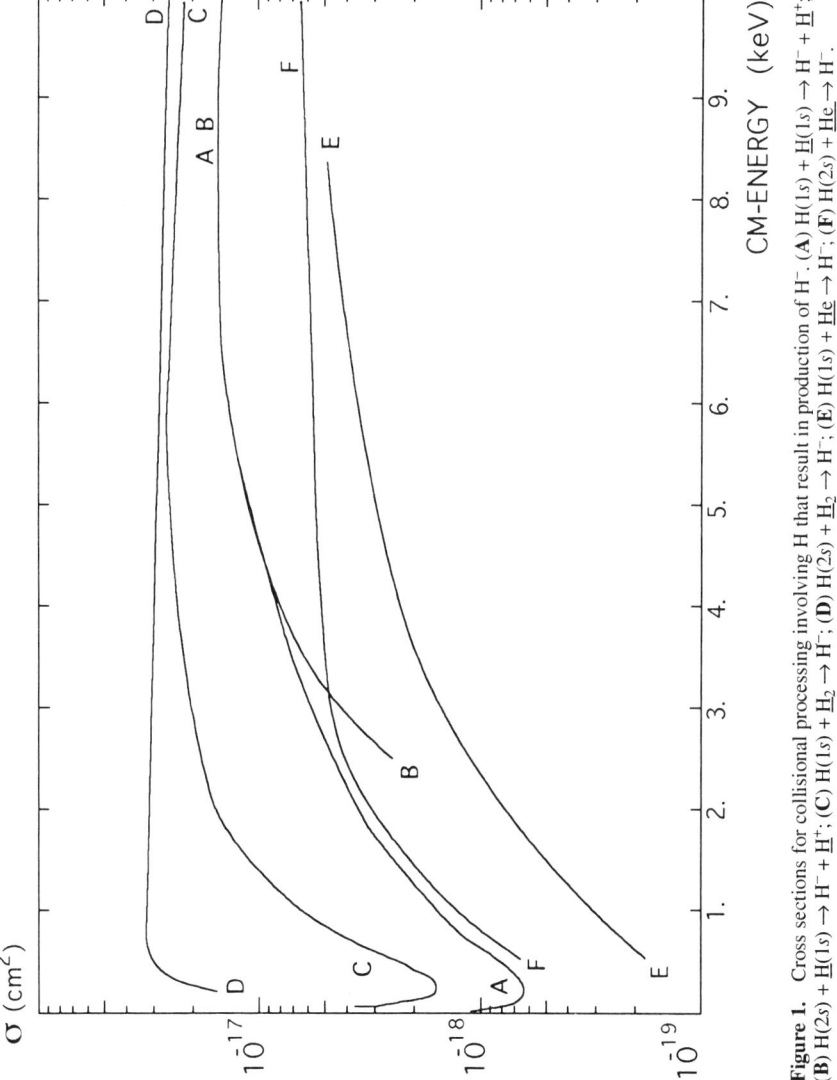

Figure 1. Cross sections for collisional processing involving H that result in production of H⁻: (**A**) $H(1s) + \underline{H}(1s) \rightarrow H^- + \underline{H}^+$; (**B**) $H(2s) + \underline{H}(1s) \rightarrow H^- + \underline{H}^+$; (**C**) $H(1s) + \underline{H}_2 \rightarrow H^-$; (**D**) $H(2s) + \underline{H}_2 \rightarrow H^-$; (**E**) $H(1s) + \underline{He} \rightarrow H^-$; (**F**) $H(2s) + \underline{He} \rightarrow H^-$.

et al.[16] for processes A, B, C, and D, McClure[17] for process A, and Roussel et al.[18] for processes C, D, E, and F. The results of Gealy and Van Zyl[15] for process A deviate from those of Hill et al.[16] below 2 keV and have not been retained because of their large experimental error.

Process B should be distinguished from the related process

$$H(2s) + \underline{H}(1s) \rightarrow H^+ + \underline{H}^-$$

which is presented in Fig. 2 (process B).

Recent measurements by Alsamour et al.[19] on the reaction

$$H(ns) + He \rightarrow H^- + He^+$$

have yielded the ratio σ_3/σ_2 of the cross sections for $n = 3$ and $n = 2$. These authors also found the ratio (σ_2/σ_1) for $n = 2$ and $n = 1$, to be five times larger than in the measurements by Roussel et al.[18] This discrepancy is probably caused by incomplete collection of the H^- ions in the case of $H(1s)$, where the angular distribution is likely to be large. The magnitude of the discrepancy indicates that the angular distribution is larger than the one accepted in the measurements of Roussel et al. for the ground state, which thus also suffered from incomplete collection of H^- ions. The data shown here for process E could be too small by as much as 50%.

3.2. Collisions of H—Production of H⁺

The cross sections for collisional process involving H that would result in the production of H^+ are presented in Fig. 2.

The cross section for process A, which includes ionization and charge transfer, is taken from the measurements of Gealy and Van Zyl,[15] Hill et al.[16] and McClure,[17] that for process B, not to be confused with process B in Fig. 1, from Fussen et al.,[20] and that for process C from Hill et al.[16]

The cross section for electron loss in H_2 (process D) is taken from the measurements by Hill et al.,[16] McClure,[17] Van Zyl et al.,[21] Stier and Barnett,[22] and Smith et al.[23] Additional data on electron loss by $H(3p)$ in H_2 and He have been reported by Cornet et al.[24]

Ionization of H in helium (process E) has been measured by Roussel et al.,[18] Smith et al.,[23] and Williams.[25] Roussel et al.[18] also measured the ionization of metastable hydrogen atoms in helium (process F) and in H_2. The cross section for ionization of H by H^+ (process G) has been taken from measurements by Shah and Gilbody[26] and Shah et al.,[27] which have been preferred to those of Fite et al.,[28] Park,[29] and Rudd et al.[30] because of their consistency with the elaborate treatment of Winter and Lin.[31] Recent measurements and calculations by Pieksma[32] over the energy range 0.5–3 keV confirm the data of Shah and Gilbody[26] and Shah et al.[27]

The cross section for electron transfer (process H) has been taken from the work of Gealy and Van Zyl,[15] Fite et al.,[28] and McClure.[33] Electron capture into the $H(2s)$ state is presented in Fig. 3.

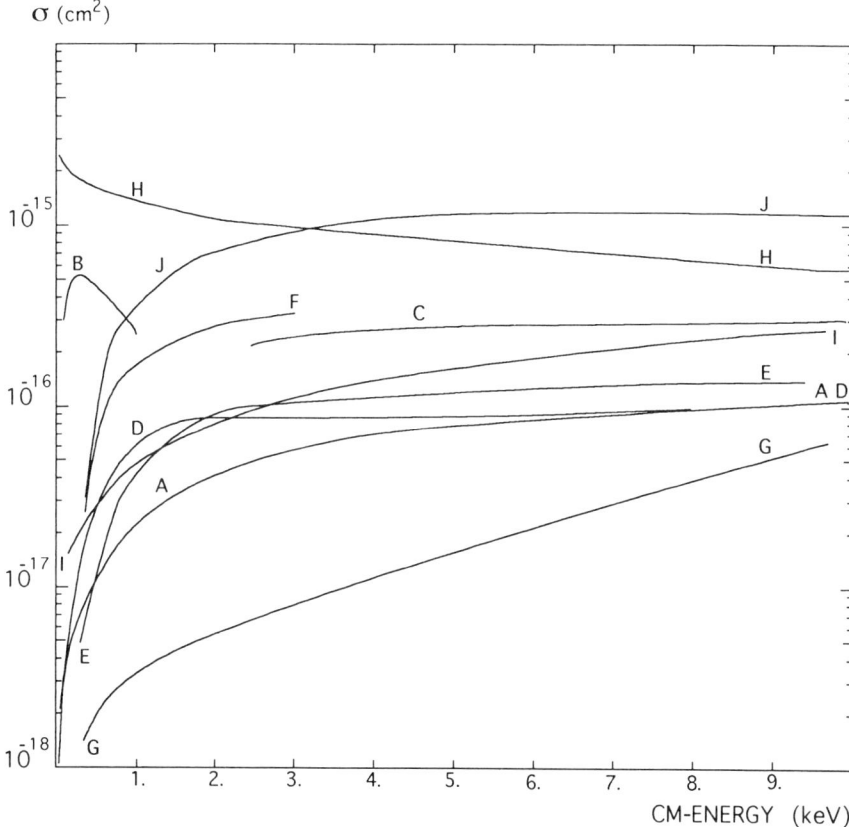

Figure 2. Cross sections for collisional processes involving H that result in production of H$^+$. (**A**) H(1s) + $\underline{\text{H}}$(1s) → H$^+$; (**B**) H(2s) + $\underline{\text{H}}$(1s) → H$^+$ + $\underline{\text{H}}^-$; (**C**) H(2s) + $\underline{\text{H}}$(1s) → H$^+$; (**D**) H(1s) + $\underline{\text{H}}_2$ → H$^+$; (**E**) H(1s) + He → H$^+$ + e^- + $\underline{\text{He}}$; (**F**) H(2s) + He → H$^+$ + e^- + $\underline{\text{He}}$; (**G**) H(1s) + $\underline{\text{H}}^+$ → H$^+$ + e^- + $\underline{\text{H}}^+$; (**H**) H(1s) + $\underline{\text{H}}^+$ → H$^+$ + $\underline{\text{H}}$; (**I**) H(1s) + $\underline{\text{He}}^+$ → H$^+$ + $\underline{\text{He}}$; (**J**) H(1s) + $\underline{\text{He}}^{2+}$ → H$^+$ + $\underline{\text{He}}^+$.

The cross section for charge exchange with He$^+$ ions (process I) is based on the measurements by Olson et al.[34] and Lockwood et al.[35] and can be compared with the calculation by Harel and Salin.[36] The cross section for ionization by He$^+$ ions has not been measured but is much smaller than that for charge exchange below 10 keV according to the calculation of Willis et al.[37]

The cross section for charge exchange with He^{2+} ions (process J) includes the data of Olson et al.,[34] Shah and Gilbody,[38] Nutt et al.,[39] and Fite et al.[40] but not those of Bayfield and Khayrallah,[41] which are larger than the others above 4 keV. There are no experimental data for ionization by He^{2+} below 10 keV. At high energies, measurements have been done by Shah and Gilbody[26] and Shah et al.,[42] and these are well reproduced by the theoretical calculation of Shingal and Lin.[43] Charge exchange between He^{2+} and excited H ($n = 10$–25) has been investigated experimentally by Burniaux et al.[44]

3.3. Collisions of H^+—Production of H

The cross sections for collisional processes involving H^+ that result in the production of H are presented in Fig. 3.

Formation of metastable hydrogen in charge exchange of protons with ground state hydrogen atoms (process A) has been measured by Hill et al.[45] and Morgan et al.,[46] from whose work the data presented in Fig. 3 are extracted, and also by Bayfield,[47] Chong and Fite,[48] and Morgan et al.[49] The cross section for charge

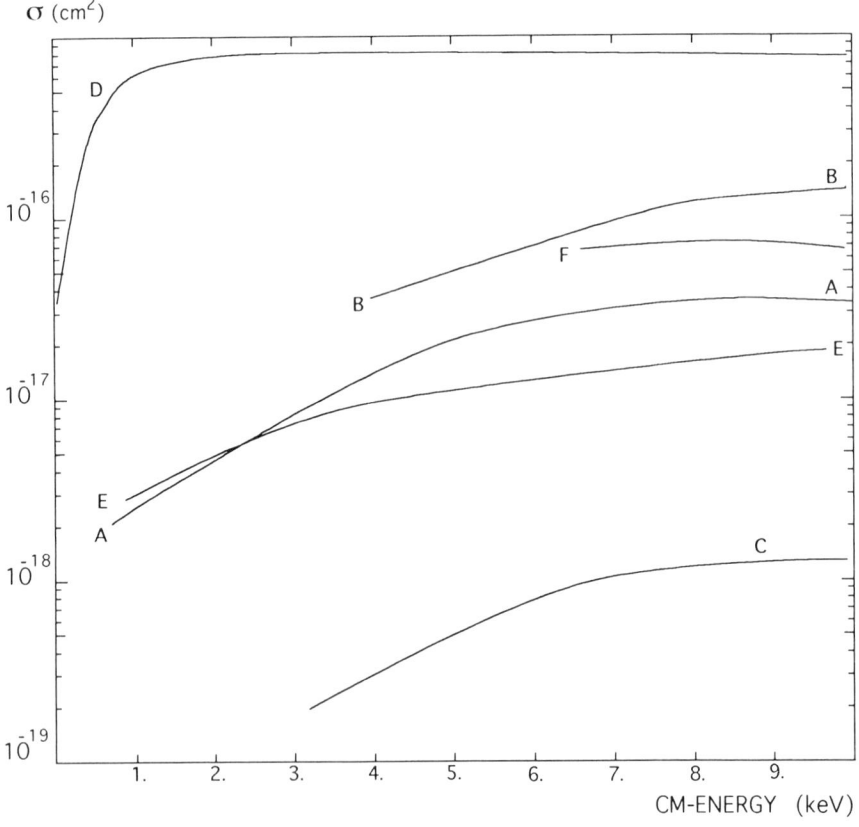

Figure 3. Cross sections for collisional processes involving H^+ that result in production of H. (**A**) $H^+ + \underline{H}(1s) \to H(2s) + \underline{H}^+$; (**B**) $H^+ + \underline{He} \to H + \underline{He}^+$; (**C**) $H^+ + \underline{He} \to H(2s) + \underline{He}^+$; (**D**) $H^+ + \underline{H_2} \to H + \underline{H_2^+}$; (**E**) $H^+ + \underline{H_2} \to H(2s) + \underline{H_2^+}$; (**F**) $H^+ + \underline{H_2} \to H + \underline{H}^+ + \underline{H}$.

exchange with helium (process B) is based on the measurements of Rudd et al.,[50] Shah et al.,[51] and Stier and Barnett.[22] The results of Crandall and Jaecks,[52] Jaecks et al.,[53] and Andreev et al.[54] for the formation of metastable hydrogen (process C) have been employed. Formation of H(2p) has also been investigated by several authors. The most recent results by Hippler et al.[12] are much smaller (by a factor of 3 at 1.6 keV) than those obtained previously by Andreev et al.,[54] Gaily et al.,[55] and Risley et al.[56]

Electron capture by protons in collisions with H_2 has been investigated by many authors.[22,33,57–65] The data given here for process D combine the results of Gealy and Van Zyl[57] up to 3 keV with those of Shah et al.[59] above 3 keV. The two sets of data agree within 10%. The cross section for formation of metastable hydrogen (process E) is that measured by Hill et al.[45] The cross section for dissociative electron transfer (process F) has been measured by Afrosimov et al.[58] and Shah et al.[59] The data presented are those of Shah et al., which are somewhat larger (30%) than those of Afrosimov et al.

The formation of H through the mutual neutralization of H^+ and H^- is discussed in Section 3.8.

3.4. Collisions of H^+—Production of H^-

Figure 4 shows the cross sections for collisional processes involving H^+ that result in the production of H^-.

Double electron capture in collisions with helium (process A) and with molecular hydrogen (process B) has been measured by Williams,[66] Toburen and Nakai,[67] and Fogel et al.[68] Process A has also been investigated by Schryber[69] and Kozlov and Bondar,[70] and process B by McClure.[71] All the data are in mutual agreement. The data presented in Fig. 4 are those of Williams.[66]

The cross section for the resonant double electron transfer between H^+ and H^- (process C) has been measured by Brouillard et al.[72] and Peart and Forrest.[73] An unusual oscillatory structure was observed in the two measurements, which are in good agreement.

3.5. Collisions of He^{2+}

The cross sections for collisions of He^{2+} are presented in Fig. 5.

Charge exchange with atomic hydrogen, which we have already encountered in the section dealing with the collisions of H, can lead to the formation of metastable He^+ (process A). The cross section for this process has been measured by Bayfield and Khayrallah[41] and by Shah and Gilbody.[38] The data given here are those of Shah and Gilbody. The data of Bayfield and Khayrallah are smaller (by a factor of 0.6) at low energy and larger (by a factor of 1.5) at high energy (10 keV). Data on the capture into other excited states have been reported by Ciric et al.[74] and Hoekstra et al.[75]

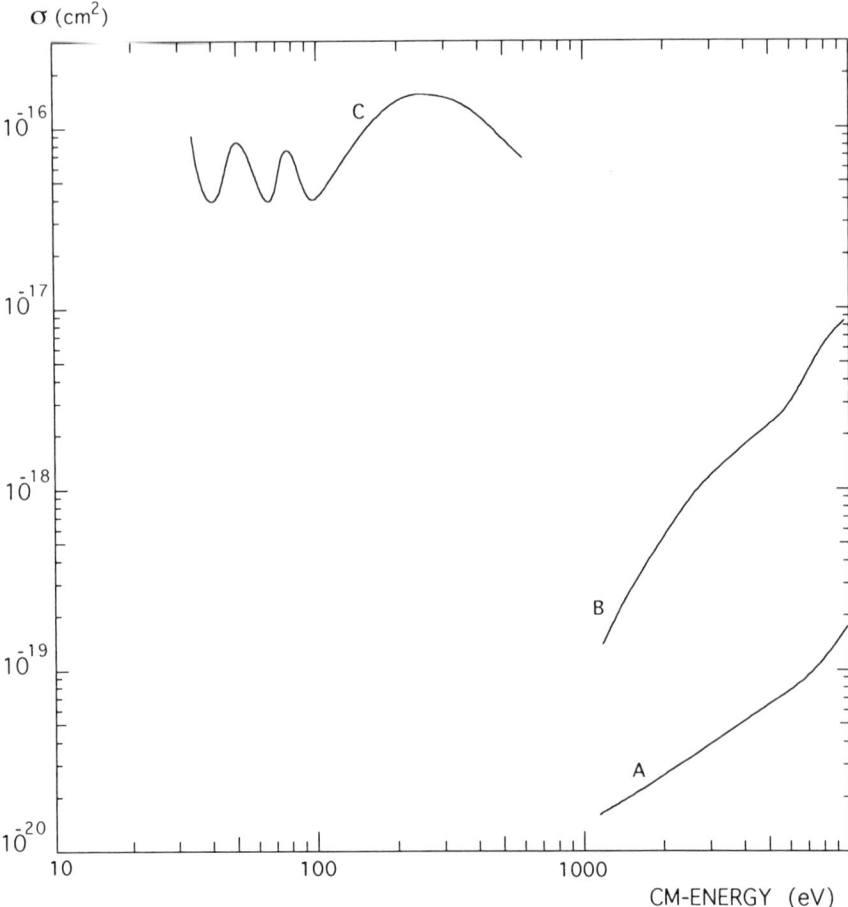

Figure 4. Cross sections for collisional processes involving H^+ that result in production of H^-. (**A**) $H^+ + \underline{He} \rightarrow H^- + \underline{He}^{2+}$; (**B**) $H^+ + \underline{H_2} \rightarrow H^- + \underline{H}^+ + \underline{H}^+$; (**C**) $H^+ + \underline{H}^- \rightarrow H^- + \underline{H}^+$.

Electron capture from H_2 (process B) has been measured by a number of authors.[34,38,39,59,76–80] The agreement is good except at the low energies, where the results of Okuno et al.,[80] which extend down to 1 eV, do not merge properly with the results of Shah and Gilbody[38] and Nutt et al.[39]

The formation of metastable He^+ (process C) has been investigated by Shah and Gilbody[38] and Khayrallah and Bayfield.[76] The cross section for the formation of ground state He^+ (process D) is obtained by subtracting from the total cross section the cross section for capture into the $2s$ state and that for capture into all the radiating excited states. This last cross section was measured by Hoekstra et al.[81]

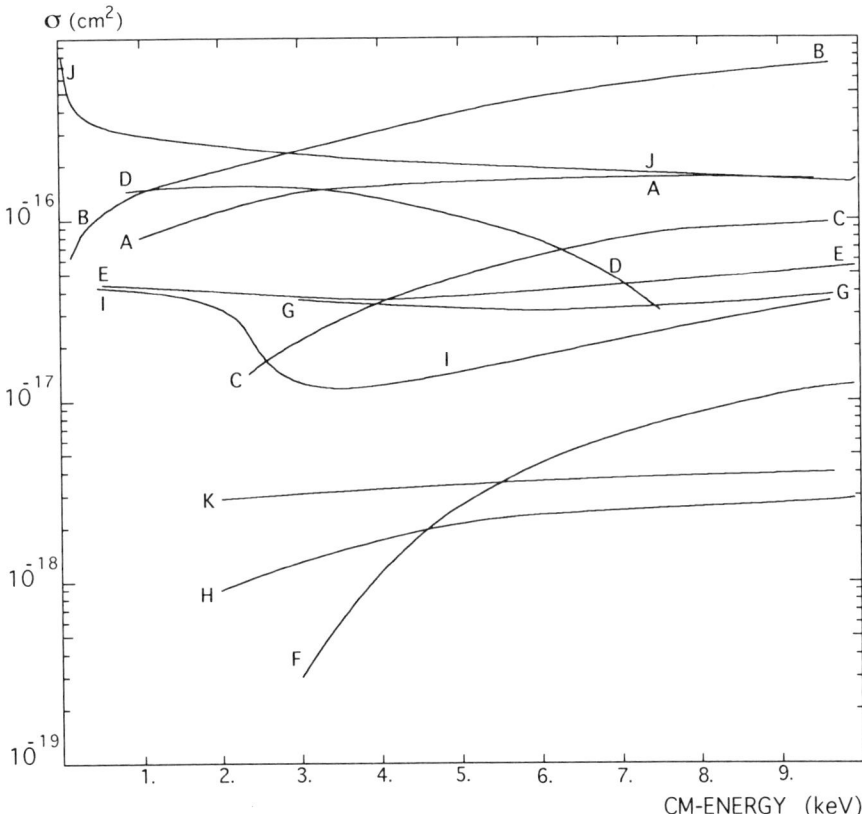

Figure 5. Cross sections for collisional processes involving He^{2+} that result in production of He^+ or He. (**A**) $He^{2+} + \underline{H} \to He(2s) + \underline{H}^+$; (**B**) $He^{2+} + \underline{H_2} \to He^+$; (**C**) $He^{2+} + \underline{H_2} \to He^+(2s)$; (**D**) $He^{2+} + \underline{H_2} \to He^+(1s)$; (**E**) $He^{2+} + \underline{He} \to He^+ + \underline{He}^+$; (**F**) $He^{2+} + \underline{He} \to He^+(1s) + \underline{He}^+(1s)$; (**G**) $He^{2+} + \underline{He} \to He^+(1s) + \underline{He}^+(n = 2, 3, 4)$; (**H**) $He^{2+} + \underline{He} \to He^+(n \geq 2) + \underline{He}^+(n \geq 2)$; (**I**) $He^{2+} + \underline{H_2} \to He + \underline{H}^+ + \underline{H}^+$; (**J**) $He^{2+} + \underline{He} \to He + \underline{He}^{2+}$; (**K**) $He^{2+} + \underline{He} \to He^+ + He^{2+} + e^-$.

Electron capture from helium (process E) has been the object of several measurements[79,80,82–86] that are in good accord except that the recent low-energy data of Kusakabe et al.[79] and Okuno et al.[80] mutually disagree (by a factor of 2) and do not merge properly with the higher energy data. Electron capture into the ground state of He^+ without excitation (process F) or with excitation (process G) of the partner has been investigated by Afrosimov et al.[85] and Bordenave-Montesquieu and Dagnac,[87] who also measured the cross section for producing two excited ions (process H). The formation of excited $He^+(n = 4)$ has been investigated by Folkerts et al.[88] Some data on the formation of metastable He^+ have been reported by Shah and Gilbody.[89] Differential cross sections have been measured by Gao et al.[90]

Double electron capture from H_2 (process I) has been investigated by Shah and Gilbody,[38] Kusakabe et al.,[79] Okuno et al.,[80] Bayfield and Khayrallah,[83] Rudd et al.,[86] and Afrosimov et al.[91] The agreement is satisfactory. The cross sections given here combine the low-energy data of Kusakabe et al. with the average of the others at higher energies.

The resonant double electron capture from He (process J) has been studied by Kusakabe et al.,[79] Okuno et al.,[80] Afrosimov et al.,[82] and Rudd et al.[86] The data presented in Fig. 5 have been taken from Rudd et al. Cross sections down to 1 eV are given by Okuno et al.[80] The transfer ionization (process K) was measured by Afrosimov et al.[82]

3.6. Collisions of He^+

Figure 6 shows the cross sections for collisional processes involving He^+.

The cross section presented for the charge exchange with protons (process A) is that measured by Rinn et al.[92] Measurements of both charge exchange and production of He^{2+}, which also includes ionization of He^+, have been done by Peart et al.,[93] Angel et al.,[94] Watts et al.,[95] Rinn et al.,[96] and Angel et al.[97] The cross section for ionization is smaller by two orders of magnitude than that for charge exchange below 10 keV.

The cross section for ionization of excited He^+ is of course large. The cross section for $He^+(n = 2)$ presented in Fig. 6 (process B) is that calculated by Krstic and Janev,[77] who were able, in the same calculation, to properly reproduce the experimental data of Rinn et al.[92] for $He^+(n = 1)$.

No experimental data are available below 8 keV for electron loss in collisions with H. The measurements of Shah et al.[78] give a cross section of 1.7×10^{-19} cm^2 at 8.4 keV.

The cross section for the resonant electron exchange with He^{2+} (process C) was measured by Jognaux et al.[98] and Peart and Dolder.[99] The measurements by Jognaux et al. extend down to 10 eV. Ionization of He^+ by He^{2+} has not been measured, but its cross section is expected to be small. The value calculated by Olson[100] at 100 keV is of the order of 10^{-18} cm^2.

Electron transfer to He^+ (process D) was investigated by Peart et al.[101] and Melchert et al.[102]

Ionization in collisions with He and H_2 has not been investigated below 10 keV. At higher energy, measurements have been done by Shah et al.,[78] Allison,[103] Pivovar et al.,[104] and Rudd et al.[105] The cross section at 10 keV is of the order of 5×10^{-19} cm^2 for He and 1×10^{-19} cm^2 for H_2. Ionization of metastable $He^+(2s)$ in collisions with He (process E) and H_2 (process F) was measured to somewhat lower energies by Shah and Gilbody.[106] Electron capture by ground state He^+ in collisions with helium (process G) was measured by Rudd et al.,[105] de Heer et al.,[107] Pivovar et al.,[108] and Barnett and Stier.[109] It was also measured in collisions with molecular hydrogen (process I) by Rudd et al.[105] Electron capture by metastable $He^+(2s)$ has

Rearrangement Processes Involving H and He

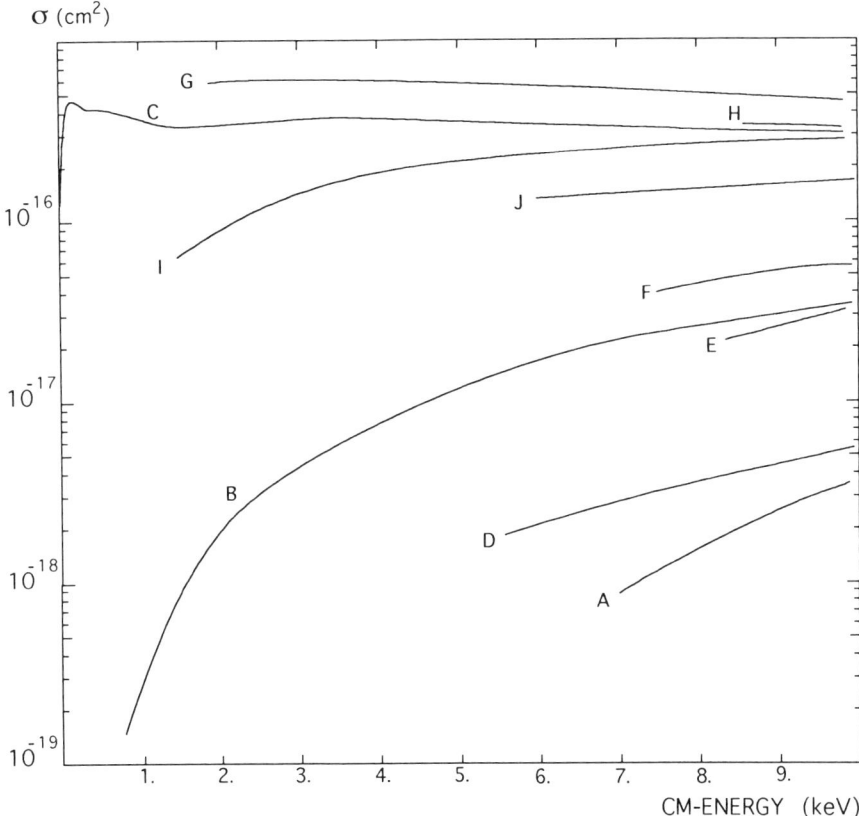

Figure 6. Cross sections for collisional processes involving He^{2+} that result in production of He^{2+} or He. (**A**) $He^+ + \underline{H}^+ \to He^{2+} + \underline{H}$; (**B**) $He^+(n=2) + \underline{H}^+ \to He^{2+} + e^- + \underline{H}^+$; (**C**) $He^+ + \underline{He}^{2+} \to He^{2+} + \underline{He}^+$; (**D**) $He^+ + \underline{He}^+ \to He^{2+} + \underline{He}$; (**E**) $He^+(2s) + \underline{He} \to He^{2+} + e^- + \underline{He}$; (**F**) $He^+(2s) + \underline{H}_2 \to He^{2+} + e^- + \underline{H}_2$; (**G**) $He^+ + \underline{He} \to He + \underline{He}^+$; (**H**) $He^+(2s) + \underline{He} \to He + \underline{He}^+$; (**I**) $He^+ + \underline{H}_2 \to He$; (**J**) $He^+(2s) + \underline{H}_2 \to He$.

been measured in helium (process H) and in molecular hydrogen (process J) by Shah and Gilbody.[106]

3.7. Collisions of He

The cross sections for collisional processes involving He are presented in Fig. 7. Charge exchange with protons was considered in Section 3.3 among the collisional processes of protons.

Ionization by protons (process A) was measured below 10 keV by Rudd et al.,[50] Afrosimov et al.,[110] and Solov'ev et al.[111] The agreement is satisfactory among the measurements and, also, with the measurements at higher energy by Shah and

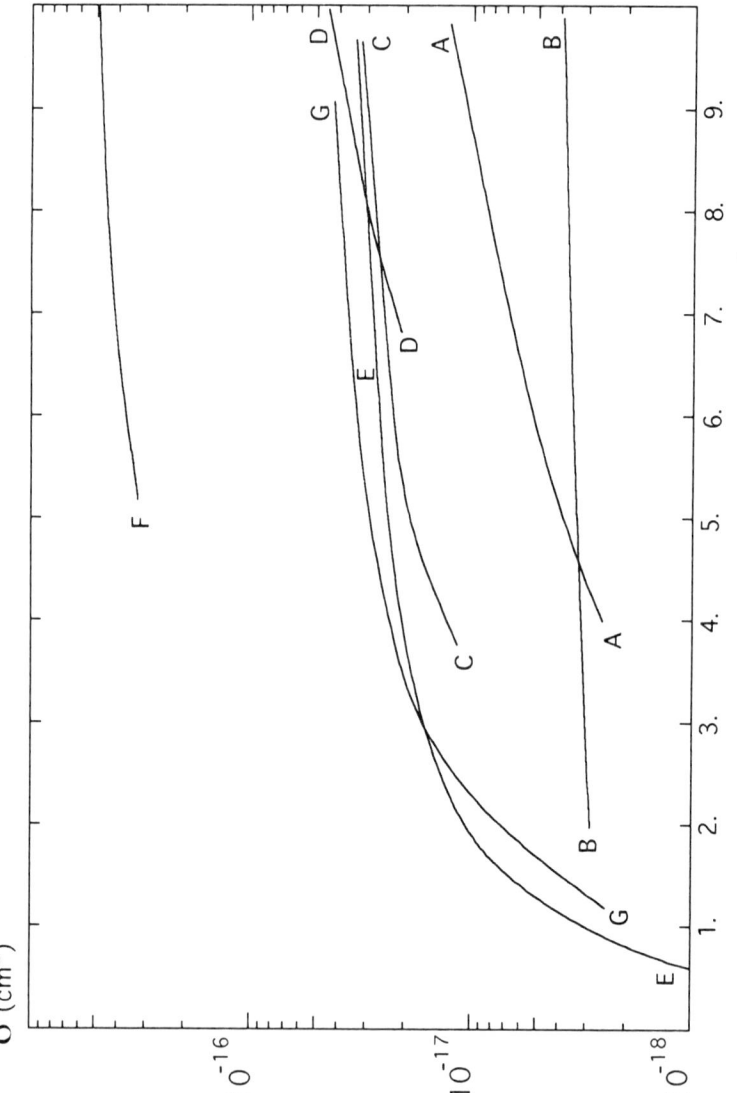

Figure 7. Cross sections for collisional processes involving He that result in production of He^+ or He^{2+}. (**A**) $He + \underline{H}^+ \to He^+ + \underline{H}^+$; (**B**) $He + \underline{He}^{2+} \to He^+ + e^- + \underline{He}^{2+}$; (**C**) $He + \underline{He}^+ \to He^+ + e^- + \underline{He}^+$; (**D**) $He + \underline{H} \to He^+ + e^- + H$. (**E**) $He + \underline{H}_2 \to He^+$; (**F**) $He(2^3S) + \underline{H}_2 \to He^+$; (**G**) $He + \underline{He} \to He^+$.

Gilbody,[112] Knudsen *et al.*,[113] and DuBois and Manson.[114] Afrosimov *et al.*[110] also measured the cross section for double ionization above 10 keV. This cross section is of the order of 10^{-19} cm^2 at 10 keV, smaller than the cross section for double electron transfer (see Section 3.3) and the cross section for transfer ionization measured, above 10 keV, by DuBois and Manson[114] and Solov'ev *et al.*[111]

Charge exchange with He^{2+} was presented in Section 3.5. Ionization by He^{2+} (process B) was measured by DuBois[115] and Afrosimov *et al.*[82] and, as pointed out by the latter authors, appears to be strongly correlated to the process of transfer ionization:

$$\text{He} + \underline{\text{He}}^{2+} \rightarrow \text{He}^{2+} + e^- + \underline{\text{He}}^+$$

The two cross sections are almost identical at low energy, except for a weak oscillation that has opposite phases. The results of Dubois[115] are smaller (by a factor of 2) than those of Afrosimov *et al.*[82] at low energy. Recent measurements by McGuire *et al.*[116] of the ratio of the cross sections for transfer ionization and charge exchange lead to values of the transfer ionization cross section that are even higher than those of Afrosimov *et al.* (40% higher at 10 keV). The data of Afrosimov *et al.* have therefore been retained here.

There are no experimental data for double ionization by He^{2+} below 10 keV, but the measurements at higher energy[82,112] indicate that the cross section is small, of the order of 5×10^{-19} cm^2 at 30 keV. Double electron transfer was considered in Section 3.5.

Charge exchange in collisions with He$^+$ was presented in Section 3.6. Ionization by He$^+$ (process C) has been measured by Rudd *et al.*,[105] de Heer *et al.*,[107] and Solov'ev *et al.*[117] The data of Rudd *et al.* deviate, at low energy, from those of the others. They are, however, retained here because the other data level off, in a doubtful manner, on the low-energy side, possibly as the result of a contamination of the helium target. Measurements at higher energy have been carried out by Pivovar *et al.*[108] and Langley *et al.*[118] There are no experimental data for double ionization by He$^+$. The lowest energy reached is 19 keV, in the measurement by DuBois *et al.*,[119] and a cross section of 10^{-18} cm^2 was reported. Other measurements are at higher energies.[120,121]

Charge exchange with atomic hydrogen has been presented in Section 3.1. Ionization by H (process D) was measured by Solov'ev *et al.*[111] and, at higher energies, by Horsdal Pedersen and Larsen[122] and Hvelplund and Andersen.[123]

Electron loss in H$_2$ (process E) was measured by Gilbody *et al.*,[124] Barnett and Stier,[109] Tawara,[125] and Tobita and Takeuchi,[126] and also, at higher energy, by Horsdal Pedersen and Hvelplund.[127] The data given in Fig. 7 are from Tobita and Takeuchi. Double electron loss in collisions with molecular hydrogen has been investigated by Hvelplund and Pedersen[128] at high energy.

Electron loss by metastable He(2^3S) in H$_2$ (process F) has been measured by Gilbody *et al.*[124] who also investigated electron loss by metastable helium in helium, from 10 keV. The cross section in helium is somewhat smaller than in H$_2$.

The cross section for electron loss by ground state atoms of helium in helium (process G), obtained by Gilbody et al.[124] and retained here, is smaller than that found previously by Barnett and Stier[109] and Fogel et al.[129] probably because of the presence of metastable atoms in the experiments of the last authors. Double electron loss in helium has only been measured at high energy.[128,130]

3.8. Collisions of H⁻

Figure 8 shows the cross sections for collisional processes involving H⁻.

The cross section for mutual neutralization in the collision of H⁻ with H⁺ (process A) was first measured by Rundel et al.,[131] Moseley et al.,[132] Gaily and

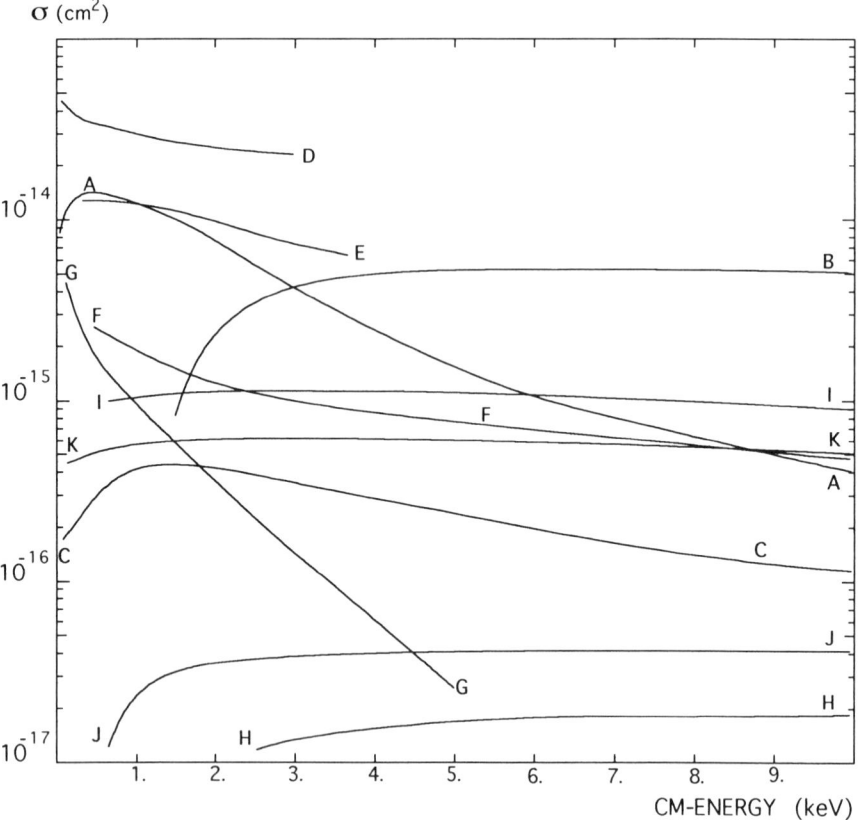

Figure 8. Cross sections for collisional processes involving H⁻. (A) H⁻ + $\underline{H}^+$ → H + $\underline{H}$; (B) H⁻ + $\underline{H}^+$ → H + e^- + $\underline{H}^+$; (C) H⁻ + $\underline{H}^+$ → H⁺ + e^- + $\underline{H}$; (D) H⁻ + $\underline{He}^{2+}$ → H + $\underline{He}$+; (E) H⁻ + $\underline{He}^+$ → H + $\underline{He}$; (F) H⁻ + $\underline{H}$ → H; (G) H⁻ + $\underline{H}$ → H + $\underline{H}^-$; (H) H⁻ + $\underline{H}$ → H⁺; (I) H⁻ + $\underline{H}_2$ → H; (J) H⁻ + $\underline{H}_2$ → H⁺; (K) H⁻ + $\underline{He}$ → H + e^- + $\underline{He}$.

Rearrangement Processes Involving H and He

Harrison,[133] and Peart et al.[134] The puzzling structure observed in those measurements has remained unexplained, and more elaborate measurements by Szücs et al.[135] do not show this structure. The results of the latter authors are regarded as the correct ones as they have been confirmed by Peart et al.[136] and agree with higher energy measurements by Schön et al.[137] and with the *ab initio* calculations of Fussen and Kubach.[138] This process leads to the formation of one excited atom in the $n = 2$ or $n = 3$ state.

The electron detachment by protons (process B) was measured by Peart et al.[139] The cross section published was obtained as the difference between the cross section to form H from H^+ and the cross section for mutual neutralization. The latter cross section was found by extrapolating the existing experimental data to higher energies, assuming an E^{-1} energy dependence. New experimental results by Schön et al.[137] have shown that such an extrapolation is not valid. The cross section presented in Fig. 8 takes the results of Schön et al. into account. It agrees well with the calculation of Fussen and Claeys.[140]

The cross section for transfer ionization (process C) is that measured by Schön et al.[141]

Electron transfer in collisions with He^{2+} (process D) was investigated by Terao et al.[142] and Peart and Bennett.[143] The results of Terao et al. are somewhat larger (25%) than those of Peart and Bennett, and an average has been taken. The measurements by Terao et al. extend down to 0.5 eV.

The transfer ionization process,

$$H^- + He^{2+} \rightarrow H^+ + e^- + He^+$$

has been measured by Cherkani et al.[144] over the energy range 0.2–1300 eV. The cross section decreases from 2.2×10^{-15} to 2.8×10^{-16} cm^2.

Electron transfer to He^+ ions (process E) has been measured by Gaily and Harrison[145] and Peart et al.[146] and remeasured by Peart et al.[136] The structures observed in the older measurements are almost unobservable in the new ones. The data given in Fig. 8 are an average of the three sets of data.

Single and double electron loss in collisions with atomic hydrogen (processes F and H, respectively) have been measured by Geddes et al.,[147] and electron transfer (process G) was measured by Hummer et al.[148] Geddes et al.[147] also measured the cross sections for single and double electron loss in molecular hydrogen (processes I and J, respectively). Their results are in good agreement with earlier measurements by Simpson and Gilbody[149] and Risley and Geballe[150] for single electron loss and by Williams[151] and Fogel et al.[152] for double electron loss.

Electron loss in helium has been investigated many times, for example, by Simpson and Gilbody,[149] Risley and Geballe,[150] and Williams.[151] Only the cross section for detachment is reported in Fig. 8 (process K), taken from Risley and Geballe. The cross section for double detachment is estimated to be 4% of that for single detachment at 500 eV and to rise to 10% of that for single detachment at 10 keV.

Detachment processes in collisions of H^- with H^- have been measured by Schulze et al.[153] but are of limited interest for fusion.

3.9. Collisions of H_2—Ionization and Dissociation

Figure 9 shows the cross sections for collisions of H_2 that result in its ionization or dissociation.

The production of H_2^+ in collisions with protons (process A) has been measured, below 10 keV, by Afrosimov et al.[58] and Shah et al.[59] The two sets of measurements are in excellent agreement. The main contribution is charge exchange, the cross section of which was presented in Fig. 3 (process D). The cross section for ionization is one order of magnitude smaller at 10 keV and even smaller below.

The same authors also obtained similar cross sections for the production of H^+ (process B). In this case, the dominating mechanism is the dissociative electron transfer shown in Fig. 3 as process F.

The simultaneous production of two protons (process C) was investigated by Yousif et al.,[154] Afrosimov et al.,[58] and, at higher energies, Shah et al.[59] The contribution of double electron transfer has been given in Fig. 4 (process B).

The cross sections for formation of H_2^+ (process D) or H^+ (process E) from H_2 colliding with atomic hydrogen have been measured by Afrosimov et al.[155] The former process is dominated by ionization, and the latter by dissociative ionization:

$$H_2 + H \rightarrow H^+ + H + e^- + H$$

The contributions of dissociative transfer,

$$H_2 + H \rightarrow H^+ + H + H^-$$

and double ionization,

$$H_2 + H \rightarrow H^+ + H^+ + 2e^- + H$$

are approximately one order of magnitude smaller.

Ionization and dissociation of H_2 in collisions with He^{2+} have been studied experimentally mainly by Afrosimov et al.[91] The cross section obtained by these authors for electron loss (process F) is in good agreement with the higher energy results of Shah et al.[59] The main channel is electron transfer,

$$H_2 + He^{2+} \rightarrow H_2^+ + He^+$$

which exceeds ionization by a factor of 20. The cross section obtained by Afrosimov et al.[91] for dissociation into H^+ and H (process G) is in less good agreement with the higher energy results of Shah et al.,[59] which are about two times larger. The main channel is dissociative transfer, which, at 10 keV, is 40 times more efficient than dissociative ionization. Double electron transfer (process I) is one order of

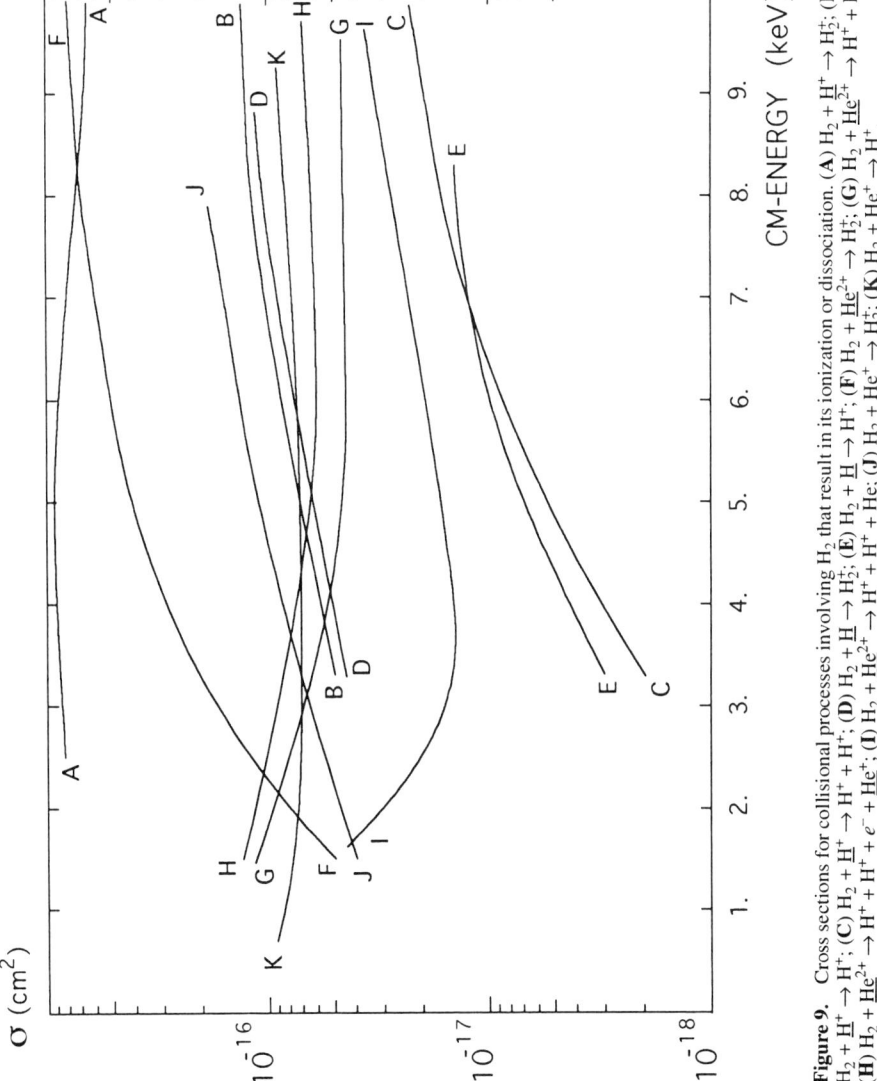

Figure 9. Cross sections for collisional processes involving H_2 that result in its ionization or dissociation. (**A**) $H_2 + H^+ \to H^+_2$; (**B**) $H_2 + \underline{H}^+ \to \underline{H}^+ + H^+ + H^+$; (**C**) $H_2 + \underline{H}^+ \to H^+ + \underline{H}$; (**D**) $H_2 + \underline{H} \to H^+$; (**E**) $H_2 + \underline{He}^{2+} \to H^+_2$; (**F**) $H_2 + \underline{He}^{2+} \to H^+_2$; (**G**) $H_2 + \underline{He}^{2+} \to H^+ + H$; (**H**) $H_2 + \underline{He}^{2+} \to H^+ + H^+ + e^- + \underline{He}^+$; (**I**) $H_2 + \underline{He}^{2+} \to H^+ + H^+ + \underline{He}$; (**J**) $H_2 + \underline{He}^+ \to H^+_2$; (**K**) $H_2 + \underline{He}^+ \to H^+$.

magnitude less important than transfer ionization (process H). The cross section for the latter process was also measured by Shah et al.,[59] who confirmed the results of Afrosimov et al.[91]

Measurements of the cross section for H_2^+ or H^+ formation from H_2 colliding with He^+ (processes J and K, respectively) have been carried out by Solov'ev et al.[156] and Browning et al.[157] The two sets of data are discrepant. For process J, the results of Solov'ev et al. are larger than those of Browning et al., by 50% at 6 keV and 25% at 9 keV. For process K, they are three to four times smaller. The data presented in Fig. 9 are those of Browning et al.

Formation of the H^+–H^- ion pair in collisions of H_2 with He has been investigated by Brouillard et al.[158]

3.10. Collisions of H_2^+—Ionization, Dissociation, and Charge Exchange

The cross sections for collisional processes involving H_2^+ that result in its ionization, dissociation, or charge exchange are presented in Fig. 10.

The formation of H^+ from H_2^+ colliding with atomic hydrogen (process A) and with molecular hydrogen (process B) has been measured by McClure[159] Process B was also investigated by Guidini,[160] Williams and Dunbar,[13] Fedorenko et al.,[161] Barnett and Ray,[162] Schmid,[163] McClure,[14,164] and Il'in et al.[165] Cross sections for collisions of H_2^+ are sensitive to its vibrational excitation, which is uncontrolled in the experiments. This explains part of the scatter observed among the reported data for process B. Another source of difficulty, at low energy, is the large scattering of dissociation fragments, which could therefore have been incompletely collected in some measurements. This has been demonstrated by McClure,[14] who had to revise his own earlier data[164] because they suffered from this problem. The data presented in Fig. 10 are those of McClure.[159] They are in good agreement with those of Schmid,[163] and also with those of Fedorenko et al.[161] except at the lowest energy, where the cross section is surprisingly found to drop by one order of magnitude in the data of Federenko et al.

Measurements distinguishing between the two ways of producing H^+,

$$H_2^+ \xrightarrow{\sigma_1} H^+ + H$$

$$H_2^+ \xrightarrow{\sigma_2} H^+ + H^+ + e^-$$

have been carried out by Guidini[160] but only above 12.5 keV. The ratio σ_1/σ_2 was found to be equal to 2.5 at 12.5 keV and to decrease at higher energies. Thus, below 10 keV, the cross section for production of H^+ is approximately equal to σ_1.

Schmid[163] also determined the cross section for production of H (process C).

Formation of H^+ and H from H_2^+ colliding with helium (processes D and E, respectively) has been investigated by Suzuki et al.[166] Process D has also been

Rearrangement Processes Involving H and He

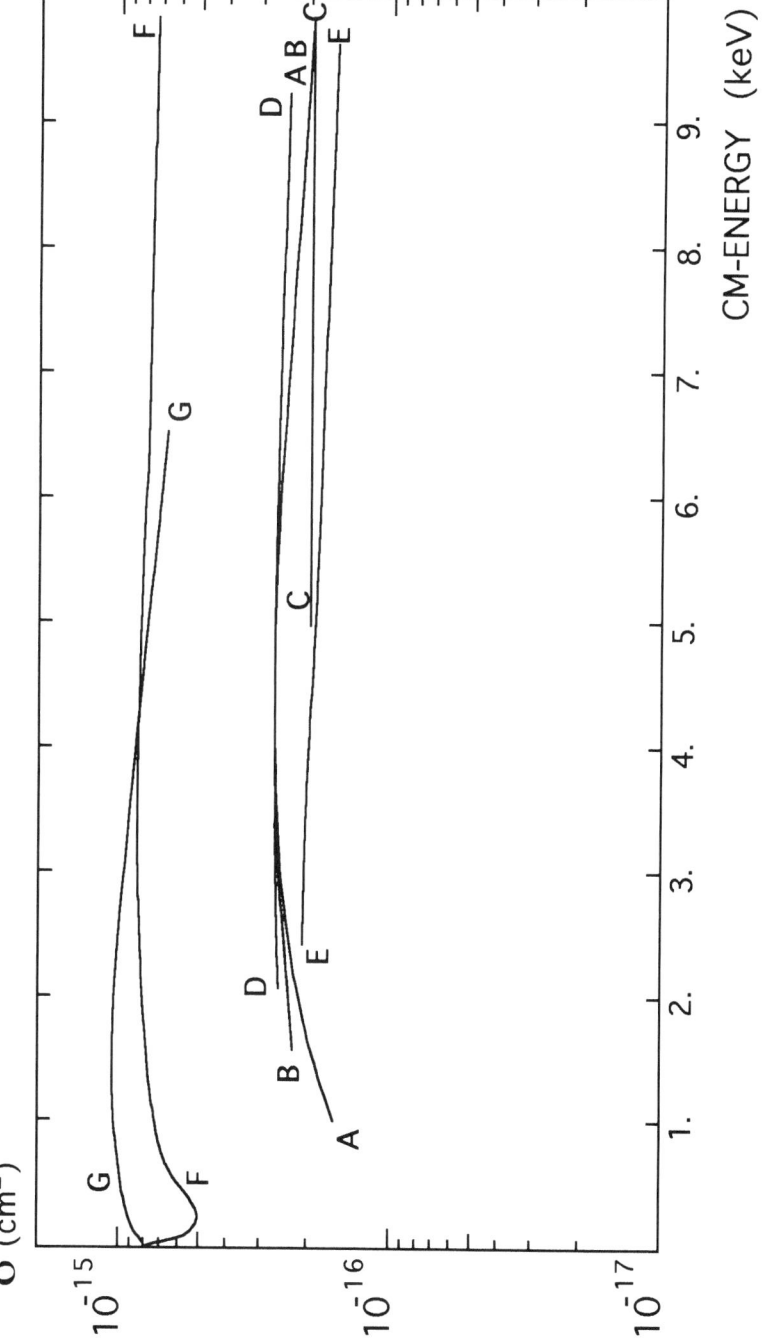

Figure 10. Cross sections for collisional processes involving H_2^+ that result in its ionization, dissociation, or charge exchange. (**A**) $H_2^+ + \underline{H} \to H^+$; (**B**) $H_2^+ + \underline{H_2}$ $\to H^+$; (**C**) $H_2^+ + \underline{H_2} \to H$; (**D**) $H_2^+ + \underline{He} \to H^+$; (**E**) $H_2^+ + \underline{He} \to H$; (**F**) $H_2^+ + \underline{H_2} \to H_2 + H_2^+$; (**G**) $H_2^+ + \underline{H} \to H_2 + \underline{H}^+$.

measured by Williams and Dunbar,[13] Fedorenko et al.[161] and Il'in et al.[165] The data shown in Fig. 10 are those of Suzuki et al. They are in good agreement with all the other data except those of Williams and Dunbar at low energy, which are smaller, probably because of an incomplete collection of H^+.

Formation of the H^+–H^- ion pair in the collision of H_2^+ with helium has been measured by Oliver et al.[167] who found an extremely small cross section (2×10^{-22} cm^2 at 3.6 keV). Experimental data on electron capture by H_2^+ are scarce.

Reliable data exist for the symmetric charge exchange with H_2 (process F). The cross section in Fig. 10 combines the results of McClure,[164] Hollricher,[168] Koopman,[169] and Rothwell et al.[170] which are in good mutual agreement. The dependence of the cross section on the vibrational state of the ion has been investigated below 500 eV by Campbell et al.[171] and Liao et al.[172] This dependence is not dramatic, but it is not simple either. It is in general agreement with the semiclassical calculation of Lee and De Pristo.[173]

The only measurement of the cross section for charge exchange with atomic hydrogen is that by Fite et al.[174] with a quoted accuracy of 20%. The measurement is relative and calibrated on the cross section for charge exchange between H^+ and H_2 at 900 eV. The cross section given in Fig. 10 (curve G) is lower by 4% as a result of the use of a more recent value of the calibration cross section. A measurement of the rate constant of this reaction at near-thermal energies has been reported by Karpas et al.[175]

3.11. Associative and Penning Ionization

We have collected in Fig. 11 the available experimental data related to associative or Penning ionization in collisions of hydrogen and helium atoms or ions. The energy range has been adapted to the low-energy character of this process and covers energies between 0.01 and 10 eV.

Associative ionization in the collision of two hydrogen atoms, one in the ground state and the other in the metastable state $2s$ (process A), has been measured by Urbain et al.[176] from threshold to 10 eV. The corresponding reaction with deuterium has recently been investigated by Karangwa et al.[177] (process B). The cross section has a pronounced oscillatory structure—different for hydrogen and deuterium—that has been explained in terms of an interference of two molecular states. The data reported here are from recent, improved measurements and are slightly different from the published ones.

Urbain et al.[178] have also carried out some measurements of the cross section for associative ionization in the collision of two metastable hydrogen atoms. They found a cross section of the order of 4×10^{-15} cm^2 at 0.06 eV that decreases as the inverse of the energy. Associative ionization in collisions between H^+ and H^- (process C) has been measured by Poulaert et al.[179] The formation of H_3^+ in the collision of a metastable hydrogen atom with a hydrogen molecule was investigated

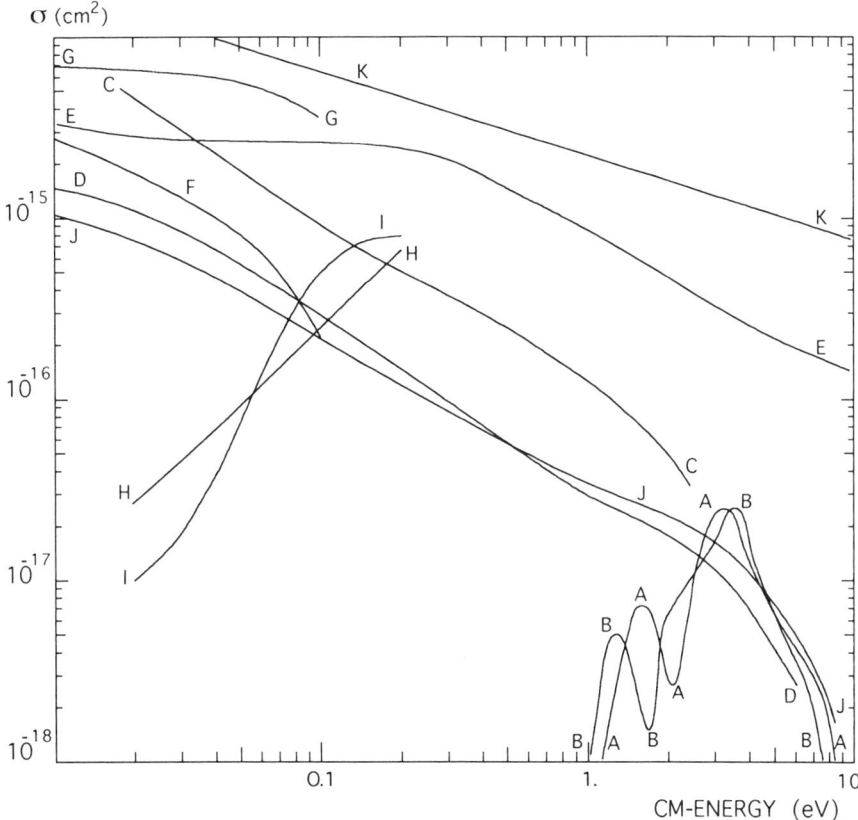

Figure 11. Cross sections for associative and Penning ionization. (**A**) $H(1s) + H(2s) \rightarrow H_2^+ + e^-$; (**B**) $D(1s) + D(2s) \rightarrow D_2^+ + e^-$; (**C**) $H^+ + H^- \rightarrow H_2^+ + e^-$; (**D**) $H + He(2^3S) \rightarrow HeH^+ + e^-$; (**E**) $H + He(2^3S) \rightarrow H^+ + e^- + He$; (**F**) $H + He(2^1S) \rightarrow HeH^+ + e^-$; (**G**) $H + He(2^1S) \rightarrow H^+ + e^- + He$; (**H**) $H_2 + He(2^3S) \rightarrow H_2^+ + e^- + He$; (**I**) $H_2 + He(2^1S) \rightarrow H_2^+ + e^- + He$; (**J**) $He^* + He^* \rightarrow He_2^+ + e^-$; (**K**) $He^* + He^* \rightarrow He^+ + e^- + He$.

by Van Volkenburgh et al.[180] who reported a cross section of the order of 1×10^{-16} cm^2 at 0.03 eV.

Associative and Penning ionization in collisions of metastable helium with atomic hydrogen (or deuterium) has been studied experimentally by Howard et al.,[181] Fort et al.,[182] Neynaber and Tang,[183] Morgner and Niehaus,[184] and Waibel et al.[185] The cross section for associative ionization with $He(2^3S)$ (process D) is taken from Waibel et al. below 0.1 eV and from Neynaber and Tang above 0.1 eV. The cross section for Penning ionization (process E) is from Waibel et al., which is in good agreement with that reported by Howard et al., but it has been extrapolated to higher energies using Neynaber and Tang's data. In the case of He(2^1S), the cross sections presented in Fig. 11 for associative ionization (process F) and for Penning ionization (process G) have been obtained using the corresponding cross sections of He(2^3S) and the results of Fort et al.[182] for the ratios

$$\frac{\sigma_{TI}(2^3S)}{\sigma_{TI}(2^1S)} \text{ and } \frac{\sigma_{AI}(2^1S)}{\sigma_{TI}(2^1S)}$$

where σ_{TI} and σ_{AI} are of the cross sections for Penning ionization and associative ionization, respectively.

The ionizing processes in collisions of metastable helium with molecular hydrogen have been investigated by Neynaber et al.,[186] Fort et al.,[187] Bregel et al.,[188] and Martin et al.[189] The cross section given in Fig. 11 for Penning ionization with He(2^3S) (process H) and with He(2^1S) (process I) are from Martin et al. and Fort et al. In both cases, the measurements are relative ones, and they have been normalized using the calculations of Martin et al. These are detailed calculations that yield the branching ratios of the different reactive channels.

Associative and Penning ionization in collisions of two metastable helium atoms have been measured by Neynaber et al.,[190] Devdariani et al.,[191] and Müller et al.[192] The cross sections presented in Fig. 11 (processes J and K) are those obtained by Neynaber et al. They relate to a statistical mixture of singlet and triplet states. Specific data for singlet–singlet, singlet–triplet, and triplet–triplet collisions have been measured or calculated by Müller et al.[192]

A review paper on Penning ionization has been recently published by Siska.[193] A discussion of the experimental techniques relevant to low-energy ionizing collisions can be found in a book chapter by Brunetti and Vecchiocattivi.[194]

REFERENCES

1. H. Tawara, *At. Data Nucl. Data Tables* **22**, 491 (1978).
2. J. S. Risley, *Comments At. Mol. Phys.* **12**, 215 (1983).
3. K. Dolder and B. Peart, *Rep. Prog. Phys.* **48**, 1283 (1985).
4. E. Salzborn, Electron Capture and Ionization in Ion–Ion Collisions, *J. Phys. Suppl. No. 1* **50** (1989).
5. C. F. Barnett, Atomic Data for Fusion, Collisions of H, H_2, He and Li Atoms and Ions with Atoms and Molecules, Oak Ridge National Laboratory, Report ORNL-6086. Vol. 1, 1990.
6. R. K. Janev and J. J. Smith, *Nuclear Fusion—Supplement to the Journal of Nuclear Fusion*, Vol. 4, International Atomic Energy Agency, Vienna (1993).
7. W. Fritsch and C. D. Lin, *Phys. Rep.* **202**, 1 (1991).
8. R. K. Janev and P. S. Krstic, *Phys. Rev. A* **46**, 5554 (1992).
9. P. S. Krstic and R. K. Janev, *Phys. Rev. A* **47**, 3894 (1993).
10. W. Fritsch, *J. Phys. B*, **28**, 3461 (1994).
11. P. Defrance, F. Brouillard, W. Claeys, and G. Van Wassenhove, *J. Phys. B.* **14**, 103 (1981).
12. R. Hippler, W. Harbich, H. Madeheim, H. Kleinpoppen, and H. O. Lutz, *Phys. Rev. A* **35**, 3139 (1987).

13. J. F. Williams and N. F. Dunbar, *Phys. Rev.* **149**, 62 (1966).
14. G. W. McClure, *Phys. Rev.* **140**, A769 (1965).
15. M. W. Gealy and B. Van Zyl, *Phys. Rev. A* **36**, 3100 (1987).
16. J. Hill, J. Geddes, and H. B. Gilbody, *J. Phys. B* **12**, 3341 (1979).
17. G. W. McClure, *Phys. Rev.* **166**, 22 (1968).
18. F. Roussel, P. Pradel, and G. Spiess, *Phys. Rev. A* **16**, 1854 (1977).
19. B. Alsamour, X. Urbain, F. Brouillard, and A. Cornet, *J. Phys. B* **26**, 1317 (1993).
20. D. Fussen, W. Claeys, A. Cornet, J. Jureta, and P. Defrance, *J. Phys. B* **15**, L715 (1982).
21. B. Van Zyl, T. Q. Le, and R. C. Amme, *J. Chem. Phys.* **74**, 314 (1981).
22. P. M. Stier and C. F. Barnett, *Phys. Rev.* **103**, 896 (1956).
23. K. A. Smith, M. D. Duncan, M. W. Geis, and R. D. Rundel, *J. Geophys. Res.* **81**, 2231 (1976).
24. A. Cornet, W. Claeys, V. Lorent, J. Jureta, and D. Fussen, *J. Phys. B* **17**, 2643 (1984).
25. J. F. Williams, *Phys. Rev.* **153**, 116 (1967).
26. M. B. Shah and H. B. Gilbody, *J. Phys. B* **14**, 2361 (1981).
27. M. B. Shah, D. S. Elliott, and H. B. Gilbody, *J. Phys. B* **20**, 2481 (1987).
28. W. L. Fite, R. F. Stebbings, D. G. Hummer, and R. T. Brackmann, *Phys. Rev.* **119**, 663 (1960).
29. J. T. Park, *Adv. At. Mol. Phys.* **19**, 67 (1983).
30. M. E. Rudd, Y.-K. Kim, D. M. Madison, and J. M. Gallagher, *Rev. Mod. Phys.* **57**, 965 (1985).
31. T. G. Winter and C. D. Lin, *Phys. Rev. A* **29**, 3071 (1984).
32. M. Pieksma, Doctoral Thesis, University of Utrecht, The Netherlands, 1993.
33. G. W. McClure, *Phys. Rev.* **148**, 47 (1966).
34. R. E. Olson, A. Salop, R. A. Phaneuf, and F. W. Meyer, *Phys. Rev. A* **16**, 1867 (1977).
35. G. J. Lockwood, G. H. Miller, and J. M. Hoffman, *Bull. Am. Phys. Soc.* **21**, 1266 (1976).
36. C. Harel and A. Salin, *J. Phys. B* **16**, 55 (1983).
37. S. L. Willis, G. Peach, M. R. C. McDowell, and J. Banerji, *J. Phys. B* **18**, 3939 (1985).
38. M. B. Shah and H. B. Gilbody, *J. Phys. B* **11**, 121 (1978).
39. W. L. Nutt, R. W. McCullough, K. Brady, M. B. Shah, and H. B. Gilbody, *J. Phys. B* **11**, 1457 (1978).
40. W. L. Fite, A. C. H. Smith, and R. F. Stebbings, *Proc. R. Soc. A* **268**, 527 (1962).
41. J. E. Bayfield and G. A. Khayrallah, *Phys. Rev. A* **12**, 869 (1975).
42. M. B. Shah, D. S. Elliott, P. McCallion, and H. B. Gilbody, *J. Phys. B* **21**, 2455 (1988).
43. R. Shingal and C. D. Lin, *J. Phys. B* **22**, L445 (1989).
44. M. Burniaux, F. Brouillard, A. Jognaux, T. R. Govers, and S. Szücs, *J. Phys. B* **10**, 2421 (1977).
45. J. Hill, J. Geddes, and H. B. Gilbody, *J. Phys. B* **12**, L341 (1979).
46. T. J. Morgan, J. Stone, and R. Mayo, *Phys. Rev. A* **22**, 1460 (1980).
47. J. E. Bayfield, *Phys. Rev.* **185**, 105 (1969).
48. Y. P. Chong and W. L. Fite, *Phys. Rev. A* **16**, 933 (1977).
49. T. J. Morgan, J. Geddes, and H. B. Gilbody, *J. Phys. B* **6**, 2118 (1973).
50. M. E. Rudd, R. D. DuBois, L. H. Toburen, C. A. Ratcliffe, and T. V Goffe, *Phys. Rev. A* **28**, 3244 (1983).
51. M. B. Shah, P. McCallion, and H. B. Gilbody, *J. Phys. B* **22**, 3037 (1989).
52. D. H. Crandall and D. H. Jaecks, *Phys. Rev. A* **4**, 2271 (1971).
53. D. H. Jaecks, R. Van Zyl, and R. Geballe, *Phys. Rev.* **137**, A340 (1965).
54. E. P. Andreev, V. A. Ankudinov, and S. V. Bobashev, *Sov. Phys. JETP* **23**, 375 (1966).
55. T. D. Gaily, D. H. Jaecks, and R. Geballe, *Phys. Rev.* **167**, 81 (1968).
56. J. S. Risley, F. J. de Heer, and C. B. Kerkdijk, *J. Phys. B* **11**, 1759 (1978).
57. M. W. Gealy and B. Van Zyl, *Phys. Rev. A* **36**, 3091 (1987).
58. V. V. Afrosimov, G. A. Leiko, Y. A. Mamaev, and M. N. Panov, *Sov. Phys. JETP* **29**, 648 (1969).
59. M. B. Shah, P. McCallion, and H. B. Gilbody, *J. Phys. B* **22**, 3983 (1989).

60. R. S. Gao, L. K. Johnson, G. J. Smith, C. L. Hakes, K. A. Smith, N. F. Lane, R. F. Stebbings, and M. Kimura, *Phys. Rev. A* **44**, 5599 (1991).
61. J. B. H. Stedeford and J. B. Hasted, *Proc. R. Soc. A* **227**, 466 (1955).
62. E. Gustafson and E. Lindholm, *Ark. Fys.* **18**, 219 (1960).
63. W. H. Cramer, *J. Chem. Phys.* **35**, 836 (1961).
64. R. Curran, T. M. Donahue, and W. H. Kasner, *Phys. Rev.* **114**, 490 (1959).
65. D. Hollricher, *Z. Phys. B* **187**, 41 (1965).
66. J. F. Williams, *Phys. Rev.* **150**, 7 (1966).
67. L. H. Toburen and M. Y. Nakai, *Phys. Rev.* **177**, 191 (1969).
68. Ya. M. Fogel, R. V. Mitin, V. F. Kozlov, and N. D. Romashko, *Sov. Phys. JETP* **35**, 390 (1959).
69. U. Schryber, *Helv. Phys. Acta A* **40**, 1023 (1967).
70. V. F. Kozlov and S. A. Bondar, *Sov. Phys. JETP* **23**, 195 (1966).
71. G. W. McClure, *Phys. Rev.* **132**, 1636 (1963).
72. F. Brouillard, W. Claeys, G. Poulaert, G. Rahmat, and G. Van Wassenhove, *J. Phys. B* **12**, 1253 (1979).
73. B. Peart and R. A. Forrest, *J. Phys. B* **12**, L23 (1979).
74. D. Ciric, D. Dijkkamp, E. Vlieg, and F. J. De Heer, *J. Phys. B* **18**, 4745 (1985).
75. R. Hoekstra, F. J. de Heer, and R. Morgenstern, *J. Phys. B* **24**, 4025 (1991).
76. G. A. Khayrallah and J. E. Bayfield, *Phys. Rev. A* **11**, 930 (1975).
77. P. S. Krstic and R. K. Janev, private communication (1993).
78. M. B. Shah, T. V. Goffe, and H. B. Gilbody, *J. Phys. B* **10**, L723 (1977).
79. T. Kusakabe, Y. Yoneda, Y. Mizumoto, and K. Katusurayama, *J. Phys. Soc. Jpn.* **59**, 1218 (1990).
80. K. Okuno, K. Soejima, and Y. Kaneko, *J. Phys. B* **25**, L105 (1992).
81. R. Hoekstra, H. O. Folkerts, J. P. M. Beijers, R. Morgenstern, and F. J. de Heer, *J. Phys. B* **27**, 2021 (1994).
82. V. V. Afrosimov, G. A. Leiko, Yu. A. Mamaev, and M. N. Panov, *Sov. Phys. JETP* **40**, 661 (1974).
83. J. E. Bayfield and G. A. Khayrallah, *Phys. Rev. A* **11**, 920 (1975).
84. K. H. Berkner, R. V. Pyle, J. W. Stearns, and J. C. Warren, *Phys. Rev.* **166**, 44 (1968).
85. V. V. Afrosimov, A. A. Basalaev, G. A. Leiko, and M. N. Panov, *Sov. Phys. JETP* **47**, 837 (1978).
86. M. E. Rudd, T. V. Goffe, and A. Itoh, *Phys. Rev. A* **32**, 2128 (1985).
87. D. Bordenave-Montesquieu and R. Dagnac, *J. Phys. B* **25**, 2573 (1992).
88. H. O. Folkerts, R. Hoekstra, L. Meng, R. E. Olson, W. Fritsch, R. Morgenstern, and H. P. Summers, *J. Phys. B* **26**, L619 (1993).
89. M. B. Shah and H. B. Gilbody, *J. Phys. B* **7**, 256 (1974).
90. R. S. Gao, C. M. Dutta, N. F. Lane, K. A. Smith, R. F. Stebbings, and M. Kimura, *Phys. Rev. A* **45**, 6388 (1992).
91. V. V. Afrosimov, G. A. Leiko, and M. N. Panov, *Sov. Phys. Tech. Phys.* **25**, 313 (1980).
92. K. Rinn, F. Melchert, and E. Salzborn, *J. Phys. B* **18**, 3783 (1985).
93. B. Peart, K. Rinn, and K. Dolder, *J. Phys. B* **16**, 1461 (1983).
94. G. C. Angel, E. C. Sewell, K. F. Dunn, and H. B. Gilbody, *J. Phys. B* **11**, L297 (1978).
95. M. F. Watts, K. F. Dunn, and H. B. Gilbody, *J. Phys. B* **19**, L355 (1986).
96. K. Rinn, F. Melchert, K. Rink, and E. Salzborn, *J. Phys. B* **19**, 3717 (1986).
97. G. C. Angel, K. F. Dunn, E. C. Sewell, and H. B. Gilbody, *J. Phys. B* **11**, L49 (1978).
98. A. Jognaux, F. Brouillard, and S. Szücs, *J. Phys. B* **11**, L669 (1978).
99. B. Peart and K. Dolder, *J. Phys. B* **12**, 4155 (1979).
100. R. E. Olson, *J. Phys. B* **11**, L227 (1978).
101. B. Peart, K. Rinn, and K. Dolder, *J. Phys. B* **16**, 2831 (1983).
102. F. Melchert, K. Rink, K. Rinn, E. Salzborn, and N. Grün, *J. Phys. B* **20**, L223 (1987).
103. S. K. Allison, *Rev. Mod. Phys.* **30**, 1137 (1958).
104. L. I. Pivovar, V. M. Tubaev, and M. T. Novikov, *Sov. Phys. JETP* **14**, 20 (1962).

105. M. E. Rudd, T. V. Goffe, A. Itoh, and R. D. DuBois, *Phys. Rev. A* **32**, 829 (1985).
106. M. B. Shah and H. B. Gilbody, *J. Phys. B*, **9**, 2685 (1976).
107. F. J. de Heer, J. Schutten, and H. Moustafa, *Physica* **32**, 1793 (1966).
108. L. I. Pivovar, Yu. Z. Levchenko, and A. N. Grigor'ev, *Sov. Phys. JETP* **27**, 699 (1968).
109. C. F. Barnett and P. M. Stier, *Phys. Rev.* **109**, 385 (1958).
110. V. V. Afrosimov, Yu. A. Mamaev, M. N. Panov, and N. V. Federenko, *Sov. Phys. Tech. Phys.* **14**, 109 (1969).
111. E. S. Solov'ev, R. N. Il'in, V. A. Oparin, and N. V. Ferorenko, *Sov. Phys. JETP* **15**, 459 (1962).
112. M. B. Shah and H. B. Gilbody, *J. Phys. B* **18**, 899 (1985).
113. H. Knudsen, L. H. Andersen, P. Hvelplund, G. Astner, H. Cederquist, H. Danared, L. Liljeby, and K. G. Rensfelt, *J. Phys. B* **17**, 3545 (1984).
114. R. D. DuBois and S. T. Manson, *Phys. Rev. A* **35**, 2007 (1987).
115. R. D. DuBois, *Phys. Rev. A* **33**, 1595 (1986).
116. J. H. McGuire, E. Salzborn, and A. Müller, *Phys. Rev. A* **35**, 3265 (1987).
117. E. S. Solov'ev, R. N. Il'in, V. A. Oparin, and N. V. Fedorendko, *Sov. Phys. JETP* **18**, 342 (1964).
118. R. S. Langley, D. W. Martin, D. S. Harmer, J. W. Hooper, and E. W. McDaniel, *Phys. Rev.* **136**, A379 (1964).
119. R. D. DuBois, L. H. Toburen, and M. E. Rudd, *Phys. Rev. A* **29**, 70 (1984).
120. R. M. Wood, A. K. Edwards, and R. L. Ezell, *Phys. Rev. A* **34**, 4415 (1986).
121. A. K. Edwards, R. M. Wood, and R. L. Ezell, *Phys. Rev. A* **32**, 1346 (1985).
122. E. Horsdal Pedersen and L. Larsen, *J. Phys. B* **24**, 4099 (1979).
123. P. Hvelplund and A. Andersen, *Phys. Ser.* **26**, 370 (1982).
124. H. B. Gilbody, K. F. Dunn, R. Browning, and C. J. Latimer, *J. Phys. B.* **3**, 1105 (1970).
125. H. Tawara, *J. Phys. Soc. Jpn.* **31**, 871 (1971).
126. K. Tobita and H. Takeuchi, *J. Phys. Soc. Jpn.* **55**, 4231 (1986).
127. E. Horsdal Pedersen and P. Hvelplund, *J. Phys. B* **7**, 132 (1974).
128. P. Hvelplund and E. H. Pedersen, *Phys. Rev. A* **9**, 2434 (1974).
129. Ya. M. Fogel, V. A. Ankudinov, and D. V. Pilipenko, *Sov. Phys.* **11**, 18 (1960).
130. S. K. Allison, *Phys. Rev.* **110**, 670 (1958).
131. R. D. Rundel, K. L. Aitken, and M. F. A. Harrison, *J. Phys. B* **2**, 954 (1969).
132. J. Moseley, W. Aberth, and J. R. Peterson, *Phys. Rev. Lett.* **24**, 435 (1970).
133. T. D. Gaily and M. F. A. Harrison, *J. Phys. B* **3**, L25 (1970).
134. B. Peart, R. Grey, and K. T. Dolder, *J. Phys. B* **9**, L369 (1976).
135. S. Szücs, M. Karemera, M. Terao, and F. Brouillard, *J. Phys. B* **17**, 1613 (1984).
136. B. Peart, M. A. Bennett, and K. Dolder, *J. Phys. B* **18**, L439 (1985).
137. W. Schön, S. Krüdener, F. Melchert, K. Rinn, M. Wagner, and E. Salzborn, *J. Phys. B* **20**, L759 (1987).
138. D. Fussen and C. Kubach, *J. Phys. B* **19**, L31 (1986).
139. B. Peart, R. Grey, and K. T. Dolder, *J. Phys. B* **9**, 3047 (1976).
140. D. Fussen and W. Claeys, *J. Phys. B* **17**, L89 (1984).
141. W. Schön, S. Krüdener, F. Melchert, K. Rinn, M. Wagner, E. Salzborn, M. Karemera, S. Szücs, M. Terao, D. Fussen, R. Janev, X. Urbain, and F. Brouillard, *Phys. Rev. Lett.* **59**, 1565 (1987).
142. M. Terao, S. Szücs, M. Cherkani, F. Brouillard, R. J. Allan, C. Harel, and A. Salin, *Europhys. Lett.* **1**, 123 (1986).
143. B. Peart and M. A. Bennett, *J. Phys. B* **19**, L321 (1986).
144. M. Cherkani, S. Szücs, M. Terao, H. Hus, and F. Brouillard, *J. Phys. B* **24**, 209 (1991).
145. T. D. Gaily and M. F. A. Harrison, *J. Phys. B* **3**, 1098 (1970).
146. B. Peart, R. Grey, and K. Dolder, *J. Phys. B* **9**, L373 (1976).
147. J. Geddes, J. Hill, M. B. Shah, T. V. Goffe, and H. B. Gilbody, *J. Phys. B* **13**, 319 (1980).
148. D. G. Hummer, R. F. Stebbings, W. L. Fite, and L. M. Branscomb, *Phys. Rev.* **119**, 668 (1960).

149. F. R. Simpson and H. B. Gilbody, *J. Phys. B* **5**, 1959 (1972).
150. J. S. Risley and R. Geballe, *Phys. Rev. A* **9**, 2485 (1974).
151. J. F. Williams, *Phys. Rev.* **154**, 9 (1967).
152. Ya. M. Fogel, V. A. Ankudinov, and R. E. Slabospitskii, *Sov. Phys. JETP* **5**, 382 (1957).
153. R. Schulze, F. Melchert, M. Hagmann, S. Krüdener, J. Krüger, E. Salzborn, C. O. Reinhold, and R. E. Olson, *J. Phys. B* **24**, L7 (1991).
154. F. B. Yousif, B. G. Lindsay, and C. J. Latimer, *J. Phys. B* **21**, 4157 (1988).
155. V. V. Afrosimov, G. A. Leiko, Yu. A. Mamaev, M. N. Panov, and N. V. Fedorenko, *Sov. Phys. JETP* **35**, 1070 (1972).
156. E. S. Solov'ev, R. N. Il'in, V. A. Oparin, and N. V. Fedorenko, *Sov. Phys. JETP* **18**, 342 (1964).
157. R. Browning, C. J. Latimer, and H. B. Gilbody, *J. Phys. B* **2**, 534 (1969).
158. F. Brouillard, W. Claeys, and J. M. Delfosse, *J. Phys. B* **8**, 1149 (1975).
159. G. W. McClure, *Phys. Rev.* **153**, 182 (1967).
160. J. Guidini, *C. R. Acad. Sci. Paris* **253**, 829 (1961).
161. N. V. Fedorenko, V. V. Afrosimov, R. N. Il'in, and D. M. Kaminker, *Sov. Phys. JETP* **9**, 267 (1959).
162. C. F. Barnett and J. A. Ray, in *Proceedings of the Third International Conference on the Physics of Electronic and Atomic Collisions* (M. R. C. McDowell, ed.), North-Holland, Amsterdam (1964), p. 743.
163. A. Schmid, Z. Phys. **161**, 550 (1961).
164. G. W. McClure, *Phys. Rev.* **130**, 1852 (1963).
165. R. N. Il'in, B. I. Kikiani, V. A. Oparin, E. S. Solov'ev, and N. V. Fedorenko, *Sov. Phys. JETP* **19**, 817 (1964).
166. Y. Suzuki, T. Kaneko, M. Tomita, and M. Sakisaka, *J. Phys. Soc. Jpn.* **55**, 3037 (1986).
167. A. Oliver, F. Brouillard, W. Claeys, and G. Poulaert, *J. Phys. B* **9**, 3295 (1976).
168. O. Hollricher, *Z. Phys.* **187**, 41, (1965).
169. D. W. Koopman, *Phys. Rev.* **154**, 79 (1967).
170. H. L. Rothwell, B. Van Zyl, and R. C. Amme, *J. Chem. Phys.* **61**, 3851 (1974).
171. F. M. Campbell, R. Browning, and C. J. Latimer, *J. Phys. B* **14**, 3493 (1981).
172. C.-L. Liao, C.-X. Liao, and C. Y. Ng, *J. Chem. Phys.* **81**, 5672 (1984).
173. C.-Y. Lee and A. E. De Priston, *J. Chem. Phys.* **80**, 1116 (1984).
174. W. L. Fite, R. T. Brackmann, and W. R. Snow, *Phys. Rev.* **112**, 1161 (1958).
175. Z. Karpas, V. Anicich, and W. T. Huntress, Jr., *J. Chem. Phys.* **70**, 2877 (1979).
176. X. Urbain, A. Cornet, F. Brouillard, and A. Giusti-Suzor, *Phys. Rev. Lett.* **66**, 1685 (1991).
177. P. C. Karangwa, X. Urbain, J. Jureta, and N. Essarroukh, *J. Phys. B*, to be published.
178. X. Urbain, A. Cornet, and J. Jureta, *J. Phys. B* **25**, L189 (1992).
179. G. Poulaert, F. Brouillard, W. Claeys, J. McGowan, and G. Van Wassenhove, *J. Phys. B* **11**, L671 (1978).
180. G. Van Volkenburgh, T. Carrington, and R. A. Young, *J. Chem. Phys.* **59**, 6035 (1973).
181. J. S. Howard, J. P. Riola, R. D. Rundel, and R. F. Stebbings, *J. Phys. B* **6**, L109 (1973).
182. J. Fort, J. J. Laucagne, A. Pesnelle, and G. Watel, *Phys. Rev. A* **18**, 2063 (1978).
183. R. H. Neynaber and S. Y. Tang, *J. Chem. Phys.* **69**, 4851 (1978).
184. H. Morgner and A. Niehaus, *J. Phys. B* **12**, 1805 (1979).
185. H. Waibel, M.-W. Ruf, and H. Hotop, *Z. Phys. D* **9**, 191 (1988).
186. R. H. Neynaber, G. D. Magnuson, and J. K. Layton, *J. Chem. Phys.* **57**, 5128 (1972).
187. J. Fort, R. Bolzinger, D. Corno, T. Ebding, and A. Pesnelle, *Phys. Rev. A* **18**, 2075 (1978).
188. T. Bregel, A. J. Yencha, M.-W. Ruf, H. Waibel, and H. Hotop, *Z. Phys. D* **13**, 51 (1989).
189. D. W. Martin, C. Weiser, R. F. Sperlein, D. L. Bernfeld, and P. E. Siska, *J. Chem. Phys.* **90**, 1564 (1989).
190. R. H. Neynaber, G. D. Magnuson, and S. Y. Tang, *J. Chem. Phys.* **68**, 5112 (1978).

191. A. Z. Devdariani, V. I. Demidov, N. B. Kolokolov, and V. Ĭ. Rubtsov, *Sov. Phys. JETP* **57**, 960 (1983).
192. M. W. Müller, A. Merz, M.-W. Ruf, H. Hotop, W. Meyer, and M. Movre, *Z. Phys. D* **21**, 89 (1991).
193. P. E. Siska, *Rev. Mod. Phys.* **65**, 337 (1993).
194. B. G. Brunetti and F. Vecchiocattivi, in *Cluster Ions* (C. Y. Ng, T. Baer, and I. Powis, eds.), John Wiley & Sons, New York (1993), pp. 359–445.

Chapter 13

Electron Capture Processes in Slow Collisions of Plasma Impurity Ions with H, H$_2$, and He

R. K. Janev, HP. Winter, and W. Fritsch

1. INTRODUCTION

Electron capture in collisions between plasma edge impurity ions and atoms is the most important heavy-particle collision process in the plasma edge region. While excitation and ionization of the main plasma edge neutrals, H, H$_2$, and He, by slow plasma impurity ions are subject to adiabatic evolution constraints, electron capture can proceed via localized nonadiabatic couplings. For ions in higher charge states, the number of nonadiabatic couplings may become very large, and the electron capture process assumes a quasi-resonant character with large (10^{-15}–10^{-14} cm^2) cross section.

In this chapter we will discuss the basic physics and the available cross section information for the one- and two-electron capture processes

$$A^{q+} + B \rightarrow A^{(q-1)+} + B^+ \tag{1}$$

$$A^{q+} + B \rightarrow A^{(q-2)+} + B^{2+} \tag{2}$$

R. K. JANEV • International Atomic Energy Agency, A-1400 Vienna, Austria. HP. WINTER • Institut für Allgemeine Physik, Technische Universität Wein, A-1040 Vienna, Austria. W. FRITSCH • Bereich Theoretische Physik, Hahn-Meitner Institute, D-14109 Berlin, Germany.

Atomic and Molecular Processes in Fusion Edge Plasmas, edited by R. K. Janev. Plenum Press, New York, 1995.

where B is any of the neutral species H, H$_2$, and He, and A^{q+} is a plasma impurity ion. Our discussion will be confined to plasma impurities generated in the interactions of plasma particles with the walls of the fusion device, notably A = Be, B, C, Si, Ti, V, Cr, Fe, Ni, Cu, Mo, and W. Oxygen will also be included as an omnipresent impurity. Electron capture processes involving hydrogen and helium ions colliding with H, H$_2$, and He are considered in Chapter 12 of this book. Our study will focus on the low-energy region, with particular attention to the collision energies below 1 keV/amu, which are pertinent to the magnetic fusion boundary plasmas. The discussion in this chapter will also include the state-selective electron capture process

$$A^{q+} + B(n_0 l_0) \to A^{(q-1)+}(nl) + B^+ \quad (3)$$

($n_0 l_0$ and nl being the principal and orbital angular momentum quantum numbers of, respectively, the initial and the final electronic state), the transfer–ionization process

$$A^{q+} + B \to A^{(q-1)+} + B^{2+} + e^- \quad (4)$$

which, at low energies, may proceed through autoionizing two-electron capture, and, to a lesser extent, the transfer–excitation process

$$A^{q+} + B \to A^{(q-1)+} + B^{+*} \quad (5)$$

and the dissociative electron capture reaction

$$A^{q+} + H_2 \to A^{(q-1)+} + H_2^{+*} \to A^{(q-1)+} + H^+ + H \quad (6)$$

where H_2^{+*} is a dissociative intermediary state.

Among all the processes considered in this chapter, one-electron capture has the largest cross section, with only a few exceptions; it will, therefore, be discussed in greater detail than the two-electron processes. One-electron capture has also been the subject of extensive experimental and theoretical studies in the past, and hence the amount of cross section information for the collision systems of interest to fusion edge plasmas is considerable. The two-electron processes given in Eqs. (2) and (4)–(6) have been much less studied so far, and the corresponding cross section information is very limited. However, in the low-energy collision region, the cross sections of these two-electron processes increase linearly with increasing ionic charge, and, therefore, they may play a significant role in studies of the physics at the plasma edge.

The large cross sections for electron capture, particularly for ions in higher charge states, render them an important factor for the plasma edge neutral-gas kinetics and for impurity recycling. The cold ionized products of the one-electron capture process (H$^+$, H$_2^+$, He$^+$) absorb plasma energy by long-range Coulomb interactions with hotter plasma electrons. After their neutralization in resonant or quasi-resonant collisions with plasma edge neutrals (H, H_2, He), they may leave the plasma as hot neutrals and thus contribute to the plasma edge cooling. The products of charge-exchange processes, because of their lower charge state, absorb plasma energy much easier than their parent ions (owing to the inverse dependence of electron–ion excitation and ionization cross sections on the ionic charge and their lower thresholds) and further enhance the desired edge plasma cooling.

The cross section information on electron capture into specific ion states, Eq. (3), is required for a quantitative understanding of the radiation from plasma impurities and is essential for certain methods of plasma diagnostics. The presently widely used charge-exchange recombination plasma spectroscopy depends entirely on the information on state-selective electron capture processes. This method is, for instance, the only one capable of diagnosing fully stripped impurities in fusion plasmas.

There exist several comprehensive reviews in the literature that discuss the physics of electron capture processes in low-energy ion–atom collisions.[1-8] These reviews are mainly of theoretical nature and describe in detail the methods used in studies of electron capture processes. Reviews describing the experimental methods for investigation of electron capture collisions are also available.[4,9-11] The cross section information for the processes in Eqs. (1)–(6), involving the plasma edge constituents as specified above, has been reviewed on several occasions.[12-17] There has also been extensive activity in the past on the compilation and critical assessment of the cross section data information for electron capture processes in the systems considered. The most comprehensive cross section compilations for these (and other) systems are contained in Refs. 18–22. Critical assessments of the total electron capture cross sections have been performed in Ref. 23 (for $A^{q+} = C^{q+}$ and O^{q+}, B = H, H_2, and He, $q = 1-Z$), Ref. 24 (for $A^{q+} = Fe^{q+}$, B = H, H_2, and He, $q = 5-26$), and Ref. 25 (for $A^{q+} = Be^{4+}$, B^{5+}, C^{6+}, and O^{8+}, B = H). For state-selective electron-capture in C^{6+} + H and O^{8+} + H collisions, the critical assessment was done in Ref. 26. These critical data evaluations have led to sets of "recommended" electron capture cross sections, represented by appropriate analytic fits that facilitate the use of cross sections in fusion plasma applications.

Experimental and theoretical investigations of electron capture processes continue to grow, and the amount of available cross section information is increasing constantly. In this chapter we will also review the results of the most recent studies on the processes considered.

In the next section, we present an overview of experimental and theoretical methods as they are used in state-of-the-art investigations of low-energy electron capture collisions. Results for total and state-selective electron capture are presented and discussed separately in subseqent sections. We close this chapter with a few remarks on the status and prospects of the data base for fusion applications.

2. METHODS IN STUDIES OF LOW-ENERGY ELECTRON CAPTURE PROCESSES

Investigations of low-energy electron capture collisions have been performed with a wide range of methods, by experimentalists and theorists alike. The requirement for reliable information on details of the collisions, such as the distribution of electron transfer over final states of the ions, has led to the development of very sophisticated measurements and theoretical descriptions which are mutually complementary. In the following two subsections we will present separate short summaries of the most viable experimental and theoretical methods.

2.1. Experimental Methods

In this section we briefly review modern experimental methods for studying inelastic collisions between slow ions A^{q+} and neutral species $B(n_0l_0)$, where B is atomic hydrogen, molecular hydrogen, or helium. We consider single-electron capture [cf. Eq. (1)] and double-electron capture. "True double-electron capture" [cf. Eq. (2)] is only relevant at low q (see Section 3.2.2), whereas double-electron capture with projectile autoionization, which is also called transfer–ionization [cf. Eq. (4)], can become, in special cases, equally important to single-electron capture. If the initial and final states of the colliding particles are well defined and (practically) all scattered reaction products are detected, we speak of "state-to-state selective total cross sections." Distinction according to different scattering angles gives access to the respective "differential cross sections." In the present context, however, the latter data are of limited interest because electron capture with cross sections of interesting sizes involves electronic transitions at relatively large internuclear distances, for which, except at rather low impact energies, projectiles will be scattered into the near-forward direction. Depending on the extent of state definition for total cross section measurements, we may further distinguish "initial-state-selective" and "final-state-selective" apparent total cross sections. More commonly, however, the former and the latter are simply named "total cross sections" and "state-selective cross sections," respectively. Especially at low impact energy, total cross sections for both single- and double-electron capture depend strongly on the initial states of both reaction partners; as a consequence, related measurements without proper initial state definition are not meaningful. On the other hand, if a primary ion species does not appear in long-lived excited states (see below), and the neutral particles are most probably in their electronic (and vibronic) ground states, the respective initial states are well defined, and final-state-selective measurements will thus provide state-to-state selectivity. Furthermore, at low impact energy, single- and double-electron capture can become equally important, and henceforth a clear separation between single-electron capture, true double capture, and autoionizing double capture requires the coincident detection of the final reaction products.

2.1.1. Preparation of Primary Collision Partners

Recently, there has been great progress in production of intense, low-energy multicharged ion beams.[10,27,28] Electron cyclotron resonance-heated plasma sources (ECRIS) can now deliver, for example, fully stripped Ar ions with electrical currents of up to several hundred nanoamperes. However, the highest charge states can be obtained with electron beam ion sources (EBIS); for example, fully stripped Xe^{54+} ions[28] up to U^{92+} ions[29] can be produced.

Whereas, in the present context, long-lived excited beam admixtures are of no practical importance for highly charged ions, possible metastable fractions of low-q ion beams must be considered carefully.[30] Neutral collision partners can be provided either as static inert (H_2, He) or dissociated (H) gas targets, or as slow or fast neutral beams. The first approach is suitable for total cross section measurements, where an ion beam crosses a target gas cell, or neutral particles are fed into an ion trap.[31] Atomic hydrogen targets can be provided inside a hot furnace, with the degree of H_2 dissociation being determined via some single- and double-electron capture reactions. For state-selective measurements by means of photon-emission- or translational energy spectroscopy (see below), a low-energy target beam can be crossed by the applied ion beam. Inert neutral particles can be effused from a nozzle into the interaction region,[32] whereas $H(1s)$ beams can be produced either by effusion from a small hole in a hot furnace, or from radio-frequency (rf)[33] or microwave discharges.[34] In merged-beams experiments, the required fast neutral beams are obtained via charge exchange from fast positive-ion beams or via photodetachment from fast negative-ion beams.[35] Fast hydrogen beams with admixtures of highly excited Rydberg states have been produced via collisional detachment from fast negative-ion beams.[36]

2.1.2. Detection of Reaction Products and Accuracy of Measurements

Depending on the technique applied, positive ions, neutral atoms, electrons, or photons have to be detected, and sometimes even several species in coincidence. Particular care is needed for absolute photon detection.[4,9] Sufficiently fast neutral particles (≥ 10 eV/amu) can be detected via particle-induced electron emission from suitable converter surfaces, for which the detection efficiency needs to be calibrated and monitored. Detection of slow neutral particles (≤ 1 eV/amu) still poses formidable experimental challenges, since only in special cases can laser-induced fluorescence, surface ionization, or highly sensitive bolometry be used.[32] The accuracy of cross section measurements is naturally linked to the experimental technique used. Principal error sources result from the quality of definition of fluxes and compositions of primary collision partners, accuracy of the collision geometry, and calibration of collection and detection efficiencies for the reaction products. Possible additional error sources are pointed out in the following description of particular experimental techniques.

2.1.3. Measurement of Total Capture Cross Sections

Apart from their own importance, absolute total capture cross sections are needed for calibrating relative data from state-selective electron capture measurements (see below). Electron capture from He and H_2 is commonly studied with static gas targets, for which the target thickness and, especially at low impact velocity, the collection efficiency for charge-exchanged scattered projectiles have to be determined carefully. Atomic hydrogen targets produced by a hot furnace[37,38] require the additional determination of the degree of dissociation (see above). At impact energies below typically $10q$ eV, a slow ion beam crossing a gas target becomes subject to incomplete collection efficiency of its charge-exchanged products, which depends rather critically on the particular reaction channels involved. This may be circumvented by the technique of merged beams (see Fig. 1, which is taken from Ref. 36).

In collision systems for which single- and double-electron capture cross sections are of comparable importance, the latter have to be studied by coincident detection of scattered projectiles and recoil particles (see Fig. 2).[39] This permits not only the separation of the single-electron, true double-electron, and autoionizing double-electron capture reaction branches but also the distinction of the latter from "direct processes" such as single- and double-ionization events. A promising new technique has emerged whereby trapping of ions in combined electric and magnetic fields can be used for measuring capture rate coefficients at rather low impact

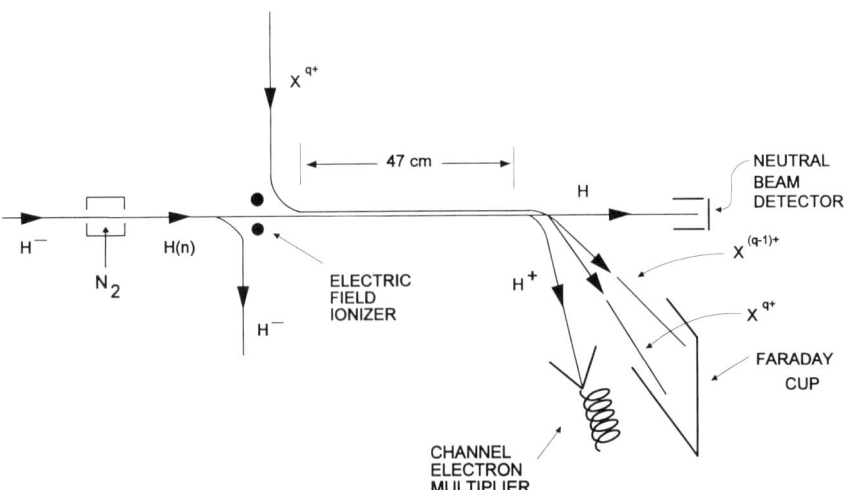

Figure 1. Schematic of an ion–atom merged-beams apparatus. (Courtesy of Dr. C. C. Havener, Oak Ridge National Laboratory.)

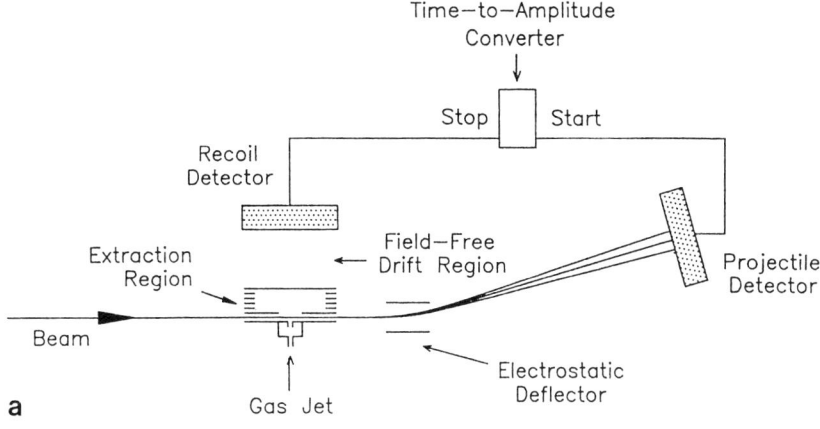

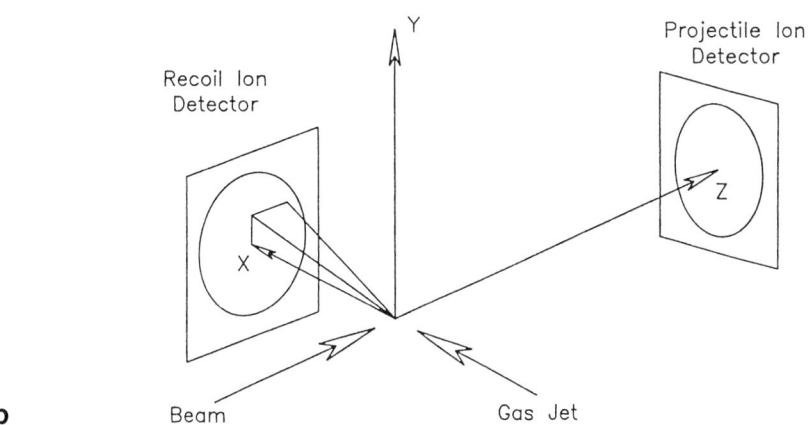

Figure 2. (a) Schematic of an apparatus for coincident measurements of the charge states of scattered projectiles and recoiled target particles. (b) Coordinate system showing interaction and collection of the collision partners. (c) Typical coincidence spectrum and its projections for Ar^{16+} + He collisions. (Courtesy of Prof. C. L. Cocke, Kansas State University).

energies (see Ref. 31 for a comprehensive review). Loading the ion trap with ions of given species (and in different charge states) is followed by admission of the neutral particles of interest into the trapping region. From the slopes of plots of the inverse ion-storage-time constants versus target-gas density, capture rate coefficients can be derived for each initial charge state. With additional laser cooling of trapped ions, such electron capture studies may be extended to ultralow collision energies.

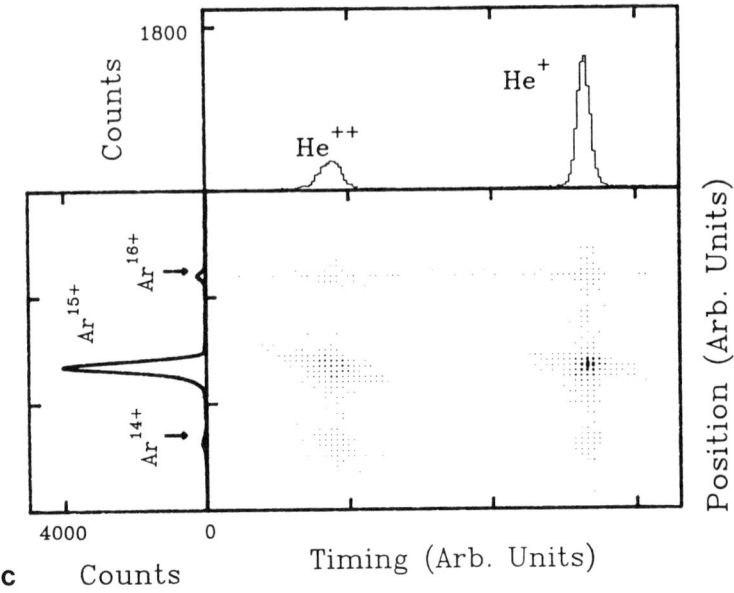

Figure 2. (continued)

2.1.4. Survey of Methods for Investigation of State-Selective Electron Capture

There are basically three experimental methods that may be used to derive state-selective electron capture cross sections, namely, translational energy spectroscopy, photon-emission spectroscopy, and electron-emission spectroscopy, all of which can also be combined with each other (see below). The different methods have their genuine advantages and shortcomings and thus generally can give access to somewhat different aspects of the inelastic collisions under study.[4]

2.1.4a. Translational Energy Spectroscopy. The method of translational energy spectroscopy relies on the kinetic energy gains or losses of projectile ions when scattered in the near-forward direction. It is a quite straightforward, rather sensitive method, and, in contrast to photon- and electron-emission spectroscopy, it also can cover capture channels involving final ground or long-lived excited states. Its state selectivity is mainly limited by the kinetic-energy resolution of electrostatic or time-of-flight analyzers applied for energy spectroscopy of the inelastically scattered, charge-exchanged ions. Special care is needed to define the collection efficiency of scattered projectiles into a given forward-scattering cone. As an example, translational energy spectroscopy has been applied at Belfast for a broad range of studies of state-selective single-electron capture from atomic and molecular hydrogen and from helium.[40–43]

The method of translational energy spectroscopy permits the preparation of specific initial states of the ion for capture studies in a rather convenient way, by using a charge-exchange cell in front of the ion monochromator, which usually precedes the main collision region. In this way, electron capture from suitable target gases into ions A^{q+} of interest can be applied for preparation of desired long-lived states $A^{(q-1)+*}$, which can then be selected for impact on target particles in the main collision region.[44] Translational energy spectroscopy has also been applied for studying single- and double-electron capture in collisions of He with slow, highly charged ions ($15 \leq q \leq 42$) from the Stockholm EBIS.[45]

2.1.4b. Photon-Emission Spectroscopy. The technique of photon-emission spectroscopy has been applied most widely by AMOLF Amsterdam and KVI Groningen. Numerous studies of single-electron capture have been conducted for impact of relatively slow ions from an ECRIS on H, H_2, and He, respectively. Photons emitted from excited projectiles, which have undergone capture reactions, are detected by means of absolutely calibrated grating spectrometers in the visible as well as the vacuum-ultraviolet (VUV) region (for the methods involved, see Refs. 4 and 9). The present status of this technique is described in Ref. 9, and Fig. 3 shows details of the most recent apparatus layout.

In comparison to translational energy spectroscopy, photon-emission spectroscopy generally permits a much higher resolution of final states, but it is considerably less sensitive because of the necessity for photon detection. The quality of its results strongly depends on the spectrometer/detector calibration and a sufficiently accurate definition of photon emission characteristics (spatial distribution, polarization), and also on the clear correlation of emitted photons to primarily excited states (see Ref. 4). Moreover, no information on capture into ground states or long-lived excited final states can be obtained. Consequently, translational and photon-emission spectroscopy should be regarded as complementary techniques. It is thus quite useful to study complicated collision systems in both ways, as has been the case, for example, for C^{3+} + H [cf. Refs. 46 (photon-emission spectroscopy) and 40 (translational energy spectroscopy)].

2.1.4c. Electron-Emission Spectroscopy. Double-electron capture from H_2 or He into multicharged ions usually results in the formation of doubly excited states. Spectroscopy of electrons from the decay of such doubly excited states gives state-selective information on the involved reaction channels.[4] Recently, several groups have conducted interesting studies in this field (e.g., at HMI Berlin[47,48] and KVI Groningen.[49,50]).

2.1.4d. Further Refined State-Selective Measurements. Still more sophisticated state-selective electron capture studies involving multicharged ion impact on two- and more-electron neutral particles have been conducted by combining translational energy spectroscopy of inelastically scattered projectiles with a deter-

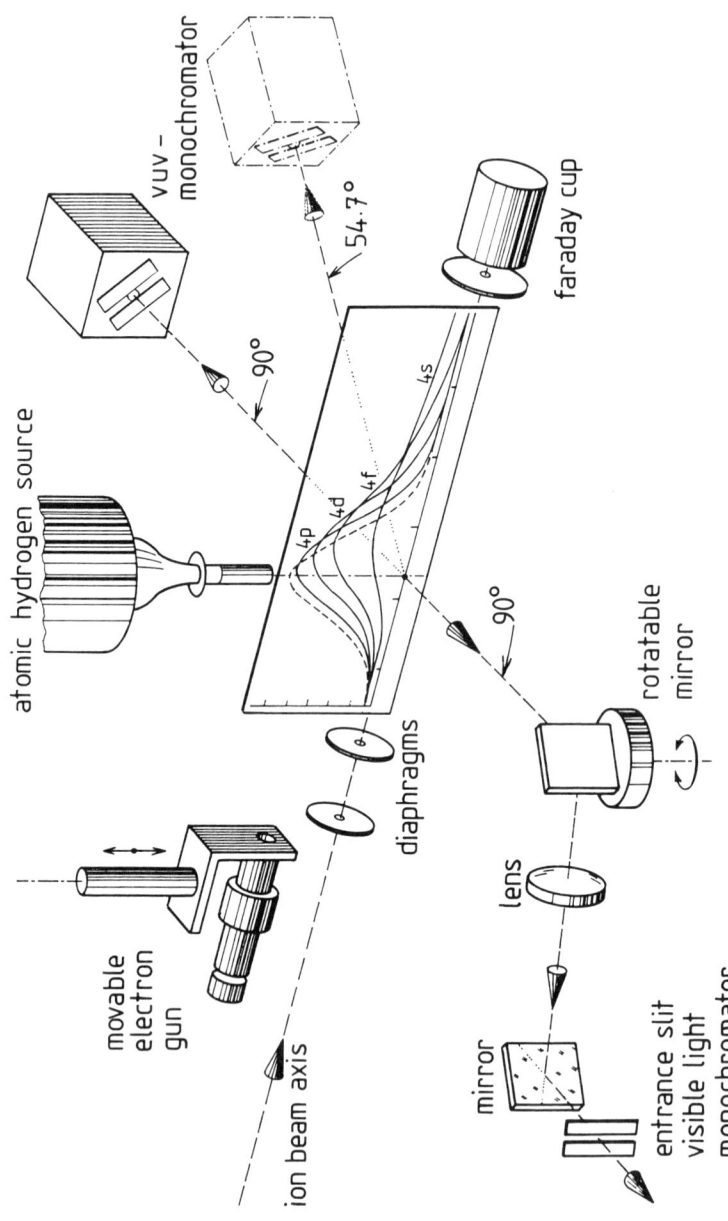

Figure 3. Schematic view of an experimental setup at KVI Groningen used for studying electron capture by ion impact on H, H$_2$, and He, by means of photon-emission spectroscopy. (Courtesy of Prof. R. Morgenstern, KVI Groningen.)

2.2. Theoretical Methods

2.2.1. General Considerations

The theoretical description of electron capture processes at low collision energies starts from an expansion of the total electron wave function of the collision system in terms of some complete basis set of functions. The time-dependent Schrödinger equation is then equivalent to an infinite system of coupled integro-differential or first-order differential equations for the expansion coefficients. The formalism can include a fully quantum-mechanical description of both the electronic and the nuclear motion (see, e.g., Ref. 1) or a quantum-mechanical description for the electron and a classical description for the nuclear motion (see, e.g., Refs. 2–8). The semiclassical description is appropriate for collision energies above about 100 eV/amu. The vast majority of theoretical work has been based on the semiclassical approximation, and we will hence concentrate, in the presentation of methods in this section, on such work.

The collision problem in the semiclassical approximation is described by the time-dependent Schrödinger equation (atomic units will be used throughout this chapter, unless explicitly stated otherwise),

$$(H - i\frac{\partial}{\partial t}) \Psi = 0 \tag{7}$$

where H is the total electron Hamiltonian of the system, and Ψ is the electron wave function. The expansion of Ψ in terms of a complete basis set $\{\psi_k\}$ of linearly independent functions,

$$\Psi = \sum_k a_k(t)\psi_k \tag{8}$$

and its substitution in Eq. (7) leads to an infinite set of first-order coupled equations for the coefficients $a_k(t)$, which in matrix notation may be written

$$i\mathbf{S}\dot{\mathbf{a}} = \mathbf{M}\mathbf{a} \tag{9}$$

Here, **a** is a column matrix formed from the expansion coefficients $a_k(t)$, and **S** and **M** are, respectively, overlap and coupling matrices with elements

$$S_{kj} = \langle \psi_j | \psi_k \rangle, \quad M_{kj} = \langle \psi_j | H - i\frac{\partial}{\partial t} | \psi_k \rangle \tag{10}$$

The terms of M_{kj} associated with the operators H and $-i(\partial/\partial t)$ are called "potential" and "dynamic" couplings, respectively. They depend on time t and on the classical interatomic trajectory $\mathbf{R}(t)$, and hence on the impact parameter b and the initial projectile velocity $v = |\dot{\mathbf{R}}(t = -\infty)|$.

The basis set is usually chosen in such a way that the unitarity condition is satisfied. If, in addition, the elements ψ_k in the basis are chosen such that Ψ satisfies the correct asymptotic boundary condition

$$a_k(-\infty) = \delta_{k0} \tag{11}$$

then the probability for a $0 \to j$ electron transition is given by

$$P_j(b,v) = |a_j(+\infty)|^2 \tag{12}$$

The cross section of the process is then determined as

$$\sigma_j(v) = 2\pi \int_0^\infty db\, b P_j(b,v) \tag{13}$$

In practical solutions of the scattering problem (Eq. 9), the basis set is truncated to a finite number of terms which can be handled computationally. It is then that the choice of the functions ψ_k becomes crucial. Two criteria are used for guidance in this choice: the basis elements ψ_k should (i) reflect the physics of the electronic motion during the collision and (ii) conserve the Galilean invariance of the system of coupled equations. The first condition leads to minimization of the couplings between the states (and thereby of the basis size), whereas the second condition, in conjunction with the condition given by Eq. (11), ensures fulfillment of the proper scattering boundary conditions.

In satisfying the first criterion, the following general ideas have been pursued. For collision velocities much smaller than the characteristic classical electron velocity in the isolated atom or ion, the electronic motion is adiabatic, and the eigenfunctions of the total electronic two-center Hamiltonian are an appropriate choice for $\{\psi_k\}$. The basis elements ψ_k are then taken as one-electron molecular orbitals (MOs) or two-electron molecular configurations. At higher collision velocities, comparable to or higher than the classical electron velocity in the isolated atom, the eigenfunctions of the atomic Hamiltonians of projectile and target constitute an appropriate choice for $\{\psi_k\}$. The elements ψ_k are in this case taken as one-electron atomic orbitals (AOs) or two-electron atomic configurations. The Galilean invariance of the scattering equations is achieved in either case by attaching to the basic MOs or AOs some phase functions, $F_k = \exp(i\alpha_k)$, which should describe the translational motion of the bound electron when the interatomic distance varies. It is obvious that for intermediate ($v \approx 1$) and high ($v \gg 1$) relative collision velocities, these electron translational factors F_k take on a plane-wave form. In the low-velocity region, however, a unique determination of F_k is not obvious.

For collisions systems with nonbare $A^{(q-1)+}$ ions, the first step in MO or AO close-coupling calculations is the choice of an appropriate ionic core potential. This potential is usually chosen in the form of a model potential or pseudopotential which, after diagonalization within the selected basis functions, reproduces the experimental energy spectrum of the $A^{(q-1)+}$ ion. This procedure determines the corresponding parameters both in the potential and in the basis functions. In both the MO- and AO-expansion methods, the basis functions are usually constructed from Slater type or scaled hydrogenic orbitals. In the case of AO expansions, the optimized orbitals may already serve as an appropriate basis for states localized on the ion, whereas in the MO method the molecular basis set is obtained by using a valence-bond configuration-interaction procedure.

Some of the specific features of the systems of coupled equations, based on the MO and AO expansions, will be given in Sections 2.2.2 and 2.2.3, respectively.

Another general approach[53–55] has recently been developed for the description of the collision dynamics in one-electron, two-nuclei systems in the low-energy region. This approach is based on the exact solution of the stationary Schrödinger equation for the electronic motion in the complex plane of interatomic separation R and on identification of the singularities of adiabatic molecular energies which are analytically continued into the complex R-plane. These singularities represent square-root branching points connecting adiabatic molecular states of the same symmetry. These branch points are called "hidden crossings" of adiabatic energies and appear as avoided crossings of the corresponding potential energy curves on the real R-axis. A remarkable feature of this approach is that the transition probability between two adiabatic states is calculated by a contour integral around the branching point connecting the energies of these states, and no calculation of matrix elements is required (at least in the strict adiabatic limit). The branching points of the adiabatic energies in the complex R-plane form characteristic series, many of which can be involved in a particular electron transfer reaction.

From the MO or AO close-coupled equations, one can derive various approximations and models by simplifying certain aspects of the collision dynamics. This will be discussed further below.

In the following brief discussion of the MO- and AO-expansion methods, we confine ourselves to the approximation of one "active" electron, moving in the effective fields of the atomic cores. The classical velocity of the "active" electron will be denoted by v_0, with $v_0 = 1$ for the hydrogen atom in its ground state.

2.2.2. Molecular-Orbital-Expansion Method

For relative collision velocities $v \ll v_0$, the molecular eigenfunctions χ_k of the electronic Hamiltonian of the system form the natural basis set. A set of stationary

MOs, however, does not satisfy the proper boundary conditions for the electron transfer problem. The elements of the basis expansion may, however, be written as

$$\psi_k = F_k(\mathbf{r}, \mathbf{R})\chi_k(\mathbf{r}, \mathbf{R}) \exp[-i \int^t E_k(R)\, dt'] \quad (14)$$

where the energy $E_k(R)$ is the eigenenergy associated with the eigenfunction $\chi_k(\mathbf{r}, \mathbf{R})$, and

$$F_k = \exp[i\mathbf{v}\mathbf{r} f_k(\mathbf{r}, \mathbf{R}) - iv^2 t/8] \quad (15)$$

is a general translational factor in a coordinate system with the origin placed at the midpoint between the two atoms. The "switching function" $f_k(\mathbf{r}, \mathbf{R})$ is fixed in the limit of large separations, where it should assume values of $\mp\frac{1}{2}$, corresponding to the plane-wave form for F_k. At small separations, a plane-wave translation factor would be inconsistent with the molecular picture of the electronic motion. Hence, different, R-dependent forms of f_k have been proposed (see Refs. 2–8), based on either intuitive arguments or on optimization schemes. In practical applications, it is ensured or at least assumed that results do not suffer critically from the exclusion of higher states and that they do not depend critically on the specific form of the chosen translation factor.

Early studies within the MO-expansion method were usually performed by setting $F_k = 1$. This version of the MO method is known as the perturbed stationary-state (PSS) approximation. Owing to nonphysical couplings at large separations, only total transfer cross sections may be deduced from such simplified descriptions.

The structure of the coupling matrix elements (Eq. 10) in the MO expansion can be seen in a more transparent way if $(\partial/\partial t)$ is calculated in a coordinate system that rotates with the interatomic axis. If any contributions of translation factors are disregarded, the couplings become

$$\langle \chi_j | \frac{\partial}{\partial t} | \chi_k \rangle = v_R \langle \chi_j | \frac{\partial}{\partial R} | \chi_k \rangle - \frac{bv}{R^2} \langle \chi_j | - iL_y | \chi_k \rangle \quad (16)$$

where v_R is the radial velocity, L_y is the component of the electron angular momentum perpendicular to the collision plane, and $-bv/R^2$ is the angular velocity of the interatomic axis. At low collision velocities, electronic transitions are seen to be induced by radial $(\partial/\partial R)$ and angular or rotational $(-iL_y)$ couplings between adiabatic molecular states. The states coupled by $\partial/\partial R$ have the same "symmetry," that is, the same magnetic quantum number m with respect to the interatomic axis ($\sigma, \pi, \ldots$ states), whereas those coupled by iL_y differ by one unit of m ($\sigma \leftrightarrow \pi$, $\pi \leftrightarrow \delta, \ldots$ couplings). The matrix elements $\langle \chi_j | \partial/\partial R | \chi_k \rangle$ are strongly peaked in the region where the corresponding adiabatic energies $E_j(R)$ and $E_k(R)$ display an avoided crossing. A molecular energy diagram therefore allows for some insight into the most important transition mechanisms in slow collisions. In the solution of the coupled equations, it is sometimes advantageous to use not the adiabatic MOs directly but some "diabatic" linear combinations of the MOs, in order to smooth the couplings at avoided crossings.

2.2.3. Atomic-Orbital-Expansion Method

When the collision velocity increases toward $v \sim v_0$, the molecular picture becomes less appropriate and the effects of electron momentum transfer become more important. The electron translation factors in this velocity region ($v \sim v_0$) attain a plane-wave form since the electronic motion is well localized around either of the atomic cores. The eigenfunctions (atomic orbitals, AOs) of the separated atoms' Hamiltonians $H_{A,B}$, in conjunction with plane-wave translation factors, form the natural basis for the expansion of the total electron wave function. The potential couplings between these basis states (see Ref. 56) do not lend themselves to any transparent interpretation; on the other hand, they are smooth functions of time and hence numerically convenient.

Most studies have been done with two-center expansions, containing "travelling" AOs (i.e., AOs with translation factors) of both projectile and target. Ionization channels or effects of the increased binding in close collisions can be included in this method if pseudostate wave functions are added to the two-center expansion. This extended two-center expansion model[57] has been sometimes termed the AO+ model. Pseudostates can be added also at a third center (at the equiforce point) between the two nuclei[58,59] (triple-center expansion).

The inclusion of united-atom pseudostates in the AO expansion method significantly improves its description of electron transfer processes in the collision velocity region below v_0. The two-center AO basis can then be visualized as almost equivalent to an MO basis constructed from a linear combination of atomic orbitals (LCAO). For the description of weak-electron-capture channels, however, a very large number of pseudostates may be required. In order to achieve a better and more economic description of molecular effects in both dominant and nondominant electron capture channels at low collision velocities, one may divide the interatomic separations into two regions: an inner region ($R \leq R_0$), in which an MO expansion is used, and an outer region ($R \geq R_0$), in which a two-center AO expansion is used.[60,61] This AO–MO matching method (similar to the **R**-matrix method in electron–atom collisions) should, in principle, describe the collision dynamics adequately in the entire energy range from low to intermediate energies provided the matching distance R_0 is chosen properly. R_0 should, in general, be channel-dependent, but in practice it is usually chosen to be the same for all channels. Current applications of the AO–MO matching method solve the coupled equations in the inner region within the PSS approximation ($F_k = 1$).

2.2.4. Other Theoretical Models

The fact that the system of coupled equations represented by Eq. (9) is identical to the time-dependent Schrödinger equation (Eq. 7) if the basis set $\{\chi_k\}$ is complete, irrespective of the character of the functions ψ_k, has led to the development of various theoretical models based on the expansion method. The basis functions in these models may still be chosen to reflect some of the characteristic features of the electronic motion, but the main idea is to gain practical advantages in solving the coupled equations with a large basis set. The results of such large-basis calculations tend to be less sensitive to the choice of translation factors, although the calculated amplitudes have to be projected onto the proper traveling atomic orbitals. The use of traveling Sturmian functions in a two-center expansion is one example of such an approach.[62,63] The Sturmian functions are closely related to scaled hydrogenic orbitals. They have the merit of being both discrete and, in their entirety, complete. Within a truncated basis, however, the Sturmian functions do not satisfy the proper boundary conditions; they are not eigenfunctions of either of the atomic Hamiltonians. The extraction of the physical amplitudes from the calculated ones needs additional projection procedures.

A set of pseudostates that reflects some of the molecular character of electronic states during slow collisions is the set of two-center Hylleraas functions.[64,65] In a rotating coordinate system, all coupling matrix elements can be evaluated analytically. These basis functions are stationary, and the calculated final-state amplitudes have to be projected onto traveling atomic orbitals as $R \to \infty$.

Other theoretical models have introduced additional, sometimes drastic, approximations to the coupled equations of the MO- or AO-expansion models. The approximations may aim at different aspects of the problem, for example, a drastic reduction of the size of the basis, the systematic omission of certain types of couplings that are considered small, or the phase interference effects (see Refs. 2–4). At intermediate collision energies, where many states of the system are strongly coupled, a classical description of the collision dynamics seems to be appropriate. The classical-trajectory Monte Carlo (CTMC) method[66] for solving the Hamilton equations for the effective three-body problem has proved to be very successful in describing the electron capture process in this energy region (see Refs. 2–4 and 8). A method based on the direct solution of the time-dependent Schrödinger equation in terms of continuum Keldysh–Volkov states has recently been developed[67] for the charge-exchange reactions and ionization, but broader applications are still missing.

3. TOTAL ELECTRON CAPTURE

In this section we present an overview and analysis of the results for total electron capture cross sections, obtained either experimentally or theoretically, for collision systems involving H, He, and H_2 neutrals and the plasma edge impurity ions that were mentioned in Section 1. The information from the general data sources[18–25] has been appropriately updated and extended to include collision systems not covered there. Because of the different electronic structures of H, He, and H_2, it is convenient to discuss the physics of electron capture and the data base for each of these electron donors separately.

3.1. Collisions with Atomic Hydrogen

Although most of the impurity ions in the plasma edge are incompletely stripped, the analysis of collisions between fully stripped ions and hydrogen atoms is of considerable interest. Since the electronic structure in these systems is comparatively simple, such studies provide a testing ground for the accuracy of various theoretical methods. These are also the systems for which various general properties of the electron capture cross section (such as scaling relations, dependence on the initial- and final-state binding energy of the electron, etc.) can be derived in the easiest way. Finally, transfer cross sections for highly charged systems are often little influenced by the structure of the ionic core; they can then be substituted by transfer cross sections for a bare projectile of equal charge.

We will hence concentrate first on collisions with fully stripped ions. Results for incompletely stripped ions are gathered further below.

3.1.1. Collisions of Fully Stripped Ions with Ground State Hydrogen

The electron capture process in the $H(1s) + A^{Z+}$ system (A^{Z+} is a fully stripped ion) at low collision energies ($E < 25$ keV/amu, $v < 1$) is determined by the strong coupling of quasimolecular states. The number N_c of strongly coupled states which give dominant contributions to the cross section depends on both the collision velocity and the ionic charge Z. For small values of Z (e.g., $Z < 8$), N_c is relatively small and increases with the collision velocity. For large values of Z, N_c is always large because of the high density ($\sim Z^2$) of the electronic states at the ion. Since both radial and angular couplings depend linearly on the velocity, the electron capture cross section is small at low velocities when N_c is small, but it increases rapidly with an increase of N_c (or Z). In general, the low-energy behavior of electron capture cross sections for low-Z ions depends on the specific two-state coupling that initiates the transfer process, on its strength and distribution along the R-axis, so that there is no uniform Z or E dependence of the cross section. For high values of Z, however (e.g., for $Z > 10$), the large density of final states available for capture ensures that the initial molecular state is strongly coupled radially with the final states, in a quasicontinuous manner that washes out all information on the particular system. The capture process becomes then quasi-resonant. In this case, the total electron capture cross section has a uniform energy behavior and varies smoothly (linearly) with Z (see, e.g., Refs. 2 and 3).

In the context of fusion edge plasmas, we will consider here only the electron transfer from H(1s) to the fully stripped impurities Be^{4+}, B^{5+}, C^{6+}, N^{7+}, and O^{8+}. Low-energy total electron capture cross section measurements for these systems are available, except for the system Be^{4+} + H, from several groups, who have employed the beam–gas target (Ref. 68 for B^{5+} and C^{6+}), crossed-beams (Ref. 69 for C^{6+} and N^{7+} and Ref. 38 for C^{6+}, N^{7+}, and O^{8+}), and beam–gas target/optical emission (Refs. 70 and 71 for C^{6+} and O^{8+}) techniques. There also exist several highly elaborate calculations for the total electron capture cross sections of these collision systems, and most of these also predict the distribution over final states. These calculations are performed either within the MO-expansion model (Refs. 72 and 73 for Be^{4+} and B^{5+}; Ref. 74 for C^{6+}; Ref. 75 for O^{8+}), within the AO-expansion model (Ref. 76 for all systems with Z = 4–8; Ref. 77 for Be^{4+} and C^{6+}; Refs. 78 and 79 for C^{6+} and O^{8+}), or within the AO–MO model (Ref. 80 for C^{6+}). The advanced adiabatic (or hidden crossing) method has also been employed for cross section calculations in Be^{4+} + H and B^{5+} + H collisions[81] and C^{6+} + H collisions.[82]

The available theoretical and experimental cross section information has been critically assessed,[23,25] with due consideration of the uncertainties in the calcula-

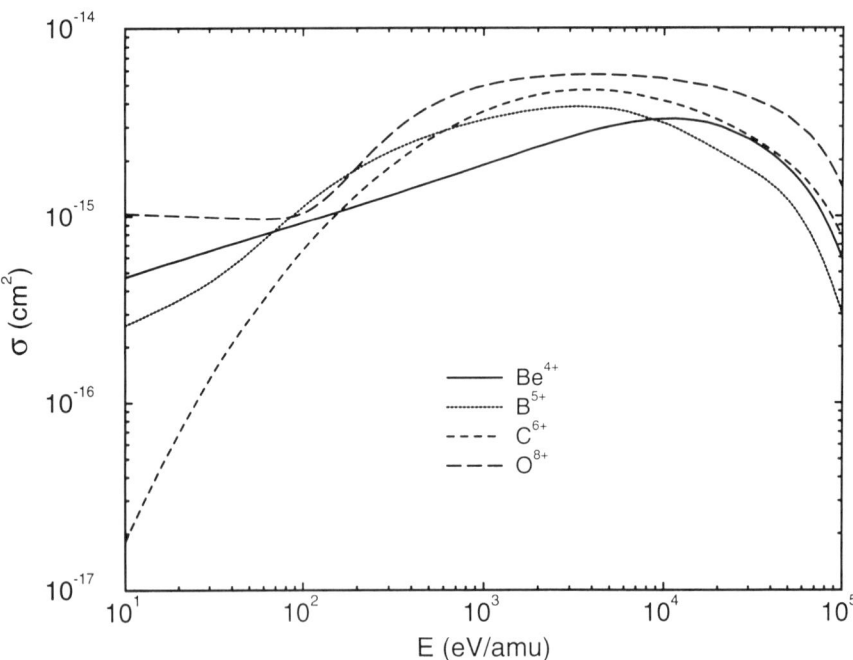

Figure 4. Recommended total electron capture cross sections for H(1s) + Be^{4+}, H(1s) + B^{5+}, H(1s) + C^{6+}, and H(1s) + O^{8+} collisions, generated from Eq. (17).

Table I. Values of the Fitting Coefficients a_i in Eq. (17)

	Be^{4+}	B^{5+}	C^{6+}	O^{8+}
a_1	19.952	31.226	418.181	54.535
a_2	0.20036	1.1442	2.1585	0.27486
a_3	$1.7295(-4)^a$	$4.8372(-8)$	$3.4808(-4)$	$1.0104(-7)$
a_4	$3.6844(-11)$	$3.0961(-10)$	$5.3333(-9)$	$2.0745(-9)$
a_5	5.0411	4.7205	4.6556	4.4416
a_6	$2.4689(-8)$	$6.2844(-7)$	0.33755	$7.6555(-3)$
a_7	4.0761	3.1297	0.81736	1.1134
a_8	0.88093	0.12556	0.27874	1.1621
a_9	0.94361	0.30098	$1.8003(-6)$	0.15826
a_{10}	0.14205	$5.9607(-2)$	$7.1033(-2)$	$3.6613(-2)$
a_{11}	-0.42973	-0.57923	0.53261	$3.9741(-2)$

$^a (-x) = \times 10^{-x}$.

tions and the data. Figure 4 shows the total electron capture cross sections for the $H(1s) + Be^{4+}$, B^{5+}, C^{6+}, and O^{8+} collision systems in the energy range of 0.01–100 keV/amu. They are believed to have an accuracy of 20–30% in the region below 1 keV/amu and about 15–20% above that energy. Above 100 keV/amu, the cross sections continue to decrease toward their asymptotic behavior, $\sigma(E) = E^{-5.5}$, in the MeV/amu region.[83] All the cross sections can be represented by the analytic fit[25]

$$\sigma(E) = a_1 \left[\frac{\exp(-a_2/E^{a_8})}{1 + a_3 E^2 + a_4 E^{a_5} + a_6 E^{a_7}} + \frac{a_9 \exp(-a_{10} E)}{E^{a_{11}}} \right] (10^{-16} \text{ cm}^2) \quad (17)$$

where the energy is expressed in keV/amu, and the values of the fitting coefficients a_i are given in Table I. The above expression describes the cross section behavior for all the considered reactions up to several MeV/amu and for C^{6+} and O^{8+} down to 1 eV/amu.

3.1.2. Collisions of Incompletely Stripped Ions with Ground State Hydrogen

Many-electron effects in collisions between an incompletely stripped ion A^{q+} and a hydrogen atom make both theoretical and experimental studies of electron capture a complex task. The complexity arises from the multitude of reaction channels that lie energetically close to the initial state. This multitude is caused by the non-Coulomb character of the ionic core potential, and, in the case of lowly charged ions, by the potentially active role of valence electrons in the ion core during the collision. Experimentally, this leads to problems in the determination of state-specific transfer cross sections (see Section 4) whereas total transfer cross sections may still be measured easily. On the theoretical side, the MO- and AO-expansion methods still apply within the approximation of one or two active electrons. There is so far no work on systems with more than two active electrons.

In discussing the available cross section information for the H(1s) + A^{q+} ($q <$ Z) collision systems, we will first consider the case of low-Z ions (Z < 10), for which data and calculated results are more abundant than for the systems with higher nuclear charge.

3.1.2a. Cross Sections for Low-Z Ions. The available cross sections for Be^{q+} + H(1s) and B^{q+} + H(1s) electron capture collisions as of 1991 have been collected in Ref. 21 and critically assessed in Ref. 84. The data situation for these collision system has not changed since this recent assessment, except for the new MO calculations[85] for B^{4+} + H. For the Be^{q+} + H systems, data are not available, and published theoretical studies are limited to a PSS calculation[86] for Be^{2+} + H and an MO close-coupling calculation[87] for Be^{3+} + H. Recent, more elaborate AO–MO calculations[73] for Be^{2+} + H support the earlier PSS calculations for this system in the region below 0.6 keV/amu. The cross section for the Be$^+$ + H system is expected to be small at low collision energies owing to the large endothermicity of all reaction channels. The low-energy electron capture cross sections for the Be^{2+} + H(1s) and Be^{3+} + H(1s) collision systems, based on Refs. 73 and 86 and Ref. 87, respectively, are given in Table II.

The availability of cross section information for the B^{q+} + H(1s) systems (q = 1–4) is greater than in the case of Be^{q+} + H(1s) systems. Experimental cross sections

Table II. Low-Energy Electron Capture Cross Sections for Be^{2+} + H(1s) and Be^{3+} + H(1s) Collisions

Be^{2+} + H(1s)		Be^{3+} + H(1s)	
E (keV/amu)	σ (10^{-16} cm^2)	E (keV. amu)	σ (10^{-16} cm^2)
0.05	1.15	0.20	0.05^a
0.08	1.52	0.30	0.15^a
0.10	1.75	0.40	0.23
0.12	1.92	0.50	0.41
0.15	2.05	0.75	0.85
0.20	2.20	1.00	1.63
0.30	2.35	1.50	3.00
0.40	2.45	2.00	3.95
0.50	2.46	2.50	4.53
0.60	2.45	3.00	5.95
0.80	2.42	3.50	6.62
1.00	2.35	4.00	7.41
1.20	2.30	4.5	8.11
2.00	2.20	5.0	9.2
3.00	2.12	6.0	10.2
5.00	2.02^a	8.0	12.0^a
8.00	1.85^a	10.0	13.8^a
10.00	1.78^a	12.0	14.0^a

^{a}Extrapolated values.

Electron Capture Processes

are available for the $B^+ + H$ system[88] (only for $E \geq 10$ keV/amu) and the $B^{2+} + H$, $B^{3+} + H$, and $B^{4+} + H$ systems[68,88–92] for energies down to 2 keV/amu (B^{3+} and B^{4+}) or 0.3 keV/amu (B^{2+}). Close-coupling MO-calculations have been performed for $B^{3+} + H$ collisions[73,86] and $B^{4+} + H$ collisions,[85] the results of which agree well with the data in the energy range of their overlap. Figure 5 shows the cross sections for these reactions, assessed on the basis of experimental and theoretical results.

After the critical assessment[23] of total electron capture cross sections for $C^{q+} + H$ ($q = 1$–5) and $O^{q+} + H$ ($q = 1$–7) systems in 1987, new low-energy cross section measurements were performed only for the $C^{3+} + H$ collisions,[93] $C^{4+} + H$ collisions,[41,94] and $O^{5+} + H$ collisions.[35] New multistate close-coupling calculations with the MO-expansion method were performed for the $C^{3+} + H$ system,[95,96] the $C^{4+} + H$ system,[97] the $C^{5+} + H$ system,[98] the $O^{4+} + H$ system,[99] the $O^{5+} + H$ system[85,100] and the $O^{6+} + H$ system.[101] A reassessment of all the available cross sections for these systems was done[102] in 1990, and the "recommended" cross sections were fitted to nonpolynomial analytic expressions with appropriate asymptotic behavior. The low-energy cross sections for the $C^{q+} + H$ ($q = 1$–5) and $O^{q+} +$

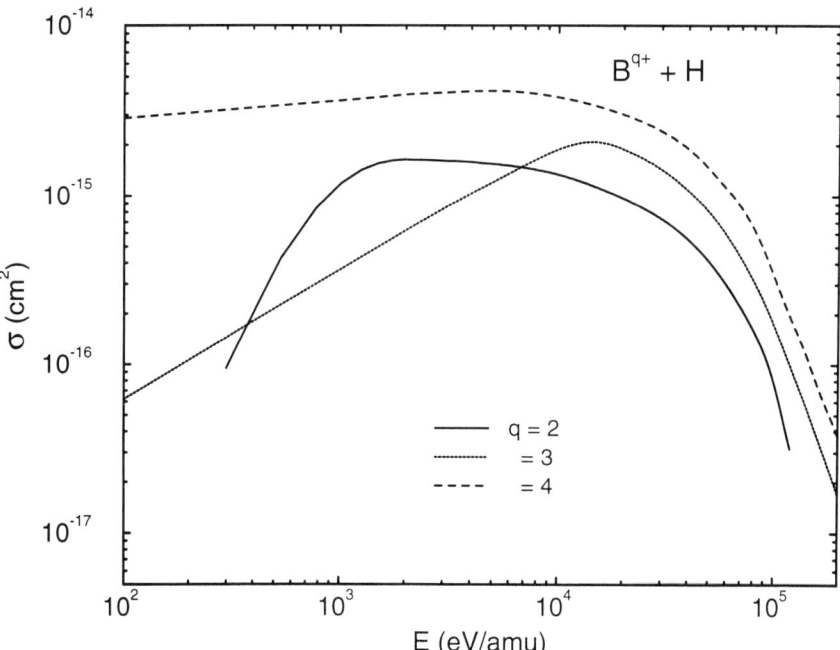

Figure 5. Total electron capture cross sections for $B^{q+} + H$ collisions ($q = 2$–4) based on a critical assessment of the experimental[68,88–92] and theoretical[73,85,86] results.

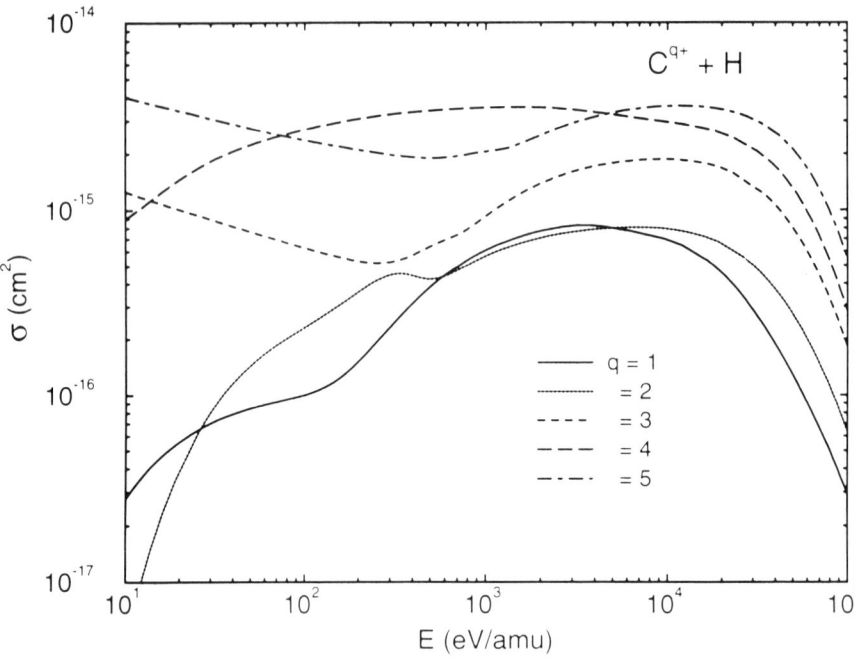

Figure 6. Recommended electron capture cross sections for C^{q+} + H collisions (q = 1–5) generated from Eq. (18) with the coefficients in Table III.

H (q = 1–3, 6, 7) systems are given in Figs. 6 and 7, respectively. All the cross sections can be represented by a unifying expression,[102]

$$\tilde{\sigma} = \frac{\sigma}{q} = b_1 \left[\frac{\exp(-b_2/\tilde{E})}{1 + b_3\tilde{E}^2 + b_4\tilde{E}^{4.5}} + b_5\frac{\exp(-b_6\tilde{E})}{\tilde{E}^{b_7}} + \frac{b_8 \exp(-b_9/\tilde{E})}{1 + b_{10}\tilde{E}^{b_{11}}} \right] (10^{-16} \text{ cm}^2) \quad (18)$$

with

$$\tilde{E} = E(\text{keV/amu})/q^{0.5}$$

and with the coefficients b_i given in Tables III and IV for C^{q+} + H and O^{q+} + H, respectively. We note that the reaction $O^+ + H \rightarrow O + H^+$ is "accidentally" resonant. Its cross section displays a behavior that is indicative of resonant transfer; that is, it increases logarithmically with decreasing collision energy.

Electron capture in N^{q+} + H(1s) collision systems has also been the subject of considerable interest. The corresponding cross section information (available as of 1988) has been collected and discussed in Ref. 19. The more recent experimental studies of these systems include merged-beams cross section measurements of the

Electron Capture Processes

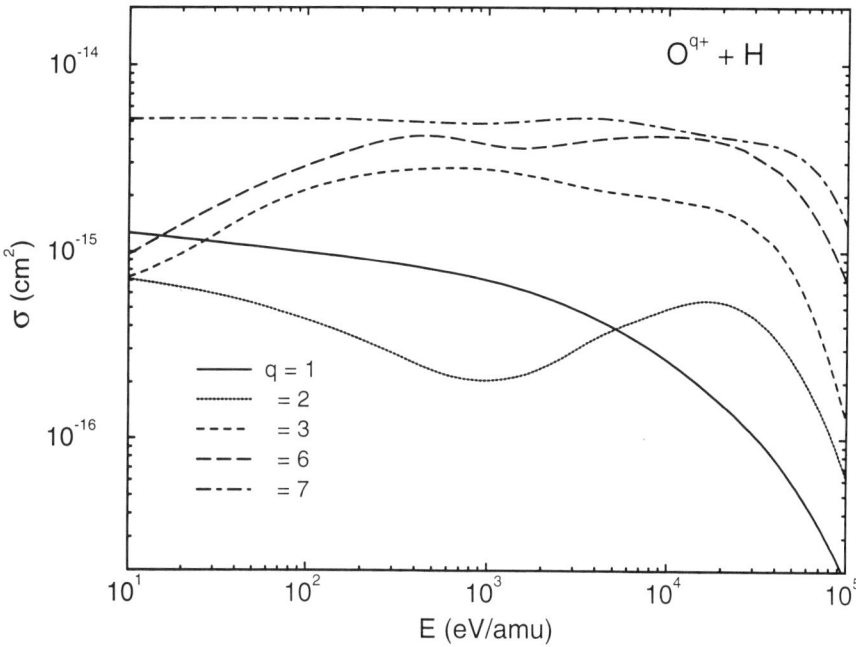

Figure 7. Recommended electron capture cross sections for O^{q+} + H collisions (q = 1–3, 6, 7) generated from Eq. (18) with the coefficients in Table IV.

Table III. Values of Parameters b_i in Eq. (18) for the C^{q+} + H Collision System (q = 1–5)

	$q = 1$	$q = 2$	$q = 3$	$q = 4$	$q = 5$
b_1	8.9260	4.4734	6.9910	7.4984	7.3591
b_2	5.8686(−1)[a]	3.2814(−1)	6.2566(−1)	9.8946(−3)	8.6154(−1)
b_3	2.2526(−3)	1.6433(−3)	1.7196(−3)	1.7233(−3)	1.3438(−3)
b_4	6.7538(−9)	2.3315(−8)	6.3293(−8)	1.2164(−7)	1.3421 (−7)
b_5	0.00	3.2033(−5)	8.3615(−2)	9.7523(−3)	3.0250(−1)
b_6	0.00	3.0081	2.3434(−2)	2.0617(−2)	4.8974(−2)
b_7	0.00	4.5424(−1)	3.6076(−1)	5.2822(−1)	2.3682(−1)
b_8	1.2567(−1)	3.8745(−1)	6.6220(−2)	2.2089(−1)	9.6709(−2)
b_9	1.3961(−2)	3.0991(−2)	4.7540(−4)	9.2267(−2)	8.6636(−2)
b_{10}	4.1377(−4)	1.8201(+3)	2.4358(+3)	1.4985(−1)	2.0738(+2)
b_{11}	5.00	5.9870	7.9403	2.51180	7.4858

[a] $(\pm x) = \times 10^{\pm x}$.

Table IV. Values of Parameters b_i in Eq. (18) for the O^{q+} + H Collision System (q = 1–7)

	$q=1$	$q=2$	$q=3$	$q=4$	$q=5$	$q=6$	$q=7$
b_1	2.4119	5.2609	6.2282	5.7292	6.5751	7.5967	7.5492
b_2	9.1143	5.2649	3.6900(–2)	1.6085(–2)	1.5087(–3)	2.060(–1)	2.2945
b_3	9.7879(–4)^a	1.7914(–3)	1.2437(–3)	1.0009(–3)	1.3414(–3)	2.2275(–3)	8.7182(–4)
b_4	1.3805 (–9)	2.7170(–8)	1.2590(–7)	9.1698(–8)	9.1059(–8)	1.0709(–7)	1.2084 (–7)
b_5	3.3180	4.4296(–1)	1.08121(–1)	9.0992(–2)	9.8415(–1)	5.1640	1.2139
b_6	1.2903(–1)	5.2110	4.7829(–2)	3.1993	1.8546(+3)	6.4620	3.6549(–1)
b_7	1.0008(–1)	9.2806(–2)	1.9064(–1)	5.9917(–2)	2.0830(–1)	–5.7621(–1)	–9.3774(–2)
b_8	0.00	2.3430(–1)	5.2076(–1)	3.3652(–1)	2.5838	0.00	0.00
b_9	0.00	1.7301(–1)	9.6184(–3)	6.4769(–3)	1.7796(–3)	0.00	0.00
b_{10}	0.00	7.3918(–3)	6.7256(–1)	2.3074(–1)	1.1206(+2)	0.00	0.00
b_{11}	0.00	3.4170	2.2217	2.5707	1.3372	0.00	0.00

$^a(\pm x) = \times 10^{\pm x.}$

N^{5+} + H system[35] and the N^{3+} + H, N^{4+} + H, and N^{5+} + H systems.[103] Cross section calculations within the MO-expansion method (in some cases including also its quantal version) have recently been performed for the N^{4+} + H system[104,105] and the N^{5+} + H systems.[101,106]

3.1.2b. Cross Sections for Medium- and High-Z Ions. The low-energy electron capture cross sections for collisions of medium- and high-Z fusion ions (Si, Ti, V, Cr, Fe, Ni, Cu, Mo, Ta, and W) with atomic hydrogen are rather scarce, particularly for the low-ionic-charge states. The information on these collision systems was reviewed[107] in 1991. The situation has not significantly improved in the meantime. For total transfer, the information may still be considered satisfactory, except for the lowest charge states or the lowest energies, since the effect of the ionic core on total transfer may be considered small. Hence, total transfer cross sections from equicharged systems can often be adopted as a substitute for missing information. On the other hand, the effect of the specific ionic core is clearly not small at very low energy or for lowly charged ions (q =1–2). Equicharged systems are also of little help when state-selective transfer cross sections are required.

Experimental cross sections exist[88] for Ti^{2+} + H in the energy region above 3 keV/amu and for Fe^{q+} + H in the energy ranges 0.01–0.095 keV/amu ($3 \le q \le 14$),[108] 0.3–2.2 keV/amu (q = 5, 6),[109] 27–290 keV/amu ($4 \le q \le 13$),[110] and 60–161 keV/amu ($4 \le q \le 15$).[111] An assessment of the data for Fe^{q+} + H (as well as those for Fe^{q+} + H_2 and Fe^{q+} + He) has been done in Ref. 24. At a limited number of collision energies in the intermediate- to high-energy region, experimental cross sections are also available for Si^{q+} (q = 4–9),[112] Mo^{q+} (q = 4–18),[111] and W^{q+} (q = 4–15)[111] ions colliding with H.

Electron Capture Processes

Only a limited number of multistate close-coupling calculations have been performed for these systems. They include MO calculations for Si^{4+} + H collisions[99] and Ti^{4+} + H collisions[113,114] and AO calculations for Si^{q+} + H collisions (q = 4, 6, 8–14),[115] Ti^{4+} + H, Cr^{6+} + H, and Fe^{8+} + H collisions,[116] and Si^{4+} + H and Ni^{10+} + H collisions.[117] Extensive CTMC calculations in the energy region above 10 keV/amu have been performed for collisions of $Ti^{(4-11)+}$, $Cr^{(4-15)+}$, $Fe^{(4-26)+}$, and $Ni^{(4-17)+}$ with atomic hydrogen[118] by using appropriate model potentials for the electron–ion interaction. Similar calculations for $Fe^{(1-26)+}$ + H collisions at energies above 50 keV/amu have been performed.[119] The Keldysh semiclassical method[67] has recently also been employed[120] to calculate the charge-exchange cross sections of the Ti^{q+}, V^{q+}, Kr^{q+}, Mo^{q+}, and W^{q+} (q = 3–10) ions colliding with H(1s) for energies above 10 keV/amu. The results of these calculations are consistent with those of the AO close-coupling method for the same collision systems in the energy range where both methods apply.

In Fig. 8 the results of AO-expansion calculations for the Si^{4+} + H and Ti^{4+} + H systems[116,117] and the Si^{6+} + H and Cr^{6+} + H systems[115,116] are shown, together with data for Ar^{4+} projectiles,[109] Fe^{4+} projectiles,[110,111] and Fe^{6+} projectiles.[109-111] This figure shows that in the low-energy region the electron capture cross sections

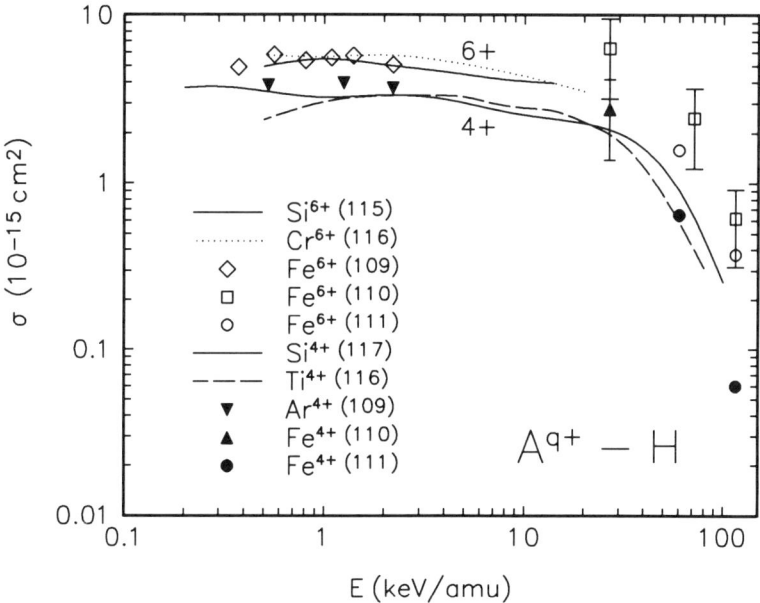

Figure 8. Total transfer cross sections in A^{q+} + H collisions (A^{q+} = Si^{4+}, Ti^{4+}, Fe^{4+}, Si^{6+}, Cr^{6+}, and Fe^{6+}). Symbols represent experimental data, and curves represent calculated cross sections. The references from which these results have been drawn are indicated on the figure.

for many-electron ions with $q \geq 4$ become weakly dependent on the collision energy and on the ionic core structure. This feature of the electron capture cross section results from the high density of ionic states that are strongly coupled with the initial state during the slow evolution of the collision system and from the fact that the mean energy of ionic states to which electron capture takes place dominantly is far from the energy of the ionic ground state. With an increase of ionic charge, these properties of the electron capture process become more pronounced. It is then that a description in terms of classical models (such as the CTMC and the classical over-barrier transition model[83]) becomes justifiable.

The simplified models of the electron capture mechanism for high-q ions at low and intermediate collision energies reveal a scaling property of the total electron capture cross section,[121]

$$\sigma/q = f(E/q^{0.5}) \tag{19}$$

where the function $f(x)$ is insensitive to the ionic species for $q \geq 5$ and the scaled energies $E/q^{0.5} \leq 25$ keV/amu. A weak ionic core dependence in the scaled cross section $\tilde{\sigma} = \sigma/q$ appears only for $E/q^{0.5} > 50$ keV/amu because, with increasing energy, electron capture increasingly goes to the lower excited and the ground state of the projectile. This core dependence of $\tilde{\sigma}$ can be largely eliminated by introducing[122] the reduced energy $\tilde{E} = E/q^{3/7} \simeq E/q^{0.43}$. All the available experimental cross section data for the incompletely stripped ions in charge states $q \geq 5$ colliding with H(1s) can be represented, within an r.m.s. deviation of 20%, by the expression[102]

$$\tilde{\sigma} = \frac{\sigma}{q} = \frac{C_1 \ln(C_2/\tilde{E} + C_3)}{1 + C_4\tilde{E} + C_5\tilde{E}^{3.5} + C_6\tilde{E}^{5.4}} (10^{-16} \text{ cm}^2) \tag{20}$$

with

$$\tilde{E} = \frac{E(\text{keV/amu})}{q^{3/7}}$$

$$C_1 = 0.7336, \; C_2 = 2.9391 \times 10^4, \; C_3 = 41.865$$

$$C_4 = 7.1023 \times 10^{-3}, \; C_5 = 3.4749 \times 10^{-6}, \; C_6 = 1.1832 \times 10^{-10}$$

The scaling relation given by Eq. (20) applies also for ions in charge states $q = 3$ and 4 except for reduced energies $\tilde{E}$ below 0.1 keV/amu.

3.1.3. Electron Capture from Excited Hydrogen States

The cross section information on electron capture from excited hydrogen atoms by plasma impurities is required in the calculations of edge plasma radiative power losses due to impurities and neutral hydrogen, as well as in calculations of the attenuation of neutral hydrogen beams injected into the plasma for heating or diagnostic purposes. Measurements of the cross section for electron capture from excited atomic hydrogen atoms have been performed only for He^{2+} projectiles[123] at low collision energies. Extensive CTMC cross section calculations have been done[119,124] for the H$^*(n)$ + A^{q+} systems (n = 1–20, A^{q+} = H$^+$, He^{2+}, Be^{5+}, Ne^{10+}; n is the principal quantum number) in the intermediate energy range. Because an extremely large number of ionic states is available for electron capture when $n \geq 2$ and $q \geq 2$, the classical description of the process can be considered appropriate even at low energies. The invariance of classical equations of motion in a Coulomb field under scale transformation of the length and the charge immediately leads to a qn^4 scaling of the classical electron capture cross section. Simple models, based on the classical dynamics, for example, the over-barrier transition model[83] or the Bohr–Lindhard model,[125] show that for $v < q^{0.25}/n$ the capture cross section should be almost energy independent, whereas for $v > q^{0.25}/n$ it should display a v^{-7} behavior. A detailed theoretical consideration of the classical scaling properties at both low and intermediate energies[122] has led to the following expression for the scaled cross section:

$$\tilde{\sigma} = \frac{\sigma}{n^4 q} = \frac{7.04B}{\tilde{E}^{3.5}(1+C\tilde{E}^2)}\left[1 - \exp\left(-\frac{2\tilde{E}^{3.5}(1+C\tilde{E}^2)}{3B}\right)\right](10^{-16}\text{ cm}^2) \quad (21)$$

with

$$\tilde{E} = \frac{n^2 E(\text{keV/amu})}{q^{0.5}}$$

$$B = 1.283 \times 10^5, \ C = 1.9704 \times 10^{-5}$$

The above expression for $\tilde{\sigma}$, in which the two free parameters B and C were determined from the high-energy CTMC results, represents the available experimental[123] and CTMC[119,124] results up to $\tilde{E}$ = 200 keV/amu with an r.m.s. deviation of about 10%. The scaled cross section is shown in Fig. 9, together with the available data and some of the CTMC results.[119]

3.2. Collisions with Helium Atoms

In collisions between singly charged ions A$^+$ and He atoms, the large binding energy of the He electrons translates into large endothermicities of all charge-exchange reactions. Correspondingly, charge-exchange cross sections in these systems are small in slow collisions. With an increase of ionic charge state q, the A^{q+} + He electron capture reactions become exothermic. For $q \geq 4$, even two-electron transfer is exothermic. For sufficiently large values of q, one- and two-electron capture processes are of comparable strength. Two-electron capture may interfere with the capture–ionization process of Eq. (4), particularly when the two electrons are captured into doubly excited states of the A$^{(q-2)+}$ ion. Clearly, the two-electron structure of the target and the multitude of different reaction channels render theoretical and experimental studies of electron capture processes a rather complex task.

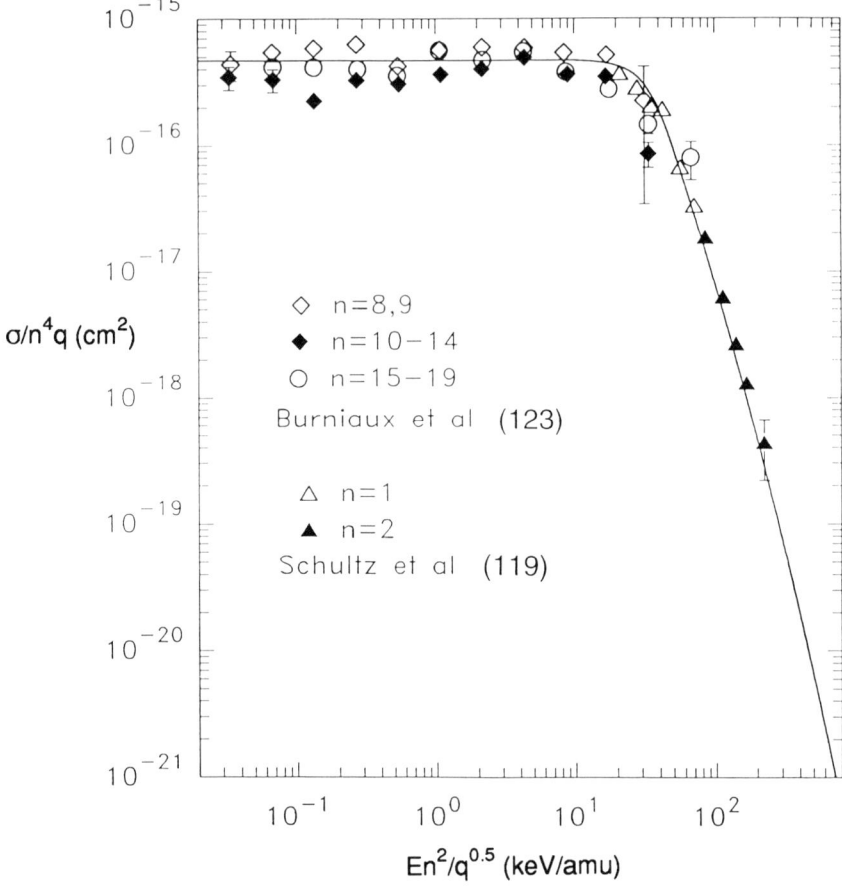

Figure 9. Scaled electron capture cross sections versus scaled collision energy for collisions of excited hydrogen atoms $H^*(n)$ with multiply charged ions ($q \geq 2$). The data[123] and theoretical results[119] are for He^{2+}. The curve is a fit of these results, as well as of the results of Ref. 124, and has been generated from Eq. (21).

3.2.1. One-Electron Capture

3.2.1a. Cross Sections for Low-Z Ions. Since the recent assessment[84] of the electron capture cross sections for the Be^{q+} + He and B^{q+} + He systems, new coupled-channel calculations have become available for Be^{4+} + He collisions,[126] B^{3+} + He collisions,[127] and B^{4+} + He collisions.[126] Experimental total electron capture cross sections for the Be^{q+} + He collision systems are still missing, and accurate theoretical calculations in the energy region above 0.25 keV/amu are available only for the Be^{4+} projectile.[126,128–130]

Electron Capture Processes

For the B^{q+} + He collision systems, low-energy cross section data exist in limited energy ranges for B^{2+} projectiles,[91] B^{3+} projectiles,[90,91,131,132] B^{4+} projectiles,[91,132] and B^{5+} projectiles.[128] We also mention the early MO calculations[133] for the B^{3+} + He system, which cover the range of 0.03–7 keV/amu. The calculated results and the data for all these systems are consistent with each other within the uncertainties of the data. We note that previous data[134] for B^{2+} + He in the energy range 0.6–4 keV/amu are about an order of magnitude larger than the more recent ones.[91]

A critical assessment of the one-electron capture data for the C^{q+} + He and O^{q+} + He ($q = 1-Z$) systems as of 1987 has been done in Ref. 23, with an update as of 1989 being given in Refs. 135 and 136. The most recent measurements of one-electron capture cross sections for these systems include the projectiles C^+ (Ref. 137; energy range 0.05–1 keV/amu), C^{2+} (Ref. 138), C^{3+} (Ref. 139), C^{4+} (Ref. 140), C^{5+} (Ref. 141), and C^{6+} (Ref. 142). The earlier extensive close-coupling calculations for the C^{q+} + He systems (such as those reported in Ref. 143 for C^{4+} + He and in Refs. 143–146 for C^{6+} + He) have recently been further expanded in Ref. 147 (for C^{4+} + He) and Ref. 148 (for C^{6+} + He). For the O^{q+} + He collision systems, recent low-energy measurements include those reported for O^+ in Refs. 137 and 149, for O^{2+}–O^{5+} in Ref. 150, and for O^{2+} in Ref. 151, whereas calculations were performed for O^+ + He collisions[152] and O^{8+} + He collisions[144,145,148] by the MO- or AO-expansion methods and for O^{8+} + He collisions[153] by the CTMC method. The new results are consistent with the earlier assessment[23] of one-electron capture in C^{q+} + He and O^{q+} + He systems, except for the C^{2+} + He case at very low energies. The updated (as of 1990) assessment of the cross sections for these reactions[102] has led to "recommended" cross sections which, in the reduced variables σ/q and $E/q^{0.5}$, can be represented, except for C^{2+} and C^{3+}, by the following expression:

$$\tilde{\sigma} = \frac{\sigma}{q} = a_1 \left[\frac{\exp(-a_2/\tilde{E})}{1 + a_3\tilde{E}^2 + a_4\tilde{E}^{4.5}} + a_5 \frac{\exp(-a_6\tilde{E})}{\tilde{E}^{a_7}} + a_8 \frac{\exp(-a_9/\tilde{E})}{1 + a_{10}\tilde{E}^{a_{11}}} \right] (10^{-16} \text{ cm}^2) \quad (22)$$

with

$$\tilde{E} = E(\text{keV/amu})/q^{0.5}$$

and for the ions C^{2+} and C^{3+} by

$$\tilde{\sigma} = \frac{\sigma}{q} = a_1 \left[\frac{\exp(-a_2/\tilde{E}^{a_7})}{1 + a_3\tilde{E}^2 + a_4\tilde{E}^{4.5} + a_5\tilde{E}^{a_6}} \right] (10^{-16} \text{ cm}^2) \quad (23)$$

The fitting coefficients a_i for C^{q+} and O^{q+} ions are given in Tables V and VI, respectively. The cross sections generated by Eqs. (22) and (23) are given in Figs. 10 and 11, in the energy range 0.01–100 keV/amu.

3.2.1b. Cross Sections for Medium- and High-Z Ions. For the high-Z ions, that is, ions of Si and metallic impurities, there are no electron capture data in the

Table V. Values of Parameters a_i in Eqs. (22) and (23) for the C^{q+} + He Collision System (q = 2–6)

	Eq. (23)		Eq. (22) (a_8 = 0)		
	$q = 2$	$q = 3$	$q = 4$	$q = 5$	$q = 6$
a_1	5.1813(+2)^a	1.2816(+1)	1.5541	4.8154(–1)	3.9888
a_2	2.5192	2.9946(–1)	2.5384	2.7704(–2)	5.7775
a_3	7.1435(–2)	1.9454(–3)	2.6213(–4)	8.9048(–5)	5.5332(–4)
a_4	1.2475(–7)	4.4980(–9)	7.0036(–10)	2.8440(–10)	2.1801(–9)
a_5	1.3939(+1)	7.6707(–1)	4.2845 (–1)	5.7335	5.4368(–1)
a_6	5.9128(–1)	6.4577(–1)	4.4078(–1)	3.4023(–2)	3.1149(–1)
a_7	1.6005(–1)	3.1491(–1)	–7.4781(–1)	–1.0934(–1)	–3.8811(–1)

$^a(\pm x) = \times 10^{\pm x}$.

low-energy region and for low charge states. Theoretical calculations include applications of the close-coupling methods to Si^{4+} (Refs. 117 and 154; energy range 10^{-4}–10^3 eV/amu) and Ti^{4+} (Ref. 117; 2–120 keV/amu) and application of the CTMC method to Fe^{q+} (q = 1–26)[119] at energies above 50 keV/amu. The cross sections of the Si^{4+} + He and Ti^{4+} + He electron capture reactions in the energy range 1–100 keV/amu are given in Fig. 12. It is apparent from the figure that for E < 25 keV/amu, the energy dependence of the cross sections is very weak, as is their sensitivity to the ionic species. As in the case of the H target (see Section 3.1), the cross sections are less sensitive to the projectile core or to energy variations with increasing ionic charge state. Indeed, the cross section measurements for Kr^{q+} + He (q = 7–25)[155] at E = $1q$ keV/amu, I^{q+} + He (q = 10–41)[156] at E = $10q$ eV/amu, and Xe^{q+} + He (q = 11–31)[157] at E = $30q$ eV/amu have confirmed this behavior and shown a linear dependence of the cross sections on the ionic charge at low collision energies. This behavior of A^{q+} + He electron capture was also observed[121] earlier for other projectiles in charge state $q \geq 6$, and so was the scaling cross section relationship $\sigma/q = f(E/q^{0.5})$. To within a 70% r.m.s. deviation, all the available experimental one-electron capture cross sections for these reactions can be represented in the form[24] ($q \geq 5$)

$$\tilde{\sigma} = \frac{\sigma}{q} = \frac{A \ln(B/\tilde{E})}{1 + C\tilde{E}^2 + D\tilde{E}^{4.5}} \, (10^{-16} \, \text{cm}^2) \quad (24)$$

with

$$\tilde{E} = E(\text{keV/amu})/q^{0.5}$$

$A = 0.1818$, $B = 1.856 \times 10^6$, $C = 2.753 \times 10^{-4}$, $D = 1.370 \times 10^{-9}$

With the same accuracy, the reduced cross section $\tilde{\sigma}$ in the reduced energy region $\tilde{E}$ < 1 keV/amu can be represented by the value 4×10^{-16} cm^2.

Table VI. Values of Parameters a_i in Eq. (22) for the O^{q+} + He Collision Systems (q = 1-8)

	$q=1$	$q=2$	$q=3$	$q=4$	$q=5$	$q=6$	$q=7$	$q=8$
a_1	7.2476(−1)	9.3909(−1)	5.2275(−1)	1.7302	1.3051	8.4779(−1)	8.8410(−1)	5.5866
a_2	4.4399(−1)	1.3724(−1)	1.1001(−3)	1.8243(−1)	2.1166(−2)	1.2570(−2)	1.4844(−2)	9.8470
a_3	1.4258(−4)	1.8449(−4)	2.0862(−4)	3.9405(−4)	1.7584(−4)	3.6525(−4)	4.4390(−4)	7.3621(−4)
a_4	3.1028(−11)	7.3794(−10)	2.4724(−10)	7.0075(−10)	1.1084(−9)	1.0325(−9)	8.6020(−10)	3.1632(−9)
a_5	1.3995	3.8463(+1)	2.1701(−2)	1.0801(−2)	1.8617	1.3497	1.5200	9.3942(−1)
a_6	6.8817(−2)	7.9608	1.1689(+2)	1.6463(−2)	1.7999(−2)	2.1201(−2)	2.3914(−2)	2.6556(−1)
a_7	−2.3280(−1)	−4.2419(−1)	6.1949(−1)	4.5898(−1)	2.0482(−1)	−1.5049(−1)	−2.1553(−1)	−2.6790(−1)
a_8	0.00	2.9398	1.8915	2.5637(−1)	0.00	0.00	0.00	0.00
a_9	0.00	1.3755(−1)	8.1097(−1)	6.5667(−3)	0.00	0.00	0.00	0.00
a_{10}	0.00	1.6106(−3)	8.1847(−8)	1.2642(−7)	0.00	0.00	0.00	0.00
a_{11}	0.00	2.9069	4.1416	6.6189	0.00	0.00	0.00	0.00

$^a(\pm x) = \times 10^{\pm x}$.

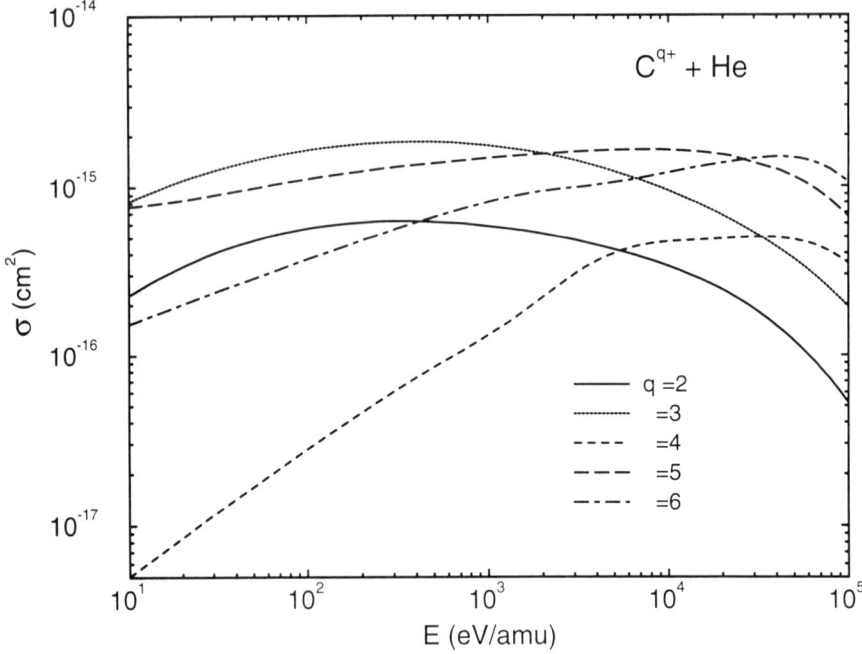

Figure 10. Total one-electron capture cross sections for C^{q+} + He collisions ($q = 2$–6) generated from Eq. (23) for $q = 2$ and 3 and from Eq. (22) for $q = 4$–6 with the coefficients in Table V.

3.2.2. Two-Electron Processes

The two-electron processes in the A^{q+} + He collision system include double-electron capture (Eq. 2), transfer–ionization (Eq. 4), and transfer–excitation (Eq. 5). The probabilities of these processes at low collision energies, and their relative magnitudes, strongly depend on the charge state of the ion and, for the low-q ions (e.g., $q < 6$), also on the ionic core. For ions in high-q states, the low-energy cross sections of two-electron processes are generally smaller than the cross section for one-electron capture. For low-q ions, this is not always the case. For instance, the two-electron capture cross section in C^{4+} + He collision in the energy region below 2 keV/amu is considerably larger than the one-electron capture cross section.[90,143] This is an example of the surprises that may occur in slow collisions when the energy separation between MOs becomes the governing issue for transitions. Another case is the observation that transfer–excitation cross sections, in slow collisions, may well be much larger than the cross section for single excitation of the He target (see Section 3.2.1b).

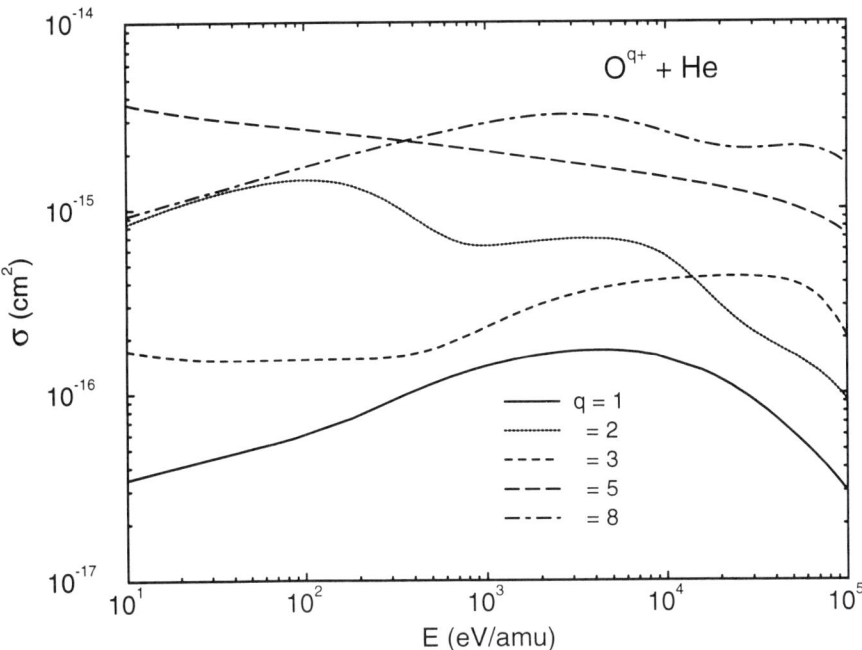

Figure 11. Total one-electron capture cross sections for O^{q+} + He collisions (q = 1–3, 5, 8) generated from Eq. (22) with the coefficients in Table VI.

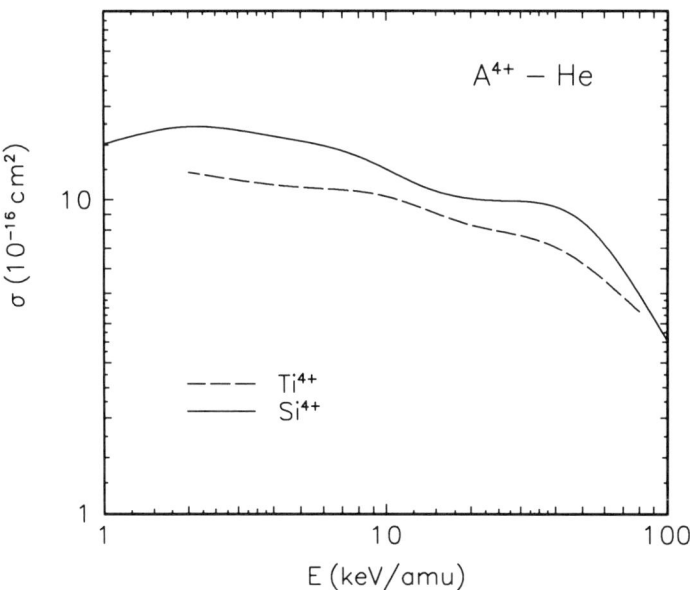

Figure 12. Calculated total transfer cross sections in Si^{4+} + He and in Ti^{4+} + He collisions.[117]

For all two-electron processes at low collision energies, two-electron capture states play an important role in the collision dynamics when q is large. These states lie in the continuum of the $A^{(q-1)+}$ ion, and Auger stabilization results, experimentally but usually not in the close-coupling calculations, in transfer–ionization as the final channel. The nonradiative decay of these states is the dominant decay mode up to very high values of q ($q \approx 30$ for a He target). At small interatomic separations, the doubly excited states $A^{(q-2)+**}$ may also contribute to the population of other channels, such as the transfer–excitation channels. If the direct population of transfer–excitation channels is weak, one would expect the cross section for transfer–excitation to be smaller than that for two-electron capture. For low-charge ions colliding with He, on the other hand, the collision dynamics strongly depends on the coupling of individual MOs, and hence general statements do not apply.

We will now consider the two-electron transitions and the status of corresponding cross section data bases for the plasma edge impurity ions in more detail.

3.2.2a. Two-Electron Capture and Transfer–Ionization. With a total binding energy of 79.0 eV, the favorable transfer of the two He electrons to an ion with atomic number $Z \geq 4$ requires ionic charge states $q \geq 3$. For the $q = 3$ ions of low-Z impurities (Be, B, C, O), reaction exothermicities are such that two-electron capture occurs still with only small cross sections. For the $q = 4$ low-Z ions, only the C^{4+} ion favors two-electron transfer to the projectile ground state in C^{4+} + He collisions. For the other systems (e.g., Be^{4+} and B^{4+}), ground and singly excited final states (e.g., $1s^2$ for Be^{2+} and $1s^2 2s$ for B^{2+}) are energetically and dynamically less favorable for two-electron transfer than the doubly excited states (such as $2s^2$, $2s2p$, and $2p^2$ in the case of Be^{2+}). For ions in still higher charge states, the two captured electrons normally populate doubly excited (nl, $n'l'$) states, with both n and n' increasing with increase of q. In the population of these states, intermediate one-electron capture states play an increasingly important role with increasing q, which leads to the classical concept of two-electron transfer as two sequential one-electron over-barrier transitions.

The one- and two-electron capture cross sections in C^{4+} + He collisions are shown in Fig. 13. The two-electron capture cross section in the low-energy region is based on both data and calculations.[90,143,147] Extensive MO or AO coupled-channel calculations have been performed for the two-electron capture cross sections in Be^{4+} + He,[126,129] B^{4+} + He,[126] C^{6+} + He,[158,159] N^{7+} + He[160,161] and O^{8+} + He[159,160] collisions. The doubly excited states created by two-electron capture in these systems decay by autoionization and have been the subject of numerous experimental investigations, involving mainly B^{5+}, C^{6+}, N^7, and O^{8+} ions, by either translational energy spectroscopy or Auger electron spectroscopy. Information on the cross section for transfer–ionization can be derived from these measurements, but data are available only for a very limited number of energies, and their uncertainties are usually high. A more consistent set of data on one- and two-electron capture processes was obtained for the Xe^{q+} + He collision system ($q =$

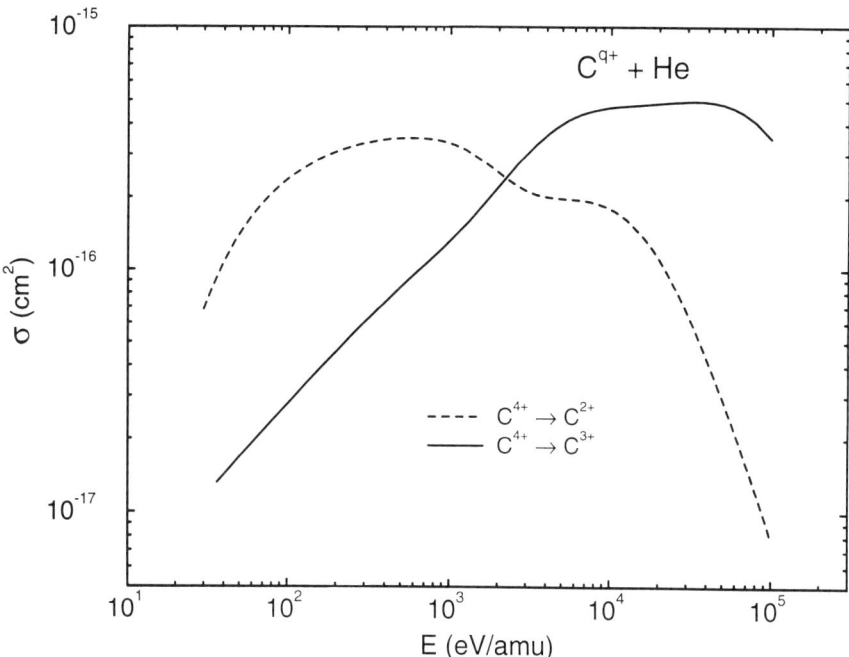

Figure 13. Total cross sections for one- and two-electron capture in C^{4+} + He collisions. The two-electron capture cross section is based on a critical assessment of the results of Refs. 90, 143, and 147, and the one-electron capture cross section is the same as in Fig. 10.

$11-31)^{157}$ at $E = 30q$ eV/amu, including differentiation between the very weak double-charge-exchange channels and the strong transfer–ionization channel. The transfer–ionization cross section from these low-energy measurements shows a linear dependence on the value of the ionic charge q, which is also confirmed by calculations for the low-q ions.

Cross section calculations for two-electron capture and transfer–ionization processes in the intermediate energy region were performed within the CTMC method[162] for C^{6+}, O^{8+}, Ne^{10+}, and Si^{14+} projectiles and within the AO close-coupling method[159] for fully stripped ions with $Z = 2-9$. The results of these calculations show that the cross sections for both double capture and transfer–ionization obey a $\sigma/q = f(E/q^{0.5})$ scaling, which is consistent with a classical picture for these processes.[163] The sum of the cross sections for two-electron capture, σ_{2c}, and transfer–ionization, σ_{ti}, can be represented over a wide energy range in the scaled form[163]

$$\tilde{\sigma}_{2c+ti} = \frac{\sigma_{2c} + \sigma_{ti}}{q} = \frac{AB \times 10^{-16} \text{ cm}^2}{B + C\tilde{E}^\alpha + A\tilde{E}^4} \quad (25)$$

with

$$\tilde{E} = E(\text{keV/amu})/100q^{0.5}$$

where A, B, C, and α are fitting parameters determined from data[157] and CTMC[162] results:

$$A = 1.37, B = 0.64, C = 1.77, \alpha = 0.75$$

The expression in Eq. (25) is valid for $q \geq 6$ and represents the data with an r.m.s. deviation of about 10%. For $\tilde{E} \gtrsim 50$, σ_{ti}/q becomes increasingly larger than σ_{2c}/q, owing to the faster decrease ($\sim E^{-4.5}$) of the latter with $\tilde{E}$.

3.2.2b. Transfer–Excitation. Transfer–excitation processes have been very little investigated, particularly for the plasma impurity ions and the energy region considered here. At low and intermediate collision energies, a transfer–excitation study has recently been performed within the AO close-coupling description for the He^{2+} + He system[164] and the C^{6+} + He system.[165] There exists another study of low-energy transfer–excitation for the Xe^{q+} systems within the classical over-barrier model.[166] The transfer–excitation cross sections σ_{TE} from these calculations can be shown to lie on a common curve:

$$\frac{\sigma_c + \sigma_{TE}}{\sigma_{TE}} = \frac{40}{E^{0.3}q^{0.4}} \quad (26)$$

where the energy E is in units of keV/amu, and σ_c denotes the total transfer cross section. The same phenomenological scaling relation has been found before for transfer–ionization processes,[167] for which it links a large set of data.

At the time of this writing, the scaling relation given in Eq. (26) is not yet well established. Notably, at energies higher than some 10 keV/amu there are no results available. There the relation is based solely on an analogy to transfer–ionization. On the other hand, measurements are difficult and so are calculations of transfer–excitation processes. Hence, any further progress would be expected to occur only at a slow pace.

3.3. Collisions with Molecular Hydrogen

3.3.1. General Considerations

The internal degrees of freedom, that is, the rotation and vibration, of a molecule make the dynamics of ion–molecule electron capture collisions more complex than in the case of ion–atom collisions. For collision energies considerably larger than the rotational energy of H_2, the molecular rotation does not have any noticeable effect on the electron transfer dynamics. The molecular vibrations, however, influence the collision dynamics by modifying the strength of nonadiabatic couplings of electronic states through the Franck–Condon vibrational overlap factors of the initial and final states and by introducing shifts in the diagonal elements of the interaction matrix. These effects are particularly important for slow collisions, when the collision time is comparable to or larger than the characteristic time of vibrational transitions. At higher energies, the transitions among all the vibrational states are strong, and the effects of vibrational motion on the nonadiabatic couplings disappear because of the sum rule for the Franck–Condon factors.

The presence of dissociative, excited electronic states in the molecules and molecular ions also introduces new elements in the collision dynamics. Two-electron transitions may populate such states and lead to dissociation of the molecular reaction products. Two-electron capture processes may also significantly contribute to transfer–ionization at low collision energies. Because of these structural and dynamical features of ion–molecule collisions, theoretical and experimental studies of electron capture processes are more demanding than in the case of ion–atom collisions, at least when information on the final state of the molecular ion is required. The most important electron capture processes in the $A^{q+} + H_2$ system are one-electron capture,

$$A^{q+} + H_2(v) \rightarrow A^{(q-1)+} + H_2^+(v') \tag{27}$$

one-electron dissociative capture,

$$A^{q+} + H_2(v) \rightarrow A^{(q-1)+} + H_2^{+*} \rightarrow A^{(q-1)+} + H^+ + H(n) \tag{28}$$

two-electron capture,

$$A^{q+} + H_2(v) \rightarrow A^{(q-2)+} + 2H^+ \tag{29}$$

and transfer–ionization,

$$A^{q+} + H_2 \rightarrow A^{(q-1)+} + 2H^+ + e^- \tag{30}$$

where v and v' denote vibrational quantum numbers.

The process of one-electron dissociative electron capture (Eq. 28) is essentially a two-electron transfer–excitation process which for low q and low energies may contribute considerably to the inclusive one-electron capture cross section. This process is very sensitive to the initial vibrational state of H_2. The processes of two-electron capture (Eq. 29) and transfer–ionization (Eq. 30) are analogous to the corresponding two-electron processes in $A^{q+} + He$ collisions. The experimental cross sections for the above reactions usually refer to H_2 in its ground vibrational state. The molecular ion produced in the one-electron capture reaction (Eq. 27) is then overwhelmingly in the $v' = 2$–4 states (resulting from vertical Franck–Condon transitions).

We will now discuss in more detail the data base of total cross sections for the processes represented by Eqs. (27)–(30).

3.3.2. One-Electron Capture

3.3.2a. Collisions with Low-Z Ions. The cross section data for total one-electron capture in collisions of Be^{q+} and B^{q+} ions with H_2 have been critically assessed in Ref. 84, while for the C^{q+} and O^{q+} ions such an assessment was done in Ref. 23. The low-energy one-electron capture cross section data for $Be^{q+} + H_2$ are fairly sparse and include an early cross section measurement[134] for Be^+, the more recent measurements[168] for Be^{2+}–Be^{4+} over a very limited energy range below 1 keV/amu, and the extensive AO–MO close-coupling calculations[73] for Be^{2+}–Be^{4+} in the energy range 0.1–10 keV/amu. The theoretical cross sections for these systems agree well (within 10%) with the data and are given in Fig. 14.

The data for $B^{q+} + H_2$ electron capture reactions are more abundant and include all charge states: B^+ (Ref. 68; energy range 10–150 keV/amu), B^{2+} (Refs. 68, 88, 90, and 91; 0.3–200 keV/amu), B^{3+} and B^{4+} (Refs. 68, 90, and 91; 1–200 keV/amu), and B^{5+} (Refs. 68 and 90; 5–200 keV/amu). The data are consistent among themselves, except for those of Ref. 91 for B^{2+}, which appear to be too low by 50%. The recent AO–MO close-coupling calculations[73] for $B^{3+} + H_2$ and $B^{5+} + H_2$ in the

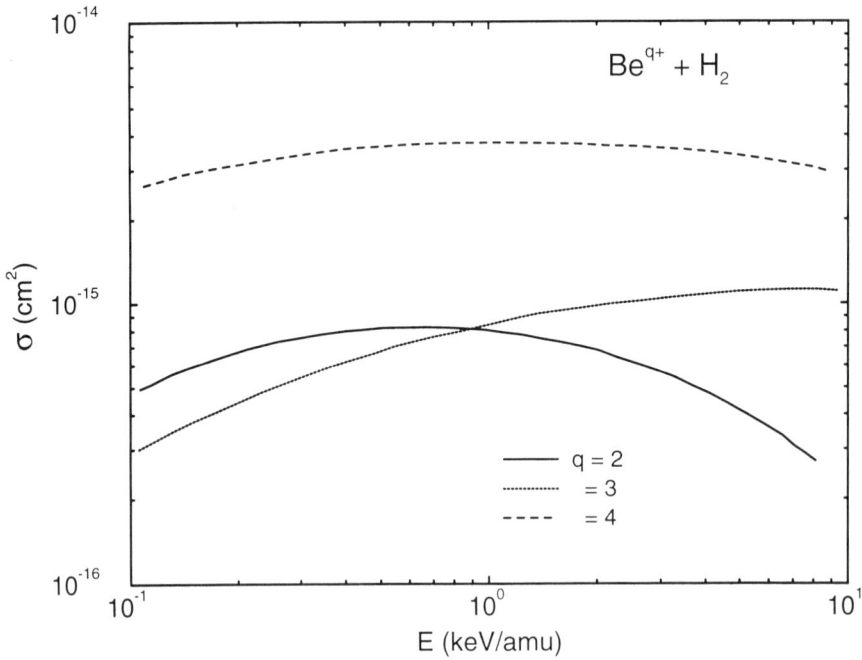

Figure 14. Total one-electron capture cross sections for $Be^{q+} + H_2$ collisions (q = 2–4) from the AO–MO close-coupling calculations. (Courtesy of Dr. M. Kimura.[73])

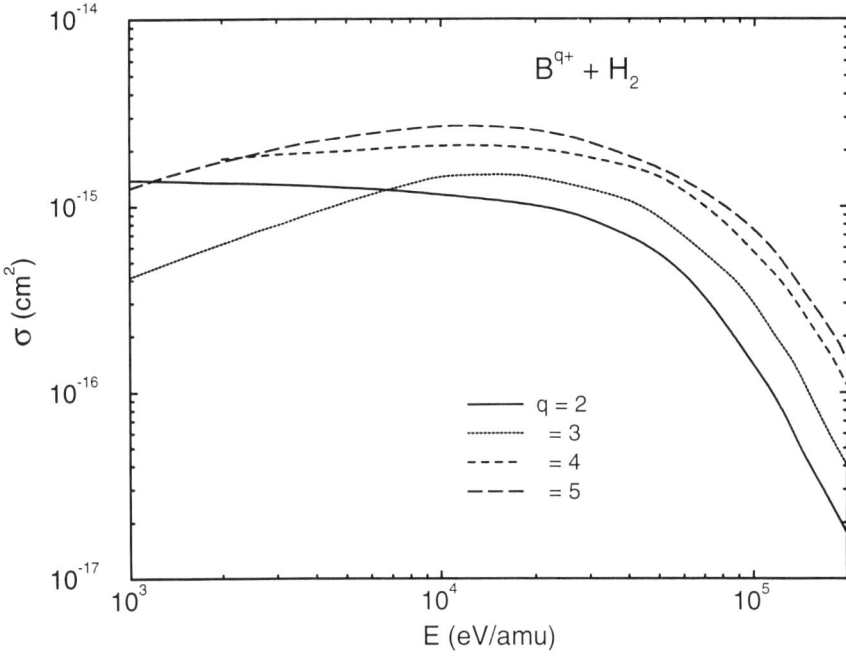

Figure 15. Total one-electron capture cross sections for $B^{q+} + H_2$ collisions ($q = 2$–5) based on a critical assessment of existing experimental and theoretical results (see text).

energy range 0.1–10 keV/amu agree well with the data[90] in the overlapping energy range. The evaluated electron transfer cross sections for the $B^{q+} + H_2$ collisions ($q = 2$–5), based on all these sources, are given in Fig. 15. This figure indicates that electron capture in the $B^{2+} + H_2$ system has a quasi-resonant character in the energy range considered. Further studies of this reaction below 0.3 keV/amu are very desirable.

The critical assessment of one-electron capture cross sections for $C^{q+} + H_2$ and $O^{q+} + H_2$ collisions in Ref. 23 still provides a comprehensive source of evaluated data, including the low-energy region. More recent total single-capture cross section measurements for these systems include the projectiles $C^+(^2P, ^4P)$ (Ref. 169; energy range 0.8–40 eV/amu), C^{3+} (Ref. 139; 125–333 keV/amu), C^{4+} (Ref. 170; 0.3–666 eV/amu), $O^+(^4S)$ (Ref. 171; 6–60 eV/amu), O^+ ($^4S, ^2D, ^2P$) (Ref. 172; 1–30 eV/amu), O^{5+} (Ref. 139; 125–333 keV/amu), and $O^{8+} + D_2$ (Ref. 173; $E > 0.5$ keV/amu). Because of difficulties in the description of the two-electron motion in H_2, close-coupling calculations for these systems have been limited to the $C^{4+} + H_2$ (AO-expansion[174]) and $C^{6+} + H_2$ (MO-expansion[175]) systems. CTMC cross section calculations have been performed[153] for $C^{6+} + H_2$ and $O^{8+} + H_2$ at $E = 50$ keV/amu and $E = 100$ keV/amu. There has been no recent reassessment of the total electron capture cross sections for the $C^{q+} + H_2$ and $O^{q+} + H_2$ systems with inclusion of the new information. The "recommended" cross sections for $C^+ + H_2$ and $O^+ + H_2$ in Ref. 23, which at very low energies include a contribution from dissociative electron capture, should now be appropriately modified to account for the new data of Ref. 169 and Refs. 171 and 172, respectively.

3.3.2b. Collisions with Medium- and High-Z Ions. There are only very few experimental studies of one-electron capture processes involving H_2 and the particular medium- and high-Z ions of interest for fusion purposes, particularly in the low-energy region. Only the $Fe^{q+} + H_2$ collision system has been investigated to a greater extent, with data available for the $q = 3$–15 ions in the 0.01–160 keV/amu energy range[108,109,111] and for the $q = 3$–25 ions at four energies within the range of 100–3400 keV/amu.[176] Fragmentary data, at one or a few collision energies, exist also for the $Si^{q+} + H_2$ ($q = 2$–9), $Mo^{q+} + H_2$ ($q = 4$–18), and $W^{q+} + H_2$ ($q = 4$–15) systems in the energy range between 50 and 160 keV/amu. Calculations for these systems include the AO coupled-channel calculations[117] for $Ti^{4+} + H_2$ and $Fe^{8+} + H_2$ in the energy range 1–100 keV/amu and the CTMC calculations[119] for $Fe^{q+} + H_2$ ($q = 1$–26) in the energy range 50–500 keV/amu. As in the case of $A^{q+} + H$ and $A^{q+} + He$ systems with $q \geq 5$, the high density of ionic states available for single-electron capture makes the low- and intermediate-energy cross section insensitive to the ionic core and species. This feature of the cross sections persists also at high energies because of the important role of momentum transfer in the collision dynamics. Indeed, all the experimental cross section data for $Fe^{q+} + H_2$ ($q = 4$–25), when plotted in the reduced form $\sigma/q = f(E/q^{0.5})$, are found to lie on a universal curve (within an r.m.s. deviation of 15%) which can be represented by the expression[24]

$$\tilde{\sigma} = \frac{\sigma}{q} + \frac{A \ln(B/\tilde{E})}{1 + C\tilde{E}^2 + D\tilde{E}^{4.5}} (10^{-16} \text{cm}^2) \qquad (31)$$

with

$$\tilde{E} = E(\text{keV/amu})/q^{0.5}$$

$$A = 0.5707, B = 5.283 \times 10^4, C = 7.800 \times 10^{-4}, D = 2.721 \times 10^{-8}$$

The data on $Si^{q+} + H_2$, $Mo^{q+} + H_2$, and $W^{q+} + H_2$ mentioned above, as well as the recent theoretical results on $Ti^{4+} + H_2$,[117] $Fe^{8+} + H_2$,[117] and $Fe^{q+} + H_2$ ($q = 4$–26),[119] are all represented by Eq. (31) within their estimated uncertainties (15–20%).

3.3.2c. Two-Electron Processes. Two-electron transition processes in $A^{q+} + H_2$ collisions have been studied much less than single capture. The coincident detection of three or more reaction products and the description of the coupling with the continuum, for the nuclear and/or electronic motion, make experimental and theoretical investigations of these processes extremely difficult. In a very limited energy range below 1 keV/amu, such measurements have been done for the $B^{3+} + H_2$ and $B^{4+} + H_2$ collision systems.[91] They show large double-capture cross sections for the $B^{3+} + H_2$ system, which are a factor of 2 smaller than the single-capture cross section. For the $B^{4+} + H_2$ system, however, double-capture cross sections are negligible. The cross section for two-electron capture in $C^{4+} + H_2$ collisions has also been measured recently[170] in the energy range 0.3–666 eV/amu. For projectiles in higher charge states, two-electron capture usually populates doubly excited states of the product ion, resulting in transfer–ionization. An experimental study of the autoionizing two-electron capture in $O^{6+} + H_2$ collisions has recently been done[177] at the collision energy of 3.8 keV/amu, and a similar theoretical study has been performed for the $O^{8+} + H_2$ system.[178] At still higher values of the projectile charge states (e.g., $q > 10$), the autoionizing two-electron capture (transfer ionization) can be described by the classical model as two successive one-electron over-barrier transitions in a similar way as for the $A^{q+} + He$ systems.[83] Since the second step of this process determines its overall probability, and the capture "radius" for the second traversing electron is $R_2 = cq^{1/2}/I_2(H_2)$, where c is a constant of the order of 1 and $I_2(H_2)$ is the second ionization potential of H_2, the cross section for two-electron removal, that is, two-electron capture and transfer–ionization, at low energies may be estimated as $\pi R_2^2 = \pi c^2 q/I_2^2(H_2)$.

At intermediate and high energies, that is, above some 30 keV/amu, transfer–ionization (Eq. 30) is the dominant two-electron process. Within a classical picture,[163] this high-energy process also takes place in two steps (ionization first and then capture of the second electron), with the second step determining the overall probability. Using the classical scaling relation for the electron capture process,[83]

$$\frac{I_2^2}{q} \sigma = f(E/I_2 q^{0.5}) \tag{32}$$

one can relate the total cross section for two-electron capture and transfer–ionization for the H_2 target to the corresponding cross section for He:

$$\frac{\sigma_{2c+ti}^{H_2}(\tilde{E}_1)}{q} = \left[\frac{I_2(He)}{I_2(H_2)}\right]^2 \frac{\sigma_{2c+ti}^{He}(\tilde{E}_2)}{q} \tag{33}$$

with

$$\tilde{E}_1 = \frac{E}{I_2(H_2)q^{0.5}} \quad \text{and} \quad \tilde{E}_2 = \frac{E}{I_2(He)q^{0.5}}$$

where $\sigma_{2c+ti}^{He}(E/q^{0.5})/q$ is given by Eq. (25), and $I_2(X)$ is the second ionization potential of target X.

Dissociative electron capture (Eq. 28) is a two-electron process in which a dissociative $np\sigma_u$ ($n = 2, 3, \ldots$) state is excited simultaneously with the capture of one of the electrons. This process takes place for impact parameters smaller than $q^{0.5}/I_1(H_2)$ (the single-electron "capture radius") and is essentially determined by the dipole $1s\sigma_g \to np\sigma_u$ transition in the H_2^+ ion. The strong $1s\sigma_g \to 2p\sigma_u$ transition leads to dissociation to $H^+ + H(1s)$ products.

Low-energy dissociative electron capture has been investigated so far only for $He^{2+} + H_2$ collisions[179] at 1–25 keV/amu. Below some 5 keV/amu, the cross section for this process appears to become larger than that for single-electron capture. In earlier electron capture cross section measurements for $C^+ + H_2$ and $O^+ + H_2$ collisions,[180] a secondary peak in the total single-capture cross section was observed in the energy region around 20 eV/amu. This peak was attributed to dissociative electron capture, but no detailed analysis of the reaction products was made. Recent experimental studies of $O^+ + H_2$ one-electron capture collisions[171,172] have revealed that the single-electron capture cross section for $O^+(^4S) + H_2$ does *not* show any structure in the energy range 1–30 eV/amu, whereas the electron capture cross sections for the excited metastable ions $O^+(^2D, ^2P)$ colliding with H_2 are, in this energy range, about an order of magnitude larger than that for the ground state $O^+(^4S)$ ion. The secondary, low-energy peak in the O+ + H_2 cross section measurements of Ref. 180 is, therefore, related to the presence of a metastable fraction in the ion beam.

4. STATE-SELECTIVE ELECTRON CAPTURE

4.1. General Considerations: *n*-Distributions and *l*-Distributions

The population of individual final states in slow electron capture collisions depends on the complex interplay between transitions from the initial state to these final states *and* on transitions within the set of final states. Hence, in a theoretical determination of state-selective electron capture, large basis sets need to be employed in any close-coupling scheme. For large charge states q of the projectile A^{q+}, typically $q \geq 12$, or for high n quantum numbers of the final states, typically $n \geq 12$, acceptable basis sizes (at most a few hundred states) cannot be met. For systems which can be studied with current implementations of the close-coupling method, such studies must be performed for each system separately since small changes in the potential translate into possibly large variations in the partial transfer cross sections.

For experimental studies of single-electron capture into fully stripped ions, because of the energy degeneracy of H-like final states, both translational energy and photon-emission spectroscopy (cf. Sections 2.1.4a and 2.1.4b) usually can deliver information on n-state population only. However, in some investigations using photon-emission spectroscopy, the difference in lifetimes of energy-degenerate l states within the same principal quantum number n could be exploited for measuring l-state populations as well. This remains feasible, however, only as long as the lifetimes are sufficiently large, that is, for H⁺ and He²⁺ projectiles.[181] In investigations of single-electron capture from incompletely stripped ions, the state resolution of the particular experimental technique generally determines to what extent, for a given collision system, the nl-state population can be studied.

Both theoretical and experimental investigations of one-electron capture show that state selectivity of the capture process at low energies is governed by the condition of an energy quasi-resonance of initial and final states. In the case of fully stripped ions, the population of l substates within a given n shell is determined by rotational mixing of states. For electron capture to fully stripped ions in charge states $q \geq 5$, energy-resonance considerations lead to the following expression for the principal quantum number n_m of the dominantly populated level[4]:

$$n_m \approx n_0 2^{1/4} q^{3/4} \qquad (34)$$

where $n_0 = \sqrt{(-2E_0)}$ is the effective principal quantum number of the initial state (binding energy E_0). The independence of n_m on collision velocity holds up to $v = q^{1/4}$. At very low collision velocities ($v \ll q^{1/4}$), only one or two levels are dominantly populated, and the n-distribution is sharply peaked. With increasing v, the number of dominantly populated levels slowly increases, and the distribution $P_n = \sigma_n/\sigma$ (σ_n is the cross section for capture into an n shell, and σ is the total electron capture cross section) acquires an asymmetric Gaussian form.

The l-distribution ($P_{nl} = \sigma_{nl}/\sigma_n$) of the captured electrons in a given n shell does not show systematic features and depends significantly on both v and q. At very low collision velocities, when rotational transitions in the united-atom limit do not play any considerable role and the initial symmetry is preserved throughout the collision, the l populations of captured electrons are determined by the projection of the final molecular state onto the atomic-orbital momentum states. For an initial σ state, this gives[4]

$$P_{nl} = \frac{(2l+1)(n-1)!}{(n+l)!(n-l-1)!} \qquad (35)$$

With an increase of collision velocity, the rotation of the internuclear axis induces transitions between the Stark states at large distances, and the strong mixing of these states leads to a statistical population of l substates,

$$P_{nl} = (2l+1)/n^2 \qquad (36)$$

The intermediate Stark mixing, coupled at higher velocities with rotational transitions in the united-atom limit, may significantly modify the statistical l-distribution.

For the case of capture to incompletely stripped ions, the n, l selectivity is governed only by the energy-resonance condition.

4.2. Collisions with Hydrogen Atoms

4.2.1. Collisions with Fully Stripped Ions

There is so far only a very small amount of data on the systems of interest. There have been measurements, based on photon-emission spectroscopy, for He^{2+} projectiles[181] in the impact energy range of 2–13 keV/amu, from which absolute emission cross sections could be derived and compared with information synthesized from AO calculations.[182] In this way, reasonably good agreement between theory and experiment has been demonstrated for single-electron capture into He^+ $2p$, $n = 3$, and $n = 4$ states. There are so far no other data for state-selective single-electron capture in these systems.

Theoretical studies of total electron transfer usually start from a consideration of individual physical channels (see the discussion in Section 2.2.1). Hence, all theoretical work cited in Section 3.1.1 on total transfer cross sections in $Be^{4+}, \ldots, O^{8+}$ + H collisions contains some information on partial transfer cross sections as well, typically for the l-distribution for the particular "dominantly populated" n shell and sometimes also for the next higher n shell. The most advanced studies of n- and l-distributions are the following: the AO study[77] ($n = 3$–8, $E = 4$–100 keV/amu) and the superpromotion study[81] ($n = 1$–6, $E = 0.2$–100 keV/amu) for Be^{4+}; the AO study[76] ($n = 3$–5, $E = 0.1$–20 keV/amu) and the superpromotion study[81] ($n = 1$–8, $E = 0.2$–30 keV/amu) for B^{5+}; and the AO study[76] ($n = 4$–6, $E = 0.1$–25 keV/amu) and the AO study[78] ($n = 7$ and 8, $E = 4$–25 keV/amu) for N^{7+} projectiles. For C^{6+} and O^{8+} projectiles, existing theoretical predictions for partial transfer cross sections have been critically assessed[26] for, respectively, $n = 3$–8 and $n = 4$–9 shells, at energies $E = 0.2$–1000 keV/amu for the respective dominant n shell, above $E = 5$ keV/amu for the next shell, and above $E = 20$ keV/amu for the higher-n shells. For each of the transitions, an analytic expression for the "recommended cross section" has been given, which fits well the predictions of various calculations.

Figure 16 shows the calculated transfer cross sections into projectile $n = 3$–7 states in Be^{4+} + H collisions, from an AO calculation[77] and a calculation within the superpromotion model.[81] A combination of the results at low energies from the superpromotion model and the results at higher energies from the AO study should be taken as a realistic prediction of partial transfer cross sections. At the time of this writing, it is not clear whether the structures in the AO results between 5 and 15 keV/amu are an artifact of the calculations, or whether they are real and reflect some subtle interaction between these transfer channels and the $n = 2$ excitation channel.

4.2.2. Collisions with Incompletely Stripped Ions

So far, experimental investigations of state-selective single-electron capture have been conducted for impact of C^{2+},[183] C^{3+},[40,46] C^{4+},[94] N^{5+},[184] O^{2+},[185] O^{3+},[186] O^{6+},[184] Ne^{3+},[186] Ar^{4+}, Ar^{5+}, and Ar^{6+},[187] Fe^{3+},[43] and Fe^{4+} ions.[43] Some of the studies

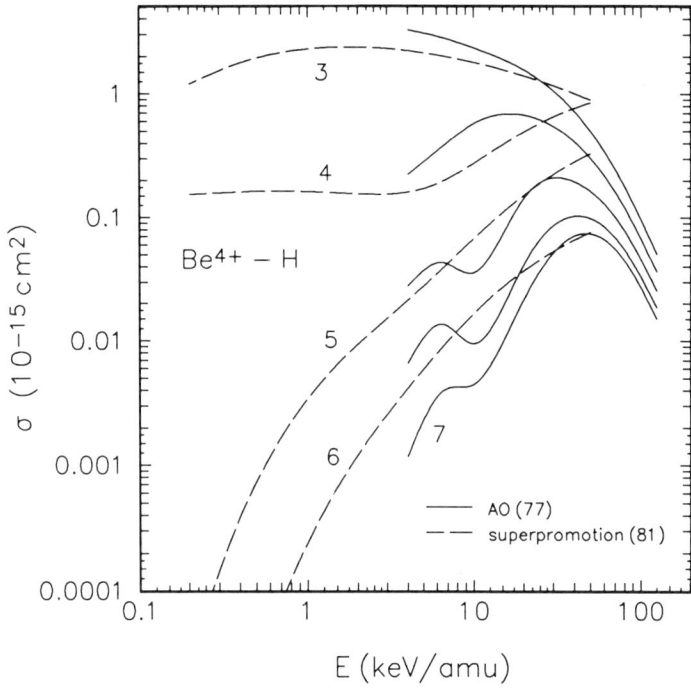

Figure 16. Calculated electron transfer cross sections into $n = 3$–7 shells of Be^{3+} in $Be^{4+} + H$ collisions, from an AO calculation[77] (—) and from a calculation within the superpromotion model[81] (- - -). The values of n are labeled on the curves.

that are based on translational energy spectroscopy have recently been reviewed in Ref. 11. Absolute cross sections for single-electron capture have been derived for some of these systems, namely, C^{3+},[40,46] C^{4+},[94] O^{3+},[186] and Ne^{3+},[186] for a varying number of the final states.

Most of the theoretical work cited in Section 3.1.2 on total transfer also provides information on state-selective capture for, typically, the l states in one or two n shells of the projectile. The MO study[188] for Be^{2+} ions predicts partial transfer cross sections into states in the $n = 2$ shell at 0.1–8 keV/amu energies. Similarly, there is state-selective information on Be^{3+} projectiles[87] ($n = 2, 3$), on B^{3+} projectiles[188] ($n = 2$, 0.1–8 keV/amu), on B^{4+} projectiles[85] ($2p$, $n = 3$, $4s$, $4f$, 10^{-4}–10 keV/amu), on C^{3+} projectiles[95,96] (Ref. 95 for $3s$ and $3p$ states, 0.01–5 keV/amu; Ref. 96 for $n = 2, 3$, and $n = 3$ states, 0.04–9 keV/amu), on C^{4+} projectiles[97,101,189] (Ref. 97 for $n = 3$, 0.001–1.2 keV/amu; Ref. 101 for $n = 3$, 0.25–16 keV/amu; Ref. 189 for $n = 3$ and $n = 4$, 0.1–20 keV/amu), on C^{5+} projectiles[98] ($n = 3$ and $n = 4$, 0.01–3 keV/amu), on N^{4+} projectiles[104,105] (Ref. 104 for $n = 3$, 0.001–10 keV/amu; Ref. 105 for $n = 3$, 10^{-5}–0.6 keV/amu), on N^{5+} projectiles[101,106] (Ref. 101 for $n =$

4, 1–7 keV/amu; Ref. 106 for 0.6–10 keV/amu), on O^{4+} projectiles[99] ($n = 3$, 0.001–1 keV/amu), on O^{5+} projectiles[100] (4s, 4p, 4d, 0.001–1 keV/amu), and on O^{6+} projectiles[101] ($n = 4$, 0.25–16 keV/amu).

There is substantially less information on state-selective transfer for medium- and high-Z ions, all of it coming from theory. Close-coupling calculations have predicted cross sections for Si^{4+} projectiles[99,117] (Ref. 99 for 3s, 4d states, 0.001–1 keV/amu; Ref. 117 for $n = 3$–5, 0.2–100 keV/amu), for a set of Si^{q+} projectiles[115] ($q = 4, 6, 8$–14, two or three n shells for each q, 0.5–14 keV/amu), for Ti^{4+} projectiles[114,116] (Ref. 114 for 4s, 4p, 0.008–1.1 keV/amu; Ref. 116 for 3s, $n = 4$, 5s, 5p, 0.5–80 keV/amu), for Cr^{6+} and Fe^{8+} projectiles[116] ($n = 4$–6, 0.5–80 keV/amu), and for Ni^{10+} projectiles[117] ($n = 4$–8, 1.5–80 keV/amu).

The cross sections from these sources tend to agree mutually, in the few cases where energy ranges overlap. There has been no attempt to represent these results in an analytic form.

Figure 17 shows, as an example, partial transfer cross sections to various channels in C^{3+} + H collisions. The low-energy data[40] from translation energy spectroscopy for both triplet and singlet 2s3s, 2s3p, and 2s3d states join smoothly

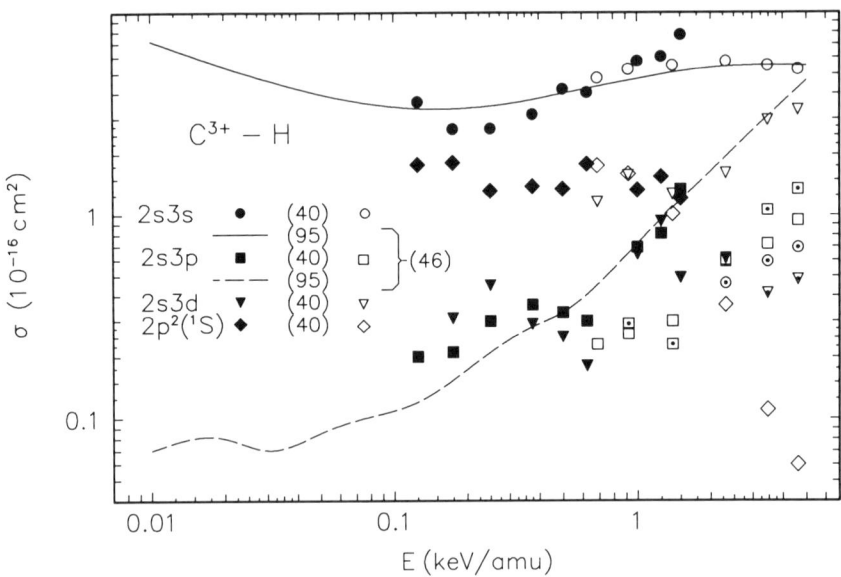

Figure 17. Electron transfer cross sections in C^{3+} + He collisions, to the 2s3s state, the 2s3p state, the 2s3d state, and the $2p^2(^1S)$ state, from Ref. 40 (closed symbols for the combined singlet and triplet states) and Ref. 46 (open symbols for the triplet states, open symbols with a dot at the center for the singlet states). The calculated results[95] are for the $2s3s(^3S)$ state (—) and for the $2s3p(^3P)$ state (- - -).

with other data[46] from photon-emission spectroscopy at higher energies, which in turn differentiate between singlet and triplet states. The calculated[95] cross section for the $2s3s(^3S)$ state agrees nicely with the data, but the cross section for the $2s3p(^3P)$ state from the calculation is larger than found by experiment. The cross section to the $2p^2(^1S)$ state is rather large in slow collisions even though its population involves two electrons. This is caused by an approximate energy matching between the initial state and this particular final state.

4.3. Collisions with He Atoms

For collisions of multicharged ions with He and H_2, there is only a very small amount of data on state-selective single-electron capture. For capture from He, measurements have been made for the following systems C^{2+},[138] C^{4+},[184] C^{5+},[190] C^{6+},[142] N^{4+},[42] N^{5+},[184,191] N^{6+},[190] O^{6+},[142,184] Ne^{6+},[192] Ne^{8+},[191] Fe^{3+},[43] and Fe^{4+}.[43] Cross section measurements for impact of protons have recently been reviewed in

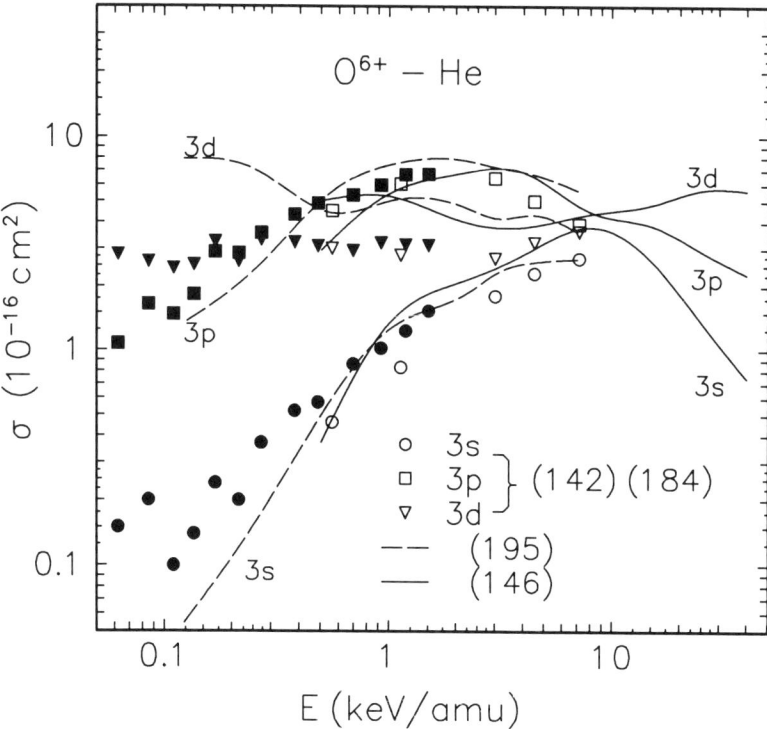

Figure 18. Electron transfer in O^{6+} + He collisions, to the $3s$ state, the $3p$ state, and the $3d$ state, from Ref. 184 (open symbols) and Ref. 142 (closed symbols). Calculated results are from AO calculations[146] (—) and from MO calculations[195] - - -).

greater detail.[193] Absolute cross sections are available only for impact of C^{4+},[184] N^{4+},[42] N^{5+},[184,191] O^{6+},[142,184] Ne^{6+},[192] and Ne^{8+}.[191]

In close-coupling calculations, partial transfer cross sections have been determined for Be^{3+} projectiles[127] ($n = 2$, 0.06–36 keV/amu), for Be^{4+} and B^{4+} projectiles[126] ($n = 2$, 2–30 keV/amu), for B^{3+} projectiles[133] ($n = 2$, 0.01–1 keV/amu), for C^{6+} projectiles[145, 146,194] (Ref. 146 for $n = 3$ and $n = 4$, 0.5–40 keV/amu; Ref. 194 for $n = 3$–6, 8–150 keV/amu; Ref. 145 for $n = 2$ and $n = 3$, 10–500 keV/amu), for O^+ projectiles[152] (^4S, $^2D + {}^2P$, 0.06–0.6 keV/amu), for O^{6+} projectiles[146,195] (Ref. 146 for $n = 3$ and $n = 4$, 0.5–40 keV/amu; Ref. 195 for $n = 3$ and $n = 4$, 0.14–7 keV/amu), and for Si^{4+} and Ti^{4+} projectiles[117] ($n = 3$–5, 1–100 keV/amu). In CTMC calculations,[153] the n-distribution for C^{6+} and O^{8+} ion impact has been determined at 50 and 100 keV/amu with n ranging up to a value of 9. In a close-coupling study on collisions of He with Be^{4+} and B^{4+} projectiles,[126] also two-electron transfer cross sections to the $2l2l'$ configurations have been determined.

In Fig. 18 the l dependence, within the $n = 3$ projectile shell, of cross sections in O^{6+} + He collisions is shown. Results from experiment,[142,184] with photon-emis-

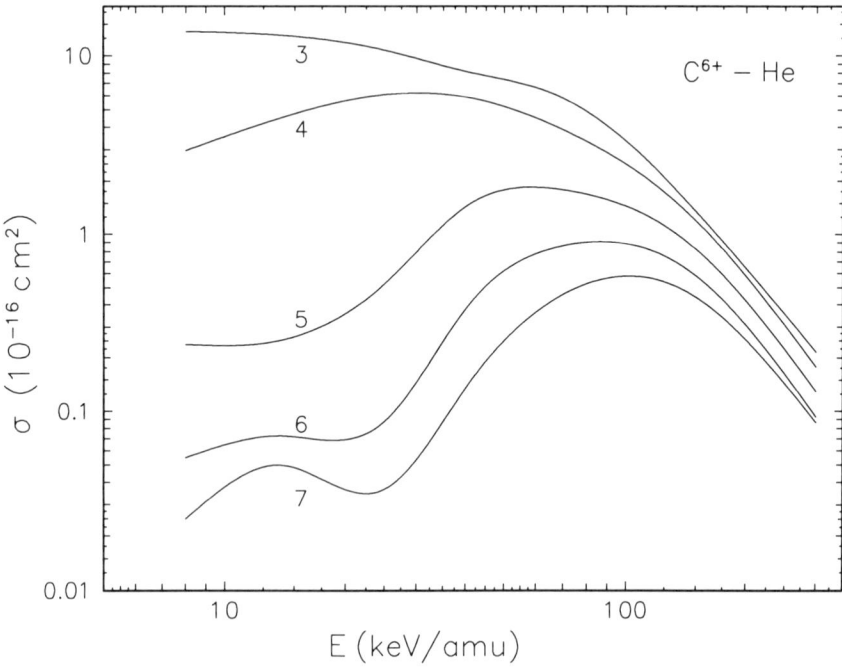

Figure 19. Calculated[194] electron transfer cross sections into $n = 3$–7 shells of C^{5+} in C^{6+} + He collisions. The values of n are labeled on the curves.

sion spectroscopy, and from calculations[146,195] are seen to agree mutually except at the low energies, where the calculation[195] appears to deviate from the data.

In Fig. 19 we show the calculated[194] n dependence of transfer cross sections in C^{6+} + He collisions. No other information is available on capture to the higher-n levels which are important for plasma diagnostics.

4.4. Collisions with H_2 Molecules

For single-electron capture from H_2, state-selective cross section measurements have been made for C^{2+},[138] C^{3+},[46] C^{4+},[41,94,184] C^{5+},[190] N^{5+},[184] N^{6+},[190] and O^{6+},[184] whereas absolute cross sections are only available for C^{3+},[46] C^{4+},[41,94,184] C^{5+},[190] N^{5+},[184] and O^{6+}.[184]

For the species of interest, close-coupling calculations have been performed for Be^{2+}, Be^{4+}, B^{3+}, and B^{5+} ions[188] (0.1–8 keV/amu), for C^{4+} ions[174,196] (Ref. 196 for $n = 3$, 5×10^{-4}–0.2 keV/amu; Ref. 174 for $n = 3$, 0.5–30 keV/amu), and for Ti^{4+} and Fe^{8+} projectiles[117] ($n = 3$–5 and $n = 4$–6, respectively, 2–80 keV/amu) only. In CTMC calculations,[153] the n-distribution for C^{6+} and O^{8+} ion impact has been determined at 50 and 100 keV/amu with n ranging up to a value of 16.

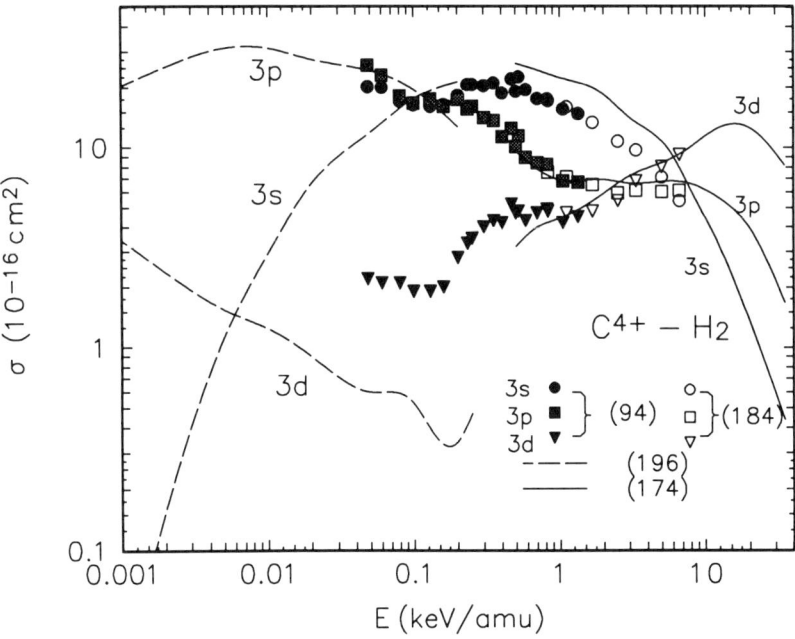

Figure 20. Electron transfer cross section in C^{4+} + H_2 collisions, to the $3s$ state, the $3p$ state, and the $3d$ state, from Ref. 184 (open symbols) and Ref. 94 (closed symbols). Calculated results are from an MO calculation[196] (- - -) and from an AO calculation (—).[174]

In Fig. 20 we show calculated[174,196] and measured[94,184] transfer cross sections to l substates in the $n = 3$ shell for $C^{4+} + H_2$ collisions. The data[41] from translational energy spectroscopy, not included in the figure, agree with the data[94,184] from photon-emission spectroscopy. All results are seen to agree reasonably well, except those for the $3d$ state at low energy. The MO calculation[196] includes only σ states. An improved calculation, with inclusion of rotational couplings, would be needed for a better understanding of the discrepancy.

5. CONCLUSION

The intense research activity during the last 10 to 15 years on the collision physics of multiply charged ions has generated a considerable amount of information on electron capture processes in collisions of plasma impurity ions with H, H_2, and He. The overwhelming majority of theoretical and experimental electron capture studies at low energies have been performed for single-electron capture processes involving H, H_2, He, and low-Z impurity ions. Total one-electron capture cross sections for hydrogen-atom–fully-stripped-ion systems with $Z \leq 10$ are presently known with a high degree of accuracy down to energies of 50–100 eV/amu. For fully stripped ions in higher charge states, the scaling relationships discussed in Section 3 can provide reliable predictions for the total single-electron capture cross sections.

For collision systems involving more than one electron (He and H_2 targets, incompletely stripped ions), the information on one-electron capture cross sections in the low-energy region is still incomplete. This is particularly true for collisions involving intermediate- and high-Z impurity ions (with the exception of Fe^{q+} with $q \geq 3$). Scaling relationships for $q \geq 5$ could provide cross section estimates for one-electron capture from H_2 and He targets with a lower reliability than in the case of a H target. Collision systems involving ions in low charge states ($q \leq 4$) have to be studied individually, because each system has its own complex collision dynamics. For energies below about 50–100 eV/amu, reliable theoretical cross section information for electron capture processes can be obtained only by using a fully quantum-mechanical description.

The cross section information on state-selective electron capture is even less complete. The reliability of calculated state-selective electron capture cross sections is high only for the dominantly populated channels, provided the basis set used in close-coupling calculations is large. The requirements on the basis size to accurately describe the capture into nondominant channels become particularly severe as the ion charge increases. Recent progress in experimental methods for state-selective electron capture studies (cf. Section 2.1), complemented by similar progress in computational capabilities, provides the basis for more vigorous studies in the future.

Two-electron processes (double capture, transfer–ionization, transfer–excitation) in slow collisions of plasma impurity ions with H, H_2, and He have been very little investigated. The existing information indicates that, for a given collision system, their cross sections are generally smaller than that for one-electron transfer. The cross section for transfer–ionization can become comparable to that for single-electron capture only when the ionic charge is high. More extensive studies of these processes are, however, very desirable because they nevertheless have significant impact on the impurity and neutral gas kinetics in the plasma edge.

REFERENCES

1. J. B. Delos, *Rev. Mod. Phys.* **53**, 287 (1981).
2. R. K. Janev and L. P. Presnyakov, *Phys. Rep.* **70**, 1 (1981).
3. B. H. Bransden and R. K. Janev, *Adv. At. Mol. Phys.* **19**, 1 (1983).
4. R. K. Janev and H. P. Winter, *Phys. Rep.* **117**, 265 (1985).
5. M. Kimura and N. F. Lane, *Adv. At. Mol. Opt. Phys.* **26**, 79 (1990).
6. W. Fritsch and C. D. Lin, *Phys. Rep.* **202**, 1 (1991).
7. W. Fritsch, in *Review of Fundamental Processes and Applications of Atoms and Ions* (C. D. Lin, ed.) World Scientific, Singapore (1993), p. 239.
8. B. H. Bransden and M. R. C. McDowell, *Charge Exchange and the Theory of Ion–Atom Collisions*, Oxford University Press, Oxford (1992).
9. R. Hoekstra, F. J. de Heer, and R. Morgenstern, *Z. Phys. D* **21**, 81 (1991).
10. C. L. Cocke in *Review of Fundamental Processes and Applications of Atoms and Ions* (C. D. Lin, ed.), World Scientific, Singapore (1993), p. 111.
11. H. B. Gilbody, *Adv. At. Mol. Opt. Phys.* **32**, 149 (1994).
12. H. Tawara and R. A. Phaneuf, *Comments At. Mol. Phys.* **21**, 177 (1988).
13. R. K. Janev, M. F. A. Harrison, and H. W. Drawin, *Nucl. Fusion* **29**, 109 (1989).
14. *Physica Scripta*, Topical Issue, **T28** (1989).
15. *Physica Scripta*, Topical Issue, **T37** (1991).
16. *Atomic and Plasma-Material Interaction Data for Fusion (Nucl. Fusion, Supplement)* **3** (1992).
17. *Nucl. Fusion, Special Supplement* (1987).
18. H. Tawara, Total and Partial Cross Sections for Electron Capture of C^{q+} ($q = 6$–2) and O^{q+} ($q = 8$–2) Ions in Collisions with H, H_2 and He, Institute of Plasma Physics, Nagoya University, Report IPPJ-AM-56, 1987.
19. H. Tawara, T. Kato, and Y. Nakai, *At. Data Nucl. Data Tables* **32**, 235 (1985).
20. W. K. Wu, B. A. Huber, and K. Wiesemann, *At. Data Nucl. Data Tables* **40**, 58 (1988); **42**, 157 (1989).
21. H. Tawara, Total and Partial Cross Sections of Electron Transfer Processes of Be^{q+} and B^{q+} Ions in Collisions with H, H_2 and He Gas Targets—Status in 1991, National Institute for Fusion Science, Nagoya, Research Report NIFS-DATA-12, 1991.
22. H. Tawara, Bibliography on Electron Transfer Processes in Ion–Ion/Atom/Molecule Collisions—Updated 1993, National Institute for Fusion Science, Nagoya, Research Report NIFS-DATA-20, 1993.
23. R. K. Janev, R. A. Phaneuf, and H. Hunter, *At. Data Nucl. Data Tables* **40**, 249 (1988).
24. R. A. Phaneuf, R. K. Janev, and H. Hunter, *Nucl. Fusion, Special Supplement* **1987**, 7.
25. R. K. Janev and J. J. Smith, *Atomic and Plasma-Material Interaction Data for Fusion (Nucl. Fusion, Supplement)*, **4**, 1 (1993).
26. R. K. Janev, R. A. Phaneuf, H. Tawara, and T. Shirai, *At. Data Nucl. Data Tables* **55**, 201 (1993).

27. I. G. Brown (ed.), *The Physics and Technology of Ion Sources*, John Wiley & Sons, New York (1989).
28. V. B. Kutner, *Rev. Sci. Instrum.* **65**, 1039 (1994).
29. R. E. Marrs, S. R. Elliott, and D. A. Knapp, *Phys. Rev. Lett.* **72**, 4082 (1994).
30. F. Aumayr and HP. Winter, *Phys. Scr.* **T28**, 96 (1989).
31. D. A. Church, *Phys. Rep.* **228**, 253 (1993).
32. G. Scoles (ed.), *Atomic and Molecular Beam Methods*, Vol. 1, Oxford University Press, New York (1988).
33. J. Slevin and W. Stirling, *Rev. Sci. Instrum.* **52**, 1780 (1981).
34. R. F. McCullough, J. Geddes, A. Donnelly, M. Liehr, M. P. Hughes, and H. B. Gilbody, *Meas. Sci. Technol.* **4**, 79 (1993).
35. C. C. Havener, M. S. Huq, H. F. Krause, P. A. Schulz, and R. A. Phaneuf, *Phys. Rev. A* **39**, 1725 (1989).
36. C. C. Havener, M. A. Haque, A. C. H. Smith, X. Urbain, and P. A. Zeijlmans van Emmichoven, in *Proceedings of the VIth International Conference on the Physics of Highly Charged Ions* (P. Richard et al., eds.), *AIP Conf. Proc.*, No. 274, American Institute of Physics, New York (1993), p. 43.
37. W. L. Nutt, R. W. McCullough, K. Brady, M. B. Shah, and H. B. Gilbody, *J. Phys. B* **11**, 1457 (1978).
38. F. W. Meyer, A. M. Howald, C. C. Havener, and R. A. Phaneuf, *Phys. Rev. A* **32**, 3310 (1985).
39. W. Wu, J. P. Giese, I. Ben-Itzhak, C. L. Cocke, P. Richard, M. Stöckli, R. Ali, H. Schöne, and R. E. Olson, *Phys. Rev. A* **48**, 3617 (1993).
40. F. G. Wilkie, R. W. McCullough, and H. B. Gilbody, *J. Phys. B* **19**, 239 (1986).
41. T. K. McLaughlin, R. W. McCullough, and H. B. Gilbody, *J. Phys. B* **25**, 1257 (1992).
42. T. K. McLaughlin, H. Tanuma, J. Hodkinson, R. W. McCullough, and H. B. Gilbody, *J. Phys. B* **26**, 3871 (1993).
43. T. K. McLaughlin, J. Hodkinson, H. Tawara, R. W. McCullough, and H. B. Gilbody, *J. Phys. B* **26**, 3587 (1993).
44. B. A. Huber and H. J. Kahlert, *J. Phys. B* **16**, 4655 (1983).
45. H. Cederquist, E. Beebe, C. Biedermann, Å. Engström, H. Gao, R. Hutton, J. C. Levin, L. Liljeby, T. Quinteros, N. Selberg, and P. Sigrey, *J. Phys. B* **25**, L69 (1992).
46. D. Čirič, A. Brazuk, D. Dijkkamp, F. J. de Heer, and HP. Winter, *J. Phys. B* **18**, 3629 (1985).
47. N. Stolterfoht, K. Sommer, J. K. Swenson, C. C. Havener, and F. W. Meyer, *Phys. Rev. A* **42**, 5396 (1990).
48. N. Stolterfoht, K. Sommer, F. Fremont, X. Husson, D. Lecler, J. K. Swenson, C. C. Havener, and F. J. Meyer, in *Proceedings of the VIth International Conference on the Physics of Highly Charged Ions* (P. Richard et al., eds.), *AIP Conf. Proc.*, No. 274, American Institute of Physics, New York (1993), p. 53.
49. M. Mack, J. H. Nijland, P. van der Straaten, A. Niehaus, and R. Morgenstern, *Phys. Rev. A* **39**, 3846 (1989).
50. J. H. Posthumus, P. Lukey, and R. Morgenstern, *J. Phys. B* **25**, 987 (1992).
51. M. Barat and P. Roncin, *J. Phys. B* **25**, 2205 (1992).
52. P. Roncin, M. N. Gaboriaud, Z. Szilagy, and M. Barat, in *The Physics of Electronic and Atomic Collisions* (T. Andersen et al., ed.), *AIP Conf. Proc.*, No. 295. American Institute of Physics, New York (1993), p. 537.
53. E. A. Solov'ev, *Zh. Eksp. Teor. Fiz.* **81**, 1681 (1981) [Engl. transl.: *Sov. Phys. JETP* **54**, 893 (1981)].
54. S. Yu. Ovchinnikov and E. A. Solov'ev, *Zh. Eksp. Teor. Fiz.* **90**, 921 (1986) [Engl. transl.: *Sov. Phys. JETP* **63**, 538 (1986)].
55. E. A. Solov'ev, *Usp. Fiz. Nauk* **157**, 437 (1989) [Engl. transl.: *Soc. Phys. Usp.* **32**, 228 (1989)].

56. W. Fritsch, *J. Phys. B* **15**, L389 (1982).
57. W. Fritsch and C. D. Lin, *Phys. Rev. A* **27**, 294 (1983).
58. C. D. Lin, T. G. Winter, and W. Fritsch, *Phys. Rev. A* **25**, 2395 (1982).
59. T. G. Winter and C. D. Lin, *Phys. Rev. A* **29**, 567 (1984).
60. M. Kimura and C. D. Lin, *Phys. Rev. A* **31**, 590 (1985).
61. T. G. Winter and N. F. Lane, *Phys. Rev. A* **31**, 2698 (1985).
62. I. M. Cheshire, *Proc. Phys. Soc.* **92**, 862 (1967).
63. R. Shakeshaft, *J. Phys. B* **8**, 1114 (1975).
64. H. J. Lüdde and R. M. Dreizler, *J. Phys. B* **14**, 2191 (1981).
65. H. J. Lüdde and R. M. Dreizler, *J. Phys. B* **15**, 2703 (1982).
66. R. E. Olson and A. Salop, *Phys. Rev. A* **16**, 531 (1977).
67. L. P. Presnyakov and D. B. Uskov, *Zh. Eksp. Teor. Fiz.* **86**, 882 (1984) [Engl. transl.: *Sov. Phys.-JETP* **59**, 515 (1984)].
68. T. V. Goffe, M. B. Shah, and H. B. Gilbody, *J. Phys. B* **12**, 3763 (1979).
69. R. A. Phaneuf, I. Alvarez, F. W. Meyer, and D. H. Crandall, *Phys. Rev. A* **26**, 1892 (1982).
70. D. Dijkkamp, D. Čirič, and F. J. de Heer, *Phys. Rev. Lett.* **54**, 1004 (1985).
71. R. Hoekstra, D. Čirič, F. J. de Heer, and R. Morgenstern, *Phys. Scr.* **T28**, 81 (1989).
72. M. Kimura and W. R. Thorson, *J. Phys. B* **16**, 1471 (1983).
73. M. Kimura, private communication, 1992.
74. T. A. Green, E. J. Shipsey, and J. C. Browne, *Phys. Rev. A* **25**, 1364 (1982).
75. E. J. Shipsey, T. A. Green, and J. C. Browne, *Phys. Rev. A* **27**, 821 (1983).
76. W. Fritsch and C. D. Lin, *Phys. Rev. A* **29**, 3039 (1984).
77. W. Fritsch, in *Proceedings of the VIth International Conference on the Physics of Highly Charged Ions* (P. Richard et al., eds.), American Institute of Physics, New York (1993), p. 24.
78. W. Fritsch, *Phys. Rev. A* **30**, 3324 (1984).
79. W. Fritsch, *J. de Phys. Colloq. C1*, **50**, C1–87 (1989).
80. M. Kimura and C. D. Lin, *Phys. Rev. A* **32**, 1357 (1985).
81. P. S. Krstic, M. Radmilovic, and R. K. Janev, *Atomic and Plasma-Material Interaction Data for Fusion* (*Nucl. Fusion, Supplement*) **3**, 113 (1992).
82. K. Richter and E. A. Solov'ev, *Phys. Rev. A* **48**, 432 (1993).
83. R. K. Janev, L. P. Presnyakov, and V. P. Shevelko, *Physics of Highly Charged Ions*, Springer-Verlag, Berlin (1985).
84. R. A. Phaneuf, R. K. Janev, H. Tawara, M. Kimura, P. S. Krstic, G. Peach, and M. A. Mazing, *Atomic and Plasma-Material Interaction Data for Fusion* (*Nucl. Fusion, Supplement*) **3**, 105 (1993).
85. N. Shimakura, S. Suzuki, and M. Kimura, *Phys. Rev. A* **47**, 3930 (1993).
86. A. E. Wetmore, H. R. Cole, and R. E. Olson, *J. Phys. B* **19**, 1515 (1986).
87. N. Shimakura, *J. Phys. B* **21**, 2485 (1988).
88. R. W. McCullough, W. L. Nutt, and H. B. Gilbody, *J. Phys. B* **12**, 4159 (1979).
89. D. H. Crandall, R. A. Phaneuf, and F. W. Meyer, *Phys. Rev. A* **19**, 504 (1979).
90. D. H. Crandall, *Phys. Rev. A* **16**, 958 (1977).
91. L. D. Gardner, J. E. Bayfield, P. M. Koch, I. A. Sellin, D. J. Pegg, R. S. Peterson, M. L. Mallory, and D. H. Crandall, *Phys. Rev. A* **20**, 766 (1979).
92. L. D. Gardner, J. E. Bayfield, P. M. Koch, I. A. Sellin, D. J. Pegg, R. S. Peterson, and D. H. Crandall, *Phys. Rev. A* **21**, 1397 (1980).
93. V. A. Belyaev, M. M. Dubrovin, and A. N. Khlopin, *Sov. J. Plasma Phys.* **17**, 337 (1991).
94. R. Hoekstra, J. P. M. Beijers, A. R. Schlatmann, R. Morgenstern, and F. J. de Heer, *Phys. Rev. A* **41**, 4800 (1990).
95. L. Opradolce, L. Benmeuraiem, R. McCarroll, and R. D. Piacentini, *J. Phys. B* **21**, 503 (1988).
96. L. F. Errea, B. Herrero, L. Mendez, and A. Riera, *J. Phys. B* **24**, 4061 (1991).

97. M. Gargaud, R. McCarrol, and P. Valiron, *J. Phys. B* **20**, 1555 (1987).
98. N. Shimakura, S. Koizumi, S. Suzuki, and M. Kimura, *Phys. Rev. A* **45**, 7876 (1992).
99. M. Gargaud and R. McCarroll, *J. Phys. B* **21**, 513 (1988).
100. L. R. Anderson, M. Gargaud, and R. McCarroll, *J. Phys. B* **24**, 2073 (1991).
101. C. Harel and H. Jouin, *J. Phys. B* **21**, 859 (1988).
102. R. F. Janev and J. J. Smith, International Atomic Energy Agency ALADDIN Database Files (1990).
103. M. S. Huq, C. C. Havener, and R. A. Phaneuf, *Phys. Rev. A* **40**, 1811 (1989).
104. N. Shimakura, M. Itoh, and M. Kimura, *Phys. Rev. A* **45**, 267 (1992).
105. B. Zygelman, D. L. Cooper, M. J. Ford, A. Dalgarno, J. Gerratt, and M. Riamondi, *Phys. Rev. A* **46**, 3846 (1992).
106. N. Shimakura and M. Kimura, *Phys. Rev. A* **44**, 1659 (1991).
107. W. Fritsch, H. B. Gilbody, R. E. Olson, H. Cederquist, R. K. Janev, K. Katsonis, and G. Yudin, *Phys. Scr.* **T 37**, 11 (1991).
108. R. A. Phaneuf, *Phys. Rev. A* **28**, 1310 (1983).
109. D. H. Crandall, R. A. Phaneuf, and F. W. Meyer, *Phys. Rev. A* **22**, 379 (1980).
110. L. D. Gardner, J. E. Bayfield, P. M. Koch, H. J. Kim, and P. H. Stelson, *Phys. Rev. A* **16**, 1414 (1977).
111. F. W. Meyer, R. A. Phaneuf, H. J. Kim, P. Hvelplund, and P. H. Stelson, *Phys. Rev. A* **19**, 515 (1979).
112. H. J. Kim, R. A. Phaneuf, and F. W. Meyer, *Phys. Rev. A* **17**, 854 (1978).
113. H. Sato, M. Kimura, A. E. Wetmore, and R. E. Olson, *J. Phys. B* **16**, 3037 (1983).
114. M. Gargaud, R. McCarroll, and L. Opradolce, *J. Phys. B* **21**, 521 (1988).
115. W. Fritsch and H. Tawara, *Nucl. Fusion* **30**, 373 (1990).
116. W. Fritsch, *Phys. Scr.* **T 37**, 75 (1991).
117. W. Fritsch, *Atomic and Plasma-Material Interaction Data for Fusion* (*Nucl. Fusion, Supplement*) **6**, (1995) to be published.
118. K. Katsonis, G. Maynard, and R. K. Janev, *Phys. Scr.* **T 37** 80 (1991).
119. D. R. Schultz, L. Meng, C. O. Reinhold, and R. E. Olson, *Phys. Scr.* **T 37**, 80 (1991).
120. D. B. Uskov, J. Botero, R. K. Janev, and L. P. Presnyakov, Cross Sections for Electron Capture and Ionization in Collision of Fusion Plasma Impurity Ions with Atomic Hydrogen, International Atomic Energy Agency, Vienna, Report NDC(NDS)-291, 1993.
121. R. K. Janev and P. Hvelplund, *Comments At. Mol. Phys.* **11**, 75 (1981).
122. R. K. Janev, *Phys. Lett. A* **160**, 67 (1991).
123. N. Burniaux, F. Brouillard, A. Jognaux, T. R. Govers, and S. Szucs, *J. Phys. B.* **10**, 2421 (1977).
124. R. E. Olson, *J. Phys. B* **13**, 483 (1980).
125. N. Bohr and B. Lindhard, *K. Dan. Vidensk. Selsk. Mat. Fys. Medd.* **28**, 1 (1954).
126. W. Fritsch and C. D. Lin, *Phys. Rev. A* **45**, 6411 (1992).
127. J. P. Hansen, A. Dubois, and S. E. Nielsen, *Phys. Rev. A* **45**, 184 (1992).
128. H. Suzuki, Y. Kajikawa, N. Toshima, H. Ryufuku, and T. Watanabe, *Phys. Rev. A* **29**, 525 (1984).
129. F. Martin, A. Riera, and M. Yanez, *Phys. Rev A* **34**, 4675 (1986).
130. R. E. Olson, *Phys Rev. A* **18**, 2464 (1978).
131. J. H. Zwally and P. G. Cable, *Phys. Rev. A* **4**, 2301 (1971).
132. T. Iwai, Y. Kaneko, M. Kimura, N. Kobayashi, S. Ohtani, K. Okuno, S. Takagi, H. Tawara, and S. Tsurubuchi, *Phys. Rev. A* **26**, 105 (1982).
133. E. J. Shipsey, J. C. Browne, and R. E. Olson, *Phys. Rev. A* **15**, 2166 (1977).
134. C. W. Sherwin, *Phys. Rev.* **57**, 814 (1940).
135. H. B. Gilbody, *Phys. Scr.* **T 28**, 49 (1989).
136. H. Tawara and W. Fritsch, *Phys. Scr.* **T 28**, 58 (1989).
137. T. Kusakabe, Y. Mizumoto, K. Katsurayama, and H. Tawara, *J. Phys. Soc. Jpn.* **59**, 1987 (1990).
138. E. Unterreiter, J. Schweinzer, and HP. Winter, *J. Phys. B* **24**, 1003 (1991).

139. E. C. Montenegro, G. M. Sigaud, and W. E. Meyerhof, *Phys. Rev. A* **45**, 1575 (1992).
140. P. Roncin, M. N. Gaboriaud, L. Guillemot, H. Laurent, S. Ohtani, and M. Barat, *J. Phys. B* **23**, 1215 (1990).
141. R. Parameswaran, C. P. Bhalla, B. P. Walch, and B. D. DePaola, *Phys. Rev. A* **43**, 5929 (1991).
142. J. P. M. Beijers, R. Hoekstra, A. R. Schlatmann, R. Morgenstern, and F. J. de Heer, *J. Phys. B* **25**, 463 (1992).
143. M. Kimura and R. E. Olson, *J. Phys. B* **17**, L713 (1984).
144. M. Kimura and N. F. Lane, *Phys. Rev. A* **35**, 70 (1987).
145. A. Jain, C. D. Lin, and W. Fritsch, *Phys. Rev. A* **34**, 3676 (1986).
146. W. Fritsch and C. D. Lin, *J. Phys. B* **19**, 2683 (1986).
147. J. P. Hansen, *J. Phys. B* **25**, L17 (1992).
148. Z. Chen, R. Shingal, and C. D. Lin, *J. Phys. B* **24**, 4251 (1991).
149. E. Wolfrum, J. Schweinzer, and H. Winter, *Phys. Rev. A* **45**, R4218 (1992).
150. J. P. Bansgaard, P. Hvelplund, J. O. P. Pedersen, L. R. Andersson, and A. Bárány, *Phys. Scr.* **T28**, 91 (1989).
151. M. S. Huq, R. L. Champion, and D. L. Doverspike, *Phys. Rev. A* **37**, 2349 (1988).
152. M. Kimura, J. P. Gu, Y. Li, G. Hirsch, and R. J. Buenker, *Phys. Rev.* **49**, 3131 (1994).
153. L. Meng, C. O. Reinhold, and R. E. Olson, *Phys. Rev A* **42**, 5286 (1990).
154. L. Opradolce, R. McCarroll, and P. Valiron, *Astron. Astrophys.* **148**, 229 (1985).
155. T. Iwai, Y. Kaneko, M. Kimura, N. Kobayashi, A. Matsumoto S. Ohtani, K. Okuno, H. Tawara, and S. Tsurubuchi, *J. Phys. B* **17** L95 (1984).
156. H. Tawara, T. Iwai, Y. Kaneko, M. Kimura, N. Kobayashi, A. Matsumoto, S. Ohtani, K. Takagi, and S. Tsurubuchi, *Nucl. Instrum. Methods. Phys. Res. B* **9**, 432 (1985).
157. H. Andersson, G. Astner, and H. Cederquist, *J. Phys. B* **21**, L187 (1988).
158. C. Harel, H. Jouin, and B. Pons, *J. Phys. B* **24**, L425 (1991).
159. R. Shingal and C. D. Lin, *J. Phys. B* **24**, 251 (1991).
160. C. Harel and H. Jouin, *J. Phys. B* **25**, 221 (1992).
161. V. K. Nikulin and A. V. Samojlov, *J. Phys. B* **22**, L201 (1989).
162. A. E. Wetmore and R. E. Olson, *Phys. Rev. A* **38**, 5563 (1988).
163. R. K. Janev, *Atomic and Plasma-Material Interaction Data for Fusion (Nucl. Fusion, Supplement)* **3**, 71 (1992).
164. W. Fritsch, *J. Phys. B* **27**, 3461 (1994).
165. W. Fritsch, *Phys. Lett. A* **192**, 369 (1994).
166. H. Cederquist, *Phys. Rev. A* **43**, 2306 (1991).
167. J. A. Tanis. M. W. Clark, R. Price S. M. Ferguson, and R. E. Olson, *Nucl. Instrum. Methods* **B23**, 167 (1987).
168. S. Takagi, S. Ohtani, K. Kadota, and J. Fujita, *J. Phys. Soc. Jpn.* **52**, 3759 (1983).
169. Y. Xu, T. F. Moran, and E. W. Thomas, *Phys. Rev. A* **41**, 1408 (1990).
170. K. Soejima, C. T. Latimer, K. Okuno, N. Kobayashi, and Y. Kaneko, *J. Phys. B* **25**, 3009 (1992).
171. A. D. Irvine and C. T. Latimer, *J. Phys. B* **24**, L145 (1991).
172. Y. Xu, E. W. Thomas, and T. F. Moran, *J. Phys. B* **23**, 1235 (1990).
173. S. Cheng, C. L. Cocke, E. Y. Kamber, C. C. Hsu, and S. L. Varghese, *Phys. Rev. A* **42**, 214 (1990).
174. W. Fritsch, *Phys. Rev. A* **46**, 3910 (1992).
175. M. Kimura, *Phys. Rev. A* **33**, 4440 (1986).
176. K. H. Berkner, W. G. Graham, R. V. Pyle, A. S. Schlachter, and J. W. Stearns, *Phys. Rev. A* **23**, 2891 (1981).
177. M. G. Suraud, S. Bliman, D. Hitz, J. E. Rubensson, J. Nordgren, J. J. Bonnet, M. Bonnefoy, M. Chassevent, A. Fleury, M. Cornille, E. J. Knystautas, and A. Barany, *J. Phys B* **25**, 2363 (1992).
178. H. Bachau, P. Roncin, and C. Harel, *J. Phys. B* **25**, L109 (1992).

179. R. Hoekstra, H. O. Folkerts, J. P. M. Beijers, R. Morgenstern, and F. J. de Heer, *J. Phys. B* **27**, 2021 (1994).
180. W. L. Nutt, R. W. McCullough, and H. B. Gilbody, *J. Phys. B* **12**, L157 (1979).
181. R. Hoekstra, F. J. de Heer, and R. Morgenstern, *J. Phys. B* **24**, 4025 (1991).
182. W. Fritsch, *Phys. Rev A* **38**, 2664 (1988).
183. R. W. McCullough, F. G. Wilkie, and H. O. Gilbody, *J. Phys. B* **17**, 1373 (1984).
184. D. Dijkkamp, D. Čirič, E. Vlieg, A. de Boer, and F. J. de Heer, *J. Phys. B* **18**, 4763 (1985).
185. T. K. McLaughlin, S. M. Wilson, R. W. McCullough, and H. B. Gilbody, *J. Phys. B* **23**, 737 (1990).
186. S. M. Wilson, R. W. McCullough, and H. B. Gilbody, *J. Phys. B* **21** 1027 (1988).
187. R. W. McCullough, S. M. Wilson, and H. B. Gilbody, *J. Phys. B* **20**, 2031 (1987).
188. N. Shimakura and M. Kimura, private communication.
189. W. Fritsch and C. D. Lin, *J. Phys. B* **17**, 31 (1984).
190. M. G. Suraud, R. Hoekstra, F. J. de Heer, J. J. Bonnet, and R. Morgenstern, *J. Phys. B* **24**, 2543 (1991).
191. J. P. M. Beijers, R. Hoekstra, and R. Morgenstern, *Phys. Rev. A* **49**, 363 (1994).
192. J. P. M. Beijers, R. Hoekstra, R. Morgenstern, and F. J. de Heer, *J. Phys. B* **25**, 4851 (1992).
193. R. Hoekstra, H. P. Summers, and F. J. de Heer, *Atomic and Plasma-Material Interaction Data for Fusion* (*Nucl. Fusion, Supplement*), **3**, 63 (1992).
194. W. Fritsch, *Nucl. Instrum. Methods Phys. Res. B*, **98**, 246 (1995).
195. N. Shimakura, H. Sato, M. Kimura, and T. Watanabe, *J. Phys. B* **20**, 1801 (1987).
196. M. Gargaud and R. McCarroll, *J. Phys. B* **18**, 463 (1985).

Chapter 14

Reactive Ion–Molecule Collisions Involving Hydrogen and Helium

F. Linder, R. K. Janev, and J. Botero

1. INTRODUCTION

Experiments on the present generation of large tokamaks have shown that the processes taking place at the plasma edge play an important role for the overall plasma performance and may even have a decisive influence on the central plasma parameters. These observations have stimulated increased interest in understanding the physics of the edge plasma in a more fundamental way.[1-6]

The edge plasma is characterized by temperatures ranging from a few electron volts to several hundred electron volts. The particle densities are in the range of 10^{12}–10^{15} cm^{-3}. The dominant constituents will always be the hydrogen isotopes. In an ignited D–T plasma, helium also must be considered as an important species whose concentration may be on the order of 10% in the edge region. In addition to these primary constituents, there is a large number of possible impurities. These include carbon and oxygen compounds, hydrocarbons and carbon oxides, various metallic (and related) impurities originating from structural materials (Ti, V, Cr, Fe, Ni, Cu, Al, Ta, Mo, W, Be, B, Si), and several diagnostic species (Li, Ne, Ar). The

---

F. LINDER • Department of Physics, University of Kaiserslautern, D-67653 Kaiserslautern, Germany.
R. K. JANEV and J. BOTERO • International Atomic Energy Agency, A-1400 Vienna, Austria.
Atomic and Molecular Processes in Fusion Edge Plasmas, edited by R. K. Janev. Plenum Press, New York, 1995.

relative abundance of these impurities in the edge plasma is typically between 0.1 and 10%. Detailed estimates based on present-day fusion reactor designs can be found in Ref. 2.

The range of collision processes relevant to edge plasma studies is extremely wide. Because of the low plasma temperature, the density of neutral particles is rather high, and there are significant amounts of molecular species as well. This, combined with the large variety of constituents, makes the collision physics of the edge plasma very complex.

In this chapter we will be concerned only with the ion–molecule collision processes in the plasma edge. The electron–atom (ion), electron–molecule (molecular ion), and ion–atom collision processes are discussed in other chapters of this volume. At the collision energies pertinent to the plasma edge, the most important ion–molecule collision processes are those which involve change of the internal energy of the molecular species (energy transfer via rotational and vibrational excitation) and particle rearrangement of the collision system (electron transfer and heavy-particle interchange reactions). This chapter will concentrate on the particle rearrangement processes:

Electron transfer:

$$A^+ + BC \rightarrow A + BC^+ \quad (1)$$

$$XY^+ + BC \rightarrow XY + BC^+ \quad (2)$$

Particle interchange:

$$A^+ + BC \rightarrow \begin{cases} AB^+ + C & (3) \\ AB + C^+ & (4) \end{cases}$$

$$XY^+ + BC \rightarrow \begin{cases} XYB^+ + C & (5) \\ XYB + C^+ & (6) \\ X + YBC^+ & (7) \\ X^+ + YBC & (8) \end{cases}$$

in which the molecular species in the entrance channel may also be in vibrationally excited states. Attention will also be given to the dissociative reaction channels

$$A^+ + BC \rightarrow \begin{cases} A^+ + B + C & (9) \\ A + B^+ + C & (10) \\ A + B + C^+ & (11) \end{cases}$$

$$XY^+ + BC \rightarrow \begin{cases} XY^+ + B + C & (12) \\ X + YB^+ + C \quad \text{(etc.)} & (13) \end{cases}$$

when their cross sections at low energies are large. Our consideration of processes (1)–(13) will be limited to collision systems involving only the primary plasma edge constituents; that is, A, B, C, X, and Y in Eqs. (1)–(13) are H, D, or He. The information on the above reactions involving tritium is extremely sparse. Emphasis in the considerations here will be given to the collision energy region below 1 keV and to those reactions which have large cross sections in this energy region. In analyzing the available cross section information for processes (1)–(13) in hydrogen and hydrogen–helium systems, we will concentrate mainly on the experimental sources, with a strong preference for the results provided by ion beam studies. In order to set the criteria for selection of the most accurate cross section information on reactions (1)–(13) and for a critical assessment of the selected information, we first discuss (in Section 2) the experimental methods used so far in the cross section measurements of the reactions considered. The selected and critically assessed information for the total and state selective cross sections of reactive collisions in hydrogen and hydrogen–helium systems is presented in Sections 3 and 4, respectively. The conclusions from the data analysis for processes (1)–(13) are given in Section 5.

We should note that for a number of the reactions considered in this chapter, cross section data assessments have been previously performed in Refs. 7–10. In many cases, however, the present reassessment leads to preferred ("recommended") cross sections that differ (sometimes considerably) from the previous recommendations. Some of the ion–molecule electron transfer reactions considered in this chapter are also discussed in Chapter 12, but at somewhat higher collision energies. We also mention that a detailed discussion of the experimental methods used in the studies of reactions (1)–(13) can be found in a recent review.[11]

2. EXPERIMENTAL METHODS

In low-energy beam experiments, ion–molecule reactions are usually studied by detecting the slow product ions. At higher energies, it is also possible to detect the fast neutrals resulting from processes such as charge transfer reactions. The traditional method in this field is the ion-beam–gas-cell (IBGC) method, which largely originated from mass spectrometry. A mass-selected ion beam of variable energy is injected into a gas-filled reaction chamber, and the resulting product ions are measured using a variety of detection methods. We first discuss the determination of total cross sections in these measurements. For completeness, we mention that a gas beam is sometimes used instead of a reaction chamber; however, this is of minor importance for the following discussion.

For measurements of total cross sections in IBGC experiments, there are two basically different methods which must be distinguished. In the first method, sometimes called the condenser method, the total of all slow product ions is collected on some plate, grid, or cylinder structure, for which very different geometries are used. Electric fields, sometimes in conjunction with a magnetic field, are applied in order to ensure saturation in the collection efficiency. The fact that no mass analysis is provided in this method certainly poses a problem. However, for processes where there is no doubt about the identity of the product ions, for example, in the case of simple charge transfer reactions, this method can give absolute total cross sections that are quite reliable. In the following, we denote this method by SID (slow ion detection) as opposed to FND (fast neutral detection).

In the second method, often called tandem mass spectrometry (TMS), a second mass spectrometer is used to analyze the product ions by their masses. Two different geometries are typically used in TMS instruments: the transverse geometry, in which the product ions are extracted perpendicularly to the direction of the primary beam, and the longitudinal geometry, in which the extraction occurs along the axis of the primary beam. The first version discriminates against products with appreciable momentum transfer in the forward direction, whereas the reverse is the case in the second configuration. In favorable cases, depending on the kinematics, the reaction mechanism, and the type of instrument used, the collection efficiency may come close to unity. In general, however, the energy and angular distributions of the product ions are unknown *a priori* so that relatively large uncertainties can arise from insufficient and/or inaccurately known collection efficiencies. Absolute total cross sections obtained with this method should therefore be regarded with caution.

This problem has been overcome to a large extent with the development of the guided-beam (GB) technique. Not only can this method be used for very low collision energies, but it also has a high sensitivity and, most importantly, a guaranteed product collection efficiency of nearly unity. Absolute total cross sections can therefore be measured with high accuracy, and the overall properties of this technique make it the ideal method for this type of measurement in the study of low-energy ion–molecule reactions. A full description of this technique, including recent advances and many applications, has been given by Gerlich.[12]

Another powerful method in this field is the merged-beams (MB) technique. By merging two fast beams coaxially and by varying their relative velocity, a broad range of center-of-mass collision energies down to extremely low energies becomes accessible, and a high center-of-mass energy resolution can be obtained. The reaction products are contained in a relatively narrow cone of laboratory angles and can therefore be collected with high efficiency, thus enabling the determination of total reaction cross sections. For absolute measurements, a careful study of the spatial beam overlap along the reaction path is necessary.

A further important and unique feature of the MB technique is the fact that the neutral reactants are usually formed from the corresponding ions by charge transfer. This allows the production of neutral reactant beams for a wide range of chemical species. In many cases, it is equally easy to produce beams of chemically unstable species (e.g., atomic and molecular radicals) as it is to produce beams of closed-shell stable molecules (e.g., H_2). On the other hand, it is important to realize that reactant molecules prepared in this way are very likely to be formed in vibrationally excited states. This has to be kept in mind when cross section data obtained by the MB technique are compared with results from other experiments.

Within certain limits, all the above methods can also be used to obtain information on the energy and angular distributions of the reaction products. However, the method that provides the most detailed information in this respect is the crossed-beams (CB) technique. In present-day CB experiments, a mass- and energy-selected ion beam is crossed with a supersonic nozzle beam of the target gas. Both beams are well collimated and can be made nearly monoenergetic. Because of the supersonic nozzle expansion, the internal state distribution of the neutral reactant molecules corresponds to a very low temperature. The kinematically well-defined conditions of a CB experiment permit measurements with high resolution in energy and angle. The product analysis is performed using a rotatable detector with mass and energy analysis. The energy resolution is generally sufficient to perform state-resolved measurements. In favorable cases, individual rotational transitions can be resolved.

Simultaneously with improvements in resolution, the development of the CB technique has reached the stage where it is now possible to carry out measurements down to collision energies on the order of 0.1 eV, so that the full energy range relevant to edge plasma studies can be covered. Elastic and inelastic collisions as well as reactive collisions can be studied. The measurements provide information that is both very detailed and very complete. The data include product- and state-specific differential cross sections, angle-integrated partial cross sections, and, finally, total cross sections summed over all product states. However, it is important to stress that cross sections have to be determined in absolute units in order to be useful in practical applications. This point has often been neglected in CB measurements. In the following, we briefly outline several possibilities for making absolute measurements in CB experiments.

In a scattering experiment, the scattered intensity I_s is given by the relation

$$I_s = \frac{d\sigma}{d\Omega} n l I_0 \Delta\Omega \tag{14}$$

in which $d\sigma/d\Omega$ is the differential cross section, n is the target gas density, l is the extension of the scattering volume in the direction of the incident beam, I_0 is the intensity of the incident beam, and $\Delta\Omega$ is the solid angle element spanned by the detector. Direct determination of absolute cross sections based on this relation is generally difficult, as in order to obtain an accurate measure for the effective value of the quantity $n l I_0 \Delta\Omega$, the absolute gas beam density, the overlap integral of the two beams, and the detector efficiency must be known. Therefore, one has to rely on other methods. One possible procedure is the following. All product- and state-specific differential cross sections are first measured in relative units, and then the data are integrated over angles, summed over product states, and finally normalized to a total cross section that is known in absolute units from other measurements. In this way, the whole set of detailed differential cross sections can be put on an absolute scale. In obtaining the absolute total cross sections needed for normalization, it is often possible to combine information from different sources covering the whole energy range from thermal energies up to the kilo-electron-volt region. A detailed example of a case in which this method has been applied can be found in Ref. 13.

Another well-known possibility is normalization with respect to theory. The experimental data, again in the form of state-specific differential cross sections, are first measured in relative units. Then a detailed comparison with theory is performed. If good agreement is achieved regarding all details of the data, one can be confident that reliable absolute cross sections are obtained by normalizing the experimental data with respect to theory. It should be emphasized, however, that the application of this method is more problematic for the processes considered here than, for example, for atomic collisions at higher energies, where relatively simple scattering approximations can often be used to normalize the experimental data. At low collision energies and, in particular, for molecular systems, a much more elaborate theoretical treatment is needed. The applicability of this method is therefore expected to be limited to relatively simple collision systems.

The third and probably most attractive method consists of using simple ion–atom scattering systems as secondary standards for absolute cross section measurements in CB experiments. The principle of the method is to compare the scattering intensity I_S of the collision system under study with that of a suitable reference system whose cross section is well known. Examples of such systems are H^+ + He and He^+ + He. These systems have been thoroughly studied both by experiment and theory, are easy to handle experimentally, and are therefore convenient to use for this purpose. Under properly chosen conditions, which have to be verified in the individual cases, the beam overlap integrals and the detector efficiencies of the two systems will cancel, to a good approximation, and the differential cross section in question can be obtained in absolute units through the relation

$$\left(\frac{d\sigma}{d\Omega}\right)_X = \frac{I_{SX}}{I_{SR}} \frac{n_R}{n_X} \left(\frac{d\sigma}{d\Omega}\right)_R \tag{15}$$

where the index R indicates the reference system and the index X the system under study. The relative target gas densities n_R and n_X are quite accurately known in the present case, since the technology of supersonic nozzle beams is highly developed and the properties of these beams are well understood.[14] Similar techniques have been developed in electron scattering and are now routinely used for absolute cross section measurements.[15,16]

Concluding the discussion of CB experiments, we can say that all types of processes (elastic, inelastic, reactive) can be studied with this method in the relevant energy range. We have emphasized the importance of making absolute measurements. All classes of data needed for edge plasma modeling can be obtained. It must be realized, however, that these measurements are very time-consuming so that, in practice, the number of such studies will be limited. The majority of processes will be studied by other techniques which may give less detailed information, but which allow total cross sections and rate coefficients to be determined for a large number of systems in an efficient way.

This overview of experimental methods is far from being complete. We have pointed out some characteristics of those methods which are of main importance in the following discussion of results. A very important class of experiments, which we have not discussed here, are measurements with state-selected reactants. Several methods, all based on photoionization, have been developed for this purpose: state-selective (single-photon) photoionization (SSPI), photoelectron–photoion coincidence (PEPICO), and resonance-enhanced multiphoton ionization (REMPI). Recent reviews of these techniques, with emphasis on applications to the study of ion–molecule reactions, have been presented by Ng,[17,18] Koyano and Tanaka,[19] and Anderson.[20] Results of state-selective measurements for collision systems relevant to the present subject are discussed in Section 4. Data with state-resolved products have been obtained mainly by time-of-flight analysis or by the use of electrostatic energy selectors. Optical methods, which can be very powerful in this respect, are of less importance for the present systems.

Finally, we mention that we have intentionally refrained from giving extensive references in this section, since all experimental methods used are stated explicitly in the following discussion of original work.

3. TOTAL CROSS SECTIONS FOR PARTICLE REARRANGEMENT COLLISIONS

The particle rearrangement ion–molecule collisions include both electron transfer and particle interchange reaction channels, and these are usually strongly coupled. In homonuclear collision systems, such as $H^+ + H_2$, the electron transfer and atom-exchange reaction channels cannot be distinguished unless the reaction products are energy-analyzed. Electron transfer and (heavy) particle interchange processes can also take place simultaneously (e.g., in the reaction $A^+ + BC \rightarrow AB + C^+$), but such reactions are usually classified as particle exchange. As mentioned in Section 1, both electron transfer and particle interchange reactions may lead to dissociation of the final molecular product, and such dissociative processes will also be included in the analyses of this section. In view of the large number of electron and heavy-particle rearrangement reactions in hydrogenic and hydrogen–helium collision systems, the electron transfer and particle interchange reactions will be discussed separately. A further classification of the reactions will be made according to the type of reactants.

3.1. Electron Transfer Reactions

3.1.1. Hydrogen Ion–Molecule Collision Systems

A survey of the total cross section measurements for the electron transfer reactions in hydrogen–hydrogen ion–molecule collision systems (including isotopic variants) in the energy region below 1 keV is given in Table I. The table provides information on the center-of-mass energy (E_{cm}) range in which a specific cross section measurement has been performed and on the method applied. As discussed in Section 2, the IBGC–TMS measurements are generally not reliable because of problems with the collection efficiency. For the electron transfer reactions considered here, the IBGC–SID method can be regarded as providing quite reliable results. This view is supported by the good agreement of the IBGC–SID results with those obtained by the crossed-beams and guided-beams methods, whose accuracy is believed to be about 10%, or better.

Table I. Experimental Total Cross Section Measurements of Electron Transfer Reactions in Hydrogen Ion–Molecule Systems

Reaction	Energy range, E_{cm} (eV)	Method[a]	Reference(s)
$H^+ + H_2 \rightarrow H + H_2^+$	2–10	IBGC (TMS)	21
	42–1300	CB (FND)	22
	33–270	IBGC (SID)	23
	45–700	IBGC (SID)	24
$H^+ + D_2 \rightarrow H + D_2^+$	2–8	IBGC (TMS)	21
	2–12	GB	25, 26
	2–80	IBGC (TMS)	27
$D^+ + HD \rightarrow D + HD^+$	2–9	IBGC (TMS)	21
	2–6	IBGC (TMS)	28
$D^+ + D_2 \rightarrow D + D_2^+$	2–10	IBGC (TMS)	21
	2–6	IBGC (TMS)	28
	2–11	GB	26
	43–270	IBGC (SID)	29
$H_2^+ + H \rightarrow H_2 + H^+$	33–4600	CB (SID, TMS)	30
$H_2^+ + H_2 \rightarrow H_2 + H_2^+$	2–200	IBGC (SID)	23
	5–2200	IBGC (SID)	31
	25–1000	IBGC (SID)	32
	35–500	IBGC (SID)	24, 33
	50–2200	IBGC (SID)	34
$H_2^+ + D_2 \rightarrow H_2 + D_2^+$	170–670	IBGC (SID)	33
$D_2^+ + H_2 \rightarrow D_2 + H_2^+$	85–330	IBGC (SID)	33
$D_2^+ + D_2 \rightarrow D_2 + D_2^+$	2–200	IBGC (SID)	29
	25–1000	IBGC (SID)	32, 35

[a] Abbreviations: IBGC, Ion-beam–gas-cell; TMS, tandem mass spectrometry; CB, crossed-beams; FND, fast neutral detection; SID, slow ion detection; GB, guided-beams.

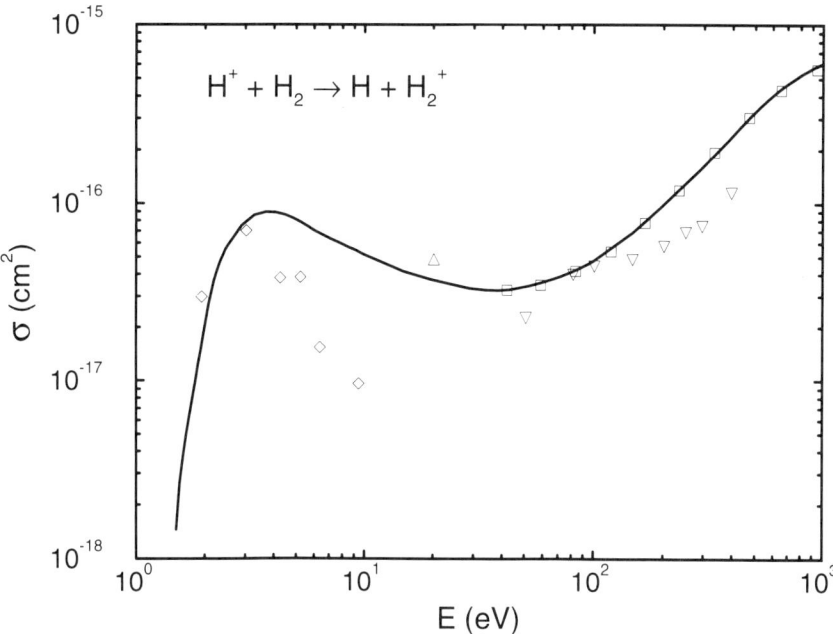

Figure 1. Total cross section for the $H^+ + H_2 \rightarrow H + H_2^+$ reaction. The symbols represent the experimental data of Holliday et al.[21] ($\diamond$), Gealy and Van Zyl[22] ($\square$), and Cramer[23] ($\triangledown$) and the single theoretical value of Niedner et al.[74] ($\triangle$). The solid curve is the "preferred" cross section for this reaction.

Figures 1 and 2 show the results of cross section measurements for the reactions $H^+ + H_2 \rightarrow H + H_2^+$ and $D^+ \rightarrow D + D_2^+$ obtained by various methods. The IBGC–SID results reported by Cramer[23] for the electron transfer in $H^+ + H_2$ collision system (Fig. 1) are in good agreement with the CB (FND) measurements of Gealy and Van Zyl[22] in the overlapping energy range. On the other hand, the GB electron transfer results of Schlier et al.[26] for the $D^+ + D_2$ system (Fig. 2) provide an illustration of the accuracy level of the IBGC–TMS results.[21,28] The solid lines in these figures represent our "preferred" cross sections for these reactions, based on the present assessment of the accuracy of the available experimental data. We note that in two previous assessments of the cross section for the $H^+ + H_2 \rightarrow H + H_2^+$ reaction,[7,9] the preferred cross section in the energy range 3–10 eV was based on the IBGC–TMS data of Ref. 21 and a smooth interpolation between these data and the CB (FND) data of Ref. 22 at 50 eV. The evaluation by Phelps[8] of the data for this reaction, however, ignores the IBGC–TMS data of Ref. 21 in the range 3–10 eV and links smoothly the IBGC–TMS cross section value at $E = 3$ eV with the CB (FND) cross section value at $E = 42$ eV. In the present assessment, the interpolation problem was solved by assuming that the energy behavior of the cross section for the $H^+ + H_2$

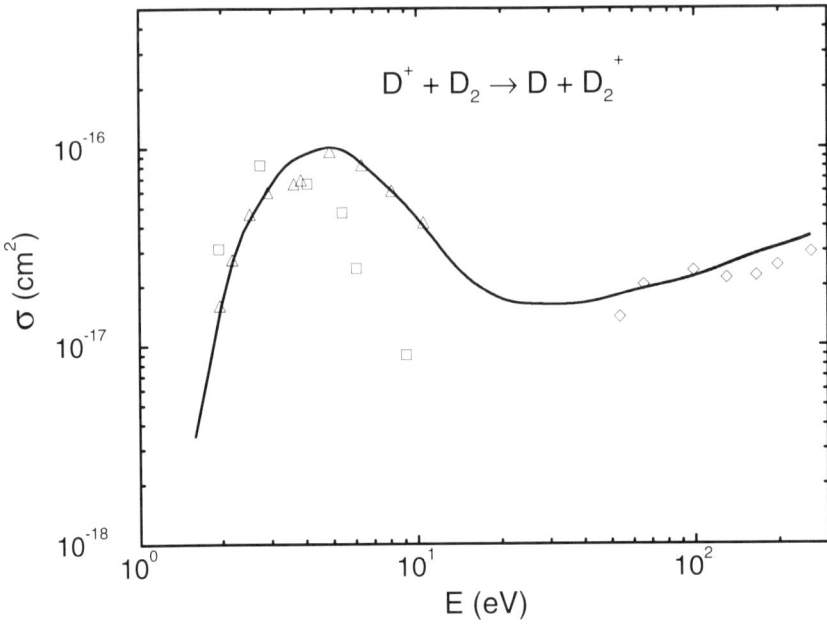

Figure 2. Total cross section for the $D^+ + D_2 \rightarrow D + D_2^+$ reaction. The symbols represent the experimental data of Holliday et al.[21] ($\square$), Schlier et al.[26] ($\triangle$), and Cramer and Marcus[29] ($\diamond$). The solid curve is the "preferred" cross section.

$\rightarrow H + H_2^+$ reaction in the 3- to 40-eV energy range is similar to that for the $D^+ + D_2 \rightarrow D + D_2^+$ reaction. This approach is supported by the theoretical cross section value of Ref. 74 at $E_{cm} = 20$ eV (also shown in Fig. 1), which is only 20% higher than our interpolated value.

The assessment of experimental total electron transfer cross sections for reactions involving a molecular ion is difficult because of the effects of ion internal (vibrational, rotational) energy on the electron capture process. Depending on the way in which they are produced (e.g., by charge exchange, electron impact ionization, etc.), molecular ions may have different vibrational state distributions, differently weighted state-selective electron capture transitions (see Section 4), and, consequently, different total electron transfer cross sections with the same target species. Therefore, it is to a certain extent inappropriate to compare experimental results obtained with differently prepared reactants. This remark is valid also for other ion–molecule rearrangement processes.

The most extensively studied electron transfer reaction involving a molecular ion is the $H_2^+ + H_2 \rightarrow H_2 + H_2^+$ reaction. The results of cross section measurements performed with the IBGC–SID method are all in fair agreement with each other.

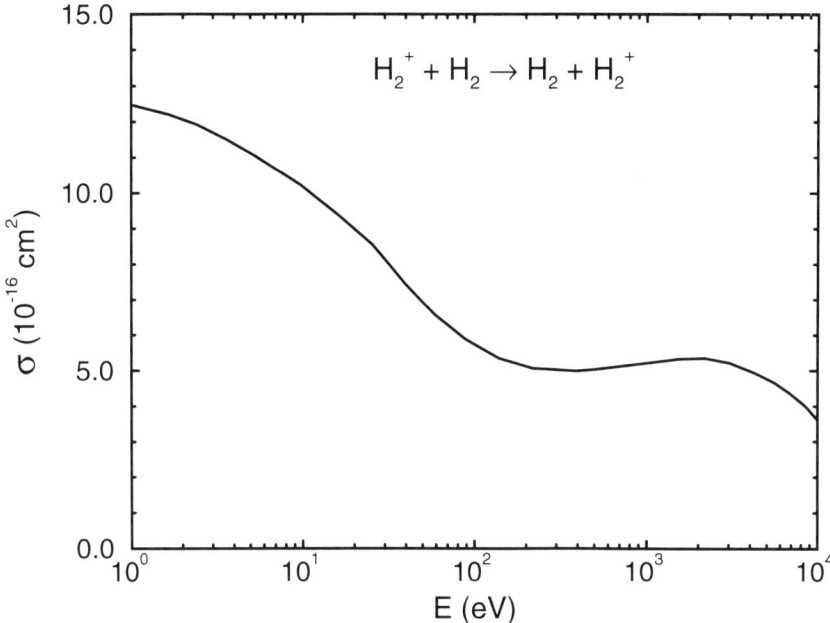

Figure 3. Total cross section for the $H_2^+ + H_2 \to H_2 + H_2^+$ electron transfer reaction based on the critical data assessment of Barnett.[9]

The cross section assessed for this reaction in the energy range 1 eV–10 keV is shown in Fig. 3 and coincides with that recommended in Ref. 9. We note that the present assessment is also supported by the state-selective cross section measurements discussed in Section 4.1, assuming a Franck–Condon population of the vibrational states of $H_2^+(v)$. The question of whether the cross section in Fig. 3 will start to decrease in the energy range below ~1 eV owing to the coupling of electron transfer with the strong particle interchange channel $H_2^+ + H_2 \to H_3^+ + H$ has at present no experimental answer.

The isotopic variants of the $H_2^+ + H_2 \to H_2 + H_2^+$ reaction have not been much investigated at low collision energies. The mentioned effect of vibrational state distribution on the electron capture process implies that a simple isotopic scaling of the total cross section for these reactions cannot be expected.

3.1.2. Hydrogen–Helium Ion–Molecule Collision Systems

The large potential energy (~ 24.5 eV) stored in the $He^+ + H_2$ (HD, D_2) system makes it possible that during the capture transition of one of the molecular electrons the other one is excited, leading to minimum change in the total electronic energy of the system. Under such conditions, the single-electron capture process is sup-

Table II. Experimental Total Cross Section Measurements of Electron Transfer Reactions in Helium–Hydrogen Ion–Molecule Systems

Reaction	Energy range, E_{cm} (eV)	Method[a]	Reference
$He^+ + H_2 \rightarrow He + H_2^+$	0.03–57	IBGC (TMS)	36
	50–300	IBGC (TMS)	37
$He^+ + H_2 \rightarrow He + H + H^+$	0.1–50	IBGC (TMS)	38
	6.7–16	IBGC (TMS)	39
	33–800	IBGC (SID)	34
	50–300	IBGC (TMS)	37
	330–1000	IBGC (FND)	40
$He^+ + D_2 \rightarrow He + D + D^+$	0.1–50	IBGC (TMS)	38
	0.2–7.2	IBGC (TMS)	41
$He^+ + HD \rightarrow He + D + H^+$	5–20	IBGC (TMS)	39
$He^+ + HD \rightarrow He + D^+ + H$	5–20	IBGC (TMS)	39

[a]For definitions of abbreviations, see footnote to Table I.

pressed in favor of the more probable two-electron capture–dissociative excitation (or dissociative electron transfer) process. The limited number of experimental studies of these processes in hydrogen–helium ion–molecule systems (see Table II) confirm this physical picture. We note that electron capture in the H_2^+ + He collision is highly endothermic and, therefore, an adiabatically improbable process.

Most of the total cross section measurements for electron capture processes in $He^+ + H_2$ (HD, D_2) systems have been performed by the IBGC–TMS method. The only IBGC–SID cross section measurement is that performed for the $He^+ + H_2 \rightarrow He + H + H^+$ dissociative electron capture reaction.[34] The available experimental data for this reaction are shown in Fig. 4. We note that the data of Jones *et al.*[38] have been normalized to the data of Rozett and Koski[39] in the overlapping energy region. This leads, however, to a severe disagreement with the measurements of Stedeford and Hasted[34] in the 30- to 50-eV energy range. The preferred cross section for this reaction, based on the results of Rozett and Koski[39] and on IBGC–SID measurements,[34] is also shown in the figure. This recommendation is supported by the Lyman-α, β and Balmer-α, β emission cross sections associated with the excited atomic hydrogen reaction product.[11] We note that the preferred cross section suggested by the present assessment practically coincides with the recommendation in Ref. 7, but drastically differs from that given in Ref. 9. We further note that the uncertainty of the cross section in the energy region below ~ 50 eV still remains very high.

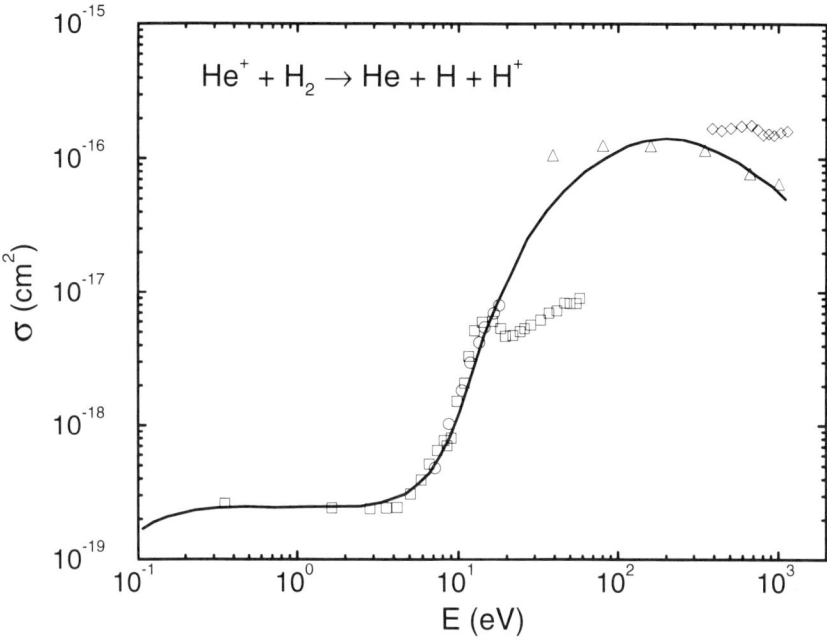

Figure 4. Total cross section for the dissociative electron capture reaction $He^+ + H_2 \rightarrow He + H + H^+$. The symbols represent the experimental data of Stedeford and Hasted[34] ($\triangle$), Jones et al.[38] ($\square$), Rozett and Koski[39] ($\circ$), and Moran and Conrads[40] ($\diamond$). The solid curve is the "preferred" cross section.

3.2. Particle Interchange Reactions

3.2.1. General Remarks

The magnitude and energy behavior of the cross sections of particle interchange ion–molecule reactions at low collision energies strongly depend on the reaction exothermicity, $\Delta\varepsilon$ ($\Delta\varepsilon$ is defined as the difference between the binding energies of the "active" particle in its final and in its initial state). The exothermic reactions ($\Delta\varepsilon > 0$) have no energy threshold, and in the energy region $E_{cm} < \Delta\varepsilon$, their cross section increases with decreasing collision energy approximately as $E^{-1/2}$ (according to the Langevin polarization capture mechanism for these reactions). For collision energies several times larger than $\Delta\varepsilon$, the cross section of an exothermic particle interchange reaction starts to decrease with the increase of collision energy much more rapidly than $E^{-1/2}$. The endothermic reactions ($\Delta\varepsilon < 0$) have a threshold ($E_{th} = |\Delta\varepsilon|$), their cross section increases rapidly (exponentially) to its maximum value (at about $E_m \approx (1-2) |\Delta\varepsilon|^2$) and after that decreases slowly with increasing collision energy. Such cross section behavior is typical of reactions governed by a nonadiabatic transition mechanism.

For collision energies above the binding energy of the "active" particle in its final state, the dissociative channels in particle interchange reactions become open. The cross section of dissociative particle interchange reactions attains its maximum at energies a few times larger than the dissociation threshold and then slowly decreases with increasing collision energy. The cross section maximum of the dissociative channel of exothermic particle interchange reactions lies in the energy region where the cross section of the nondissociative channel already rapidly decreases with energy, and in this region dissociation is the dominant reaction channel.

As discussed in Section 3.1, the internal energy state of reactants (e.g., their vibrational excitation) may have a significant influence on the reaction dynamics. This influence is particularly expressed in the particle interchange reactions. A reaction $A^+ + BC \rightarrow AB^+ + C$ that is endothermic for $BC(v = 0)$ may become exothermic for a vibrationally excited state $BC(v)$ having a smaller dissociation energy than AB^+. Since the probability for particle interchange increases with exothermicity, it is obvious that the vibrational distribution of $BC(v)$ will have a significant effect on the cross section magnitude. If the reaction $A^+ + BC \rightarrow AB^+ + C$ is exothermic even for $BC(v = 0)$, then for excited molecules $BC(v > 0)$ its exothermicity increases, and the distribution of $BC(v)$ over the vibrational states has again a dramatic effect on the cross section magnitude. For sufficiently large exothermicities, the reaction product AB^+ can be formed also in a vibrationally excited state.

3.2.2. Hydrogen Ion–Molecule Collision Systems

A survey of the total cross section measurements for particle interchange reactions in hydrogenic ion–molecule collision systems is given in Table III. As in the case of electron transfer reactions in these collision systems, most of the measurements have been performed by using the IBGC–TMS method. Our presentation here will concentrate, however, mainly on the results obtained with the more reliable guided- and merged-beams methods.

In Fig. 5 the experimental cross sections for the $H^+ + D_2 \rightarrow D^+ + HD$ (or $H + D$) and $H^+ + D_2 \rightarrow HD^+ + D$ reactions are shown in the energy range where merged- or guided-beams measurements were made. The reaction leading to $HD^+ + D$ products is endothermic by 1.85 eV if D_2 is in its ground vibrational state. The apparent threshold at a lower collision energy for this reaction seen in its cross section in Fig. 5 indicates that either the D_2 molecules in the experiment of Schlier et al.[26] were in the first two excited states (the apparent threshold corresponds to D_2 in the $v = 2$ state) or that the data in this energy region have large experimental uncertainties. The $H^+ + D_2 \rightarrow D^+ + HD$ reaction is endothermic by only 0.043 eV when D_2 is in its ground vibrational state. If D_2 is in its first vibrational excited state, the reaction becomes exothermic with $\Delta\varepsilon = 0.328$ eV. [For D_2 ($v = 2$), the

Reactive Ion-Molecule Collisions

Table III. Experimental Total Cross Section Measurements of Particle Interchange Reactions in Hydrogenic Ion–Molecule Systems

Reaction	Energy range, E_{cm} (eV)	Method[a]	Reference(s)
$H^+ + D_2 \to D + HD^+$	2–8	IBGC (TMS)	21, 42
	2–12	GB	25, 26
	2–80	IBGC (TMS)	27
$H^+ + D_2 \to D^+ + HD$ (or $H + D$)	2–8	IBGC (TMS)	21, 28, 42
	0.3–12	GB	25, 26
	0.04–0.3	From rates	43
	0.3–80	IBGC (TMS)	27
$D^+ + H_2 \to H + HD^+$	2–8	IBGC (TMS)	21
	2–9	GB	26
$D^+ + H_2 \to H^+ + HD$ (or $H + D$)	0.04–0.3	From rates	43
	1–8	IBGC (TMS)	21
	2–9	GB	26
$D^+ + HD \to H + D_2^+$	2–9	IBGC (TMS)	21
	2–6	IBGC (TMS)	28, 42
$D^+ + HD \to H^+ + D_2$ (or $D + D$)	1.5–6	IBGC (TMS)	21
	0.8–6	IBGC (TMS)	28, 42
$H_2^+ + D \to HD^+ + H$	0.05–5	MB	44
$H_2^+ + H_2 \to H_3^+ + H$	0.01–7	MB	45
	0.1–5	MB	46
	0.1–15	SSPI–GB	47[b]
	0.1–1	PEPICO–SCC	48[b]
	0.2–3	SCC	49
	0.3–5	IBGC (TMS)	50
	0.7–8.5	IBGC (TMS)	51
	0.7–7.5	SSMS	52
$HD^+ + D_2 \to HD_2^+ + D$	0.01–8	MB	53
$HD^+ + D_2 \to D_3^+ + H$	0.01–8	MB	53
$D_2^+ + HD \to HD_2^+ + D$	0.01–8	MB	53
$D_2^+ + HD \to D_3^+ + H$	0.01–8	MB	53
$D_2^+ + D_2 \to D_3^+ + D$	0.3–8	ICR, SCC	9
$H_3^+ + D_2 \to HD_2^+ + H_2$	0.01–11	MB	54
$D_3^+ + H_2 \to DH_2^+ + D_2$	0.001–1	MB	12
	0.13–10	GB	75

[a] Abbreviations: MB, Merged-beams; SSPI, state-selective photoionization; PEPICO, photoelectron–photoion coincidence; SCC, single collision chamber; SSMS, single-stage mass spectrometer; ICR, ion cyclotron resonance. For definitions of other abbreviations, see footnote to Table I.
[b] State-selective measurements with $H_2^+(v)$.

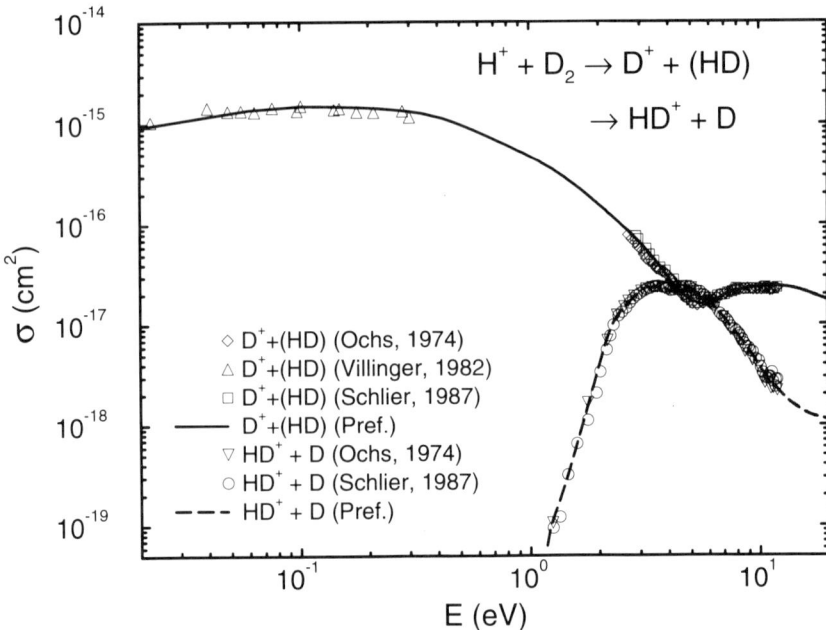

Figure 5. Total cross sections for the particle interchange reactions $H^+ + D_2 \to D^+ + HD$ (or H + D) (—) and $H^+ + D_2 \to HD^+ + D$ (– – –). The experimental data are from Villinger et al.[43] (△), Ochs and Teloy,[25] (◇, ▽), and Schlier et al.[26] (□, ○). The high-energy maximum of the $D^+ + (HD)$ cross section is attributed to the contribution of the dissociative $D^+ + H + D$ channel.

reaction exothermicity is 0.683 eV.] The large experimental cross section values in the energy range 0.02–0.3 eV reported by Villinger et al.[43] (see Fig. 5) indicate that D_2 molecules in this experiment must have been vibrationally excited. For energies above 4.5 eV, the dissociative particle interchange reaction $H^+ + D_2 \to D^+ + H + D$ becomes energetically possible, and the second maximum at about 10 eV in the cross section of the $H^+ + D_2 \to D^+ + (HD)$ reaction shown in Fig. 5 is to be attributed to the dissociative channel.

Figure 6 shows the experimental cross sections for the reactions $D^+ + H_2 \to H^+ + HD$ (or H + D) and $D^+ + H_2 \to HD^+ + H$ reported by Schlier et al.[26] and Villinger et al.[43] For H_2 in its ground vibrational state, the $H^+ + HD$ channel is exothermic ($\Delta\varepsilon = 0.035$), while the $HD^+ + H$ channel is endothermic ($\Delta\varepsilon = -1.79$). The threshold behavior of the endothermic reaction cross section indicates that H_2 molecules in the experiment of Schlier et al.[26] were in their ground vibrational state. The experimental cross section values obtained by Villinger et al.[43] for the exothermic reaction $D^+ + H_2 \to H^+ + HD$ closely follow the Langevin $E^{-0.5}$ energy behavior in the energy range 0.05–0.3 eV. In the energy region below 0.05 eV, they clearly

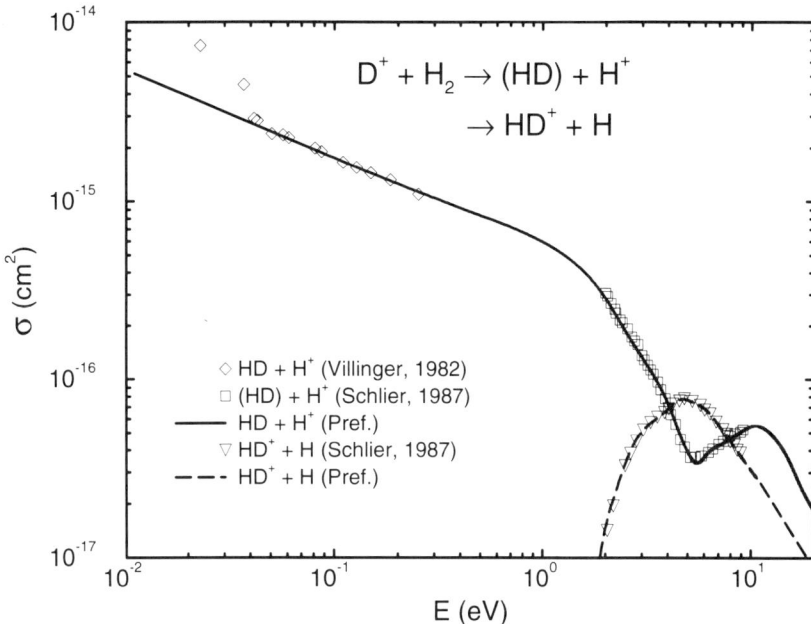

Figure 6. Total cross sections for the $D^+ + H_2 \rightarrow HD$ (or $H + D$) $+ H^+$ (—) and $D^+ + H_2 \rightarrow HD^+ + H$ (– – –) reactions. The symbols represent the experimental data of Villinger et al.[43] ($\diamond$) and Schlier et al.[26] ($\square$, $\triangledown$). The high-energy part of the (HD) $+ H^+$ cross section is mainly due to the contribution of the dissociative $H + D + H^+$ channel.

depart from this behavior. For collision energies above 4.5 eV, the dissociative channel of the $D^+ + H_2 \rightarrow H^+ + (HD)$ reaction becomes open, and the cross section for this reaction above 5.5 eV is dominated by this reaction channel.

Figure 7 shows the merged-beams cross section for the reaction $H_2^+ + D \rightarrow HD^+ + H$ reported by Wendell and Pol.[44] For the H_2^+ ion in its ground vibrational state, this reaction is exothermic by only 0.06 eV. The high values of the cross section for this reaction in the region below 1 eV indicate that H_2^+ ions might have been vibrationally excited in this experiment. The energy dependence of the cross section in this region is given by $E^{-0.46}$.

The cross section measurements for the $H_2^+ + H_2 \rightarrow H_3^+ + H$ particle interchange reaction are fairly abundant (see Table III). This reaction may proceed by both proton transfer

$$\underline{H_2^+} + H_2 \rightarrow (H^+H_2) + \underline{H} + 2.3 \text{ eV} \tag{16}$$

and atom transfer

$$\underline{H_2^+} + H_2 \rightarrow (\underline{H_2^+}H) + H + 0.5 \text{ eV} \tag{17}$$

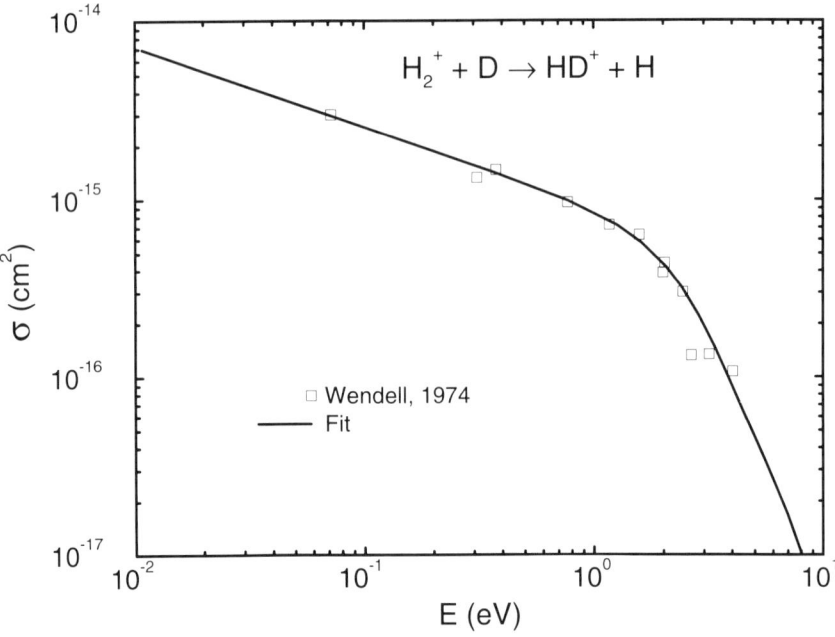

Figure 7. Total cross section for the $H_2^+ + D \rightarrow HD^+ + H$ reaction. The solid curve is a fit to the experimental data of Wendell and Pol[44] with appropriate extrapolation at high energies.

If both H_2^+ and H_2 are in their ground vibrational states, the exothermicity of reaction (16) is considerably larger than that for reaction (17) and, correspondingly, the cross section for reaction (16) is considerably larger than for reaction (17). If H_2^+ is kept in its ground (or in any fixed) vibrational state and H_2 is in a vibrationally excited state, the exothermicity for the proton-exchange reaction is reduced, while that for the atom-exchange reaction becomes larger. Therefore, for a fixed vibrational distribution of $H_2^+(v)$, an increase in the vibrational excitation of H_2 leads to suppression of reaction (16) and enhancement of reaction (17). In the case in which H_2 is in its ground (or in any other fixed) vibrational state and H_2^+ is in a vibrationally excited state, the situation is reversed: reaction (16) is enhanced while reaction (17) is inhibited. In different experiments the reactants H_2^+ and H_2 may be produced with different vibrational state distributions, and the cross section for the $H_3^+ + H$ reaction channel may vary by a factor of up to 3–4. In merged-beams experiments, where H_2 is formed from H_2^+ by charge exchange and the H_2 molecules are vibrationally excited (predominantly in $v = 1$–2), the measured cross section is expected to be smaller than for the ground state reactants. In the experiments in which the H_2^+ ions are formed by electron impact ionization (as is the case in those using the IBGC

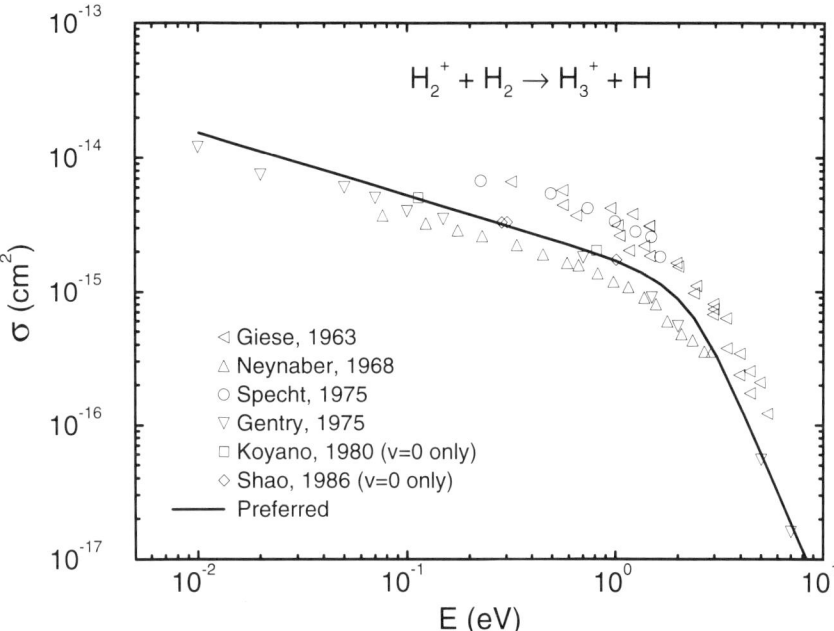

Figure 8. Total cross section for the $H_2^+ + H_2 \to H_3^+ + H$ reaction. The symbols represent experimental data from Neynaber and Trujillo,[46] Gentry et al.,[45] Specht et al.,[49] Giese and Maier,[50] Shao and Ng[47] [for $H_2^+(v=0)$], and Koyano and Tanaka[48] [for $H_2^+(v=0)$]. The solid curve is the preferred cross section.

method, for instance), the population of excited vibrational states of H_2^+ leads to cross sections larger than in the case of $H_2^+ (v = 0)$. This is illustrated in Fig. 8, where the merged-beams cross section data of Refs. 45 and 46 and the IBGC–TMS data of Refs. 49 and 50 for this reaction are shown. The "preferred" cross section for this reaction, shown by the solid curve in Fig. 8, is constructed on the basis of state-selective [$H_2^+ (v = 0)$] measurements reported in Refs. 47 and 48. With respect to the preferred cross section, the merged-beams data in the region below 0.8 eV appear to be 30–50% too low.

We note that the merged-beams data of Gentry et al.[45] follow an $E^{-0.46}$ dependence in the energy region below 0.8 eV while for energies above 4 eV they follow approximately an $E^{-3.5}$ dependence. The preferred cross section retains these asymptotic behaviors.

In Fig. 9 the merged-beams cross section results of Douglas et al.[53] for the exothermic reactions $HD^+ + D_2 \to HD_2^+ + D$ and $HD^+ + D_2 \to D_3^+ + H$ are shown. Because D_2 in these reactions is most probably in vibrationally excited states, the cross section for the $D_3^+ + H$ channel (D^+ transfer) is smaller and that for the HD_2^+

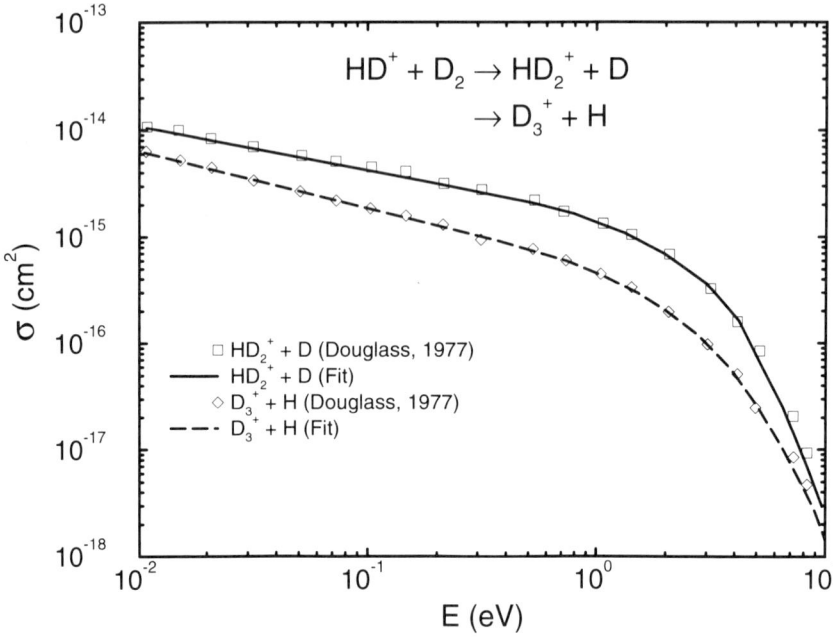

Figure 9. Total cross sections for the reactions $HD^+ + D_2 \to HD_2^+ + D$ and $HD^+ + D_2 \to D_3^+ + H$ from the merged-beams experiments of Douglass et al.[53]

+ D channel is larger than the corresponding cross sections for the reactions with $D_2(v = 0)$. Figure 10 shows the merged-beams cross sections for the exothermic reactions $D_2^+ + HD \to HD_2^+ + D$ and $D_2^+ + HD \to D_3^+ + H$ taken from Ref. 53. The cross sections in Figs. 9 and 10 all follow an approximate $E^{-0.45\pm0.02}$ dependence at low energies.

In Fig. 11 the experimental cross sections for the reactions $H_3^+ + D_2 \to HD_2^+ + H_2$ and $D_3^+ + H_2 \to H_2D^+ + D_2$ are shown, taken from Ref. 54 and Refs. 12 and 75, respectively. For ground state reactants, the first of these reactions is slightly exothermic (by ~0.07 eV), whereas the second one is endothermic by only 0.029 eV. However, both H_3^+ and D_3^+ are usually produced in vibrationally excited states (with H_3^+ predominantly in $v = 3$–4, and D_3^+ predominantly in $v = 5$–7 states) and the experimentally measured reactions both have significant exothermicities. The cross section for the $D_3^+ + H_2 \to H_2D^+ + D_2$ reaction shown in Fig. 11 is for thermalized D_3^+ ions (at $T \approx 350$ K). For "hot" D_3^+ ions with an internal energy of about 2 eV, the cross section for this reaction becomes larger by a factor of about 2.[12]

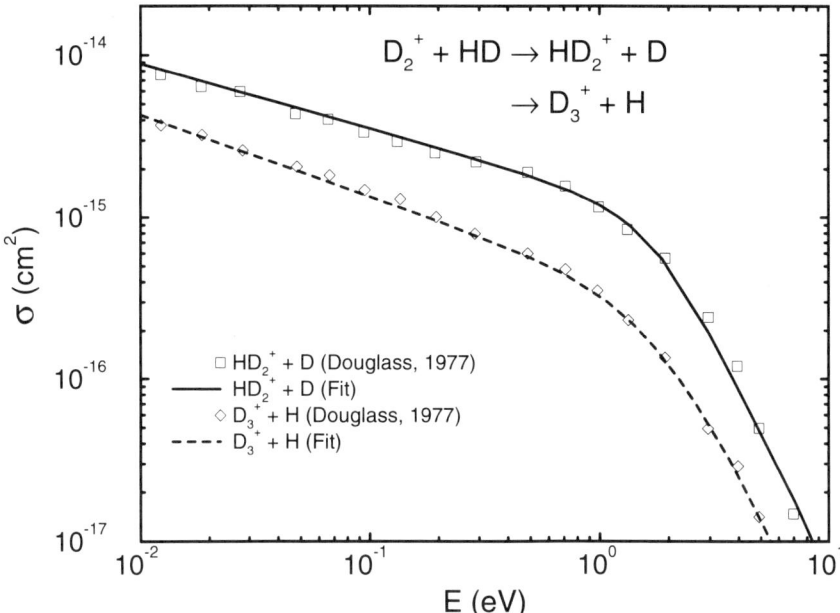

Figure 10. Total cross sections for the reactions $D_2^+ + HD \rightarrow HD_2^+ + D$ and $D_2^+ + HD \rightarrow D_3^+ + H$ from the merged-beams experiments of Douglas et al.[53]

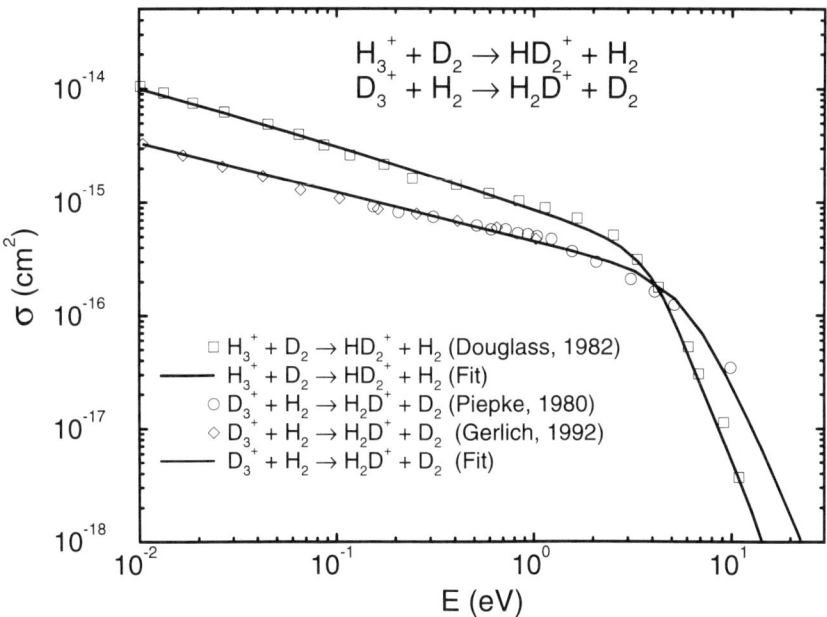

Figure 11. Total cross sections for the $H_3^+ + D_2 \rightarrow HD_2^+ + H_2$ and $D_3^+ + H_2 \rightarrow H_2D^+ + D_2$ reactions. The data for $H_3^+ + D_2$ are from Douglass et al.,[54] and those for $D_3^+ + H_2$ are from Gerlich[12] and Piepke.[75] The solid curves represent least-squares fits to the data.

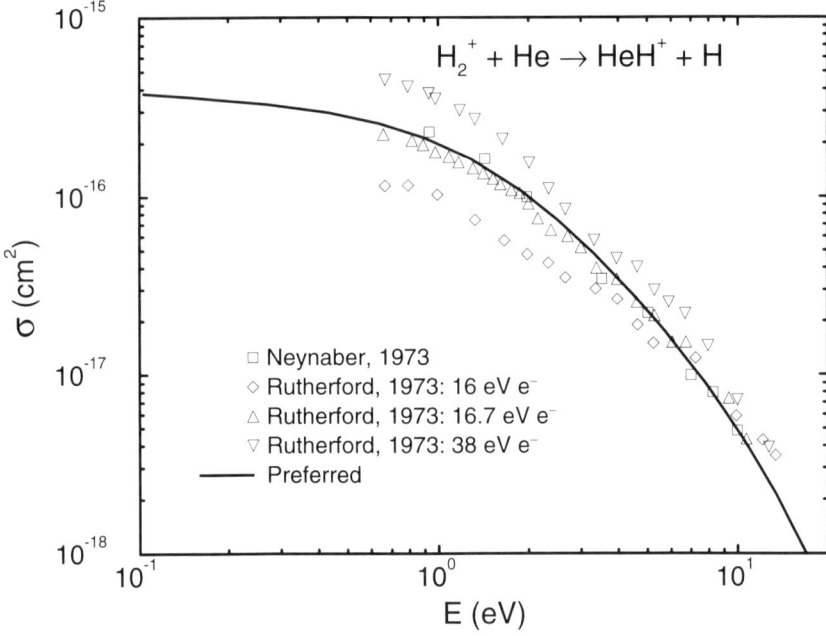

Figure 12. Total cross section for the $H_2^+ + He \rightarrow HeH^+ + H$ reaction. The symbols represent the merged-beams data of Neynaber and Magnuson[55] and the crossed-beams data of Rutherford and Vroom[56] for H_2^+ produced by electron impact at three different collision energies. The solid curve is obtained as a weighted sum of state-selective cross sections (see text).

Table IV. Experimental Total Cross Section Measurements of Particle Interchange Reactions in Helium–Hydrogen Ion–Molecule Systems

Reaction	Energy range, E_{cm} (eV)	Method[a]	Reference
$H_2^+ + He \rightarrow He H^+ + H$	0.05–12	MB	55
	0.7–3	IBGC (TMS)	50
	0.7–14	CB (TMS)	56
	1–6	GB	57
$He^+ + H_2 \rightarrow He H^+ + H$	0.1–30	IBGC (TMS)	38
	0.1–50	GB	59
$He^+ + D_2 \rightarrow He D^+ + D$	0.1–30	IBGC (TMS)	38
	0.1–50	GB	59
$He^+ + HD \rightarrow He H^+ + D$	0.1–50	GB	59
$He^+ + HD \rightarrow He D^+ + H$	0.1–50	GB	59
$HeH^+ + H \rightarrow H_2^+ + He$	0.2–4	CB (TMS)	56
$HeH^+ + H_2 \rightarrow H_3^+ + He$	0.21–4.2	CB (TMS)	58

[a] For definitions of abbreviations, see footnotes to Table I and III.

3.2.3. Hydrogen–Helium Ion–Molecule Collision Systems

Because of the small dissociation energy of HeH$^+$ and HeD$^+$ ions (about 1.8 eV), all of the ground state particle interchange reactions in hydrogen-helium systems leading to HeH$^+$ and HeD$^+$ products are endothermic and generally characterized by small cross sections at low energies. A survey of the cross section measurements for these and other particle interchange reactions in hydrogen–helium systems is given in Table IV.

The most extensively studied reaction, $H_2^+ + He \rightarrow HeH^+ + H$,[50,55-57] is endothermic by about 0.85 eV when H_2^+ is in its ground vibrational state. All measured total cross sections for this reaction (the beam measurements of Refs. 55–57 and the IBGC–TMS measurement of Ref. 50) show large cross sections even in the energy region below 0.8 eV, indicating high vibrational population of H_2^+ ions in these experiments. [The reaction becomes exothermic for $H_2^+(v \geq 3)$.] The total cross sections for this reaction obtained from various measurements are shown in Fig. 12. The influence of the vibrational distribution of $H_2^+(v)$ on the total reaction cross section is clearly demonstrated by the three data sets from the crossed-beams experiments of Rutherford and Vroom[56] using H_2^+ ions obtained by ionization with

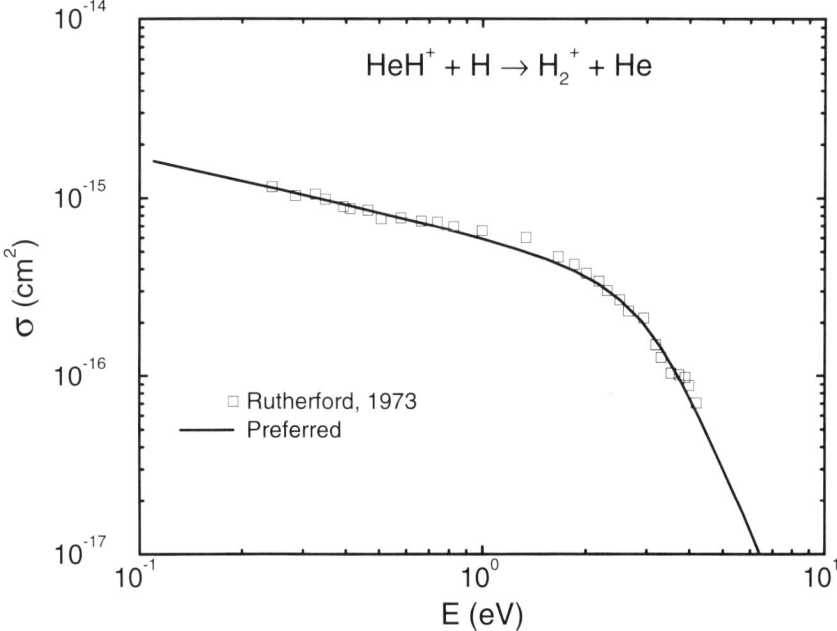

Figure 13. Total cross section for the HeH$^+$ + H → H_2^+ + He reaction. The symbols represent the crossed-beams data of Rutherford and Vroom,[56] and the solid curve is their least-squares fit (with appropriate extrapolations).

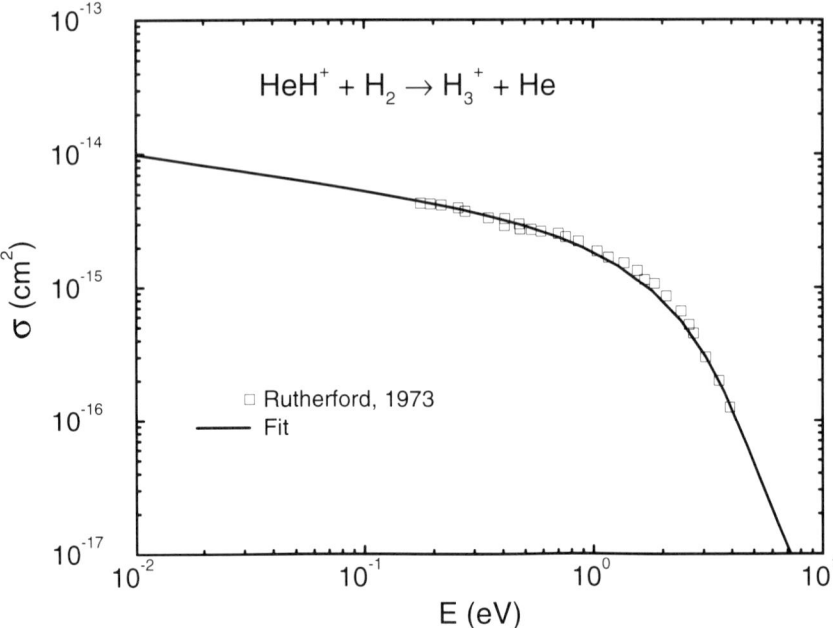

Figure 14. Total cross section for the HeH$^+$ + H$_2$ → H$_3^+$ + He reaction. The symbols represent the crossed-beams data of Rutherford and Vroom,[58] and the solid curve is their least-squares fit (with appropriate extrapolations).

electrons of energy 16, 16.7, and 38 eV. The solid curve in Fig. 12 represents the cross section for this reaction obtained as a weighted sum of the state-selective cross sections $\sigma(v)$ (v = 0–5) presented in Section 4.2 [and a plausible estimate of the contribution from $\sigma(v \geq 6)$], with the weighting factors (i.e., the relative vibrational populations) taken from Ref. 60.

The cross section for the reverse HeH$^+$ + H → H$_2^+$ + He exothermic reaction ($\Delta\varepsilon$ = 0.85 eV), measured by the crossed-beams method,[56] is shown in Fig. 13. The low-energy part of this cross section exhibits an $E^{-0.45}$ dependence. Its high-energy part is an extrapolation, consistent with the high-energy behavior of other particle interchange reactions. The cross section of the reaction HeH$^+$ + H$_2$ → H$_3^+$ + He ($\Delta\varepsilon$ = 3.15 eV) is shown in Fig. 14, the data being taken from Ref. 58.

The cross sections for the endothermic ($\Delta\varepsilon \approx$ –2.7 eV) reactions He$^+$ + H$_2$ → HeH$^+$ + H and He$^+$ + D$_2$ → HeD$^+$ + D have been measured by using both the guided-beams[59] and IBGC–TMS[38] techniques. The latter measurements are relative, and the results have been normalized to the absolute values of Ref. 59. The cross sections of these reactions are shown in Fig. 15. In the threshold region, they

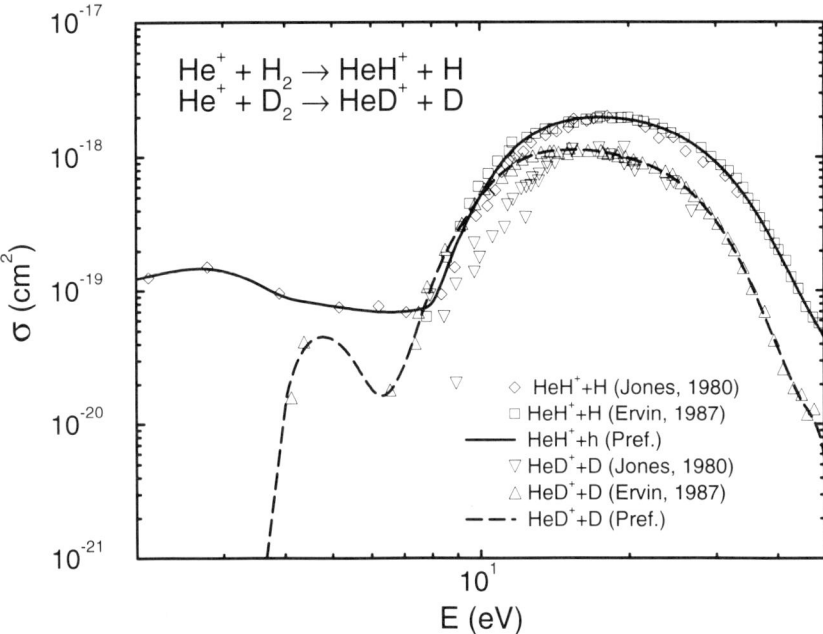

Figure 15. Total cross sections for the $He^+ + H_2 \to HeH^+ + H$ and $He^+ + D_2 \to HeD^+ + D$ reactions. The symbols represent the experimental data of Ervin and Armentrout[59] and Jones et al.[38] The curves represent the "preferred" cross sections for these reactions.

exhibit a nontypical behaviour. The cross sections for the similar endothermic reactions ($\Delta\varepsilon \simeq -2.7$ eV) $He^+ + HD \to HeD^+ + H$ and $He^+ + HD \to HeH^+ + D$, measured in Ref. 59, are given in Fig. 16.

3.2.4. Analytic Fits for the Total Cross Sections of Exothermic Particle Interchange Reactions

The total cross sections of all exothermic particle interchange reactions considered in the preceding two subsections show a similar energy behavior at both low and high energies. Therefore, they can all be represented by a unique general analytic expression with a limited number of free parameters. Taking into account the inverse-power energy dependence of the cross sections at both low and high energies, a suitable form of the analytic cross section expression is

$$\sigma = \frac{a_1[1 - \exp(-a_2 E^{a_3})]}{E^{a_4} + a_5 E^{a_6}} \; (\times 10^{-15} \text{ cm}^2) \tag{18}$$

in which a_1–a_6 are constants, and E is the center-of-mass collision energy in electron volts. The values of the parameters a_1–a_6 have been obtained by fitting Eq. (18) to

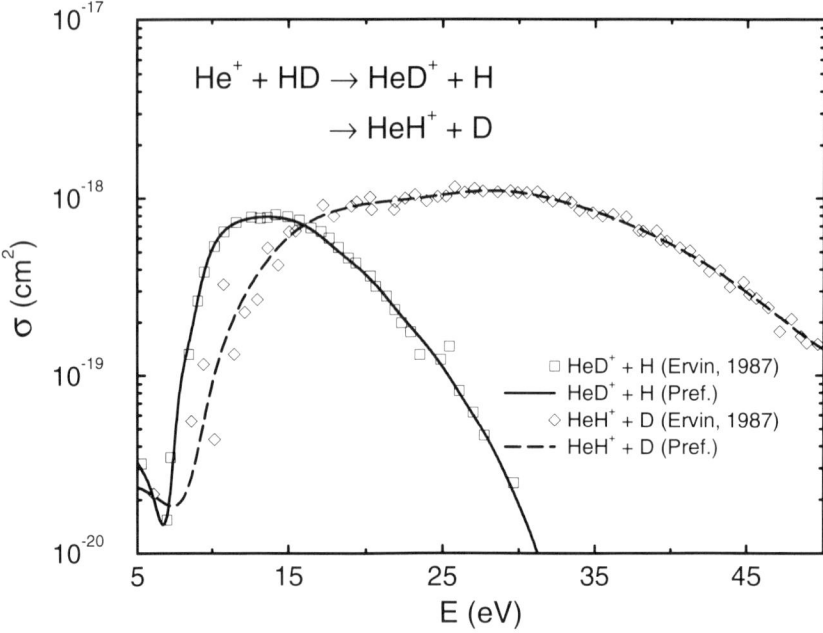

Figure 16. Total cross sections for the He$^+$ + HD → HeD$^+$ + H and He$^+$ + HD → HeH$^+$ + D reactions, based on the guided-beams experimental data of Ervin and Armentrout.[59]

Table V. Values of Fitting Parameters a_i in the Analytic Expression (Eq. 18) for the Total Cross Sections of Exothermic Particle Interchange Reactions

Reaction	a_1	a_2	a_3	a_4	a_5	a_6
D$^+$ + H$_2$ → HD + H$^+$	0.566	—	—	0.4839	0.0314	3.986
H$_2^+$ + D → HD$^+$ + H	5.712	0.1607	2.551	2.998	—	—
H$_2^+$ + H$_2$ → H$_3^+$ + H	18.08	0.0992	3.073	3.540	—	—
HD$^+$ + D$_2$ → HD$_2^+$ + D	1.63	1.97	−1.202	0.412	0.002	4.78
HD$^+$ + D$_2$ → D$_3^+$ + H	0.5484	1.902	−1.15	0.532	0.0078	3.691
D$_2^+$ + HD → HD$_2^+$ + D	3.971	0.357	2.372	2.77	—	—
D$_2^+$ + HD → D$_3^+$ + H	2.236	2.497	−0.3641	2.898	5.078	0.4986
H$_3^+$ + D$_2$ → HD$_2^+$ + H$_2$	144.3	0.00754	3.749	4.397	0.2552	4.0704
D$_3^+$ + H$_2$ → H$_2$D$^+$ + D$_2$	0.4571	—	—	0.432	0.0016	4.005
HeH$^+$ + H → H$_2^+$ + He	105.4	0.0233	3.648	4.551	3.113	4.01
H$_2^+$ + He → HeH$^+$ + H	21.09	0.01435	1.846	1.948	0.5085	3.721
HeH$^+$ + H$_2$ → H$_3^+$ + He	48.09	0.06098	2.867	3.132	0.560	4.541

the experimental data, or to the "preferred" cross sections, of the reactions considered in Sections 3.2.2 and 3.2.3 and are presented in Table V. The r.m.s. deviation of the fits is below 2%, which is much smaller than the uncertainties in the data (10–20% at the best). We note that the exothermicity of the reactions listed in Table V is an effective one that appears under the specific experimental conditions (i.e., it accounts for an unknown vibrational population of the reactants).

4. STATE-SELECTIVE CROSS SECTION MEASUREMENTS

The studies of ion–molecule reactions with state-selected reactants and reaction products are generally aimed at revealing the details of reaction dynamics. Therefore, these studies are usually restricted to a narrow range of collision energies, and the measured cross sections are normally presented in relative units.

Table VI. Experimental Cross Section Measurements of State-Selective Charge Transfer and Particle Interchange Reactions in Hydrogen Ion–Molecule Systems

Reaction	Energy range, E_{cm} (eV)	Method[a]	Reference
$H_2^+ (v, J) + H_2 \rightarrow H_2 + H_2^+$			
$v = 0; J = 0, 1, 2$	2–4	SSPI–CB	61
$H_2^+ (v_1) + H_2 \rightarrow H_2 + H_2^+ (v_2)$			
$v_1 = 0, 1; v_2 = 0, 1, 2$	2–16	SSPI–CB	62
$H_2^+ (v) + H_2 \rightarrow H_2 + H_2^+$			
$v = 0$–4	2–200	SSPI–CB	61
$v = 0$–5	4–500	PEPICO–CB	63
$v = 0$–10	4–16	PEPICO–CB	64
$D_2^+ (v) + H_2 \rightarrow D_2 + H_2^+$			
$v = 0$	0.2–3	SSPI–GB	65
$v = 0$–10	4	PEPICO–CB	64
$H_2^+ (v) + H_2 \rightarrow H_3^+ + H$			
$v = 0$–8	~0.04	PEPICO–SCC	66
$v = 0$–4	0.04–15	SSPI–GB	47
$v = 0$–3	0.1–1	PEPICO–SCC	48
$v = 0$–2	1.5–5.3	REMPI–CB	67
$H_2^+ (v) + D_2 \rightarrow HD_2^+ + H$			
$v = 0$–4	0.2–6	SSPI–GB	65
$D_2^+ (v) + H_2 \rightarrow HD_2^+ + H$			
$v = 0$–4	0.2–6	SSPI–GB	65
$D_2^+ (v) + D_2 \rightarrow D_3^+ + D$			
$v = 0$–12	~0.04	PEPICO–SCC	66

[a] For definitions of abbreviations, see footnotes to Tables I and III; REMPI = resonance-enhanced multiphoton ionization.

In the context of ion–molecule reactions, the term "state-selected" is usually used for measurements in which the internal state of at least one of the reactants is specified. When the internal states of both the reactants and the reaction products are specified, the measurements are termed "state-to-state selective." In most measurements, the "selectivity" refers to vibrational states only. Experimental studies of ion–molecule reactions with selected ro-vibrational states are extremely sparse. An account of the state-selective and state-to-state selective ion–molecule reaction studies performed in the past two decades can be found in Refs. 12 and 18–20. These references provide also detailed discussions of the experimental methods currently used in these reaction studies.

4.1. Reactions in Hydrogen Ion–Molecule Systems

A survey of the available experimental cross section measurements for the charge transfer and particle interchange reactions in hydrogen ion–molecule systems is given in Table VI. As seen from this table, most of the experimental investigations of these reactions have been performed by the state-selective photoionization (SSPI) and photoelectron–photoion coincidence (PEPICO) techniques.

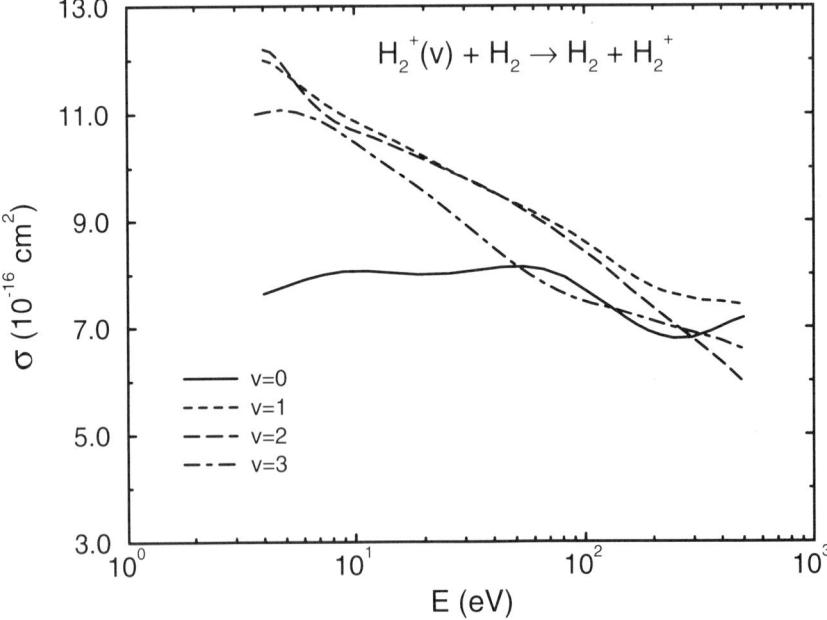

Figure 17. Cross sections for the state-selective electron transfer reactions $H_2^+(v) + H_2 \to H_2 + H_2^+$ based on the evaluation of Barnett.[9]

Reactive Ion–Molecule Collisions

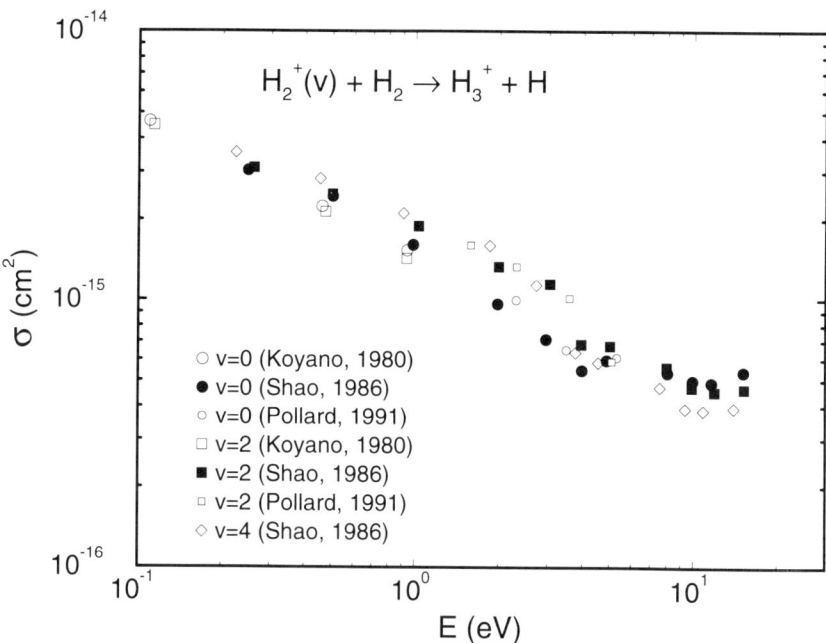

Figure 18. Cross sections for state-selective particle interchange reactions $H_2^+(v) + H_2 \rightarrow H_3^+ + H$ ($v = 0, 2, 4$). The symbols represent the experimental data of Koyano and Tanaka,[48] Shao and Ng,[47] and Pollard et al.[67]

The state-selective electron transfer in the $H_2^+ + H_2$ collision system has been studied most extensively. There is, however, only one experiment[61] in which rotationally state-selected H_2^+ ions have been used. The measured cross section shows no noticeable dependence on the rotational state of the ion. All other experiments are concerned with the vibrational state dependence of the electron transfer cross section. The measurements show a quite pronounced dependence of the cross section on the vibrational state of H_2^+, the cross section varying by up to a factor of 4 in the range $v = 0$–10.[64] In addition, the state-selective electron capture cross sections vary differently with the collision energy. This results from the fact that the quasi-resonant condition for the process for different initial vibrational states is reached at different collision energies, leading to different positions of the cross section maximum on the collision energy scale. It should also be noted that the results of different groups are not fully consistent with each other. In Fig. 17, we show the state-selective cross sections $\sigma(v)$, $v = 0$–3, from Ref. 9 to illustrate their magnitude and variation with the collision energy.

The state-selective cross sections for the particle interchange reaction $H_2^+(v) + H_2 \rightarrow H_3^+ + H$ have been extensively studied[47,48,66,67] and show a relatively weak

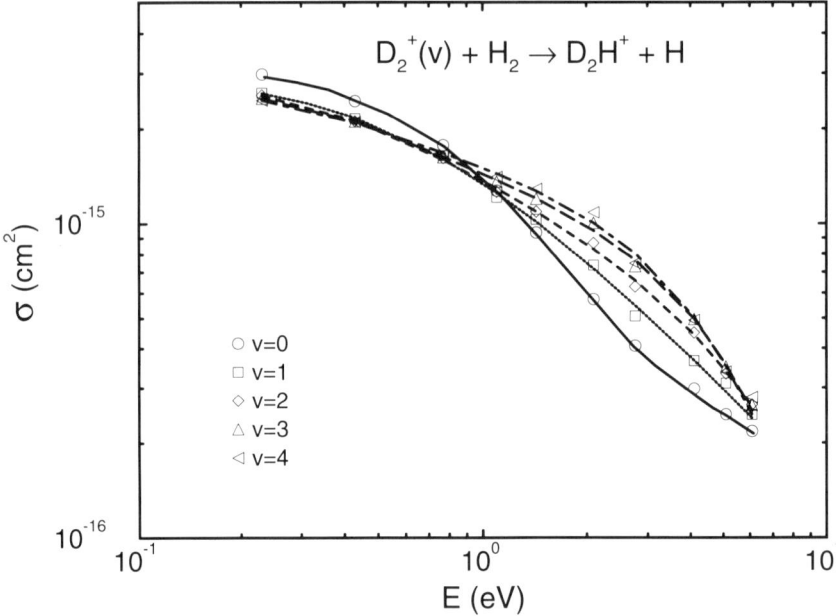

Figure 19. Cross sections for state-selective particle interchange reactions $D_2^+(v) + H_2 \rightarrow D_2H^+ + H$ ($v = 0$–4). The data are from Anderson et al.,[65] and the curves serve to guide the eye.

dependence on the vibrational state of the H_2^+ ion (see Fig. 18; only the $v = 0$, $v = 2$, and $v = 4$ cross sections are shown for the sake of clarity). This may be a consequence of the fact that the proton and atom transfer channels of this reaction have opposite dependences on the vibrational state of $H_2^+(v)$. The situation is similar also in the case of other isotopic variants of this reaction and is illustrated in Fig. 19 for the reaction $D_2^+(v) + H_2 \rightarrow D_2H^+ + H$.

The collision-induced dissociation (CID) processes in hydrogenic ion–molecule systems show a very strong dependence on the initial vibrational state of the molecular ion. For the $H_2^+(v) + H_2$ system, the CID cross section at 4 eV increases by a factor of 7 in the range $v = 0$–6.[64]

4.2. Reactions in Hydrogen-Helium Ion-Molecule Systems

The available state-selective cross section measurements for the reactive collisions of hydrogen–helium ion–molecule systems, including a number of dissociative processes, are given in Table VII. The reaction $H_2^+(v) + He \rightarrow HeH^+ + H$, with values of v up to 8, has been the most extensively studied.[66,68–71] The state-selective cross sections $\sigma(v)$ ($v = 0$–5) for this reaction are shown in Fig. 20

Table VII. Experimental Cross Section Measurements of State-Selective Reactive Collisions in Hydrogen–Helium Ion–Molecule Systems

Reaction	Energy range, E_{cm} (eV)	Method[a]	Reference(s)
$H_2^+ (v) + He \rightarrow HeH^+ + H$			
$v = 0$–8	~0.04	PEPICO–SCC	66
$v = 0$–5	0.04–7	SSPI–SCC	68, 69
$v = 0$–4	1–8	SSPI–GB	70
$v = 0$–6	3.1	PEPICO–CB	71
$HD^+(v) + He \rightarrow HeH^+ + D$			
$v = 0$–4	1–8	SSPI–GB	70
$HD^+(v) + He \rightarrow HeD^+ + H$			
$v = 0$–4	1–8	SSPI–GB	70
$H_2^+ (v) + He \rightarrow H^+ + H + He$			
$v = 0$–5	0.04–7	SSPI–SCC	68, 69
$v = 0$–6	3.1	PEPICO–CB	71
$He^+ + H_2 \rightarrow He + H^+ + H^*(nl)$	5–50	IBGC–OES	38
	10–230	IBGC–OES	72
	10–670	IBGC–OES	73
$He^+ + D_2 \rightarrow He + D^+ + D^*(nl)$	5–80	IBGC–OES	38

[a] For definitions of abbreviations, see footnote to Table VI; IBGC, ion-beam–gas-cell; OES, optical emission spectroscopy.

and are based on the data of Refs. 68–71. For $v = 0$–3, the reaction is endothermic, and the corresponding cross sections show a distinct threshold. For $v \geq 4$, the reaction becomes exothermic with an approximate $E^{-0.5}$ dependence of the cross section at low energies. These state-selective cross sections, weighted according to the vibrational state distribution from Ref. 60, have been used to generate the "preferred" total cross section for this reaction given in Fig. 12.

The state-selective cross sections for the isotopic variants of the $H_2^+(v) + He \rightarrow HeH^+ + H$ reaction display similar features. The significant vibrational state dependence of the state-selective reactive cross sections in the $H_2^+(v) + He$ collision system is also present in the cross sections for dissociative channels. For a collision energy of 3 eV, the collision-induced dissociative cross section increases by a factor of 24 in the range $v = 0$–6.[71]

For the dissociative electron transfer reactions in the $He^+ + H_2$ and $He^+ + D_2$ systems, state-selective measurements have been performed[38,72,73] with respect to the excited products $H^*(nl)$ and $D^*(nl)$. These measurements, however, have yielded only total line emission cross sections and do not include the "dark" channels $H(1s)$ and $H(2s)$.

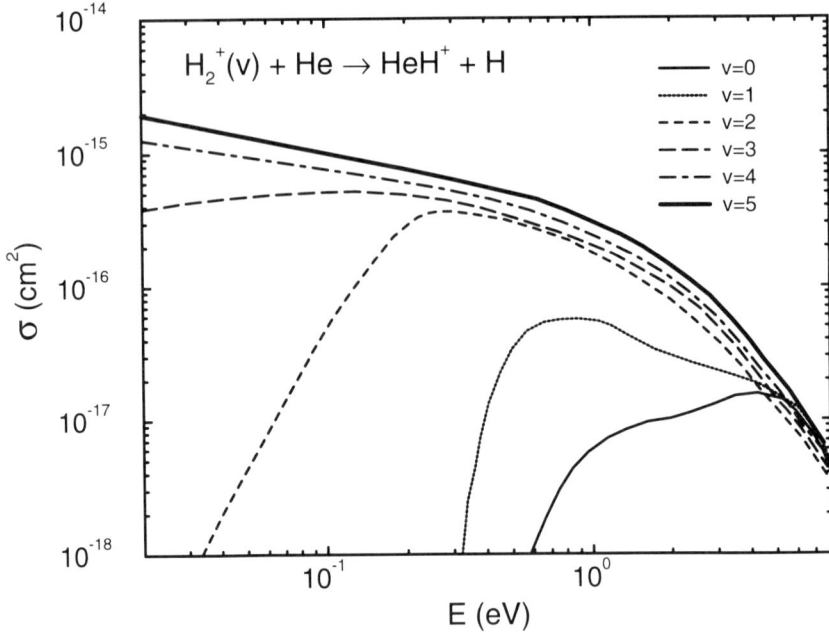

Figure 20. Cross sections for state-selective particle interchange reactions $H_2^+(v) + He \rightarrow HeH^+ + H$ ($v = 0\text{–}5$), based on the assessed data of Refs. 68–71.

4.3. Energy and Angular Distribution of Reaction Products

An important aspect of the analysis of reactive ion–molecule collisions is the energy and angular distribution of the reaction products. Measurements of these distributions can provide an insight into the collision dynamics and reveal the underlying reaction mechanisms in various collision systems. Such measurements have been performed for the majority of particle interchange reactions in the hydrogenic and hydrogen–helium systems discussed here and have recently been reviewed by Reinig et al.[11] Most of the available information on the reaction product distributions is in relative units. Converting this information to an absolute scale requires normalization either to certain accurate theoretical results or to an absolute experimental total cross section (if the information on energy and angular distribution is sufficiently complete).

The use of product energy and angular distributions to elucidate the reaction mechanisms in various collision systems can be most successful if it is coupled with an appropriate dynamical analysis of the evolution of the collision system on the potential energy surfaces. It is obvious that owing to the strong effects of the internal energy of reactants on the collision dynamics, the energy and angular distributions of reaction products will be highly sensitive to the vibrational state distributions of reactants. Therefore, extreme caution should be exercised in using the existing information on the product energy and angular distributions from reactive collisions in plasma edge modeling applications.

5. SUMMARY AND CONCLUSIONS

In this chapter we have reviewed the available experimental cross section information for the particle rearrangement processes (electron transfer and particle interchange) in low-energy ion–molecule collisions involving hydrogen and helium species. The cross section information presented is based mainly on the results of crossed-, merged-, and guided-beams experiments, which are currently considered as having the highest reliability. We have chosen to present the total and state-selective cross section data for only the most important particle rearrangement reactions in hydrogenic and hydrogen–helium ion–molecule collision systems, characterized by large cross sections in the low-energy region (below 200 eV). For this reason, the emphasis in our considerations was given to the exothermic particle interchange reactions with cross sections typically greater than 10^{-15} cm^2 for energies below 1–2 eV. For the total cross sections that we have presented for these reactions, we have also provided simple analytical fits. Because of its fragmentary nature, the cross section information on state-selective processes has been less exhaustively reviewed in this chapter; instead, we refer to the recent critical evaluation by Barnett[9] and to the original data sources (Tables VI and VII).

As discussed in Section 3, the total cross sections for particle rearrangement processes in the collision systems considered are very sensitive to the vibrational state populations of the reactants. This makes the comparison of cross section results from different experiments (with differently prepared molecular reactants) meaningless to some extent and impairs the intercomparison-based assessment of the accuracy of the data. The generation of accurate total cross sections for particle rearrangement ion–molecule reactions must, therefore, be based on systematic studies with state-selected reactants and on accurately known internal state distributions of the individual reactants. Although the presently available experimental techniques provide a good basis for carrying out very accurate state-selective cross section measurements, the existing information of this kind is limited to very narrow energy ranges and does not cover all the collision systems of interest. As mentioned in Section 4, much of this information is given only in relative units. Since the experimental measurements are normally restricted to a certain range of physical parameters, it would be highly useful to parallel such measurements with analogous theoretical studies in order to extend the (normalized) data into a broader parametric space.

REFERENCES

1. H. Tawara and R. A. Phaneuf, *Comments At. Mol. Phys.* **21**, 177 (1988).
2. R. K. Janev, M. F. A. Harrison, and H. W. Drawin, *Nucl. Fusion* **29**, 109 (1989).
3. R. K. Janev, *Comments At. Mol. Phys.* **26**, 83 (1991).
4. D. Reiter, in *Atomic and Plasma-Material Interaction Processes in Controlled Thermonuclear Fusion* (R. K. Janev and H. W. Drawin, eds.), Elsevier, Amsterdam (1993), p. 243.
5. M. F. A. Harrison, in *Atomic and Plasma-Material Interaction Processes in Controlled Thermonuclear Fusion* (R. K. Janev and H. W. Drawin, eds.), Elsevier, Amsterdam (1993), p. 285.
6. R. K. Janev, in *Review of Fundamental Processes and Applications of Atoms and Ions* (C. D. Lin, ed.), World Scientific, Singapore (1993), p. 1.
7. R. K. Janev, W. D. Langer, K. Evans, and D. E. Post, *Elementary Processes in Hydrogen–Helium Plasmas*, Springer-Verlag, Heidelberg (1987).
8. A. V. Phelps, *J. Phys. Chem. Ref. Data* **19**, 653 (1990).
9. C. F. Barnett, Collisions of H, H_2, He and Li Atoms and Ions with Atoms and Molecules, Atomic Data for Fusion, Vol. 1, Oak Ridge National Laboratory, Report ORNL-6086, 1990.
10. H. Tawara, Y. Itikawa, Y. Itoh, T. Kato, H. Nishimura, S. Ohtani, H. Takagi, K. Takayanagi, and M. Yoshino, Atomic Data Involving Hydrogens Relevant to Edge Plasmas, Institute of Plasma Physics, Nagoya, Japan, Report IPPJ-AM-46, 1986.
11. P. Reinig, M. Zimmer, and F. Linder, *Atomic and Plasma-Material Interaction Data for Fusion (Nucl. Fusion, Supplement)* **2**, 95 (1992).
12. D. Gerlich, *Adv. Chem. Phys.* **82**, 1 (1992).
13. G. Bischof and F. Linder, *Z. Phys. D* **1**, 303 (1986).
14. D. R. Miller, in *Atomic and Molecular Beam Methods*, Vol. 1 (G. Scoles, ed.), Oxford University Press, Oxford (1988), p. 14.
15. S. Trajmar and D. F. Register, in *Electron–Molecule Collisions* (I. Shimamura and K. Takayanagi, eds.), Plenum, New York (1984), p. 427.
16. J. C. Nickel, P. W. Zetner, G. Shen, and S. Trajmar, *J. Phys. E: Sci. Instrum.* **22**, 730 (1989).
17. C. Y. Ng, in *Techniques for the Study of Ion–Molecule Reactions* (J. M. Farrar and W. H. Saunders, eds.), John Wiley & Sons, New York (1988), p. 417.
18. C. Y. Ng, *Adv. Chem. Phys.* **82**, 401 (1992).
19. I. Koyano and K. Tanaka, *Adv. Chem. Phys.* **82**, 263 (1992).
20. S. L. Anderson, *Adv. Chem. Phys.* **82**, 177 (1992).
21. M. G. Holliday, J. T. Muckerman, and L. Friedman, *J. Chem. Phys.* **54**, 1058 (1971).
22. M. W. Gealy and B. Van Zyl, *Phys. Rev. A* **36**, 3091 (1987).
23. W. H. Cramer, *J. Chem. Phys.* **35**, 836 (1961).
24. D. W. Koopman, *Phys. Rev.* **154**, 79 (1967).
25. G. Ochs and E. Teloy, *J. Chem. Phys.* **61**, 4930 (1974).
26. C. Schlier, U. Nowotny, and E. Teloy, *Chem. Phys.* **111**, 401 (1987).
27. W. B. Maier II, *J. Chem. Phys.* **54**, 2732 (1971).
28. J. R. Krenos and R. Wolfgang, *J. Chem Phys.* **52**, 5961 (1970).
29. W. H. Cramer and A. B. Marcus, *J. Chem. Phys.* **32**, 186 (1960).
30. W. L. Fite, R. T. Brackmann, and W. R. Snow, *Phys. Rev.* **112**, 1161 (1958).
31. H. B. Gilbody and J. B. Hasted, *Proc. R. Soc. London, Ser. A* **238**, 334 (1956).
32. H. L. Rothwell, B. Van Zyl, and R. C. Amme, *J. Chem. Phys.* **61**, 3851 (1974).
33. H. C. Hayden and R. C. Amme, *Phys. Rev.* **172**, 104 (1968).
34. J. B. H. Stedeford and J. B. Hasted, *Proc. R. Soc. London, Ser. A* **227**, 466 (1955).
35. R. N. Stocker and H. Neumann, *J. Chem. Phys.* **61**, 3852 (1974).
36. R. L. C. Wu and D. G. Hopper, *Chem. Phys.* **57**, 385 (1981).
37. E. Gustafsson and E. Lindholm, *Ark. Fys.* **18**, 219 (1960).

38. E. G. Jones, R. L. C. Wu, B. M. Hughes, T. O. Tiernan, and D. G. Hopper, *J. Chem. Phys.* **73**, 5631 (1980).
39. R. W. Rozett and W. S. Koski, *J. Chem. Phys.* **48**, 533 (1968).
40. T. F. Moran and R. J. Conrads, *J. Chem. Phys.* **58**, 3793 (1973).
41. D. G. Hopper and R. L. C. Wu, *Chem. Phys. Lett.* **81**, 230 (1981).
42. J. R. Krenos, R. K. Preston, R. Wolfgang, and J. C. Tully, *J. Chem. Phys.* **60**, 1634 (1974).
43. H. Villinger, M. J. Henchman, and W. Lindinger, *J. Chem. Phys.* **76**, 1590 (1982).
44. K. L. Wendell and P. K. Pol, *J. Chem. Phys.* **61**, 2059 (1974).
45. W. R. Gentry, D. J. McClure, and C. H. Douglass, *Rev. Sci. Instrum.* **46**, 367 (1975).
46. R. H. Neynaber and S. M. Trujillo, *Phys. Rev.* **167**, 63 (1968).
47. J. P. Shao and C. Y. Ng, *J. Chem. Phys.* **84**, 4317 (1986).
48. I. Koyano and K. Tanaka, *J. Chem. Phys.* **72**, 4858 (1980).
49. L. T. Specht, K. D. Foster, and E. E. Muschlitz, *J. Chem. Phys.* **63**, 1582 (1975).
50. C. F. Giese and W. B. Maier II, *J. Chem. Phys.* **39**, 739 (1963).
51. D. W. Vance and T. L. Bailey, *J. Chem. Phys.* **44**, 486 (1966).
52. T. F. Moran and J. R. Roberts, *J. Chem. Phys.* **49**, 3411 (1968).
53. C. H. Douglass, D. J. McClure, and W. R. Gentry, *J. Chem. Phys.* **67**, 4931 (1977).
54. C. H. Douglass, G. Ringer, and W. R. Gentry, *J. Chem. Phys.* **76**, 2423 (1982).
55. R. H. Neynaber and G. D. Magnuson, *J. Chem. Phys.* **59**, 825 (1973).
56. J. A. Rutherford and D. A. Vroom, *J. Chem. Phys.* **58**, 4076 (1973).
57. C. Schlier, cited in: F. Schneider, U. Havemann, L. Zülicke, and Z. Herman, *Chem. Phys. Lett.* **48**, 439 (1977).
58. J. A. Rutherford and D. A. Vroom, *J. Chem. Phys.* **59**, 4561 (1973).
59. K. M. Ervin and P. B. Armentrout, *J. Chem. Phys.* **86**, 6240 (1987).
60. F. Busch and G. H. Dunn, *Phys. Rev. A* **5**, 1726 (1972).
61. C. L. Liao, C. X. Liao, and C. Y. Ng, *J. Chem. Phys.* **81**, 5672 (1984).
62. C. L. Liao and C. Y. Ng, *J. Chem. Phys.* **84**, 197 (1986).
63. F. M. Campbell, R. Browning, and C. J. Latimer, *J. Phys. B* **14**, 3493 (1981).
64. P. M. Guyon, T. Baer, S. K. Cole, and R. T. Govers, *Chem. Phys.* **119**, 145 (1988).
65. S. L. Anderson, F. A. Houle, D. Gerlich, and Y. T. Lee, *J. Chem Phys.* **75**, 2153 (1981).
66. D. van Pijkeren, E. Boltjes, J. van Eck, and A. Niehaus, *Chem. Phys.* **91**, 293 (1984).
67. J. E. Pollard, L. K. Johnson, D. A. Lichtin, and R. B. Cohen, *J. Chem. Phys.* **95**, 4877 (1991).
68. W. A. Chupka and M. E. Russell, *J. Chem. Phys.* **49**, 5426 (1968).
69. W. A. Chupka, J. Berkowitz, and M. E. Russell, in: *Book of Abstracts of VIth International Conference on the Physics of Electronic and Atomic Collisions*, MIT Press, Cambridge, Massachusetts (1969), p. 71.
70. T. Turner, O. Dutuit, and Y. T. Lee, *J. Chem. Phys.* **81**, 3475 (1984).
71. T. R. Govers and P. M. Guyon, *Chem. Phys.* **113**, 425 (1987).
72. R. C. Isler and R. D. Nathan, *Phys. Rev. A* **6**, 1036 (1972).
73. G. H. Dunn, R. Geballe, and D. Pretzer, *Phys. Rev.* **128**, 2200 (1962).
74. G. Niedner, M. Noll, J. P. Toennies, and C. Schlier, *J. Chem. Phys.* **87**, 2685 (1987).
75. G. Piepke, Diploma Thesis, University of Freiburg, 1980.

Chapter 15

Particle Interchange Reactions Involving Plasma Impurity Ions and H_2, D_2, and HD

P. B. Armentrout and J. Botero

1. INTRODUCTION

The edge plasma in a tokamak nuclear fusion reactor is characterized by low plasma temperatures and high plasma densities. In the divertor region, the plasma temperature may be as low as a few electron volts, and plasma densities may be as high as 10^{17} cm^{-3}.[1] An important consequence of the low edge plasma temperature is that molecular species are present in this region, resulting either from plasma–wall interactions (e.g., hydrocarbons) or from recycling and plasma fueling (molecular hydrogen and its isotopes). In addition to the primary constituents of the plasma, for example D, T, and He (in D–T plasmas) and H, D, and He in the present tokamaks, a relatively large variety of atomic impurities (at concentrations between 0.1 and 10%) are present. The most common impurity-generating processes are particle–surface interaction processes, mainly desorption, physical sputtering, and evaporation. The main impurities in the plasma edge of most of the present-generation tokamaks and fusion reactor designs are carbon (≤10%), oxygen (≤5%),

---

P. B. ARMENTROUT • Department of Chemistry, University of Utah, Salt Lake City, Utah, 84112.
J. BOTERO • International Atomic Energy Agency, A-1400 Vienna, Austria.

Atomic and Molecular Processes in Fusion Edge Plasmas, edited by R. K. Janev. Plenum Press, New York, 1995.

various metallic (and related) impurities originating from structural materials (Ti, V, Cr, Fe, Ni, Cu, Al, Mo, Mn, Mg, B, Si, Ge, Nb, Ag; with a concentration of ≤2%), and various diagnostic species (Li, Ne, Ar). Under these plasma conditions, a wide range of atomic and molecular processes that are not important in the core plasma become relevant.

Heavy-particle interchange reactions in ion–molecule collisions refer to processes in which heavy-particle rearrangement occurs, such as

$$A^{q+} + BC \rightarrow AB^{q+} + C \qquad (1)$$

In this chapter we will cover particle interchange reactions in which the ion is a plasma impurity with a charge state $q = 1$ and the molecule is H_2, D_2, or HD. In general, there are two different types of these reactions. Endothermic reactions, which are characterized by a threshold below which the reaction does not take place (generally equivalent to the endothermicity of the reaction), and exothermic reactions, in which case the reaction often (but not always) occurs without an activation energy because of the attractive long-range interaction potential.[2,3] Of the reactions presented here, all but those of O^+ are endothermic. The kinetic energy dependence of the interchange cross section for these two types of reactions is, of course, quite different. For exothermic reactions, the cross section often follows a relatively simple $E^{-1/2}$ decrease with increasing energy (in accord with the Langevin–Giou-mousis–Stevenson prediction[4]), but deviations from this behavior abound.[5] For endothermic reactions, the cross section usually increases sharply starting at an energy close to the endothermicity of the reaction, peaks at an energy close to the molecular (H_2, D_2, or HD) dissociation limit, and decreases drastically within a few electron volts. The reason for this sharp decrease at high energies is that the dissociation channel opens; that is, the reaction

$$A^+ + BC \rightarrow A^+ + B + C \qquad (2)$$

with a higher cross section, takes place. This decline in the interchange cross section after the dissociation energy has been reached may be delayed, in terms of the energy at which it occurs, if the reaction dynamics tend to place much of the excess energy in product translation.

2. EXPERIMENTAL DESCRIPTION

2.1. General Considerations

The work described in this chapter has been performed on a guided-ion-beam tandem mass spectrometer that has undergone several changes in its 10-year history.[6–8] In this apparatus, ions exiting the ion source region are focused into a magnetic sector for mass analysis, decelerated to a desired kinetic energy, and injected into an octopole ion-beam guide that passes through a collision cell filled with the neutral reactant. Pressures of this gas are generally kept sufficiently low that only single ion–molecule collisions are probable. (Deviations from single-collision conditions are readily ascertained by examination of the pressure dependence of the product yield.) Product and reactant ions drift from the gas cell to the end of the octopole, where they are extracted and focused into a quadrupole mass filter for mass analysis. The ions are then detected by using a secondary electron scintillation ion detector[9] and counted by using standard pulse counting electronics. A computer sweeps the kinetic energy of the ion beam while monitoring the reactant ion and all product ions so that extensive signal averaging can be performed easily.

Ion intensities are converted to absolute reaction cross sections as described previously.[6] Based on reproducibility, the relative uncertainty of cross sections at different energies is within about 5% for cross sections greater than 10^{-17} cm^2 and is limited by statistical counting uncertainties for smaller cross sections. The absolute accuracy of the cross sections is limited by systematic errors in the measurement of the absolute neutral reactant pressure and our estimate of the effective path length for interaction; both these errors are estimated at about 10%. Although the use of an octopole minimizes losses of ions, the possibility of ion loss especially at very low and very high energies cannot be ruled out entirely. For the heavy-on-light mass systems discussed in this work, collection deficiencies are minimized by the forward scattering in the laboratory frame necessitated by linear momentum conservation. Under most circumstances, we estimate that the uncertainties in the absolute cross sections are within ±20%. Comparisons of the absolute cross sections measured with this instrument with calculated capture rate constants[10] and with thermal rate constant measurements[6,11,12] suggest that our cross sections generally suffer from no serious systematic errors and that their accuracy is very good.

2.2. The Octopole Ion-Beam Guide

The octopole ion beam guide, first developed by Teloy and Gerlic,[13] is the primary key to the unique capabilities of this apparatus when compared with more conventional tandem mass spectrometers. A comprehensive analysis and description of such inhomogeneous rf devices has been made recently by Gerlich.[14] In our instrument, the octopole comprises eight rods (3.2-mm diameter) held in a cylindrical array (17.2-mm bolt circle). Alternate phases of an rf potential are applied to alternate rods such that an effective potential well in the radial direction is established. Energies along the axis of the octopole are perturbed little because the potential well, which depends on the inverse sixth power of the distance from the center, is flat at the bottom and has steep sides.[6,14,15] One of the virtues of the octopole is that it enables the use of retarding field analysis to measure the absolute energy of the ion beam and its distribution. The octopole avoids problems associated with contact potentials, space-charge effects, and focusing aberrations because the analysis and interaction regions are physically the same. Further, the octopole trapping field prevents anomalous losses of ions even at very low kinetic energies. We have verified the ability of the octopole to accurately measure kinetic energies by comparison with time-of-flight experiments[6] and with calculated cross sections.[10] We estimate that the uncertainty associated with measuring the zero of energy is about 0.05 eV in the laboratory frame.

2.3. Kinetic Energy Scale and Doppler Broadening

Laboratory energies are converted to center-of-mass (CM) energies by using the stationary target assumption. Thus,

$$E_{CM} = E_{lab} \times m/(M + m) \qquad (3)$$

where m and M are the masses of the neutral and ionic reactants, respectively. At the very lowest energies, this approximation is not adequate because the ion energy distribution is being truncated. In such circumstances, the mean energy that is reported properly reflects the truncated distribution of ions, as described in detail previously.[6] This kinetic energy scale in the center-of-mass frame does not include the thermal motion of the reactant neutral, which adds another $(3/2)\gamma kT$ to the average energy, where $\gamma = M/(M + m)$. The neutral reactant molecules have a Maxwell–Boltzmann distribution of velocities at the temperature of the gas cell, usually 305 K in our work. The effects of this motion are to obscure sharp features in the true cross sections, a result that is especially obvious at a reaction threshold. In addition, the velocities of the neutral reactant molecules can be comparable to or larger than the ion velocities at very low kinetic energies. At these energies, the full distribution of neutral and ion velocities must be considered in order to fully describe the average interaction energy. The means necessary to describe this so-called Doppler broadening were first developed in detail by Chantry[16] for the case of a monoenergetic ion beam and extended to include consideration of ion energy distributions by Lifshiftz et al.[17] The width of the energy distribution [full width at half maximum (FWHM)] contributed by the neutral molecules at a center-of-mass energy E was given by Chantry[16] as approximately $(11.1\,\gamma kTE)^{1/2}$.

2.4. Ion Sources

The guided-ion-beam apparatus shown in Fig. 1 is capable of using several types of ion sources: commonly, surface ionization (SI), electron impact (EI) ionization, and two high-pressure sources—a drift cell and a flow tube source. In some studies, several of these sources are used in order to systematically vary the degree of internal excitation of the ion. The SI source[18] can be used to ionize species with low ionization energies (IEs), such as metal atoms, and is believed[19] to produce ions with a Maxwell–Boltzmann distribution of internal states at the temperature of the ionizing filament, 1900–2300 K. EI[18] can ionize and fragment any volatile

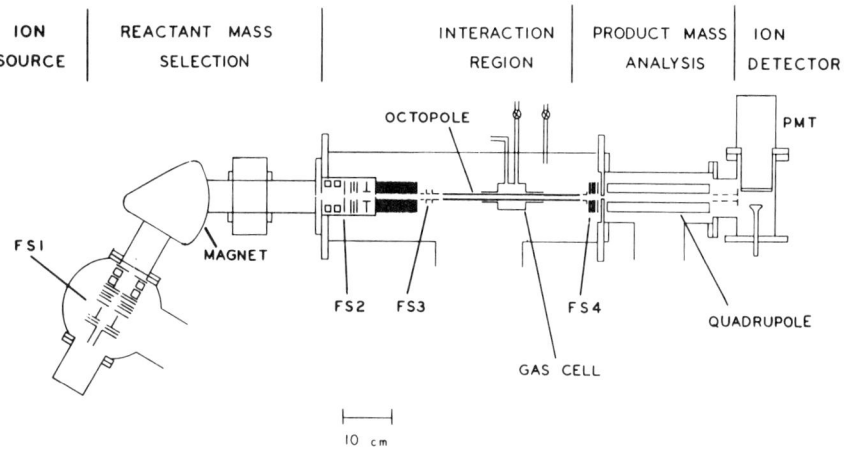

Figure 1. Schematic overview of the guided-ion-beam tandem mass spectrometer. FS1–FS4: focusing stages comprised of ion lens systems. (Reprinted with permission of VCH Publishers.)

gas but can produce internally excited species depending on the electron energy. The flow tube source, developed as a more versatile version of an earlier drift cell (DC) source,[20] is modeled after those described by Lineberger and co-workers[21] and Graul and Squires[22] and is detailed in the literature.[8] This high-pressure source (~0.5 Torr) is designed to produce atomic ions with a minimum of internal energy, that is, species thermalized to near room temperature.

3. THEORETICAL CONSIDERATIONS

The theoretical study of the particle interchange reactions considered here is still an undeveloped field. It started about 20 years ago with the trajectory surface-hopping model first suggested by Bjerre and Nikitin[23] and the multivibronic curve crossing model introduced by Bauer, Fischer, and Gilmore.[24] These two methods are based on the Landau–Zener model for the transition probability of a transition from one adiabatic curve to another. Quantum-mechanical calculations are scarce and are mostly devoted to collinear systems or to three dimensional calculations where the infinite-order sudden approximation is applied.[25] Alternative simpler theories include statistical theories, such as phase space theory (PST) or transition state theory (TST), and empirical or semiempirical approximations. Theoretical results for the reactions presented here are very scarce. In a few cases, PST results can be found in the literature.[26] Simons and co-workers[27] have examined the B^+ and Al^+ targets by *ab initio* methods.

The phase space theory for ion–molecule reactions[26,28,29,30] uses the long-range ion-induced dipole potential (Langevin model) to determine the complex-formation cross section. The complex can then decompose into products or back to reactants via "loose" transition states.[26] Total energy and angular momentum are explicitly conserved. The presumption of loose transition states means that only the molecular parameters of isolated reactants and products are required to determine the number of states available in each channel. No explicit information regarding the complex or transition states is needed for PST calculations. In its simplest form, PST does not differentiate among any of the states in the reactant or product channels. All states that are accessible while conserving both energy and momentum are included in the statistical sum of states. Sums for electronic states that are degenerate are multiplied by the appropriate degeneracy factor.

It has been shown that classical phase space theory[31,32] can be used instead of the full quantum-mechanical version for the system for which PST calculations are presented here ($C^+ + H_2$). Details of the calculation can be found elsewhere.[26] The reaction cross section is given by

$$\sigma(E) = (1/Q_{tot}) \sum_J (2J+1) g_i e^{-BJ(J+1)/kT} \sigma(E, J) \tag{4}$$

where Q_{tot} is the rotational partition function, B is the rotational constant, g_i is the nuclear spin degeneracy factor, T is the temperature of the target gas, and $\sigma(E, J)$ is the cross section for a particular rotational state J of reactant hydrogen as a function of relative energy E, given by[31]

$$\sigma(E, J) = \frac{\pi \hbar^2}{2\mu E} \frac{s}{2JG} \int_{K_-}^{K_+} dK \, 2K \, N^{ots}_{reag}(E, J, K) N^{ots}_{prod}(E_0, K)/N^{ots}_{tot}(E_0, K) \tag{5}$$

In Eq. (5), G and μ are, respectively, the total number of degenerate electronic surfaces and the reduced mass; s is the symmetry number; E and K are, respectively, the relative translational energy of the reactants and the total angular momentum; E_0 is the total system energy defined as

$$E_0 = E + E_{rot} + E_{vib} \tag{6}$$

$N^{ots}_{reag}(E, J, K)$ is the sum of orbital states of the reactants for given values of E, J, and K; $N^{ots}_{prod}(E_0, K)$ is the total sum of states of the products for given values of E and K; and $N^{ots}_{tot}(E_0, K)$ is the sum of the $N^{ots}_i(E_0, K)$ (total sum of states in channel i) over all available channels. Note that in all sums N, all accessible electronic surfaces and reaction path degeneracy have been explicitly taken into account. Finally, K_- and K_+ are the maximum and minimum values of the total angular momentum K for which the reactive flux is greater than zero.

An important aspect of the reactions studied here is the isotopic effects, for example, comparisons between the interchange cross sections for H_2, D_2, and HD. In particular, the branching ratio between reactions

$$A^+ + HD \rightarrow AH^+ + D \tag{7}$$

$$A^+ + HD \rightarrow AD^+ + H \tag{8}$$

appears to be quite sensitive to reaction dynamics and is still not fully understood. For endothermic reactions, there seem to be three different types of behavior[33]: (1) the branching ratio is nearly unity; (2) the process in which AH^+ is formed is favored by a factor of about 3; and (3) the process in which AD^+ is formed is favored by a large factor. In the first two types of reactivity, the reaction thresholds are observed to correspond to the thermodynamic thresholds, whereas in the third type, the experimental thresholds are higher than the thermodynamic thresholds and those for the reactions in Eqs. (7) and (8) differ from each other and from those for reaction with H_2 and D_2.

The first type of isotopic behavior reflects approximately a statistically behaved system. This can be illustrated by examining the density of states for the products of the reactions given in Eqs. (7) and (8). If we assume that the mass of A greatly exceeds the mass of H and D—a reasonable approximation for all cases studied here—the density of internal states favors the formation of AD^+. In the classical limit, the density of vibrational states is given by $1/\hbar\omega$, ω being the vibrational frequency. Because $\omega = (k/\mu)^{1/2}$ and the reduced mass $\mu(AH^+) \approx 1$ while $\mu(AD^+) \approx 2$, this favors the formation of AD^+ by a factor of $2^{1/2}$. The classical density of rotational states is $1/hcB$, where $B \propto 1/\mu$, and hence the formation of AD^+ is favored by a factor of 2. The density of translational states is proportional to $m^{3/2}$, where m is the reduced mass of the reactant or product channel, and $m(AD^+ + H) \approx 1$, while $m(AH^+ + D) \approx 2$. This gives an extra factor of $2^{3/2}$ in favor of AH^+. Overall, these factors approximately cancel such that the classical statistical isotope effect is about 1:1 formation of AH^+ and AD^+. This simple treatment ignores all quantum effects but does capture the essence of a statistically behaved system. More detailed calculations using PST give a branching ratio close to 1:1 for this case.[26]

In the second type of systems, the formation of AH^+ is favored by a factor that could be interpreted as a simple mass factor. In the analysis above, it is the internal density of states that favors the formation of AD^+. If only the translational degrees of freedom are taken into account, the formation of AH^+ is favored by a factor of $2^{3/2} = 2.8$. This would reflect a direct reaction in which the internal degrees of freedom are unimportant.

The third type of behavior may be explained in a case in which the energy relevant to a particle interchange reaction is not the center-of-mass (CM) energy, but a "pairwise" interaction energy. In the CM frame, the energy available for chemical change is the relative kinetic energy between the incoming atom of mass

M_A and the reactant molecule with mass $(M_B + M_C)$. The center-of-mass energy is given by Eq. (3) with $m = M_B + M_C$ and $M = M_A$. In a pairwise interaction, A is sensitive only to the potential between A and the interchanged atom B. Therefore, the pairwise energy for transfer of B from molecule BC is

$$E_{B,BC} = E_{lab} \frac{M_B}{M_A + M_B} \qquad (9)$$

In cases in which $M_A \gg M_B$ or M_C, substitution of Eq. (3) into Eq. (9) leads to the approximate relation

$$E_{B,BC} = E_{CM} \frac{M_B}{M_B + M_C} \qquad (10)$$

such that the energy available for chemical change in a pairwise reaction is always less than E_{CM}. For reactions with BC = H_2 or D_2, this pairwise mass factor $\delta = M_B/(M_B + M_C)$ is $\frac{1}{2}$. If BC = HD, then $\delta = \frac{1}{3}$ if the transferred atom is H; if the transferred atom is D, then $\delta = \frac{2}{3}$. This means that if the thermodynamic threshold occurs at E_0, the pairwise threshold will occur at E_0/δ. Therefore, the enhanced production of AD^+ may be due to the lower apparent threshold for this reaction, 1.5 E_0, as compared to that for the one in which AH^+ is formed, $3E_0$. This pairwise scheme also explains the shift in the thresholds observed for the H_2 ($2E_0$), D_2 ($2E_0$), and HD systems.

4. RESULTS

4.1. Carbon, Oxygen, and Silicon

In many presently operating fusion devices, carbon is one of the main impurities in a fusion plasma, reaching concentrations of up to 10%. Its presence in the plasma comes from the preferred use of graphite as plasma-facing material, from first-wall carbonization, and from carbon contained in first-wall alloys. The reaction

$$C^+(^2P) + H_2 [D_2, HD] \rightarrow CH^+ [CD^+] + H[D] \qquad (11)$$

is probably the best studied endothermic reaction. The endothermicity of 0.398 ± 0.003 eV is known extremely well from spectroscopic data.[20]

The cross sections for the reactions given in Eq. (11) are shown in Fig. 2.[20] The reactions are clearly endothermic, with an extended plateau followed by a decline beginning at the dissociation energy of H_2 [D_2, HD]. The apparent threshold is lower than the endothermicity, as shown more clearly in Fig. 3. This is a consequence of the thermal motion of the H_2 [D_2, HD] reactant gas, as has been demonstrated elsewhere[34] and is illustrated by comparison with experimental results of Gerlich et al.[35] in which a crossed beam of H_2 was used to drastically reduce this thermal motion. This broadening is a feature that is common to all the endothermic reactions

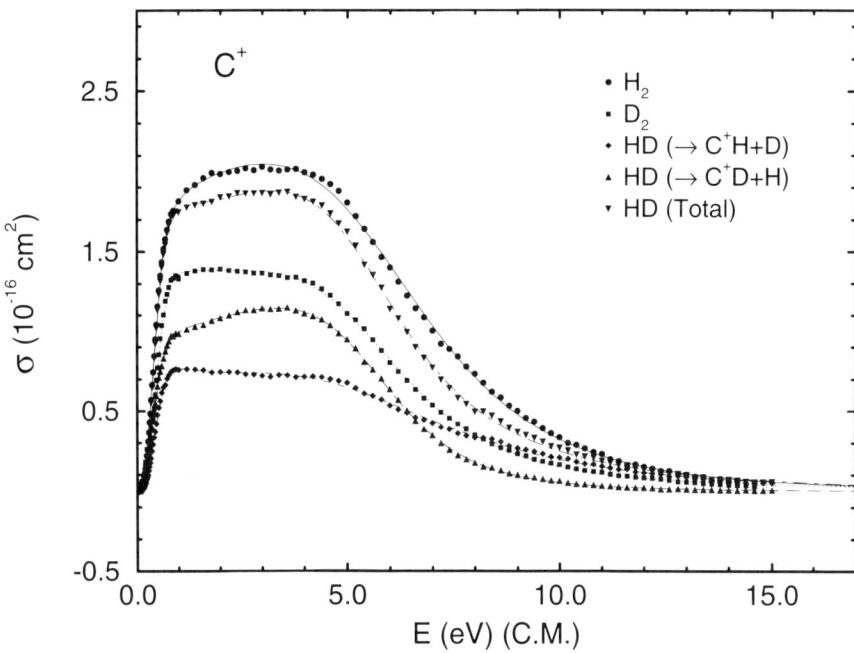

Figure 2. Cross sections for reaction of C^+ with H_2, D_2, and HD as a function of relative kinetic energy. The curves show the analytic fits with the coefficients given in Table I.

discussed here. The reaction with D_2 has a cross section about two-thirds of that of the reaction with H_2. In the reaction with HD, formation of CD^+ is favored over that of CH^+ by a factor of about 1.4 from threshold until the onset of product dissociation, indicating type 1, statistical behavior. In part, CD^+ is favored slightly because this channel has a lower zero point energy and therefore a lower threshold by 45 meV.

Phase space theory calculations for this reaction agree well (within 15%) with experimental results.[26,35] The input parameters to the PST calculation include the molecular parameters of the reactants and products, all of which are well established, and the electronic degeneracies of the reactants and products. The agreement between PST and the experimental results shown in Fig. 3 is obtained only when one-third of the reactant $C^+ + H_2$ surfaces are presumed to lead to products. This factor may be easily explained by molecular orbital analysis.[26] PST or any other theory is not able to fully explain the branching ratio observed in the HD reaction, indicating that dynamics plays an important role in this reaction.

Oxygen is a common impurity in the plasma, reaching concentrations of up to 5%, basically because of its chemical activity and omnipresence, initially in the

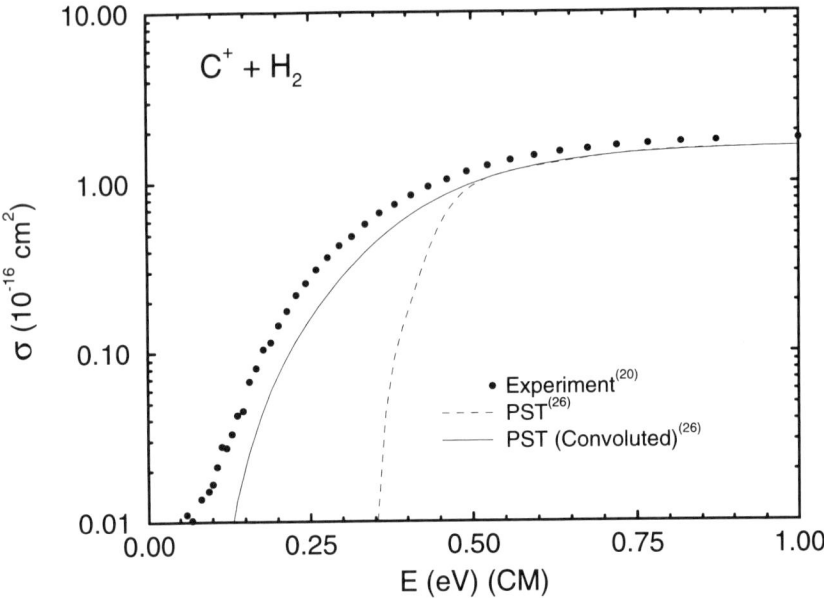

Figure 3. Cross sections for reaction of C^+ with H_2 at low relative kinetic energy. The dashed and dotted curves show phase space calculations for this process.[26,35] The solid curve gives the calculation convoluted with the experimental energy broadening.

form of adsorbed H_2O films and, after activation, in the form of other oxygen-containing compounds. One of the difficulties of examining the reactions of atomic oxygen ions is that excited electronic states are easily produced. The experimental results presented in Fig. 4 were obtained with O^+ generated in an EI/DC source.[10] In such experiments, ions are first generated by electron impact ionization and fragmentation of CO_2 and then are passed through a drift cell filled with molecular nitrogen. Excited states of O^+ rapidly react with N_2 by charge transfer, while the ground state $O^+(^4S)$ ions react very slowly. The O^+ emerging from the cell is found to have less than 0.1% of excited states. The reaction

$$O^+(^4S) + H_2 [D_2, HD] \rightarrow OH^+ [OD^+] + H [D] \qquad (12)$$

is exothermic by 0.6 eV.[36]

The cross sections for the reactions given in Eq. (12) are shown in Fig. 4.[10] They show three distinct energy regimes. At the lower energies, ~0.25 eV, the cross section decreases as $E^{-1/2}$ as predicted by the Langevin–Gioumousis–Stevenson (LGS) collision model.[4] At energies above about 0.3 eV, the cross sections deviate from the LGS prediction such that the reaction efficiency drops. In the region

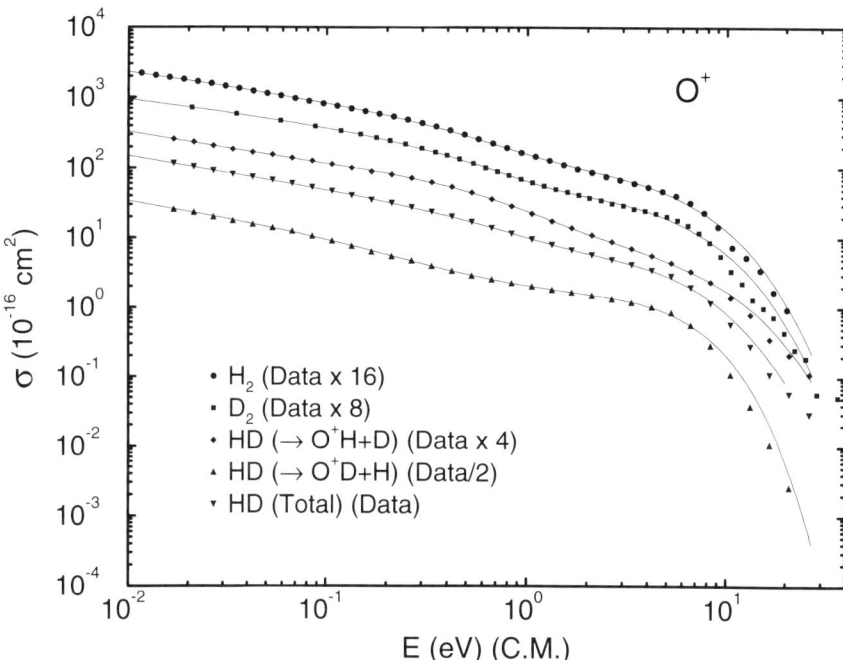

Figure 4. Cross sections for reaction of O^+ with H_2, D_2, and HD as a function of relative kinetic energy. For clarity of presentation the H_2 data and fit are multiplied by 16, the D_2 data and fit by 8, and the data and fits for the two HD channels by 4 and 0.5, respectively. The curves show the analytic fits with the coefficients given in Table I.

between 0.3 and 5 eV, the cross sections fall approximately as E^{-1}. This behavior has been explained[10,33] in terms of angular momentum conservation, because the reactants are higher in energy than the products but have a larger reduced mass than the products. This shifts the limiting transition state for reaction from the entrance channel at low energies to the exit channel at higher energies. Another effect may involve the lifetime of the ion-induced dipole bound O^+–H_2 complex formed during the reaction. At low kinetic energies, the $O^+(^4S) + H_2$ system has time to orient into the collinear configuration favored by electronic considerations, whereas at higher kinetic energies, the time available to orient decreases (or equivalently the lifetime of the O^+–H_2 complex decreases) such that the reaction efficiency falls off. At still higher energies (above 5 eV), the cross sections drop more rapidly. This is due to the dissociation of the OH^+ product ion in the reaction given in Eq. (12) to form O^+ + H, although dissociation to O + H^+ can also occur. The thermodynamic threshold for both dissociations is 4.5 eV. The observed onset of the decline in the OH^+ cross section is somewhat higher, about 6 eV, suggesting that some of the energy available

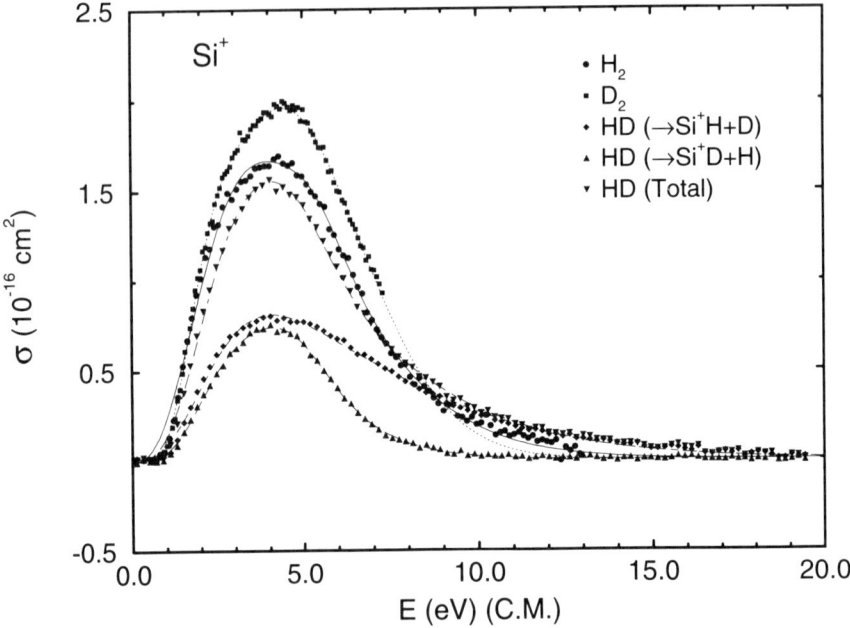

Figure 5. Cross sections for reaction of Si$^+$ with H$_2$, D$_2$, and HD as a function of relative kinetic energy. The curves show the analytic fits with the coefficients given in Table I.

to the products is preferentially placed in translation. Flesch and Ng[37] found that the H$^+$ product appears at its thermodynamic threshold, but this could occur by a dissociative charge transfer process. The HD and D$_2$ system cross sections show the same energy behavior as those of the H$_2$ system and are also of comparable magnitude.

Silicon is a relatively important impurity in the plasma, frequently with a concentration of the order of a few percent. The main source of Si impurity atoms is from erosion of silicon additives. The reaction

$$\text{Si}^+(^2P) + \text{H}_2 \,[\text{D}_2, \text{HD}] \rightarrow \text{SiH}^+ \,[\text{SiD}^+] + \text{H} \,[\text{D}] \tag{13}$$

has an endothermicity of 1.26 ± 0.03 eV. Figure 5 shows the cross sections corresponding to the reactions given in Eq. (13).[33,38] In this case, in contrast with that of C$^+$ and O$^+$, the reaction with D$_2$ has a greater cross section than that with H$_2$ (although the cross sections of these reactions are within experimental error of one another), but with similar energy behavior. The plateau present in the cross sections for C$^+$ becomes a sharper peak because the reaction threshold is higher for Si$^+$. The position of the peak correlates with the dissociation threshold of H$_2$[D$_2$]. The reaction with HD is of type 1, based on the branching ratio in the threshold region. At higher energies, the formation of SiH^+ is favored over SiD^+ formation, an observation that is also true for the C^+ and O^+ systems and for almost all cases presented here. This simply is an indication that the D atom product carries away more energy than the H atom product, a result of it being more massive.

4.2. Metals

The presence of metallic impurities in the plasma edge is due mainly to erosion of structural materials. The expected relative abundance of these impurities in a fusion reactor is of the order of 1–2% for Al, Ti, Cr, Fe, Ni, Cu, B, and V and less than 0.1% for the others. The reactions of H_2 [D_2, HD] with the cations of these elements are all endothermic, with endothermicities of the order of a few electron volts. A common feature of metal ions that makes measurements of these reactions difficult is the large number of electronic states. For instance, V^+ has eight electronic states below 2 eV (28 spin–orbit levels),[39] and these form 186 separate potential energy surfaces in the interaction with H_2. Clearly, explicit account of all these surfaces is quite difficult.

Because of the multitude of electronic states, the way in which the metal ions are produced has a very important influence on the results reported here. As noted in Section 2.4, three main ion sources are used in these studies, namely, surface ionization, electron impact, and high-pressure sources. In many cases, there exist cross section data for metal ions produced with different ion sources. Here, we present mainly data for those obtained by surface ionization, based on the fact that SI normally produces a high population of ions in the ground state, as compared to EI, for example.

Figure 6 shows the cross sections for the reactions

$$M^+ + H_2 [D_2, HD] \to MH^+ [MD^+] + H [D] \qquad (14)$$

for M = Al.[40,41] Notice that all reactions are very inefficient, with cross sections not exceeding 10^{-18} cm^2. The reaction with H_2 is more efficient than the one with D_2. The reactions have a large threshold; the threshold is different for each of the four reactions, but it is higher than the thermodynamic threshold of 3.8 eV in all four cases. This energetic behavior and the branching ratios for the HD reaction indicate that the isotopic behavior of this system is largely type 3. In the HD system, AlH^+ formation is favored by a very large factor except at the lower energies, because AlD^+ formation has a lower energy threshold and maximum. An interesting dynamical explanation for this behavior has been put forward by Simons and co-workers.[27]

Figures 7 and 8 show the cross sections for the reactions in Eq. (14) with M = Ti and M = V, from Refs. 42 and 18, respectively. The cross sections for the reactions with H_2 are larger than those for the reactions with D_2, but the differences are still

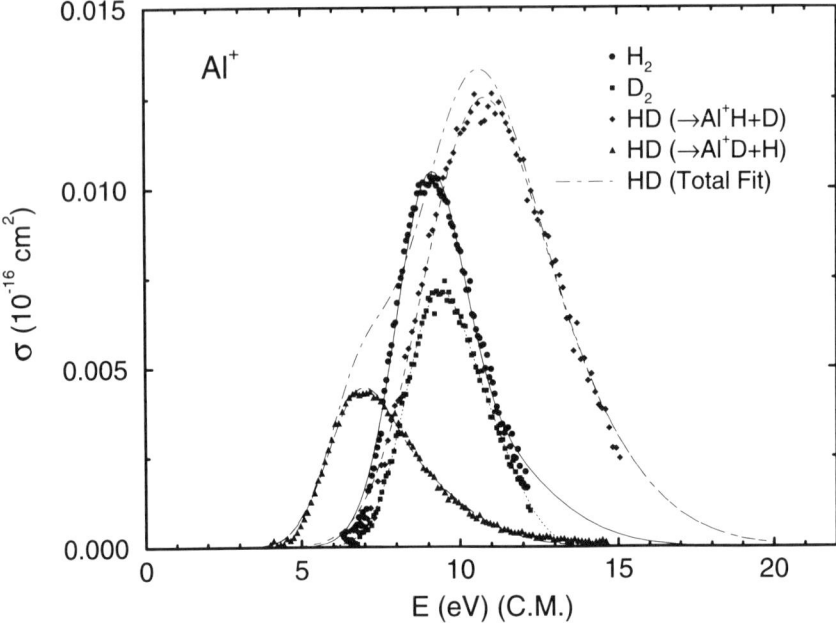

Figure 6. Cross sections for reaction of Al^+ with H_2, D_2, and HD as a function of relative kinetic energy. The curves show the analytic fits with the coefficients given in Table I.

within the experimental error. In the HD systems, the cross sections for the formation of MH^+ and MD^+ have about the same magnitudes in the threshold regions. All four reactions for both metals have similar thresholds. This indicates that these systems behave statistically (isotopic behavior type 1). At high energies in the HD reactions, the cross sections for MD^+ drop much faster than those for MH^+, for reasons described above.

Figure 9 shows the cross sections for the reactions in Eq. (14) with M = B.[40,41] Results for B^+ are rather atypical and resemble those for Al^+. The cross section for H_2 is almost a factor of two larger than that for D_2, and the HD reaction has a mixed isotopic behavior, with formation of BH^+ favored at most energies but BD^+ favored at low energies. In both this and the analogous Al^+ + HD system, the thresholds are displaced from each other and from the thresholds of the H_2 and D_2 systems. Also, the peaks in the various product cross sections are displaced from one another by several electron volts and do not correspond to the dissociation energy for H_2, D_2, or HD. Also shown in Fig. 9 are results of Ruatta et al.[43] divided by 2.75. The reason for this factor is not understood.

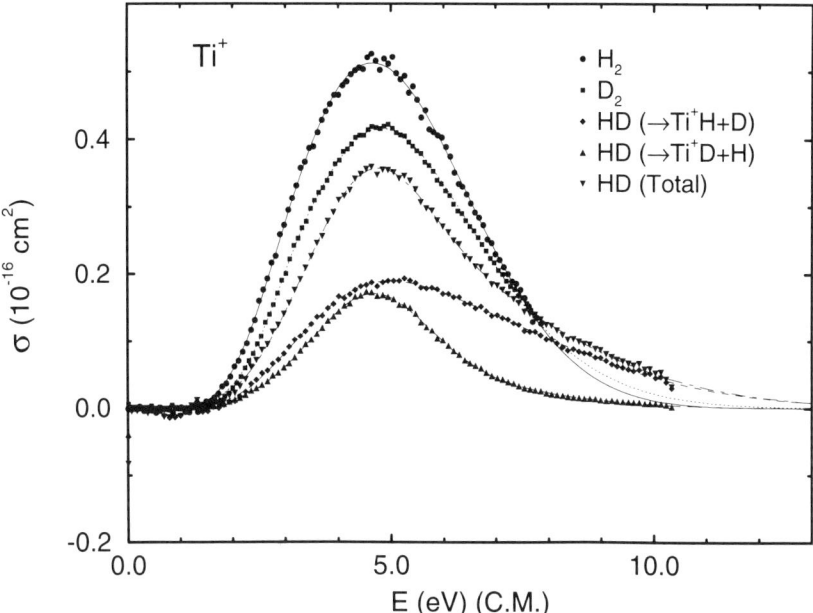

Figure 7. Cross sections for reaction of Ti$^+$ with H$_2$, D$_2$, and HD as a function of relative kinetic energy. The curves show the analytic fits with the coefficients given in Table I.

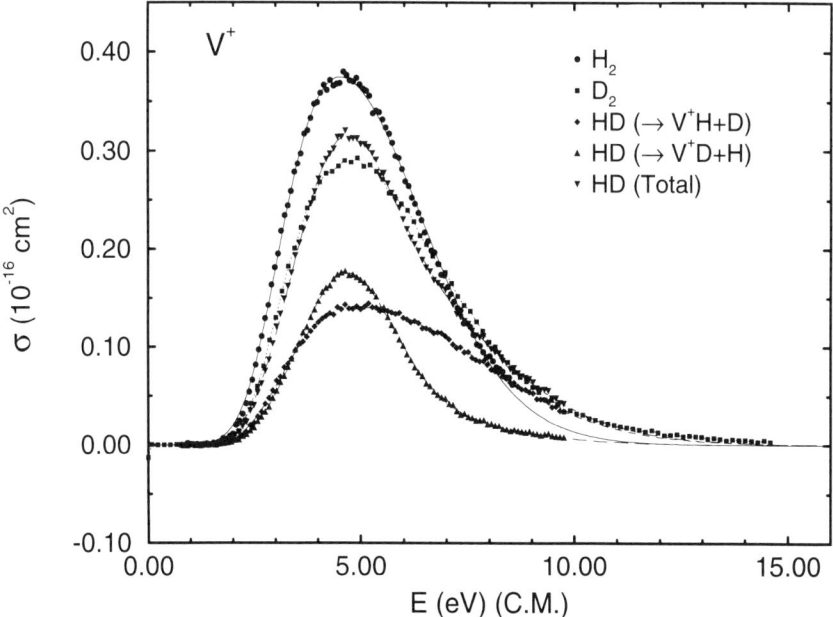

Figure 8. Cross sections for reaction of V$^+$ with H$_2$, D$_2$, and HD as a function of relative kinetic energy. The curves show the analytic fits with the coefficients given in Table I.

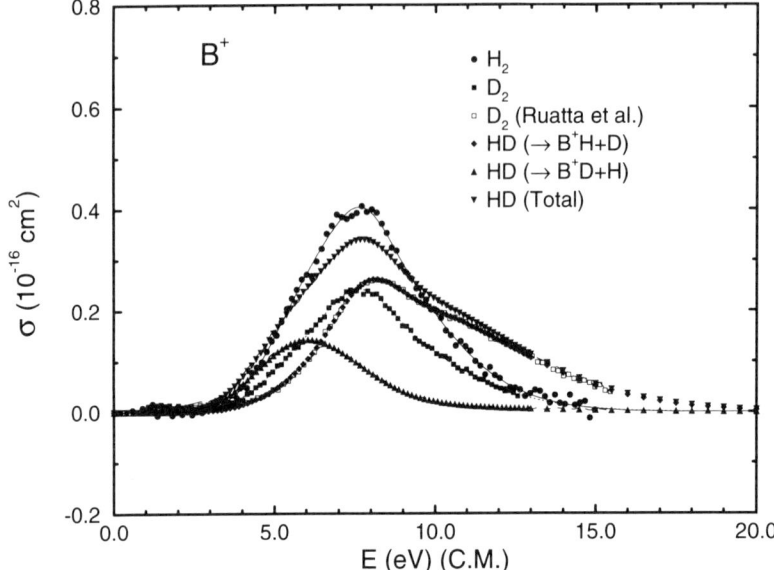

Figure 9. Cross sections for reaction of B^+ with H_2, D_2, and HD as a function of relative kinetic energy. The curves show the analytic fits with the coefficients given in Table I.

Figure 10 shows the cross sections for the reactions in Eq. (14) with M = Cr.[44] For this system, the reaction with H_2 is almost identical to the reaction with D_2, while the HD reaction has a mixed isotopic behavior most characteristic of type 3 because formation of CrD^+ is favored over formation of CrH^+. This mixed dynamic behavior accounts for the unusual double-peak structure in the latter cross section. Data appropriate for excited state Cr^+ generated by EI are also available in the literature.[44]

Figure 11 shows the cross sections for the reactions in Eq. (14) with M^+ = $Fe^+(^6D)$[45] and $Fe^+(^4F)$[45] for D_2 and HD. Notice that in this case the reactivity of the ground state [$Fe^+(^6D)$] is much lower than the reactivity of an excited state [$Fe^+(^4F)$]. The HD reaction in both cases is of type 2, formation of FeH^+ being favored by a factor of about 2.5.

Figures 12 and 13 show the cross sections for the reactions in Eq. (14) with M = Ni and M = Cu, respectively.[46] In both cases, the H_2 and D_2 reactions are similar to one another, within experimental error, and the HD reaction is of type 2, formation of MH^+ being favored by a relatively large factor (5 in the case of Ni^+ and almost 7 in the case of Cu^+ in the threshold region). The cross sections in all

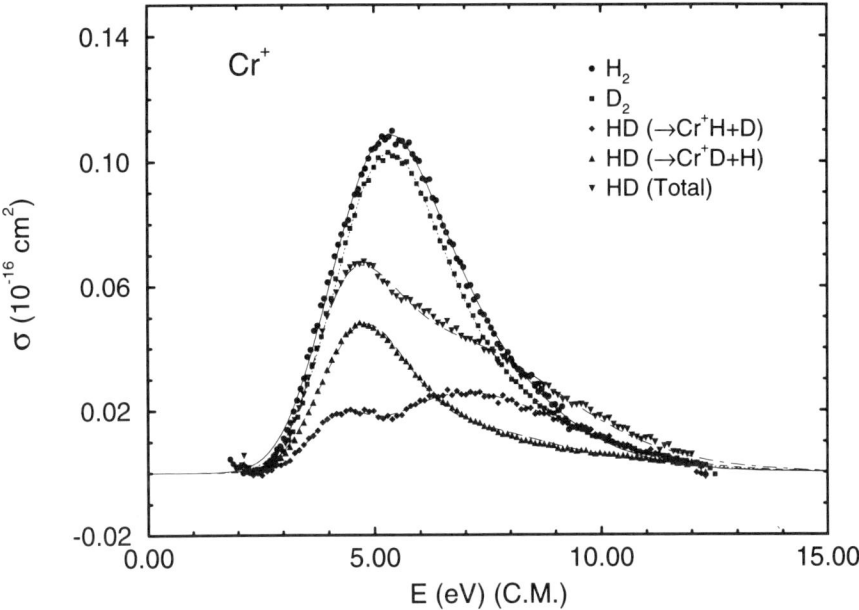

Figure 10. Cross sections for reaction of Cr^+ with H_2, D_2, and HD as a function of relative kinetic energy. The curves show the analytic fits with the coefficients given in Table I.

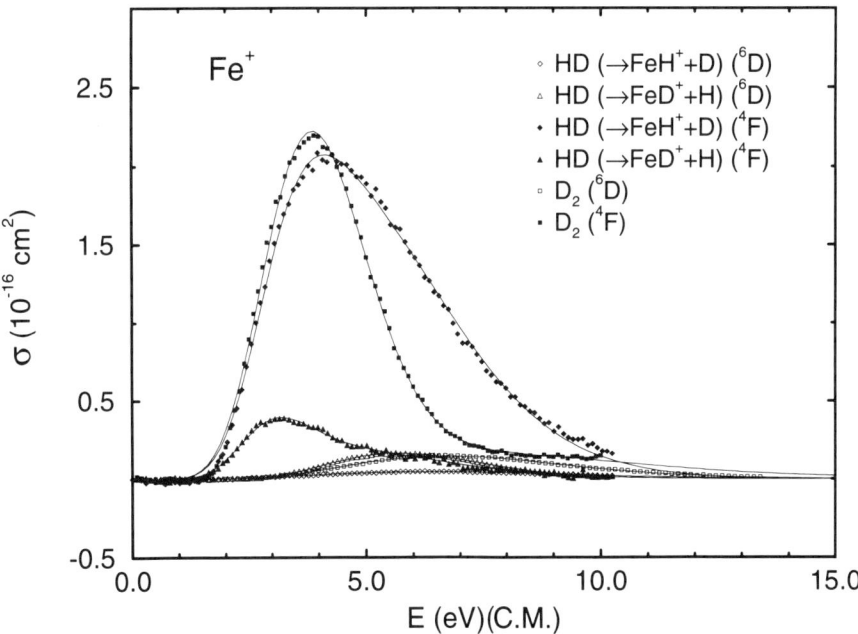

Figure 11. Cross sections for reaction of $Fe^+(^4F)$ and $Fe^+(^6D)$ with D_2 and HD as a function of relative kinetic energy. The curves show the analytic fits with the coefficients given in Table I.

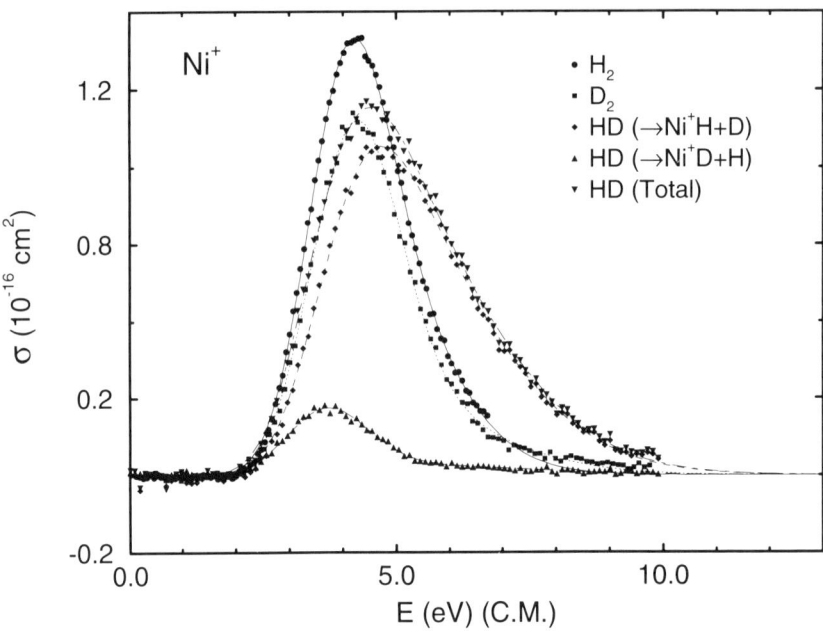

Figure 12. Cross sections for reaction of Ni^+ with H_2, D_2, and HD as a function of relative kinetic energy. The curves show the analytic fits with the coefficients given in Table I.

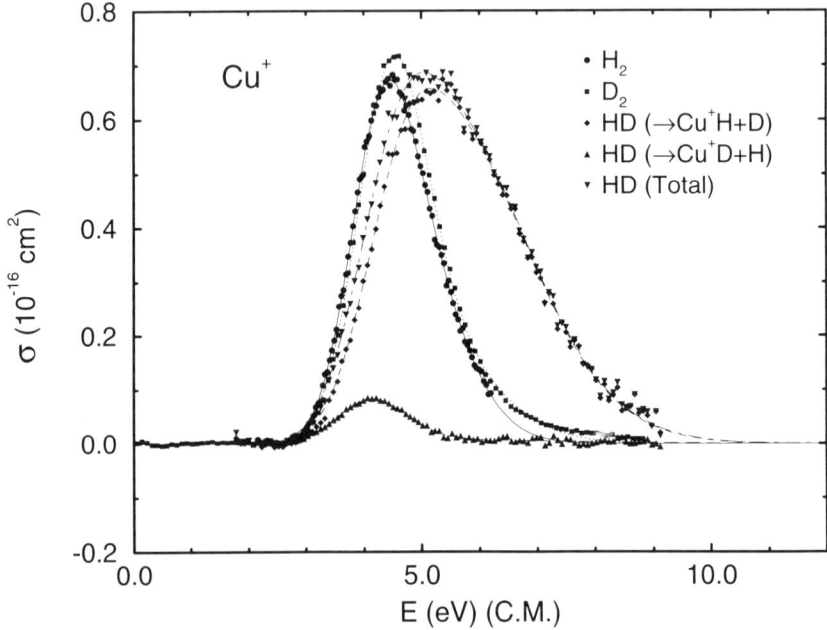

Figure 13. Cross sections for reaction of Cu^+ with H_2, D_2, and HD as a function of relative kinetic energy. The curves show the analytic fits with the coefficients given in Table I.

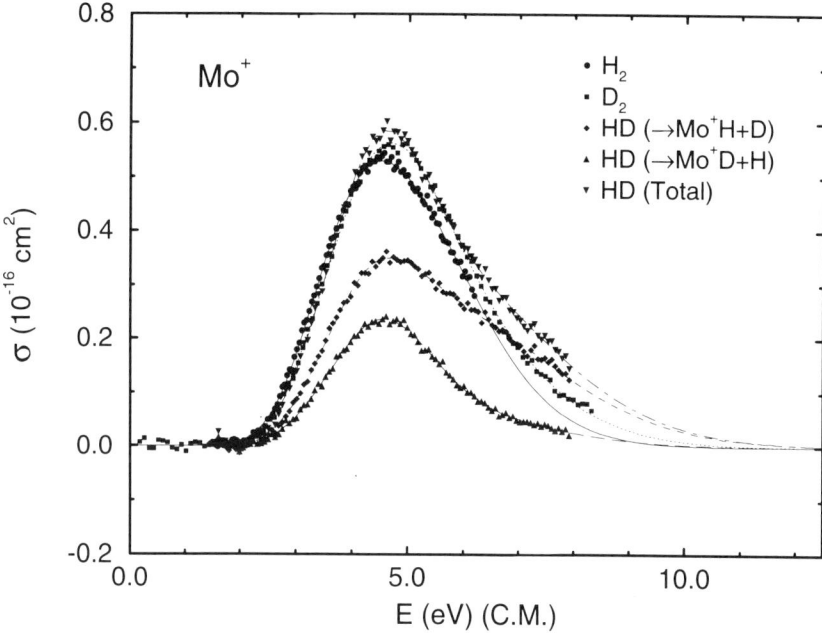

Figure 14. Cross sections for reaction of Mo^+ with H_2, D_2, and HD as a function of relative kinetic energy. The curves show the analytic fits with the coefficients given in Table I.

these reactions rise from their thermodynamic thresholds and reach maxima corresponding to the H_2, D_2, or HD bond energy (although the peaks in the MD^+ cross sections in the HD systems are somewhat lower because of competition with the MH^+ cross sections). Figure 14 shows the cross sections for M = Mo.[47] In this case, the H_2 and D_2 reactions are very similar, and the HD reaction exhibits almost statistical (type 1) behavior.

Figures 15 and 16 show data for the reactions of D_2 and HD, respectively, with Mn^+ generated by SI and EI.[48] The SI source produces 99.83% of the ions in their ^7S ground state and 0.15% in the ^5S first excited state. The behavior of the D_2 cross section is very unusual. The cross section actually rises at an energy below the thermodynamic threshold for reaction of the ^7S state of ~2.5 eV, indicating an observable contribution from the excited state. The reactivity of the ground state does not become appreciable until 5 eV, and the peak of the cross section is at an energy much higher than the thermodynamic onset of the dissociation channel, corresponding to the reaction in Eq. (2). In fact, both the apparent threshold and the peak of the product ions are at energies approximately twice the thermodynamic values, in agreement with the predictions for type 3 behavior given in Section 3.

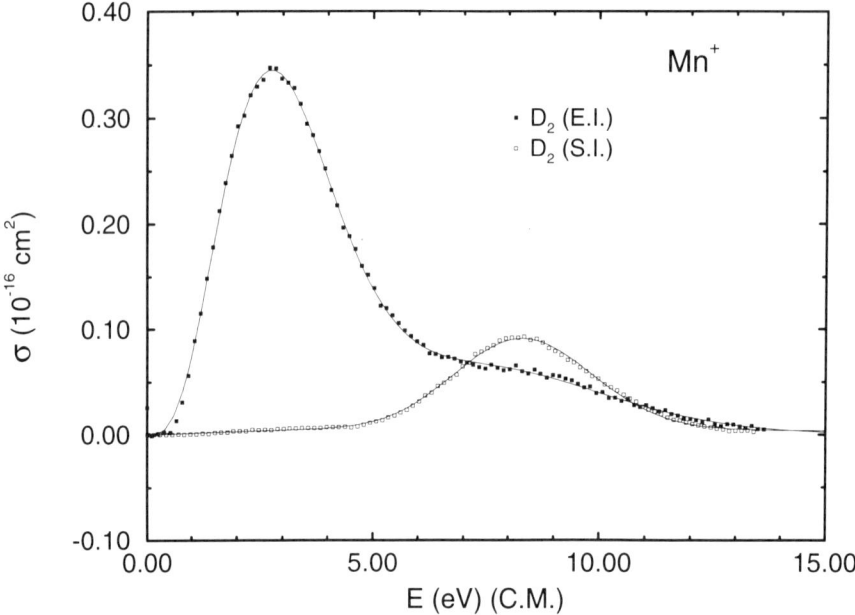

Figure 15. Cross sections for reaction of Mn^+ produced by surface ionization (S.I.) and electron impact (E.I.) with D_2 as a function of relative kinetic energy. The curves show the analytic fits with the coefficients given in Table I.

This is confirmed by the HD data, which show a large preference for formation of MnD^+ and a shift in the threshold and peak of the cross section from those observed for the D_2 system. Figures 15 and 16 also show results for reaction of D_2 and HD with Mn^+ in excited states,[48] obtained by EI ionization of $Mn_2(CO)_{10}$ at 50-eV electron energy. It is believed that the Mn^+ produced under these conditions has a population of about 50% of the ^7S ground state, 10% of the ^5S first excited state, and the remainder in other uncharacterized states. The excited states are much more reactive than the ground state of Mn^+, as is obvious from comparison of the D_2 results for the two source conditions. In contrast to the behavior of $Mn^+(^7S)$, the excited Mn^+ behaves more like most of the transition-metal ions presented here. The HD reaction is of type 2, with formation of MnH^+ being favored by a factor of about 3. Notice the double-peak structure of the MnD^+ channel. This clearly shows the presence of both ground and excited state Mn^+ ions.

Although not presented here in detailed form, kinetic energy-dependent cross sections can also be found in the literature for reaction of H_2, D_2, and HD with the rare-gas ions (He^+,[52] Ne^+,[52] Ar^+,[6] Kr^+,[53] and Xe^+,[54] spin–orbit resolved data having

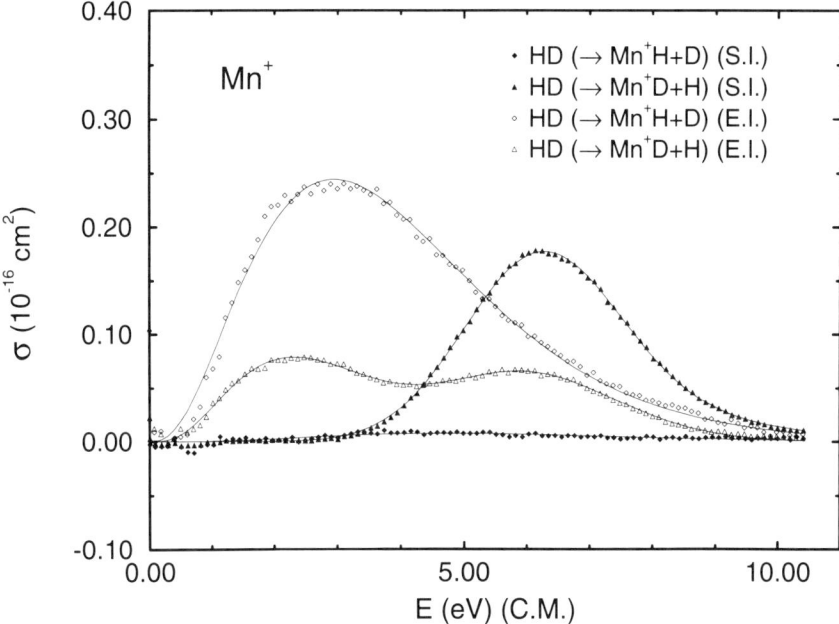

Figure 16. Cross sections for reaction of Mn^+ produced by surface ionization (S.I.) and electron impact (E.I.) with HD as a function of relative kinetic energy. The curves show the analytic fits with the coefficients given in Table I.

been obtained for the latter two systems), other atomic metal ions (Ag^+,[49] Mg^+,[50] Zn^+,[51] Ca^+,[51] Sr^+,[50] and Sc^+, Y^+, La^+, and Lu^{+55}), other atomic ions (H^+ and D^+,[14] N^+,[56,57] and S^{+58}), and several molecular ions (O_2^+,[59] N_2^+,[60] N_4^+,[61] H_2^+,[62] H_3^+ and D_3^+ [14]).

5. ANALYTIC REPRESENTATION

In many applications, plasma modeling, for example, cross sections or rate coefficients of different processes are required over a wide energy region. Analytic fits to the data serve this purpose, because with few parameters and a given function, the cross sections can be generated easily. Unlike polynomial fits, nonlinear analytic fits do not have spurious oscillations and can represent the low- and high-energy limits properly.

In this work, we have made analytic fits to all the data presented. The functions used are (E refers to relative energy in electron volts):

Table I. Parameters for the Analytic Fits

Reaction	Fit[a]	a_1	a_2	a_3	a_4	a_5	a_6	a_7	a_8
$C^+ + H_2$	2	2.4227	0.5648	0.9831	0.1369E-03	4.6645	0.6892	0.4532	-1.8112
$C^+ + D_2$	2	1.7700	0.9278	0.7499	0.7214E-04	5.2681	1.0260	0.4429	-1.3375
$C^+ + HD^b$	3	1.2963	1.3037	0.5859	0.7382E-03	3.7451	0.9493	0.8375	-1.7938
$C^+ + HD^c$	3	1.5651	0.8339	0.7394	0.1034E-04	6.4434	0.6915	0.7923	-1.9521
$O^+ + H_2$	5	6.0584	1.9028	-0.4232	2.2762	-0.4232	8.9380	0.2401	
$O^+ + D_2$	5	21.0010	3.3203	-0.3685	1.7340	0.8621	7.6810	0.2333	
$O^+ + HD^b$	5	0.4904	0.1755	0.4052	9.7028	-0.5835	13.7129	2.0795	
$O^+ + HD^c$	5	1.9387	0.4342	1.1924	2.3245	-0.5638	8.3116	10.1516	
$Si^+ + H_2$	1	0.6876E-04	4.8919	1.2857	4.7086	0.5440	0.8670	3.2978	
$Si^+ + D_2$	1	0.5697	-6.9803	26.4962	4.1046	2.2621	3.5354	8.2685	
$Si^+ + HD^b$	1	1.6141	5.0946	1.6592	-1.5019	0.1650	1.2931	4.7125	
$Si^+ + HD^c$	1	1.5127	-6.6639	23.1747	3.3913	0.3123E-01	0.5350	3.2391	
$Ge^+ + H_2$	1	0.4174	-2.6944	2.1957	13.0924	0.7389E-01	1.4102	6.2168	
$Ge^+ + D_2$	1	0.9565	-3.3142	3.3746	9.8594	0.6038E-01	1.5044	6.5809	
$Ge^+ + HD^b$	1	1.8548	-6.4204	11.0892	5.9428	0.7186E-02	1.0455	6.1512	
$Ge^+ + HD^c$	1	0.7723	-7.6205	12.8199	7.8544	23.2726	12.4083	28.0961	
$Ag^+ + H_2$	1	0.6568E-02	4.3669	0.5480	2.5636	0.2403E-11	8.2219	41.0891	
$Ag^+ + D_2$	1	31.2781	5.2694	0.7968E-01	-20.3268	0.8077E-12	12.1637	54.2531	
$Ag^+ + HD^b$	1	0.6241E-02	4.9196	0.8591	2.1010	0.2402E-10	5.1786	30.3243	
$Ag^+ + HD^c$	6	0.1256E-12	11.4063	51.2384	0.2972	7.6263	-2.0464		
$Al^+ + H_2$	1	0.3689E-03	8.8721	2.2142	1.3995	0.1324E-18	2.7312	28.2325	
$Al^+ + D_2$	4	0.6743E-02	9.3068	2.0824	0.2082E-19	18.9947			
$Al^+ + HD^b$	4	0.1512E-05	9.5038	5.5735	3.5403	0.1341E-15	1.9512	22.1544	
$Al^+ + HD^c$	1	1.2262	7.1585	1.7567	-3.1487	0.5927E-13	2.8017	22.4889	

Particle Interchange Reactions

Reaction							
$B^+ + H_2$	1	0.4247E-03	-12.8882	30.1688	10.2213	16.3517	7.7797	7.6294
$B^+ + D_2$	1	0.3757	7.7428	1.5123	-0.9222	0.2673E-05	1.4359	10.8769
$B^+ + HD^b$	1	0.2990E-04	7.4047	1.7672	3.8732	0.1214E-05	1.0626	9.8244
$B^+ + HD^c$	1	0.9230E-05	-2.1215	11.2994	8.5916	0.1615E-04	0.7957	6.3676
$Cr^+ + H_2$	1	0.7354E-05	2.6543	3.6829	6.5795	0.1816E-05	1.9784	12.3483
$Cr^+ + D_2$	1	0.9763E-06	1.7888	4.0930	8.5821	0.5181E-06	1.8249	12.2406
$Cr^+ + HD^b$	1	6.5527	4.5010	0.6620	-4.2398	0.2118E-06	1.7485	12.3046
$Cr^+ + HD^c$	1	0.2120E-06	2.2394	2.4814	9.3736	0.5945E-05	1.4206	9.1519
$Cu^+ + H_2$	1	0.1670E-01	4.1239	0.5959	2.0283	0.3612E-09	8.0077	37.7967
$Cu^+ + D_2$	1	0.1750E-02	4.1422	0.7491	3.9534	0.2205E-08	5.6379	28.8868
$Cu^+ + HD^b$	1	0.1772E-03	4.1628	0.8936	4.7420	0.5790E-07	3.9176	22.0742
$Cu^+ + HD^c$	1	32.5852	-8.9097	2.7951	38.6171	0.3700E-03	1.1177	5.3696
$Fe^+ + D_2 (^4F)$	1	0.2736	-4.6746	7.5953	8.6301	0.4540E-03	0.8444	6.0719
$Fe^+ + D_2 (^6D)$	1	5.6502	5.5811	0.7878E-04	0.7230E-01	0.4417E-04	1.3478	9.0108
$Fe^+ + HD^b (^6D)$	1	0.2051	-5.5536	6.4411	10.0888	0.2923E-02	1.4364	7.4147
$Fe^+ + HD^c (^6D)$	1	0.1091E-03	5.4707	2.8138	3.4492	0.7654E-02	1.1503	5.3401
$Mg^+ + H_2$	1	0.3424E-05	1.5051	14.9878	5.9390	0.8094E-02	2.2444	0.5252
$Mg^+ + D_2$	1	0.9140E-05	4.5121	0.6696	3.4206	0.1210E-09	2.2676	18.1563
$Mg^+ + HD^b$	1	0.1254E-02	4.2359	0.2873E-01	-0.1186	0.2146E-04	0.6322	4.7068
$Mg^+ + HD^c$	1	0.4347E-06	1.0594	7.3930	8.4011	0.3013E-05	0.6466	5.8995
$Mn^+ + D_2$	1	0.3220E-05	6.6032	5.6048	5.0548	0.1650E-02	0.1785	1.2988
$Mn^+ + HD^b$	1	122.6415	6.5709	0.9055E-01	-68.1373	0.7997E-03	0.8379	3.9994
$Mn^+ + HD^c$	1	0.4830E-04	4.8650	3.6120	4.6836	0.3127E-06	1.6829	11.9207
$Mo^+ + H_2$	1	0.1816E-04	2.1826	1.5543	7.9769	0.4972E-04	3.2083	15.6359
$Mo^+ + D_2$	1	0.1926E-04	-0.4929	4.0158	10.5236	0.3775E-04	2.5744	13.5932
$Mo^+ + HD^b$	1	0.1800E-04	2.3841	2.1239	7.5721	0.1367E-04	2.1381	12.5991
$Mo^+ + HD^c$	1	0.7533E-03	3.6189	1.8516	3.9266	0.5133E-05	2.4155	13.3818

(continued)

Table I. (continued)

Reaction	Fit[a]	a_1	a_2	a_3	a_4	a_5	a_6	a_7	a_8
$Nb^+ + H_2$	1	1.5223	−9.5323	25.3637	4.9966	−10.3680	7.1868	14.7585	
$Nb^+ + D_2$	1	0.6861	−8.4661	20.8691	5.6957	−8.1364	7.8389	16.6475	
$Nb^+ + HD$[b]	1	115.8729	4.5951	0.8161E−02	−5.5772	0.6879E−02	2.4215	10.8280	
$Nb^+ + HD$[c]	1	616.3727	4.9555	0.9216E−01	−40.8749	0.1256E−02	2.7818	12.0326	
$Ni^+ + H_2$	1	0.3563E−03	3.0002	1.4252	6.1225	0.4069E−04	3.7812	17.7161	
$Ni^+ + D_2$	1	0.4115E−03	2.9791	1.7485	6.0246	0.3029E−04	2.5064	13.2607	
$Ni^+ + HD$[b]	1	0.2360E−04	−1.1986	4.0207	12.0903	0.4559E−04	2.4243	13.3949	
$Ni^+ + HD$[c]	1	0.2484E−03	2.5367	1.4873	5.8079	0.6917E−04	1.7151	9.1565	
$Ti^+ + H_2$	1	0.2583E−02	−4.7730	14.3468	7.2519	0.7673E−02	2.9512	10.8452	
$Ti^+ + D_2$	1	0.3020E−06	2.2869	0.5557	10.9316	0.2999E−02	1.7810	8.5950	
$Ti^+ + HD$[b]	1	0.2015E−01	−6.2360	11.1255	7.8524	0.2843E−03	1.2340	7.5683	
$Ti^+ + HD$[c]	1	0.4398E−02	4.1688	2.1959	2.0990	0.1370E−02	1.4925	7.0814	
$V^+ + H_2$	1	0.4411E−03	−2.9449	3.4480	13.8489	0.7630E−03	2.0604	10.1866	
$V^+ + D_2$	1	0.2279E−02	−4.4457	5.4449	12.1460	0.6615E−03	1.6011	8.6188	
$V^+ + HD$[b]	1	0.1402E−02	−4.2247	5.6416	11.3306	0.1432E−03	1.4533	8.6303	
$V^+ + HD$[c]	1	0.5596E−03	3.3363	2.8820	4.0073	0.2565E−04	1.7036	9.7621	
$Zn^+ + H_2$	1	0.2505	4.5991	0.1561E−03	−0.3726	0.2522E−05	1.6796	12.3514	
$Zn^+ + D_2$	1	0.4839E−01	4.6978	0.8887E−04	−0.4276	0.6046E−06	1.9485	14.1202	

[a] Refers to the expressions given in Eqs. (15)–(20).
[b] $M^+ + HD \rightarrow MH^+ + D$.
[c] $M^+ + HD \rightarrow MD^+ + H$.

Particle Interchange Reactions

Expression 1:

$$\sigma = a_1 \exp\left[\frac{-(E-a_2)^2}{a_3}\right] E^{a_4} + a_5 \exp(-a_6 E) E^{a_7} \qquad (15)$$

Expression 2:

$$\sigma = a_1 \frac{\exp(-a_2/E^{a_3})}{1 + a_4 E^{a_5}} + a_6 \exp\left(\frac{-a_7}{E^2}\right) E^{a_8} \qquad (16)$$

Expression 3:

$$\sigma = a_1 \frac{\exp(-a_2/E^{a_3})}{1 + a_4 E^{a_5}} + a_6 \exp\left(\frac{-a_7}{E^{1.5}}\right) E^{a_8} \qquad (17)$$

Expression 4:

$$\sigma = a_1 \exp\left[\frac{-(E-a_2)^2}{a_3}\right](1 + a_4 E^{a_5}) \qquad (18)$$

Expression 5:

$$\sigma = a_1 \exp(-a_2 E)(E^{a_3} + a_4 E^{a_5}) + a_6 \exp(-a_7 E) \qquad (19)$$

Expression 6:

$$\sigma = a_1 \exp(-a_2 E) E^{a_3} + a_4 E^{a_5}) \exp(-a_7/E) \qquad (20)$$

Table I gives the expressions used to represent the cross sections for each reaction and the corresponding parameters in Eqs. (15)–(20).

In using these analytic fits to model systems, it should be remembered that the experimental conditions for the system being modeled may differ from the experimental conditions used to generate these data. The kinetic energy distributions of the ions and neutrals and the electronic states of the ions influence the shapes of these cross sections. Reactions taking place at higher neutral reactant densities may also lead to different reaction channels from those shown here, which correspond to single-collision conditions in all cases.

6. DISCUSSION

We have seen that many endothermic particle interchange reactions between metallic ions and H_2 (or D_2) follow a pattern: the cross section rises from an apparent threshold, peaks at an energy close to the dissociation energy of the neutral reactant, and then decreases relatively rapidly. This regularity can be seen clearly in Fig. 17, where we show the cross sections σ/σ_0 as a function of scaled energy E/E_0, where σ_0 and E_0 are the values of the cross section and the energy at the peak. For most

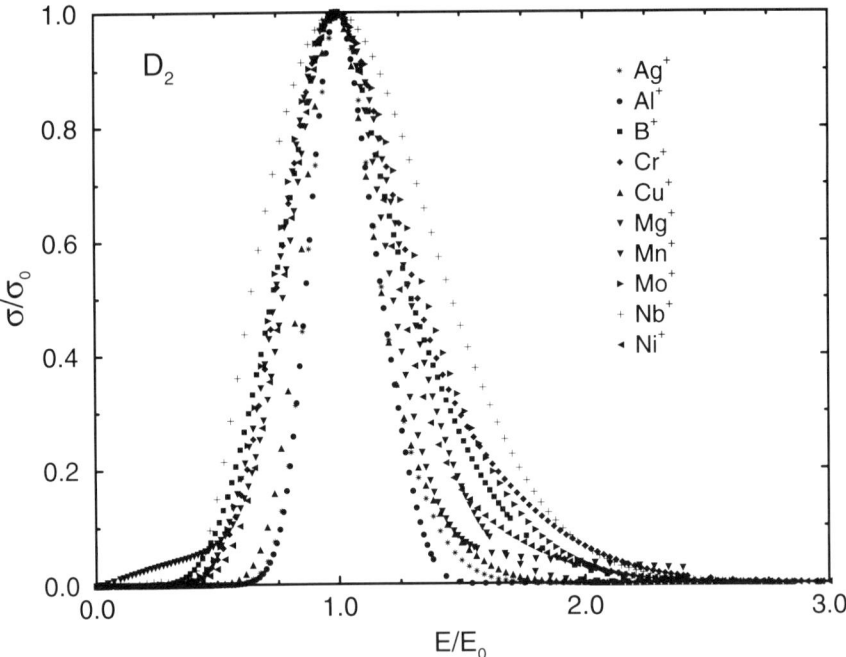

Figure 17. Scaled cross sections for reaction of metals with D_2 as a function of scaled relative kinetic energy (see text).

of these reactions, E_0 is close to the bond energy of D_2, whereas for M = Al, B, Mg, Mn, and Zn the value is displaced to higher energies.

The similarities between these various curves reflect the similarity in the overall physical processes. In the case of transition-metal ions, a more detailed comparison can be found elsewhere.[63] The apparent symmetry above and below the peak for each particular system is largely serendipitous because the molecular characteristics that control the onset of reaction (formation of $MD^+ + D$) are not the same as those that control the decline (dissociation to $M^+ + 2D$). This is evident from the data for M = C, shown in Fig. 2, which does not exhibit this type of symmetry. Some correlation between how rapidly the cross section rises from threshold and how fast it declines above its maximum is observed because the threshold for reaction is lowest when the MD^+ bond is strongest, and this also means that the MD^+ product is less likely to dissociate at higher energies.

ACKNOWLEDGMENTS. The experimental data presented here were acquired with support of the National Science Foundation. One of us (JB) would like to thank

Ms. Clara Illescas Rojas for her invaluable assistance during the preparation of the analytic fits and figures and Professor R. K. Janev for helpful discussions.

REFERENCES

1. R. K. Janev, M. F. A. Harrison, and H. W. Drawin, *Nucl. Fusion* **29**, 109 (1989).
2. V. L. Talrose, P. S. Vinogradov, and I. K. Larin, in *Gas Phase Ion Chemistry*, Vol. 1 (M. Bowers, ed.), Academic Press, New York (1979), p. 305.
3. P. B. Armentrout, in *Advances in Gas Phase Ion Chemistry*, Vol. 1 (N. G. Adams and L. M. Babcock, eds.), Jai Press, Greenwich, Connecticut (1992), p. 83.
4. G. Gioumousis and D. P. Stevenson, *J. Chem. Phys.* **29**, 294 (1958).
5. P. B. Armentrout, in *Structure/Reactivity and Thermochemistry of Ions* (P. Ausloos and S. G. Lias, eds.), D. Reidel, Dordrecht (1987), pp. 97–164.
6. K. M. Ervin and P. B. Armentrout, *J. Chem. Phys.* **83**, 166 (1985).
7. L. S. Sunderlin and P. B. Armentrout, *Chem. Phys. Lett.* **167**, 188 (1990).
8. R. H. Schultz and P. B. Armentrout, *Int. J. Mass Spectrom. Ion Processes* **107**, 29 (1991).
9. N. R. Daly, *Rev. Sci. Instrum.* **31**, 264 (1959).
10. J. D. Burley, K. M. Ervin, and P. B. Armentrout, *Int. J. Mass Spectrom. Ion Processes* **80**, 153 (1987).
11. J. D. Burley and P. B. Armentrout, *Int. J. Mass Spectrom. Ion Processes* **84**, 157 (1988).
12. E. R. Fisher and P. B. Armentrout, *J. Chem. Phys.* **94**, 1150 (1991).
13. E. Teloy and D. Gerlich, *Chem. Phys.* **4**, 417 (1974); D. Gerlich, Diplomarbeit, University of Freiburg, 1971.
14. D. Gerlich, *Adv. Chem. Phys.* **82**, 1 (1992).
15. K. M. Ervin, Ph.D. Thesis, University of California, Berkeley, 1986.
16. P. J. Chantry, *J. Chem. Phys.* **55**, 2746 (1971).
17. C. Lifshitz, R. L. C. Wu, T. O. Tiernan, and D. T. Terwilliger, *J. Chem. Phys.* **68**, 247 (1978).
18. J. L. Elkind and P. B. Armentrout, *J. Phys. Chem.* **89**, 5626 (1985).
19. L. S. Sunderlin and P. B. Armentrout, *J. Phys. Chem.* **92**, 1209 (1988).
20. K. M. Ervin and P. B. Armentrout, *J. Chem. Phys.* **84**, 6738 (1986).
21. D. G. Leopold, K. K. Murray, A. E. S. Miller, and W. C. Lineberger, *J. Chem. Phys.* **83**, 4849 (1985).
22. S. T. Graul and R. R. Squires, *Mass Spectrom. Rev.* **7**, 263 (1988).
23. A. Bjerre and E. E. Nikitin, *Chem. Phys. Lett.* **1**, 179 (1967).
24. E. Bauer, E. R. Fischer, and F. R. Gilmore, *J. Chem. Phys.* **51**, 4173 (1969); M. S. Child and M. Baer, *J. Chem. Phys.* **74**, 2832 (1981).
25. M. Baer, *Adv. Chem. Phys.* **82** (Part II), 202 (1992); M. Baer, C. Y. Ng, and D. Neuhauser, *Chem. Phys. Lett.* **93**, 4845 (1990).
26. K. M. Ervin and P. B. Armentrout, *J. Chem. Phys.* **84**, 6750 (1986).
27. M. Gutowski, M. Roberson, J. Rusho, J. Nichols, and J. Simons, *J. Chem. Phys.* **99**, 2601 (1993).
28. C. D. Light, *J. Chem. Phys.* **40**, 3221 (1964); P. Pechukas and J. C. Light, *J. Chem. Phys.* **42**, 3281 (1965).
29. E. E. Nikitin, *Teor. Eksp. Khim.* **1**, 135, 144, 248 (1965) [Engl. trans.: *Theor. Exp. Chem.* **1**, 83, 90, 275 (1975)].
30. J. Light, *J. Chem. Phys.* **43**, 3209 (1965).
31. D. A. Webb and W. J. Chesnavich, *J. Phys. Chem.* **87**, 3791 (1983).
32. W. J. Chesnavich and M. T. Bowers, *J. Chem. Phys.* **66**, 2306 (1977).
33. P. B. Armentrout, in *Isotope Effects in Chemical Reactions and Photodissociation Processes* (J. A. Kaye, ed.), *ACS Symp. Ser.* **502**, 194 (1992).

34. L. S. Sunderlin and P. B. Armentrout, *J. Chem. Phys.* **100**, 5639 (1994).
35. D. Gerlich, R. Disch, and S. Scherbarth, *J. Chem. Phys.* **87**, 350 (1987).
36. D. D Wagman, W. H. Evans, V. B. Parker, R. H. Schumm, I. Halow, S. M. Bailey, K. L. Churney, and R. L. Nuttall, *J. Phys. Chem. Ref. Data* **11**, Suppl. 2 (1982).
37. G. D. Flesch and C. Y. Ng, *J. Chem. Phys.* **94**, 2372 (1991).
38. J. L. Elkind and P. B. Armentrout, *J. Phys. Chem.* **88**, 5454 (1984).
39. J. Sugar and C. Corliss, *J. Phys. Chem. Ref. Data* **14**, Suppl. 2 (1985).
40. P. B. Armentrout, *Int. Rev. Phys. Chem.* **9**, 115 (1990).
41. J. L. Elkind and P. B. Armentrout, unpublished data (1985).
42. J. L. Elkind and P. B. Armentrout, *Int. J. Mass Spectrom. Ion Processes* **83**, 259 (1988).
43. S. Ruatta, L. Hanley, and S. L. Anderson, *J. Chem. Phys.* **91**, 226 (1989).
44. J. L. Elkind and P. B. Armentrout, *J. Chem. Phys.* **86**, 1868 (1987).
45. J. L. Elkind and P. B. Armentrout, *J. Phys. Chem.* **90**, 5736 (1986).
46. J. L. Elkind and P. B. Armentrout, *J. Phys. Chem.* **90**, 6576 (1986).
47. J. L. Elkind and P. B. Armentrout, unpublished data (1986).
48. J. L. Elkind and P. B. Armentrout, *J. Chem. Phys.* **84**, 4862 (1986).
49. Y.-M. Chen, P. B. Armentrout, and J. L. Elkind, *J. Phys. Chem.* **99**, in press (1995).
50. N. F. Dalleska, K. C. Crellin, and P. B. Armentrout, *J. Phys. Chem.* **97**, 3123 (1993).
51. R. Georgiadis and P. B. Armentrout, *J. Phys. Chem.* **92**, 7060 (1988).
52. K. M. Ervin and P. B. Armentrout, *J. Chem. Phys.* **86**, 6240 (1987).
53. K. M. Ervin and P. B. Armentrout, *J. Chem. Phys.* **85**, 6380 (1986).
54. K. M. Ervin and P. B. Armentrout, *J. Chem. Phys.* **90**, 118 (1989).
55. J. L. Elkind, L. S. Sunderlin, and P. B. Armentrout, *J. Phys. Chem.* **93**, 3151 (1989).
56. K. M. Ervin and P. B. Armentrout, *J. Chem. Phys.* **86**, 2659 (1987).
57. P. Tosi, O. Dmitriev, D. Bassi, O. Wick, and D. Gerlich, *J. Chem. Phys.* **100**, 4300 (1994).
58. G. F. Stowe, R. H. Schultz, C. A. Wight, and P. B. Armentrout, *Int. J. Mass Spectrom. Ion Processes* **100**, 177 (1990).
59. M. E. Weber, N. F. Dalleska, B. L. Tjelta, E. R. Fisher, and P. B. Armentrout, *J. Chem. Phys.* **98**, 7855 (1993).
60. R. H. Schultz and P. B. Armentrout, *J. Chem. Phys.* **96**, 1036 (1992).
61. R. H. Schultz and P. B. Armentrout, *J. Chem. Phys.* **96**, 1046 (1992).
62. C. L. Liao and C. Y. Ng, *J. Chem. Phys.* **84**, 197 (1986). J. D. Shao and C. Y. Ng, *J. Chem. Phys.* **84**, 4317 (1986).
63. J. L. Elkind and P.B. Armentrout, *J. Phys. Chem.* **91**, 2037 (1987).

Chapter 16

Electron Collision Processes Involving Hydrocarbons

Hiroyuki Tawara

1. INTRODUCTION

Currently, graphites or carbon-coated materials are most commonly used as the plasma-facing inner walls of various fusion plasma research devices because of their superior qualities at high temperatures and because low atomic number is one of the most important factors in reducing radiation losses from high-temperature main plasmas. On the other hand, they are known to be significantly eroded through interactions with atomic hydrogens in plasmas in particular circumstances.

The following three processes are believed to be the most responsible for such graphite erosion at intermediate to high temperatures:

1. Physical sputtering[1,2] due to impact of energetic heavy particles
2. Chemical sputtering[3] due to interactions of carbon atoms with active atomic hydrogens
3. Radiation-enhanced sublimation at extremely high temperatures (>1500 K)

The first and third processes usually result in the emission of mostly carbon atoms, whereas the second process results in the emission of hydrocarbon molecules such as CH_4.

HIROYUKI TAWARA • National Institute for Fusion Science, Nagoya 464-01, Japan.

Atomic and Molecular Processes in Fusion Edge Plasmas, edited by R. K. Janev. Plenum Press, New York, 1995.

1.1. Physical Sputtering

The measured total sputtering yields of graphite at room temperatures under helium impact are shown in Fig. 1. As the impact particle energy increases from the threshold of ~7 eV, the sputtering yield increases up to 0.1 at 500–1000 eV and then decreases. This dependence of the sputtering yield on the particle impact energy is typical in physical sputtering that is caused through collision cascades, and its features are well understood.[1,2] Physical sputtering yields also increase as the incident particles become heavier because their momentum transfer is large. In

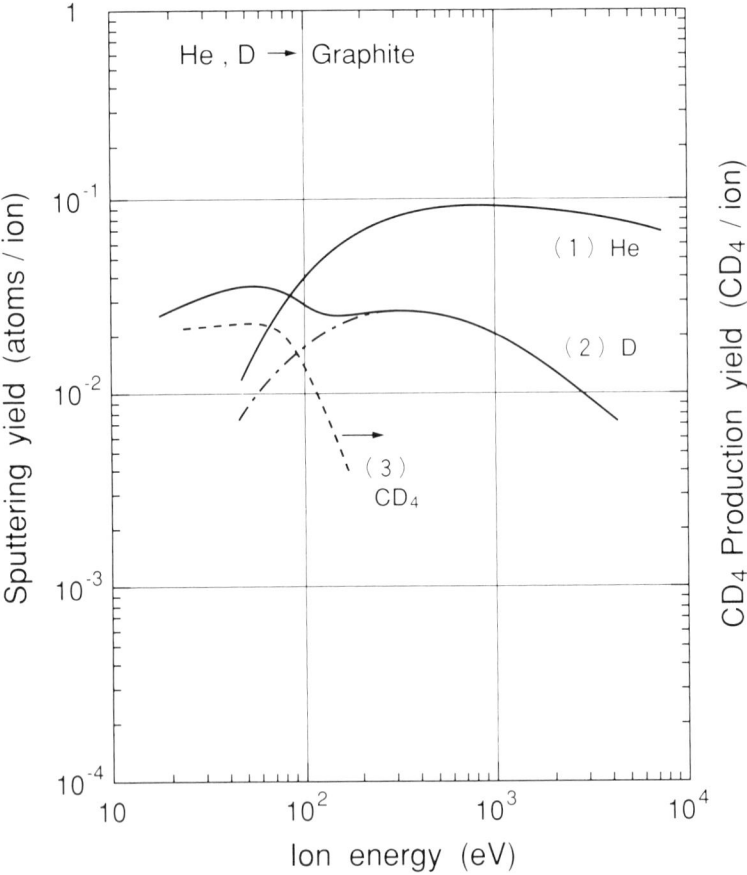

Figure 1. Sputtering yields of graphite as a function of helium (curve 1) and deuteriun (curve 2) particle impact energy, and production of CD_4 under deuterium particle impact (curve 3) at room temperatures. (Adapted from Ref. 2.)

fact, sputtering yields of graphite under carbon particle impact reach unity at 1500 K under 50-keV impact (see Fig. 2).

Another important feature seen in Fig. 1 is the fact that even when the impact energy of hydrogen particles decreases below 200 eV down threshold, the sputter-

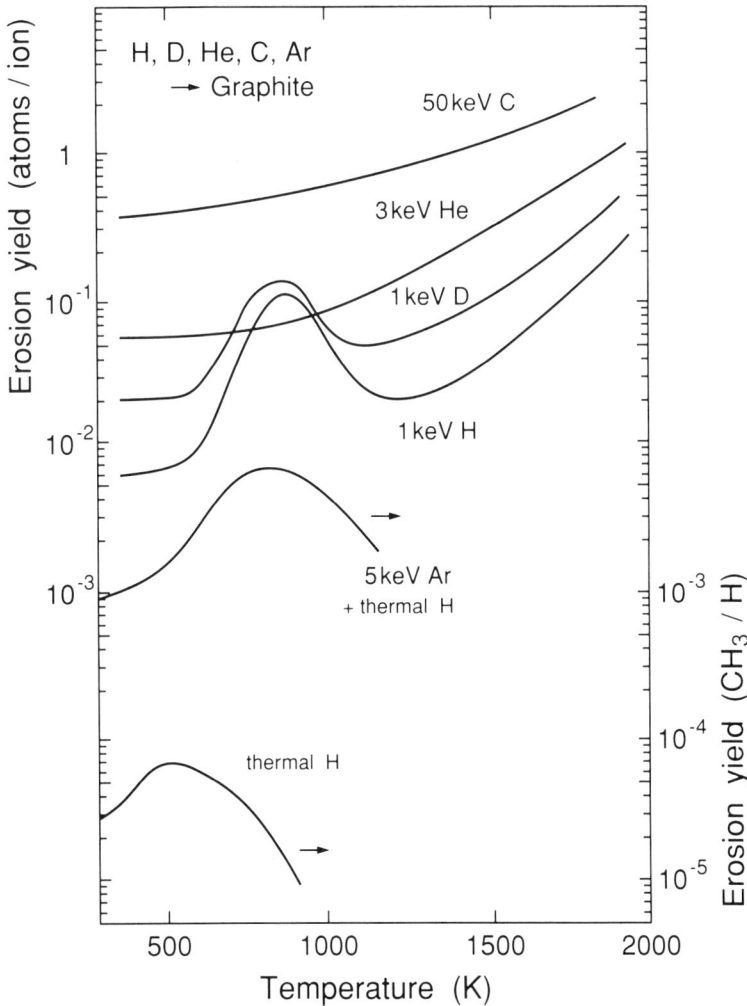

Figure 2. Sputtering yields of graphite as a function of temperature under various types of particle impact. Note that strong enhancement is observed at 700–1000 K under hydrogen particle impact, whereas no such enhancement is seen under helium and carbon particle impact. Production of CH_3 is also shown in interactions with thermal hydrogen particles and increase significantly under simultaneous Ar ion impact, suggesting strong synergistic enhancement. (Adapted from Ref. 3.)

ing yield does not decrease but stays nearly constant. This enhancement of sputtering yields under hydrogen particle impact at low energies, distinctly different from the behavior observed with other types of other particle impact, is closely related to chemical sputtering and is understood to be due to the erosion of graphite via the production of hydrocarbon molecules through chemical reaction between hydrogen atoms and carbon atoms. The production of CH_4 (or CD_4 in deuterium particle impact) molecules as a function of impact particle energy is also shown in Fig. 1. As can be clearly seen in Fig. 1, the CH_4 production decreases sharply as the hydrogen particle impact energy increases beyond 100 eV. This feature suggests that CH_4 molecules are formed not on the surfaces but deep inside the graphite and therefore is understood to be due to the fact that, as the range of the incident hydrogen particles becomes large at higher impact energies, CH_4 molecules, which are formed after the incident hydrogen particles are sufficiently thermalized inside the graphite, cannot escape from the solid surface.

1.2. Chemical Sputtering

As seen in Fig. 2, where sputtering yields of graphite are shown as a function of graphite temperature under various types of particle impact, sputtering of graphite exhibits an interesting feature under hydrogen particle impact. Sputtering by helium and carbon particle impact increases slowly as the graphite temperature increases up to 2000 K, where radiation-enhanced sublimation becomes the dominant process. On the other hand, sputtering yields under hydrogen particle impact show a broad peak at 800 K, with an enhancement factor of more than 10 at 1-keV impact. This enhancement is due to chemical reactions of carbon atoms with hydrogen atoms to form hydrocarbon molecules. Chemical sputtering is more significant at low energies, compared with physical sputtering (see Fig. 1).

Some other interesting features in such chemical sputtering from graphite under hydrogen particle impact should be pointed out. Under impact of hydrogen atoms at thermal energies, CH_3 yields are observed to be much larger than CH_4 yields. On the other hand, CH_4 becomes dominant under energetic hydrogen particle impact[4] (see Tables I and II). This again supports the fact that CH_3 radicals are formed near the surface, whereas CH_4 molecules are formed inside the solid through multiple collisions. This fact also indicates that the large hydrocarbon molecules observed are formed only near the surface. This conclusion is supported by the observation that sputtering yields of thin deposited-carbon layers, whose thickness is less than the range of the incident hydrogen particles, are almost independent of their temperature and are not much different from physical sputtering yields at room temperatures because most of the energetic incident hydrogen particles pass through the carbon layers, and thus the formation of hydrocarbon molecules is significantly hindered.

Table I. Relative Production of Hydrocarbon Molecules under Pure 5-keV D_2 Impact and under Simultaneous 5-keV D_2 + Thermal D Impact on 800 K Graphite[a]

	Relative production		
	CD_4	CD_3	CD_2
5-keV D_2	100	16	0
5-keV D_2 + thermal D[b]	13	100	6

[a]From Ref. 4.
[b]Note that the density of thermal D is 18 times that of 5-keV D_2 particles.

The observed production rates, at the temperatures of maximum sputtering yields, of various hydrocarbon molecules under hydrogen particle impact[5] are shown in Fig. 3 as a function of the impact energy. It should be noted that total erosion of graphite is equally shared by CH_4 and C_2H_n molecule formation, though the absolute yields of C_2H_n molecules are smaller than those of CH_4 molecules.

When physically sputtered carbon atoms or chemically produced hydrocarbon molecules are released from graphite surfaces and come into plasmas and finally interact with plasma constituents, they, or their collision products, should play some important roles in both cold edge plasmas (such as cooling of the edge plasmas) as well as hot plasmas in the region of the high-temperature plasma center (such as radiation losses). The penetration and transport processes of molecules and dissociation products into plasmas[6] are strongly influenced by their kinetic energy as well as their internal energy. Thus, in order to diagnose, model, and understand plasma behavior, in particular, edge plasma behavior, collision data[7,8] for hydrocarbon molecules under electron impact are requisite.

In this chapter, the present status of these data is discussed, urgent needs for additional data are identified, and new investigations are proposed. As collision cross sections have already been described in detail in recent papers,[7,8] we focus our discussion here on some new aspects of hydrocarbon molecule data under electron impact.

Table II. Relative Production of Various Hydrocarbon Molecules from Graphite at Room Temperature under Different Bombardment Conditions[a]

Incident particle	Dominant CH_x	C_2H_n/CH_x	C_3H_n/CH_x	Total chemical erosion/CH_x
Thermal H	CH_3	2–6	1–3	8–22
100-eV H	CH_4	0.3	0.1	1.9
2.5-keV H	CH_4	0.05–0.4	0.02–0.04	1.2–2.2

[a]From Ref. 4.

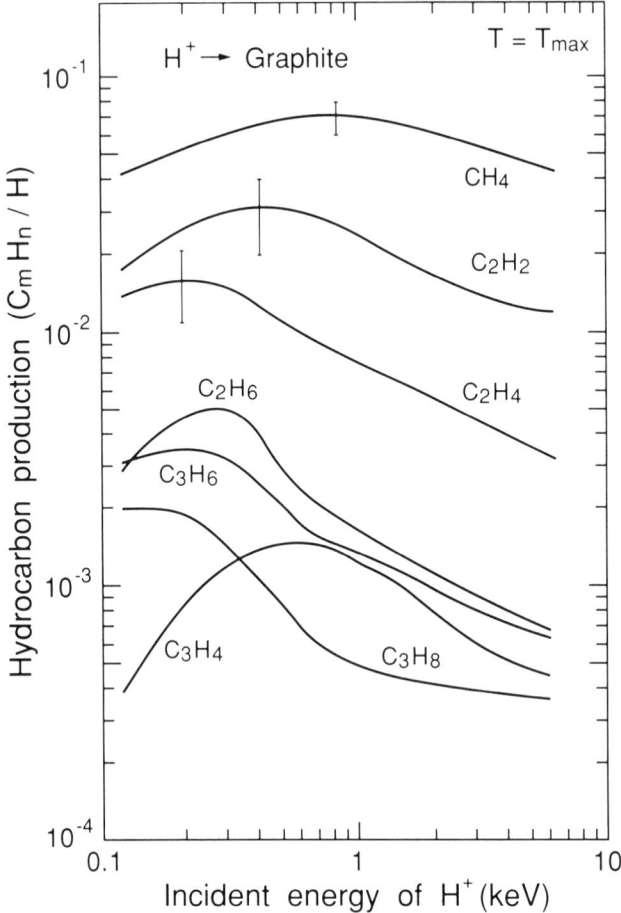

Figure 3. Production of various hydrocarbon molecules as a function of the impact energy of hydrogen particles at the temperature of maximum sputtering yields. The vertical bars indicate the uncertainties of the observed intensities. (Adapted from Ref. 5.)

2. IMPORTANT COLLISION PROCESSES INVOLVING HYDROCARBON MOLECULES

The importance and accuracies of data for particular collision processes are dependent upon how these data are used. For example, in modeling and describing the overall properties of plasmas that contain hydrocarbon molecules, systematic data for a series of collision processes over a wide range of collision energy are required, though uncertainties in these data of the order of a factor of 2, or even more, may be acceptable. On the other hand, much better accuracies are required for diagnostics of the edge plasma for some particular processes.

In fact, for proper modeling of plasmas, a large set of data for a series of collision processes is necessary, comprising total as well as partial cross sections as a function of the incident electron energy (E_0), scattering angle (θ), scattered electron energy (E_s), and molecule internal energy (E_i) prior to as well as after collisions. Typical electron–hydrocarbon molecule collision processes that play an important role in plasma modeling are listed below:

1. Electron elastic scattering (total and differential in angle and energy)
2. Electron inelastic scattering (total and differential in angle and energy)
3. Target molecule excitation (rotational, vibrational, and electronic): j- and v-distributions (j and v denote the rotational and the vibrational quantum number, respectively)
4. Target molecule dissociation: products and their kinetic energy distributions, angular distributions, and internal energy distributions
5. Target molecule ionization (including dissociative ionization; total and partial)
6. Electron attachment to target molecule (forming negative ions)
7. Dissociative recombination of target molecular ion
8. Chemical reaction or rearrangement collisions

Some of these processes and their features have already been discussed in detail in the previous chapters in this volume, which are mostly concerned with atomic species, except for Chapters 8, 9, 14, and 15.

One of the most important, yet less emphasized and not well-documented, aspects is the fact that the data for almost all collision processes involving molecules or molecular ions are strongly dependent upon their internal energy, that is, their excitation states. It is well known that the differences between the collision cross sections of species with different internal states are generally large and that the collision cross sections for such excited species are often an order of magnitude larger than those for the ground state. For example, partial cross sections for production of the various ions from two differently prepared CH_4 molecules are compared in Table III. One of the CH_4 molecular targets is a neutral gas target which is totally in the ground state, whereas the other is formed via neutralization of CH_4^+ ions through gas collisions[9] and is very likely to be in (highly) excited states. The data in Table III show that electron impact on CH_4 molecules in the ground state mostly results in production of CH_4^+ ions through direct single ionization, together with a comparable intensity of CH_3^+ ions, whereas the dominance of CH_3^+ ions among the products from CH_4 molecules in the excited states suggests that one of the hydrogen atoms (in the excited state) is easily blown away from the parent molecule.

Table III. Relative Production of CH_4^+, CH_3^+, and CH_2^+ Ions in Differently Prepared CH_4 Targets under 100-eV Electron Impact[a]

Target	Relative production		
	CH_4^+	CH_3^+	CH_2^+
Ground state CH_4 gas	100	85	15
Neutralized CH_4	100	400	200

[a]From Ref. 9.

For diagnostics, the most important and useful data are those on the optical emission that results from excitation and dissociative excitation/ionization of hydrocarbon molecules by electron impact. Two types of optical data are necessary for obtaining information on plasmas containing hydrocarbon molecules:

1. Emission from atoms and atomic ions
   (a) Photons emitted from specific transitions: These data are necessary to identify atoms and atomic ion species and to determine their intensities or numbers in plasmas.
   (b) Line profiles: The detailed profiles are useful for estimating/determining the kinetic energy of the species from the broadened line widths.
2. Emission from molecules, molecular ions, and radicals
   (a) Band emission: This type of data is needed to identify molecules and radicals and to determine their intensities.
   (b) Band structures: These data are important for determination of the internal energy (temperatures) of the species; variations in band structures can give information on environmental effects such as plasma conditions.

3. EXPERIMENTAL TECHNIQUES AND THEIR FEATURES

There are some well-established experimental techniques that can be employed in determining absolute cross sections for various collision processes involving hydrocarbon molecules. No single measurement technique is sufficient to determine these absolute cross sections. Thus, the combination of various techniques is necessary in order to obtain reasonably accurate absolute cross sections. Typical techniques and their features in determining various collision cross sections as a function of E_0, E_s, E_i and θ are described here.

1. *Incident Electron Beam Attenuation.* This technique gives total cross sections summed over all possible processes, including elastic collisions, and provides information on the magnitudes and overall features of total collision

processes. The absolute cross sections obtained are fairly accurate under well-controlled measurement conditions (to within ~2–3%) and can be used as standards to determine absolute cross sections of other collision processes.

2. *Scattered Electron Energy Loss Spectroscopy*. This is an important technique for obtaining excitation cross sections for specific processes. These cross sections depend strongly on the scattering angles {differential cross section in angle θ [DCS(θ)]} in most collision processes. These DCS have to be known over a wide range of scattering angles (0–180°). However, those at the extreme angles (near 0° and 180°) are not known in most cases, and thus the final integrated cross sections often have large uncertainties. These absolute cross sections are usually normalized with respect to those for a He gas target, but caution must be exercised in doing so because there are significant differences in target gas density distributions, as the density distribution of a gas effused through a nozzle is well known to depend strongly on the characteristics of the gas.

3. *Secondary Electrons and Their Energy Distributions*. Cross sections for production of total secondary electrons, including both scattered and ejected electrons, and ionization cross sections are important in determining plasma features. An electron energy spectroscopy technique, similar to that described in (2) above, is often used to obtain the energy distributions of electrons.

4. *Secondary Ions and Neutral Particles and Their Energy Distributions*. Accurate (1–2%) total ionization cross sections can be obtained through measurements of all secondary ions produced with a pair of condenser plates and are used as standards in the determination of absolute partial dissociative ionization cross sections. The measured partial dissociative ionization cross sections of molecules have significant uncertainties because the products are often discriminated in ion collection/acceleration systems as they often have relatively large initial kinetic energy, which hinders complete collection of the secondary ions produced (also see Section 4.1). The energy distributions of secondary ions also provide important information, in particular when ions are produced through dissociative ionization. Their energy is determined through electrostatic/magnetic analysis or time-of-flight methods. It is necessary to take into account the possibility of discrimination between ions of different charges and initial kinetic energies in these measurements. Otherwise, the observed cross sections from different measurements will have large variance.[10]

In obtaining the energy distributions of secondary neutral particles, tandem collision chamber systems are required in order to ionize them, to detect them effectively, and to measure their energy. However, it is quite difficult to estimate absolute ionization and collection efficiencies for neutral particles, especially when they are energetic, which is often the case when molecules dissociate.

5. *Photon Spectroscopy*. This technique provides two types of data:
   (a) Photon intensities: Absolute measurements of photon intensities from specific excited states of atoms, molecules, and radicals and their absolute emission cross sections have large uncertainties as it is

difficult to evaluate absolute detection efficiencies and to keep them constant over a long period of time.

(b) Line profiles: Measurement of the broadened line profiles provides a convenient means of estimating the kinetic energy of dissociated particles—ions as well as neutral particles. Often, the observed line profiles do not consist of a single line but of many overlapped lines, which are difficult to resolve in order to obtain information on a specific single peak. This technique is only applicable to excited species giving rise to photon emission.

4. PRESENT STATUS OF ELECTRON COLLISION DATA FOR HYDROCARBON MOLECULES

As mentioned in Section 1, we can confine ourselves to the description of relatively small hydrocarbon molecules up to C_3H_n (see Fig. 3). Thus, collision

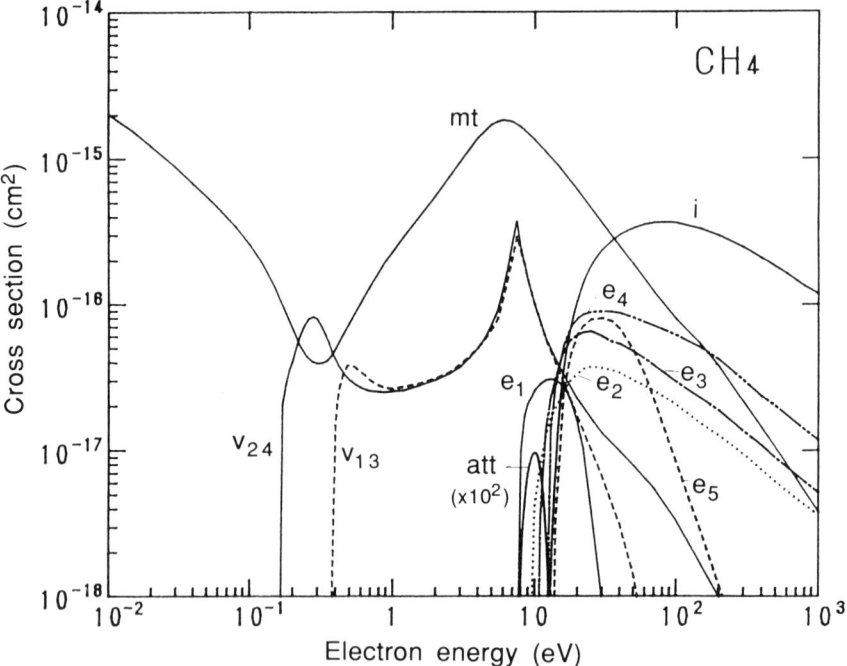

Figure 4. The evaluated cross sections for electron–CH_4 collisions as a function of the electron energy. Notation: mt, Momentum transfer process; v_j, excitation into the jth vibrational state (v_{13}: summed over the unresolved v_1 and v_3); e_j, excitation into the jth electronic excited state; att, electron attachment; i, ionization. (Adapted from Ref. 11.)

processes involving CH_4, CH_3, CH_2, CH, C_2H_6, C_2H_4, C_2H_2, C_3H_8, C_3H_6, and C_3H_4 (including isomers) are considered here. Some of these species are not stable molecules but radicals whose detailed behavior in collision processes may strongly depend on how they are formed, that is, on their internal energy. Very few experimental investigations have been performed up to now for even the most simple hydrocarbon radical, CH.

As recent surveys and evaluations of the cross sections for various collision processes involving hydrocarbon molecules[7,8,10] indicate, systematic measurements for all the processes listed above (see Section 2) are quite limited, and the relevant data are available only for relatively small hydrocarbon molecules. The most systematic compilations of the electron impact data involving hydrocarbon molecules are those published by Hayashi.[11] His evaluated cross sections for CH_4, C_2H_6, C_2H_4, C_2H_2, and C_3H_8 are shown in Figs. 4–8. Included in these figures are the cross sections for momentum transfer [defined as $2\pi \int DCS(\theta) \sin \theta (1 - \cos \theta) d\theta$], vibrational and electronic excitation (integrated over angle and energy), electron attachment, and total ionization (partial ionization cross sections will be described later in this section) of these hydrocarbon molecules as a function of the incident electron energy (the numerical values of these cross sections can be obtained upon request). From all these figures, it is clear that at ~0.1 eV, a Ramsauer

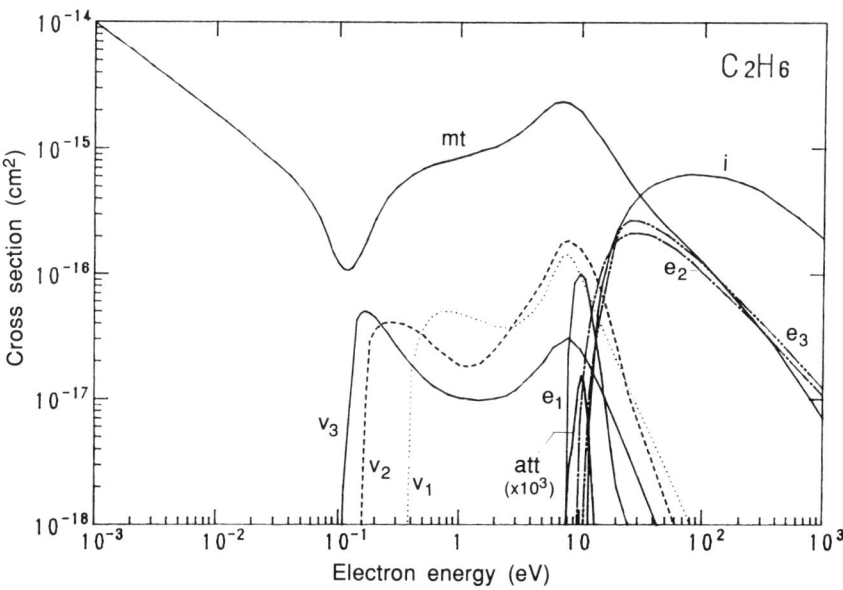

Figure 5. The evaluated cross sections for electron–C_2H_6 collisions as a function of the electron energy. The notation is the same as in Fig. 4. (Adapted from Ref. 11.)

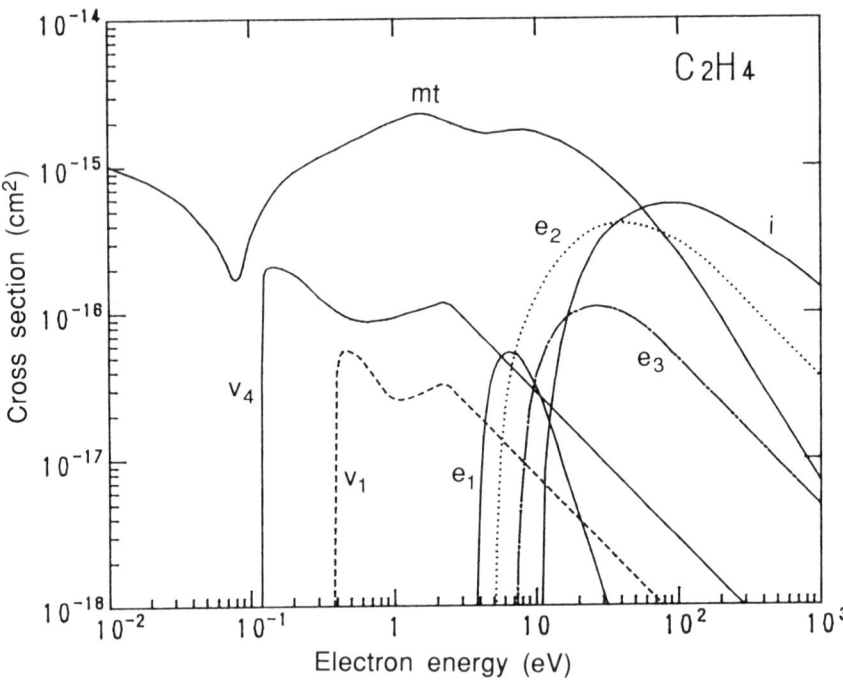

Figure 6. The evaluated cross sections for electron–C_2H_4 collisions as a function of the electron energy. The notation is the same as in Fig. 4. (Adapted from Ref. 11.)

minimum exists for most of the hydrocarbon molecules, except for C_2H_2; in the region between a few electron volts and 10 eV, vibrational excitation becomes dominant; finally, the ionization process becomes strongly dominant over other processes at electron energies above 50–100 eV.

In previous chapters of this book, detailed descriptions have been given of electron scattering and excitation by Trajmar and Kanik (Chapter 3) and of dissociative recombination on molecular ions by Mitchell (Chapter 9). Therefore, the reader should refer to these chapters. In this section, we discuss in some detail some important features of electron collision data for various processes involving hydrocarbon molecules.

4.1. Dissociation and Ionization or Ion and Neutral Particle Production

One of the relatively accurately measured collision cross sections for hydrocarbon molecules is that for ionization processes of molecules. Indeed, the cross sections for direct single ionization, in particular, the total ionization cross sections determined through so-called condenser plate techniques, seem to be sufficiently

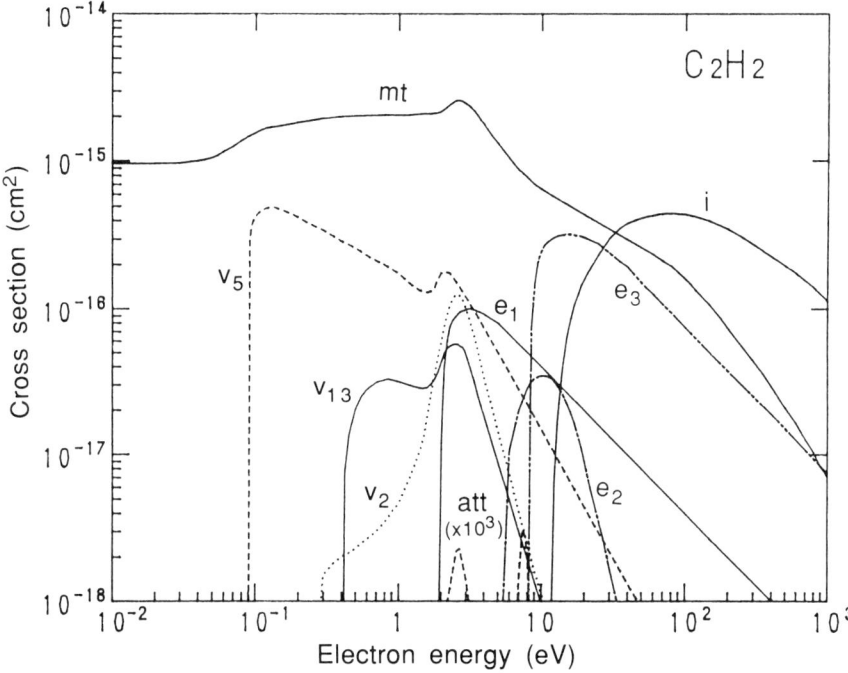

Figure 7. The evaluated cross sections for electron–C_2H_2 collisions as a function of the electron energy. The notation is the same as in Fig. 4. (Adapted from Ref. 11.)

accurate in most of the experiments (accurate to within a few percent). However, in the case of the partial cross sections for the formation of various ions by dissociative ionization, there is often significant disagreement among the results from different experiments. Discrepancies reach more than a factor of 2 or 3 in some cases, and even the electron impact energy dependence may look different, as shown by Lennon et al.[10] This is easily understood to be due to incomplete collection of product ions through the collision region on the detector as these ions have relatively large initial kinetic energies when they dissociate from the parent molecules. It has been shown that a high electric field of at least 1000 V/cm is necessary for complete collection of protons[12] (with initial kinetic energies ranging up to 10–15 eV) produced through dissociative ionization of molecular hydrogen[13] (note that an electric field of only 8–10 V/cm is sufficient to collect all the parent molecular hydrogen ions produced through direct single ionization). This field strength can be used for high-energy beams, but it is impossible to adapt this technique to low-energy electron experiments. Another important consequence of the use of a high electric field is the fact that, due to the finite size of the incident

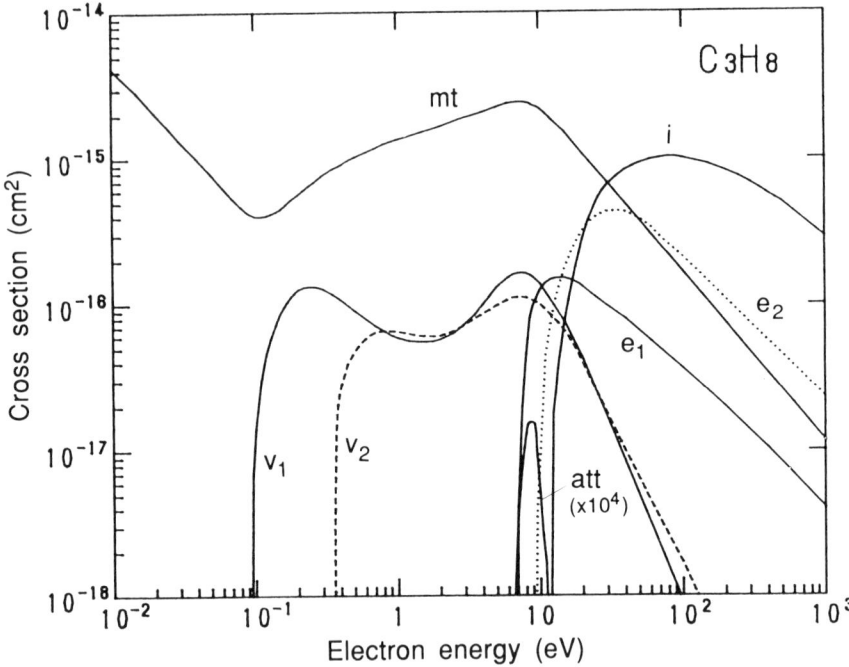

Figure 8. The evaluated cross sections for electron–C_3H_8 collisions as a function of the electron energy. The notation is the same as in Fig. 4. (Adapted from Ref. 11.)

electron beam, the resulting energy spreads of the secondary ions become so large that their original energy distributions cannot be determined accurately anymore. Recently, Grill *et al.* have tried to overcome this problem by scanning the product ions through two X–Y directions.[14]

As shown in Fig. 9, the $C_3H_8^+$ product ions formed via direct single ionization from C_3H_8,

$$e^- + C_3H_8 \rightarrow C_3H_8^+ \quad \text{(single ionization)}$$

have a single peak in their energy distributions corresponding to the near-thermal energy of about 0.04 eV. On the other hand, the energy distributions of dissociative ionization products, CH_3^+ ions for example,

$$e^- + C_3H_8 \rightarrow CH_3^+ \quad \text{(dissociative ionization)}$$

are not only broadened, due to the initial kinetic energy, but also have some additional structure, indicating the presence of several groups of ions with different kinetic energies. The widths of the broadened energy distributions observed for dissociated ions are closely related to their initial kinetic energy. Indeed, the energy

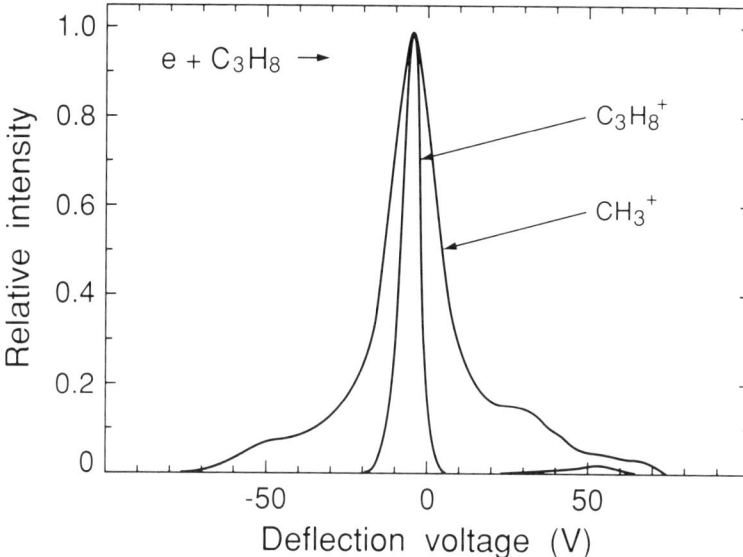

Figure 9. Energy profiles of $C_3H_8^+$ and CH_3^+ ions produced from C_3H_8 under electron impact. These profiles are obtained by moving the ion beam on a mass spectrometer entrance slit while the deflecting voltage on a pair of plates is varied. Note that the profile of the $C_3H_8^+$ ions shows a single peak, whereas that of the CH_3^+ ions consists of several additional peaks. (Adapted from Ref. 22.)

distribution of CH_3^+ ions formed from C_3H_8 molecules is found to consist of at least three components at a mean energy of 0.28, 1.55, and 7.88 eV corresponding to 31, 31, and 38%, respectively, of the total ions.

In comparison to the cross sections for ion production through dissociative ionization, those for dissociation as well as for dissociative ionization processes resulting in production of neutral species are far more difficult to measure, and thus only a limited number of such cross sections have been determined, and the accuracies of these determinations are not good. (Typically, the uncertainties are more than a factor of 2!) Total dissociation cross sections have been determined through the measurement of the pressure rise in a collision chamber due to dissociation by Winter[15] (up to 500 eV) and Perrin et al.[16] (up to 110 eV); the cross sections obtained are only upper limits and, in most cases, are not reliable as the contributions of dissociative ionization and wall effects have not been accurately evaluated.

In order to determine the partial cross sections for neutral particle production, special procedures are necessary: after the neutral species are produced in a collision chamber by electron impact, they have to be ionized by another electron beam in a second collision chamber. Indeed, there are a number of practical difficulties in determining the cross sections for the production of neutral species because of the following problems:

1. The number of product neutral species is small.
2. The neutral species produced may have relatively large kinetic energies with wide distributions, and thus their effective ionization cross sections are often different from those that would be obtained at thermal energies.
3. The angular distributions of the products are not well known, although they are unlikely to be strongly anisotropic.
4. No information is available on the excitation state or the internal energy of the products, which strongly influences their ionization cross sections in the second collision chamber.
5. The adherence of the products to walls is strongly affected by the surface conditions.

Altogether, the cross sections for the production of neutral species are far less accurate and very limited in quantity as well as in quality compared with those for production of ions.

As total ionization cross sections have been presented in Figs. 4–8, we would like to show partial cross sections for hydrocarbon molecules at 100-eV electron impact, many of which have been summarized in our recent report.[8] Some of these data have been recently obtained by employing the experimental arrangement of Grill et al.,[14] mentioned above. Relative intensities of the product ions, which are generally independent of the impact energy above 100 eV, would be useful in order to have some idea about the intensities of product species.

4.1.1. CH_4

A number of theoretical and experimental studies have been reported for CH_4 molecules, and the results have been summarized recently.[8] Some typical values of the measured cross sections for production of various ions and neutral radicals from CH_4 molecules by 100-eV electron impact are shown in Table IV. The total as well as partial ionization cross section data for neutral CH_4 gases from different experiments[17,18] seem to be in reasonably good agreement. It should, however, be noted that these cross sections are strongly dependent on the internal energy of the parent molecules (see Table III).

CH_4^+ and CH_3^+ ions, both of nearly equal intensities, amount to more than 85% of the total ions produced from CH_4. On the other hand, it may be noted from Table IV that neutral H atoms comprise roughly half of the neutral particles produced in pure dissociation processes. Some of these cross sections for neutral radical production reported by Melton and Rudolph[17] can be compared with recent measurements of partial cross sections for CH_3 and CH_2 production (from threshold

Table IV. Partial Cross Sections for Direct Single Ionization, Dissociative Ionization and Dissociation Processes of CH_4 Molecules under 100-eV Electron Impact

Product ion	Cross section (10^{-16} cm^2)		Product neutral	Cross section (10^{-16} cm^2)	
	Ref. 17	Ref. 18		Ref. 17	Ref. 19
CH_4^+	1.8	1.45	CH_3	1.2	0.54^a
CH_3^+	1.5	1.26	CH_2	0.2	
CH_2^+	0.28	0.21	CH	0.1	
CH^+	0.14	0.097	H	2.4	
C^+	0.05	0.035	H_2	0.8	
H_2^+	0.02	0.019			
H^+	0.04	0.11			

^{a}The cross sections for production of CH_2 become maximum (7×10^{-16} cm^2) at around 18 eV and decrease rapidly as the electron energy increases, whereas those for CH_3 are maximum (1.4×10^{-16} cm^2) at around 50 eV.[19]

up to 100 eV) reported by Nakano et al.,[19] who have used the cross sections for ionization of neutral radicals (CH_3) recently reported by Baiocchi et al.[9] in their analysis of the second ionization process (though there is no guarantee that CH_3 radicals in both measurements have the same internal energy) rather than those for the ground state (neutral gas) CH_4, employed by Melton and Rudolph.[17] The intensities of CH_3 radicals far exceed those of CH_2 radicals at electron impact energies above 30 eV even though the intensities of CH_3 and CH_2 radicals seem to be nearly equal just above the threshold energy region.[19] The results obtained for CH_2 radical production seem to be in reasonable agreement with a theoretical calculation by Winstead et al.[20] performed for a limited range of electron impact energy (10–30 eV).

It should be noted that the initial kinetic energy supplied to neutral and ion dissociation products from CH_4 molecules, except for hydrogen atoms, which carry a large fraction of the initial kinetic energy after dissociation, is expected to be small owing to their chemical (tetrahedral) structure, with the carbon atom at its center. Therefore, the accuracies of the cross sections for CH_4 are usually better than those of the cross sections for other hydrocarbon molecules because the dissociation products of the latter often have large kinetic energies, and thus various uncertainties are included in the final cross sections.

4.1.2. CH_3

In order to investigate experimentally the collision behavior of CH_3, this radical is formed via electron capture into the CH_3^+ ion which passes through a gas target and then is crossed by an electron beam of controlled energy. The only experiment that has been reported so far is that by Baiocchi et al.[9] at electron impact energies below 200 eV. In this type of experiment involving radicals, there is no precise information on the internal energy of the target radicals (in particular, on

Table V. Partial Cross Sections for Direct Single Ionization and Dissociative Ionization of Neutralized CD_3 and CD_2 Molecules under 100-eV Electron Impact

Target	Product ion	Cross section (10^{-16} cm^2)
CD_3	CD_3^+	1.75
	CD_2^+	1.15
CD_2	CD_2^+	1.65
	CD^+	0.64

rotational and vibrational states, even though they may be electronically in the ground state), which must certainly depend on how they are formed. This makes it difficult to compare data obtained in different experiments, insofar as the internal energy of the target radicals and its distribution are not known accurately. In the present case, another measurement[21] at low (8–14 eV) electron energies based on photodissociation of CH_3OH seems to be in relatively good agreement with that by Baiocchi et al.[9] Table V contains the cross sections at 100 eV for direct single ionization and dissociative ionization of CD_3 radicals neutralized from CD_3^+ ions.

4.1.3. CH2

The experimental procedures in this case are the same as in that of CH_3 described above. CH_2^+ ions from dissociative ionization of CH_4 are neutralized through electron capture. In the only experimental work reported, direct single ionization and dissociative ionization cross sections have been measured up to 200-eV electron impact energies[9] (see Table V for data on CD_2 radicals).

4.1.4. CH

Unfortunately, no experimental or theoretical work has been reported on ionization and dissociation of this simplest hydrocarbon molecule under electron impact.

4.1.5. C2H6

Recently, Grill et al.[22] have used a scanning technique to collect all the ions which are most likely to have different kinetic energies when they dissociate from the parent molecules. As described above, ions produced through direct single ionization, namely $C_2H_6^+$, have shown only a single peak in such scanned energy distributions (see Fig. 8, which shows similar features for C_3H_8 molecules). On the other hand, CH_3^+ ions, formed through dissociative ionization, show some broadened peaks, suggesting two or more components in the ion kinetic energy. It is noted

Table VI. Partial Cross Sections for Direct Single Ionization and Dissociative Ionization of C_2H_6 Molecules under 100-eV Electron Impact

Product ion	Cross section (10^{-16} cm^2)		Product ion	Cross section (10^{-16} cm^2)	
	Ref. 22	Ref. 23		Ref. 22	Ref. 23
$C_2H_6^+$	0.48	0.81	CH_3^+	0.26	0.13
$C_2H_5^+$	0.47	0.65	CH_2^+	0.18	0.095
$C_2H_4^+$	2.38	2.90	CH^+	0.10	0.039
$C_2H_3^+$	1.19	0.97	C^+	0.04	0.013
$C_2H_2^+$	0.81	0.64	$C_2H_5^{2+}$	0.01	
C_2H^+	0.20	0.11	$C_2H_3^{2+}$	0.0003	
C_2^+	0.05	0.02			

that the most intense ions from dissociative ionization are not the parent $C_2H_6^+$ ions (only 8%) but $C_2H_4^+$ ions (40%). The agreement between two independent measurements[22,23] seems to be generally good, except for $C_2H_6^+$ and $C_2H_5^+$ ions, which should acquire relatively small kinetic energies during dissociative ionization (see Table VI).

4.1.6. C_2H_4

Very limited measurements of partial cross sections for ion production from C_2H_4 have been reported, but no absolute values have been given. Table VII contains the cross sections that have been determined from relative values obtained by Melton[24] after normalization with respect to absolute cross sections for total ion production.[25,26]

4.1.7. C_2H_2

Only a single measurement of partial cross sections for dissociative ionization has been reported so far at electron energies ranging from 100 eV up to 2000 eV.[27] Some cross sections at 100-eV electron impact energy are given in Table VIII. No

Table VII. Partial Cross Sections for Direct Single Ionization and Dissociative Ionization of C_2H_4 Molecules under 100-eV Electron Impact[a]

Product ion	Cross section (10^{-16} cm^2)	Product ion	Cross section (10^{-16} cm^2)
$C_2H_4^+$	2.44	CH_3^+	0.010
$C_2H_3^+$	1.51	CH_2^+	0.095
$C_2H_2^+$	1.27	CH^+	0.041
C_2H^+	0.23	C^+	0.023
C_2^+	0.076	$C_2H_3^{2+}$	0.046

[a] Refs. 24 and 28.

Table VIII. Partial Cross Sections for Direct Single Ionization and Dissociative Ionization of C_2H_2 Molecules under 100-eV Electron Impact

Product ion	Cross section (10^{-16} cm^2)
$C_2H_2^+$	3.20
C_2H^+	0.65
C_2^+	0.19
CH^+	0.052
C^+	0.048
$C_2H_2^{2+}$	0.032
C_2H^{2+}	0.0003

[a]Ref. 27.

data have been reported at electron energies below 100 eV, except for total ion production cross sections.[28]

4.1.8. C_3H_8

Recent results from measurements of partial ionization cross sections by Grill et al.[14] are summarized in Table IX. The total ionization cross sections obtained by summing the partial cross sections over all ion products are slightly smaller (20%) than those recently measured with the condenser plate technique by Nishimura and Tawara.[26]

Table IX. Partial Cross Sections for Direct Single Ionization and Dissociative Ionization of C_3H_8 Molecules under 100-eV Electron Impact[a]

Product ion	Cross section (10^{-16} cm^2)	Product ion	Cross section (10^{-16} cm^2)
$C_3H_8^+$	0.544	$C_2H_5^+$	2.06
$C_3H_7^+$	0.423	$C_2H_4^+$	1.03
$C_3H_6^+$	0.125	$C_2H_3^+$	1.346
$C_3H_5^+$	0.304	$C_2H_2^+$	0.537
$C_3H_4^+$	0.076	C_2H^+	0.075
$C_3H_3^+$	0.752	C_2^+	0.014
$C_3H_2^+$	0.349	CH_3^+	0.443
C_3H^+	0.215	CH_2^+	0.160
C_3^+	0.039	CH^+	0.055
$C_3H_5^{2+}$	0.0007	C^+	0.026
$C_3H_4^{2+}$	0.0168		
$C_3H_3^{2+}$	0.0151		
$C_3H_2^{2+}$	0.0254		

[a]Ref. 14.

4.1.9. C_3H_6

Recent measurements of total scattering cross sections for both propene and cyclopropane show that a clear isomer effect exists at low energies, below 50 eV,[29] the cross sections for propene being 20–30% larger than those for cyclopropane. This is also confirmed by a recent theoretical calculation.[30] However, no partial ion and neutral species production cross sections have been reported yet.

4.1.10. C_3H_4

No investigation for C_3H_4 (including the isomers allene and methylacetylene) has been reported yet.

4.1.11. Scaling Laws

It should be pointed that scaling laws[29] can be used to estimate total ionization cross sections of various hydrocarbon molecules if these cross sections, divided by the polarizabilities of the molecules, can be expressed as a function of the electron impact energy, divided by the ionization threshold energy in a single analytic formula, analogous to the well-known Lotz formula.

4.2. Energy Distributions of Product Ion and Neutral Species

As often mentioned above, knowledge about the kinetic energy distributions of products from dissociation and dissociative ionization of molecules is important not only for a basic understanding of the relevant mechanisms but also for many applications. For example, the initial kinetic energy imparted through such collision processes plays a key role in determining the mean (collision) free path lengths and penetration–diffusion lengths of particle species in gases, plasmas, or other media. Furthermore, products with high energy, particularly if they are in excited states, are known to play a critical role in many chemical reactions, and, thus, more reactions become likely to occur if such products are formed.

The determination of the kinetic energies of ion products by means of electrostatic/magnetic analysis is straightforward. Due to the intrinsically good energy resolution, a series of different channels leading to ion production can be distinguished. Indeed, many peaks due to protons with different kinetic energies have been observed.[31] The energy distributions of protons produced by dissociative ionization of CH_4 molecules have been investigated in detail. A number of groups of protons with different kinetic energies, corresponding to 0.0, 0.5, 0.9, 2.35, and 3.97 eV, have been shown to be produced under 20- to 50-eV electron impact. In the energy distributions obtained in this electron impact energy range, the peaks due to zero-energy (thermal energy) protons are the most intense. Peaks due to protons with large kinetic energies become intense as the electron impact energy increases.

On the other hand, the determination of kinetic energies of neutral particle products is fairly complicated, in particular in cases in which the absolute efficiencies have to be known. The kinetic energies as well as energy distributions of such products can be directly determined through ionization of the neutral species by a second electron impact, followed by electromagnetic or time-of-flight analysis. However, as the ionization efficiencies depend on the velocities of the species to be ionized in the second ionization chamber, accurate determination of absolute fractions of neutral product species with different kinetic energy distributions is quite difficult.

Indirect measurement methods are often used in order to get information on the energy distributions of neutral particle products.[32,33] These are methods based upon Doppler shifts and broadened profiles of the emitted photons, which can be applied only for excited species but not for ground state particles. Systematic measurements have been reported only for hydrogen atoms $H^*(n)$ from dissociation processes. The Balmer lines associated with $n = 3 \rightarrow n = 2$ and $n = 4 \rightarrow n = 3$ transitions are most often used in determining the initial kinetic energy distributions of neutral hydrogen species in these excited states. Figure 10 shows a comparison of the Balmer-α line ($n = 3 \rightarrow n = 2$) profiles from H atoms produced in the dissociation of H_2 and CH_4 molecules by electron impact of different energies. As seen, the Balmer line profiles of H atoms from CH_4 molecules change and become broad as the electron energy increases, suggesting that the observed Balmer line consists of several components with various initial kinetic energies. Thus, the analyses become quite complicated, and it is practically impossible to clearly separate all the possible contributing components based on the observed line profiles. On the other hand, the Balmer lines from the excited hydrogen atoms dissociated from H_2 molecules are relatively simple, as expected from the energy diagram of the hydrogen molecule.[13] The analysis of the broadened Balmer line profiles of the excited hydrogen atoms dissociated from hydrocarbon molecules reveals that there are different kinetic energy components, which can likely be divided into four components corresponding to kinetic energies of 0–2, 2–6, 6–8, and 8–12 eV.[34] At low electron energies, the low-energy components are dominant, whereas at high impact energy the high-energy components become significant. It should be noted that the Balmer line profiles observed under limited spectral resolution seem to be very similar among many hydrocarbon molecules. Furthermore, the kinetic energies of neutral species other than hydrogen atoms (such as CH_3 or CH) are very difficult to analyze. Indeed, very few kinetic energy measurements have been reported for such neutral species.[35] The observed data for C atoms in Rydberg states from the dissociation of CH_4 show that their initial kinetic energy is relatively small (1–2 eV under 20- to 50-eV electron impact), as expected on the basis of the chemical structure of the parent molecule.[36] On the other hand, it is expected that heavy neutral molecules as well as ions produced from other hydro-

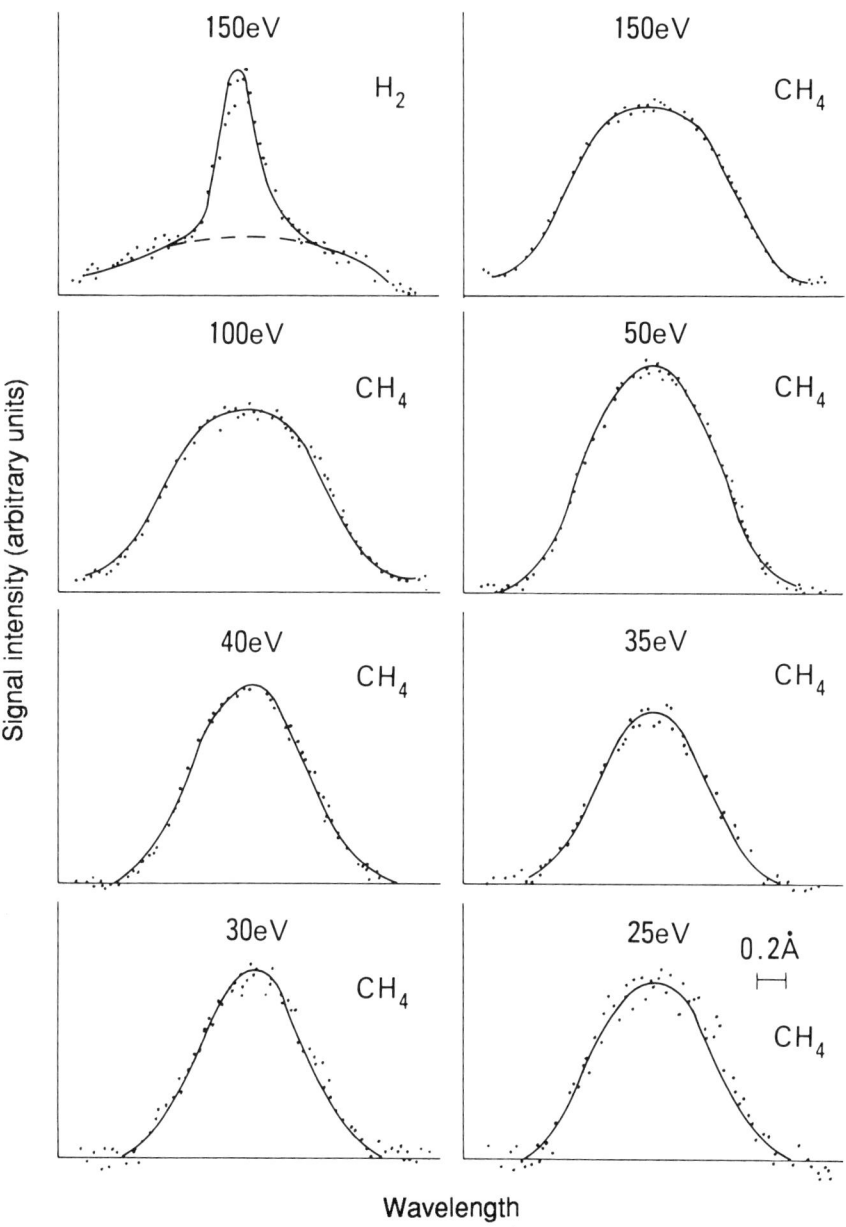

Figure 10. Comparison of Balmer-α line profiles of H* ($n = 3$) produced from H_2 and CH_4 molecules under electron impact.[8] Note that in the case of CH_4 the spectral shape becomes broad as the electron impact energy increases, whereas in the case of H_2 the line profile consists of two distinct parts (one, dominant at low electron energy, has a sharp peak corresponding to low kinetic energy, and the other, dominant at high electron energies, a broad peak corresponding to high kinetic energy). It should also be noted that each peak does not represent monoenergetic hydrogen atoms but hydrogen atoms with broad energy distributions.[13]

carbon molecules could have a large amount of kinetic energy imparted to them upon dissociation from the parent molecule. However, as yet, little is known.

It is quite important to point out that up to now no precise measurement is available at all on cross sections for the production of ground state neutral hydrogen atoms, which are expected to be significant at low-energy electron impact.

4.3. Dissociative Recombination and Dissociation/Ionization of Hydrocarbon Molecular Ions

Dissociative recombination processes[37] of hydrocarbon molecular ions in vibrationally excited states (v) under electron impact such as

$$e^- + CH^+(v) \rightarrow C + H \quad \text{(dissociative recombination)}$$

are important at very low energies. A detailed description of this process has been given by Takagi et al.[38] However, experiments of this kind, based on electron–ion crossed- or merged-beams techniques, are accompanied by a number of difficulties in preparing the ground state or, more precisely, state-specified molecular ions. The electronic as well as ro-vibrational states of the parent (target) ions used are usually not known precisely, or not known at all. The target molecular ions that have been investigated to date are limited to the simplest hydrocarbon molecules—CH_4^+, CH_3^+, CH_2^+, and CH^+—without any specification of the internal energy state. In these investigations, the collision energy covers only 0.03–1 eV. However, most of these measurements suggest a strong dependence of the cross sections on the target ion's excited (ro-vibrational or electronic) states which decrease smoothly as the collision energy increases up to 1 eV (roughly as $E_0^{-0.9}$).

In the most recent experiment,[39] molecular CH^+ ions have been stored for about 5 s in an ion storage ring facility, and thus these ions can be assumed to be mostly in the ground state, at least electronically, but no absolute cross sections have been reported yet. The measured cross sections for the stored ions, normalized to those obtained in a previous merged-beams measurement,[40] show prominent peaks at around 0.8 and 10 eV, which were not observed in the previous measurements,[34] superposed on a smooth $E_0^{-0.9}$-curve over the energy range 0.01–60 eV.

To date, no cross sections have been reported for direct dissociation and ionization of hydrocarbon molecular ions, which become important under high-energy electron impact:

$$e^- + CH^+(v) \rightarrow e^- + C^+ + H/C + H^+ \quad \text{(dissociative ionization)}.$$

In this case, the cross sections for dissociation and dissociative ionization processes have also been found to strongly depend on the internal energy of the target species, and not only the absolute values but also their electron impact energy dependence can vary for ions produced under different ion source conditions.[41] As the cross sections are known to strongly depend on the internal energy, systematic discussion of these processes is difficult insofar as the internal energy of the parent ions is not specified.

Another interesting point is that direct single ionization of molecular ions can occur under high-energy electron impact, resulting in doubly charged ions. However, most of these doubly charged molecular ions are known to be unstable and dissociate immediately after formation, except for a few cases such as CH_4^{2+} ions, which are metastable, with a lifetime of only a few microseconds, when they are formed through single ionization from CH_4^+ ions but not through direct double ionization from CH_4. A detailed description of dissociative recombination of various molecular ions is given by Mitchell in Chapter 9 of this book.

4.4. Photon Emission

In the photon-emission spectra recorded from hydrocarbon molecules under electron impact, the most intense lines are usually hydrogen lines whose profiles are often broadened owing to the initial kinetic energies afforded during dissociation processes. In addition to these hydrogen lines, relatively intense band structures due to various molecules or radicals are also seen in spectra from many hydrocarbon molecules. Typical emission cross sections for various transitions in the spectrum recorded from CH_4 are shown in Figs. 11–13. Lyman and Balmer lines all show simple smooth variations of the cross sections as a function of the electron impact energy. Naturally, the Lyman-α line is the most intense, and it varies smoothly as a function of the electron impact energy, whereas C I lines show some structure, suggesting that more than single channels contribute to these emissions. It is not clear whether such structure is present in CH band emissions observed by Aarts et al.[42] More detailed and precise observations are necessary for these band emissions.

It is interesting to note that the emission cross sections of Lyman and Balmer lines from hydrogen atoms are practically the same for various hydrocarbon molecules and are almost independent of the number of hydrogen atoms in the molecules,[8] though measurements have been made only for CH_4, C_2H_6, C_2H_4, and C_2H_2.

On the other hand, the emission cross sections of the transitions from C I seem to be proportional to the number of carbon atoms in the hydrocarbon molecule. For example, the emission cross sections for the C I ($^3P - ^3P^0$) transition line at 165.7 nm from C_2H_2 are roughly twice those from CH_4. Similarly, the emission cross sections of CH bands seem to be roughly proportional to the number of carbon atoms in the molecules, though no cross section has been reported for C_3H_n molecules yet.

Unfortunately, measurements of the absolute emission cross sections are still limited for most hydrocarbon molecules, in particular for heavy molecules. Among hydrocarbon molecules, acetylene (C_2H_2) molecules have been most widely investigated optically. Typical behaviors of the cross sections for photon emissions from

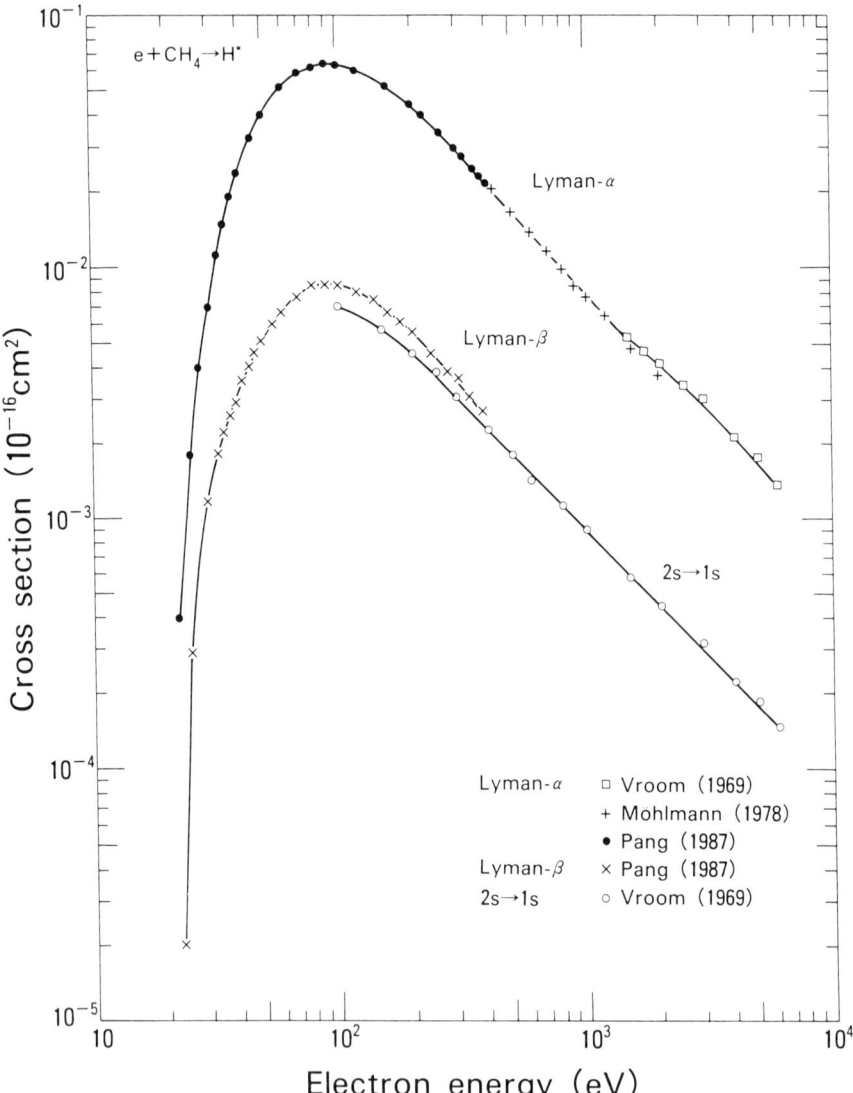

Figure 11. Photon-emission cross sections of hydrogen Lyman-α and -β and $2s \rightarrow 1s$ lines in the spectrum recorded from CH_4 under electron impact.[8]

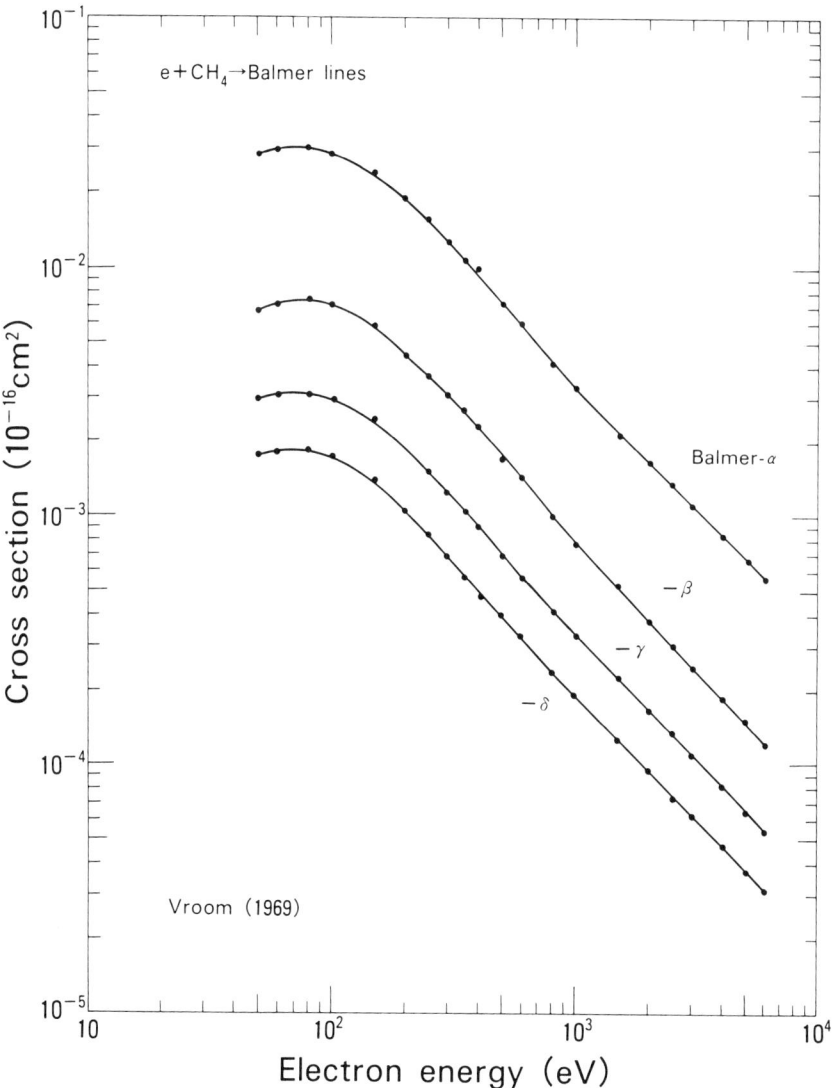

Figure 12. Photon-emission cross sections of hydrogen Balmer-α, -β, -γ, and -δ lines in the spectrum recorded from CH_4 under electron impact.[8]

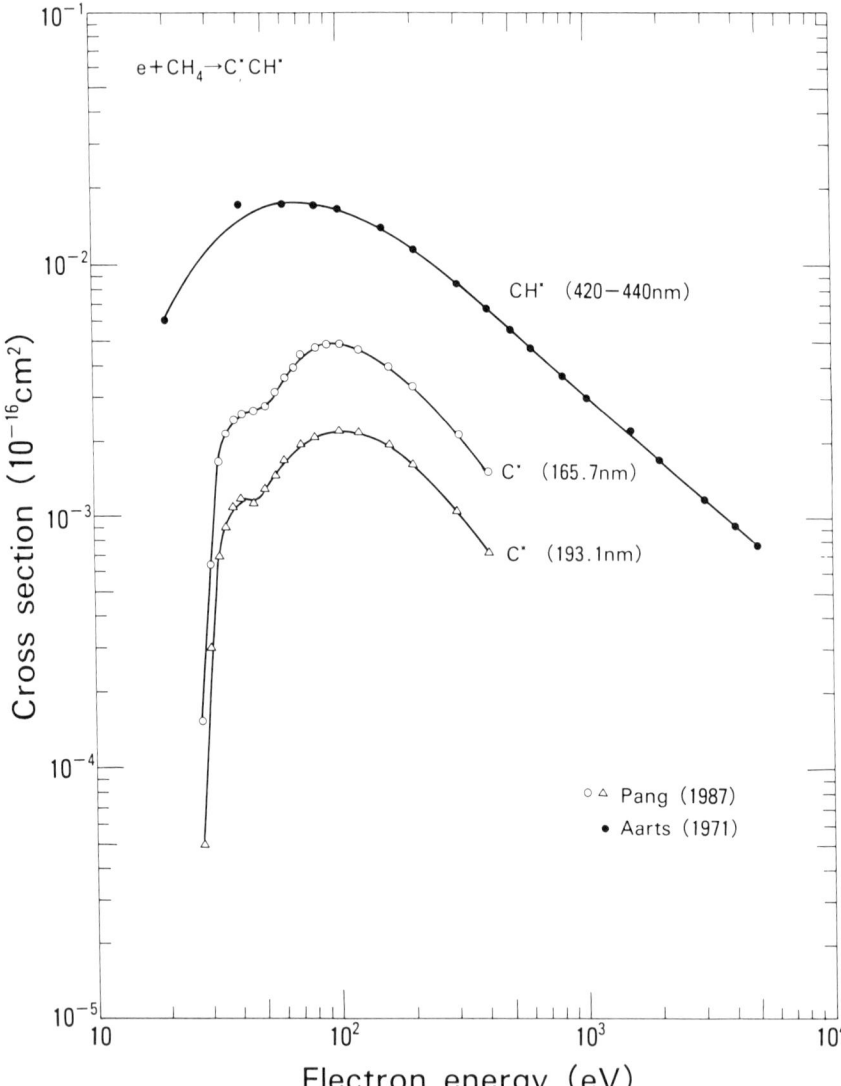

Figure 13. Photon-emission cross sections of C I and CH bands in the spectrum recorded from CH_4 under electron impact.[8]

C_2H_2 are shown in Figs. 14 and 15. The most intense lines are Lyman-α line (and probably Balmer-α line, but their cross sections have not been directly measured yet) and CH bands, followed by Lyman-β, Balmer-β, and C I lines, as seen in these figures. As in CH_4, the observed dependence[43,44] of emission cross sections of both the Lyman and Balmer lines of hydrogen atoms on the electron impact energy

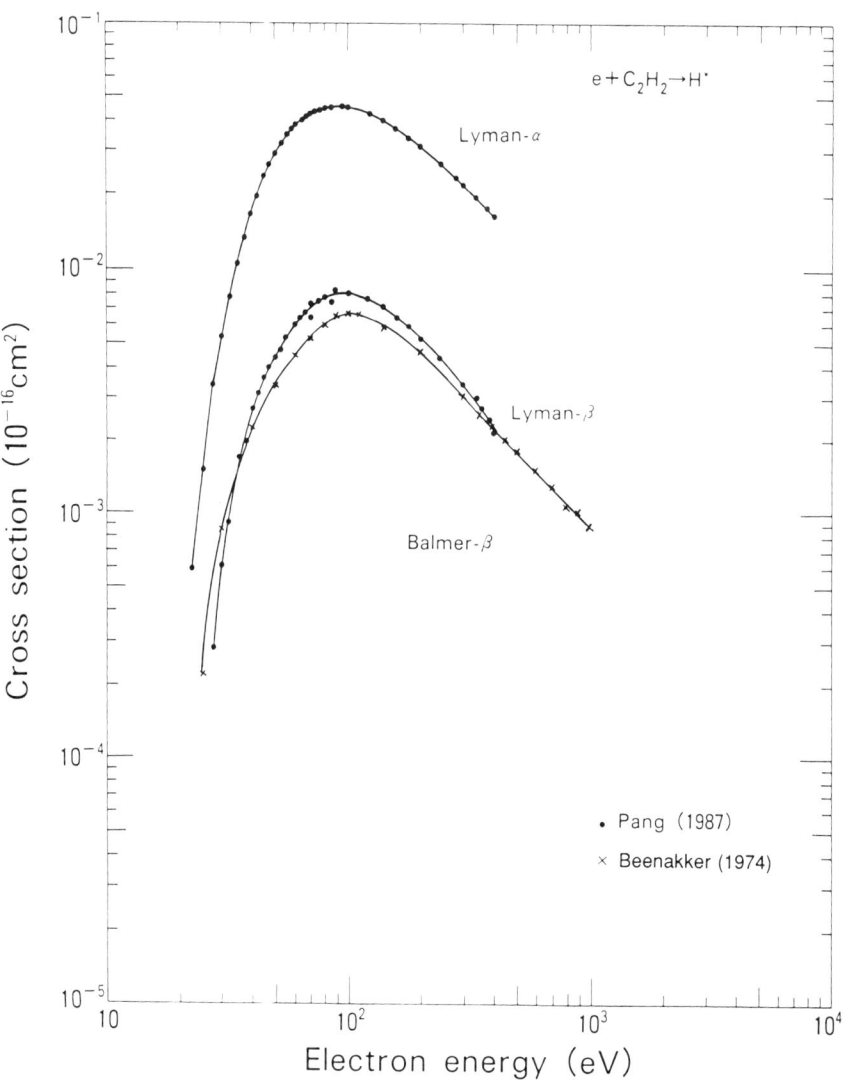

Figure 14. Photon-emission cross sections of hydrogen Lyman-α and -β and Balmer-β lines in the spectrum recorded from C_2H_2 under electron impact.[8]

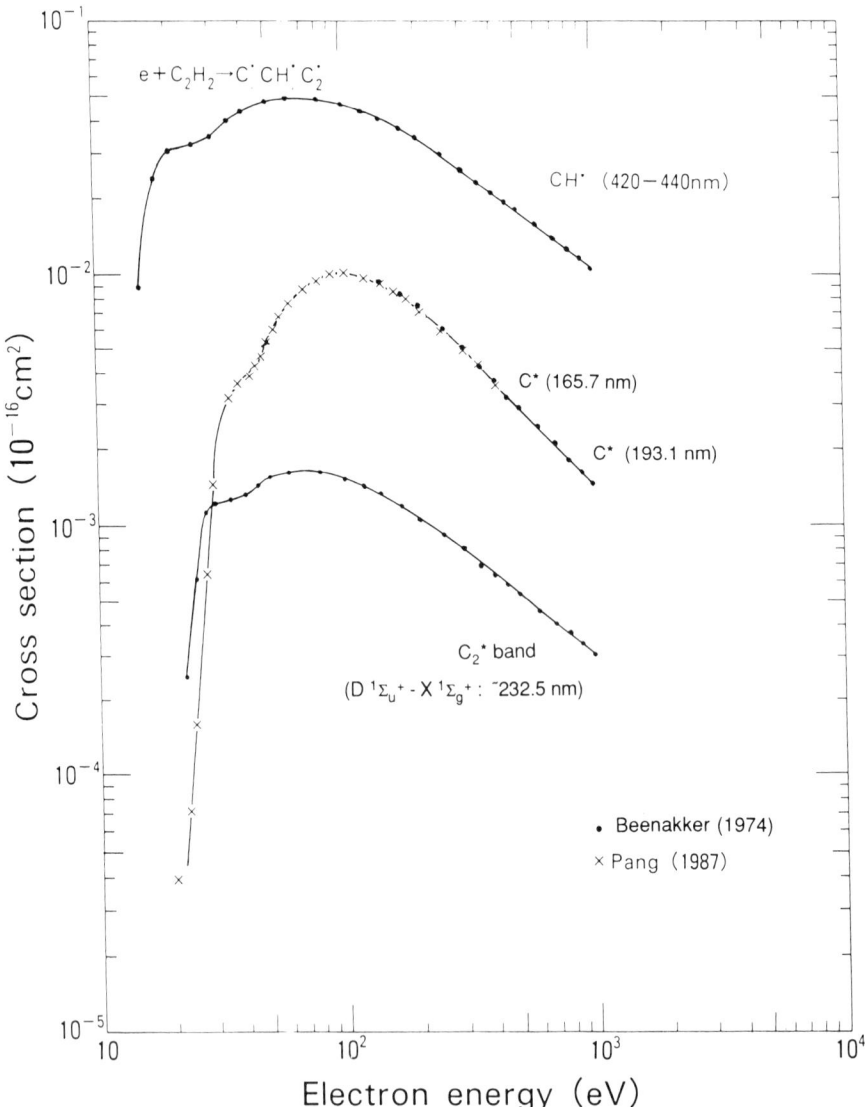

Figure 15. Photon-emission cross sections of C I, CH, and C_2 bands in the spectrum recorded from C_2H_2 under electron impact.[8]

follows simple curves, whereas that of the C I lines ($2p3s\ ^1P^0-2p^2\ ^1D$: 193.1 nm; $2p3s\ ^3P^0-2p^2\ ^3P$: 165.7 nm), the CH band ($A\ ^2\Delta-X\ ^2\Pi$: ~431.5 nm), and the C_2 bands [Mulliken system ($D\ ^1\Sigma_u^+-X^1\Sigma_g^+$): ~232.5 nm; Deslandres–d'Azambuja system ($C^1\Pi_g-A^1\Pi_u$): ~314.3 nm (the cross sections for these two band emissions are nearly equal); Swan system ($d^3\Pi_g-a^3\Pi_u$): ~434.0 nm] show some structure, indicating that at least two different channels contribute to these emissions. Note that the cross sections for both C I transitions (193.1 and 165.7 nm) are nearly equal, and branching ratios of the emission cross sections for C I ($2p3s\ ^1P^0-2p^2\ ^1D$: 193.1 nm)/C I ($2p3s\ ^1P^0-2p^2\ ^1S$: 247.9 nm) have been determined to be 0.068 ± 0.024. Also note that the emission cross sections for the C_2 band ($d^3\Pi_g-a^3\Pi_u$; Swan system) observed by Sushanin and Mishko[45] do not show such a structure.

Not only the intensities of various lines but also the variations of band structures in the photon-emission spectra are interesting features to be investigated. For example, CH bands are always observed from hydrocarbon molecules. Some features of the CH band from acetylene (C_2H_2) under electron impact can be summarized as follows. As the electron energy increases from 15 to 1000 eV, the rotational temperature decreases only slightly, while vibrational populations increase slightly.[46,47] The populations in high vibrational states increase by a factor of 2 over the 17- to 100-eV electron impact energy. It is also important to note (and also to be confirmed) that the temperatures from the CH band produced in plasmas may not be the same as those in electron impact and may depend on the plasma parameters, such as temperature and density, as described by Pospieszczyk et al.,[48] whose photon spectra produced from C_2H_2 in high- and low-temperature He plasmas are shown in Fig. 16. In the high-temperature plasma spectrum, a number of emission bands due to ro-vibrational excitation are clearly observed.

Similar behavior has been observed for C_2 ($d^3\Pi_g-a^3\Pi_u$) Swan bands (lifetime of ~119 ns). For example, the following are common features among C_2H_n ($n = 2, 4,$ and 6) molecules[49]:

1. High vibrational temperatures for larger n molecules
2. Lower vibrational temperatures at high electron impact energy
3. Boltzmann type v-distributions (decreasing smoothly for larger v, by a factor of 2 from $v = 0$ to $v = 6$).

Generally speaking, polarization of photons emitted from dissociation products of hydrocarbons is not too large, at least under usual collision conditions. However, in plasmas where strong magnetic fields are applied, polarization of photons has recently been realized to affect the estimates of electromagnetic fields inside the plasma and must be taken into account in obtaining information about these parameters.

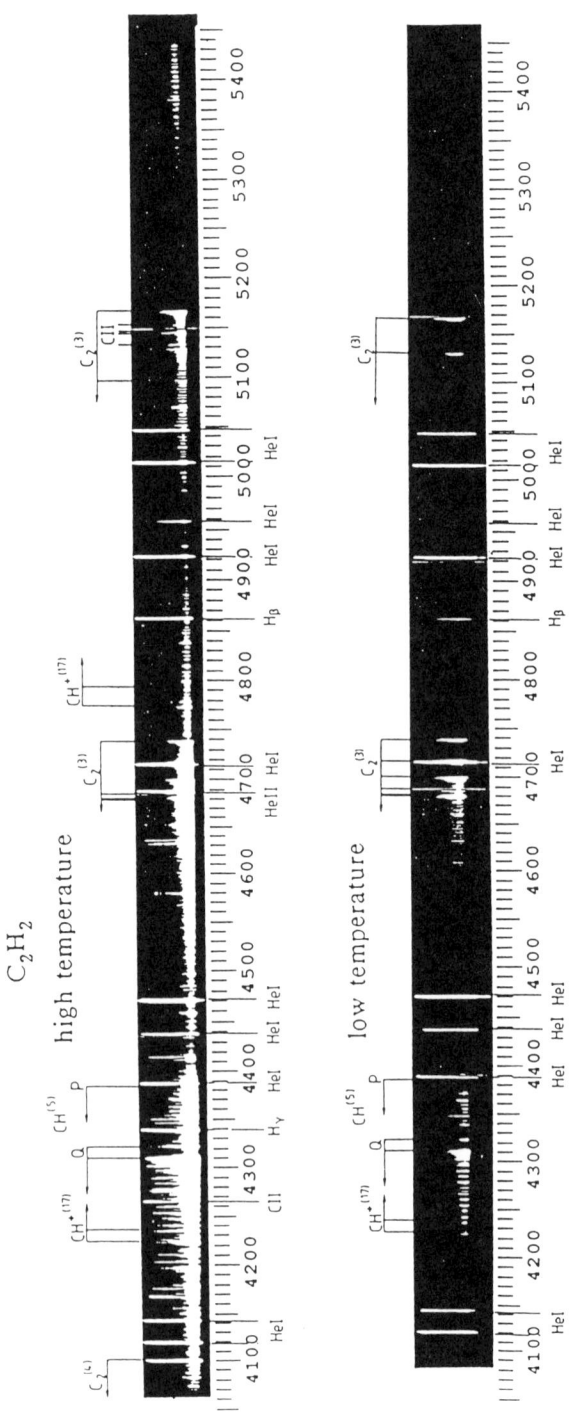

Figure 16. Photon spectra recorded from C_2H_2 in high-temperature (*top*) and low-temperature (*bottom*) He plasmas.[48]

Table X. Availability and Accuracies of Important Cross Section Data for Hydrocarbon Molecules under Electron Impact[a]

Cross section	Hydrocarbon[b]								
	CH_4	CH_3	CH_2	CH	C_2H_6	C_2H_4	C_2H_2	C_3H_8	C_3H_6[c]
Total scattering	B				B	B	b	B	b
Elastic scattering									
Total	B				b	b	b	b	
Differential	B				b	b	b	b	
Momentum transfer	B				B	B	B	B	b
Excitation									
Rotational									
Vibrational	B				c				
Electronic	B				B	B	B	B	
Dissociation									
Total	c				c				
Partial	d								
Electron attachment	A				C		B	B	
Ionization									
Total	A				A	A	A	A	A
Partial	B	c	c		B		b	B	
Photon emission									
Ion/atom	B				B	B	B		
Molecule	B					B	C		
Energy									
Ion/atom	c				c	c	c		
Molecule									
Ion recombination	c	c	c	c					
Ion dissociation	d								

[a] Accuracies: A, <10%; B, ~10–50%; C, ~50–100%; D, >100% (over wide energy range); a, <10%; b, ~10–50%; c, ~50–100%; d, >100% (over limited energy range). Blank cells indicate that no data (or practically no data) are available.
[b] All neutral gases are in the ground states, except for CH_3, CH_2, and CH, which are likely to be in highly excited states.
[c] Includes both cyclopropane and propene. (An isomer effect has recently been observed below 70 eV.[29,30])

5. SUMMARY, FURTHER DATA NEEDS, AND RECOMMENDED WORK

In diagnosing, analyzing, and modeling plasmas in fusion plasma apparatus with graphite inner walls, systematic knowledge of cross sections for a series of collisions involving hydrocarbon molecules under electron impact is required. The present availability of collision data for the various hydrocarbon molecules described above is summarized in Table X. Although a relatively large amount of data has become available recently for hydrocarbon molecules, the general trend is that, except for CH_4, many data still seem to be of relatively low accuracy or even have not yet been measured, as clearly seen in Table X. Indeed, data and information for neutral particle products are quite limited in quantity as well as in quality, and many of the reported data still have large uncertainties (e.g., a factor of 2 or more).

In particular, the optical data (on an absolute scale) are quite important in plasma diagnosis. Therefore, accurate optical emission cross sections from various atomic and molecular species are urgently required. As already described, some features of band emissions from hydrocarbon molecules such as vibrational temperatures or populations in various rotational/vibrational states can provide information on such plasma characteristics as temperatures or densities. Thus, it would be worthwhile to investigate systematically the variation of band structures under various well-defined plasma conditions and their dependence on plasma parameters and to examine in turn whether band structures could be an important tool for plasma diagnosis.

One of the most important features of electron–molecule collisions is the fact that most collision cross sections strongly depend on the internal energy of the parent molecules prior to collisions with electrons. This fact suggests that care must be taken when the existing electron (single) collision data for hydrocarbons are being applied to the analysis of the behavior of plasmas with hydrocarbon molecules, as there is no guarantee that the hydrocarbon molecules have the same internal energy in both cases.

In various fields of science and engineering, such as fusion research, a large fraction of atoms, ions, molecules, and molecular ions are very likely to be in excited states rather than in the ground state, and, therefore, their cross sections may be enhanced by orders of magnitude over the ground state data obtained in ordinal laboratory collision experiments. Therefore, it becomes quite important and urgent to investigate collision processes and to determine their cross sections under conditions in which the internal energies of the species involved are well specified, in order to obtain more reliable information that is applicable to fusion plasmas and other fields. Some techniques for preparation of state-selected species, such as laser-induced fluorescence methods, have already been widely used and can also be quite powerful methods for the present purpose.

It should also be pointed out that there is a significant as well as an urgent need of collision data for molecules involving Be and B, which are expected to play a role in reducing the radiation losses from the main plasmas. However, no systematic investigations have been performed yet for these Be- and B-containing molecules, and only limited information is available for them at present.[50]

Of course, data for other impurities, such as H_2O, CO, and CO_2, which are common in all experimental vacuum apparatuses but have not been treated in this chapter, have to be refined,[10,51] as they also suffer from problems similar to those described above for hydrocarbon molecules.

ACKNOWLEDGMENTS. The author would like to express his sincere thanks to Dr. M. Hayashi of Gaseous Electronic Institute (Nagoya) for kindly providing his original figures (Figs. 4–8), to Dr. H. Nishimura of Niigata University (Niigata) for useful discussions and suggestions, and to Dr. A. Pospieszczyk of KFA (Jülich) for his kind permission to use his optical spectra in Fig. 16.

REFERENCES

1. K. Matsunami, Y. Yamamura, Y. Itikawa, N. Itoh, Y. Kazumata, S. Miyagawa, K. Morita, R. Shimizu, and H. Tawara, *At. Data Nucl. Data Tables* **31**, 1 (1994).
2. W. Eckstein, J. Bodansky, and J. Roth, *Nucl. Fusion, Supplement* **1**, 31 (1991).
3. J. Roth, E. Vietzke, and A. A. Haasz, *Nucl. Fusion, Supplement* **1**, 63 (1991).
4. E. Vietzke, K. Flaskamp, and V. Philipps, *J. Nucl. Mater.* **128/129**, 545 (1984); E. Vietzke and V. Philipps, *Fusion Technol.* **15**, 108 (1989).
5. R. Yamada, *J. Vac. Sci. Technol.* **A5**, 305 (1987); *J. Nucl. Mater.* **145/146**, 359 (1987).
6. W. D. Langer, *Nucl. Fusion* **22**, 751 (1992).
7. A. B. Ehrhardt and W. D. Langer, Collisional Processes of Hydrocarbons in Hydrogen Plasmas, Princeton Plasma Physics Laboratory, Princeton University, Report PPPL-2477, 1987.
8. H. Tawara, Y. Itikawa, H. Nishimura, H. Tanaka, and Y. Nakamura, *Nucl. Fusion, Supplement* **2**, 41 (1992); see also Collision Data Involving Hydrocarbon Molecules, National Institute for Fusion Science, NIFS-DATA-6, 1990.
9. F. A. Baiocchi, R. C. Wentzel, and R. S. Freund, *Phys. Rev. Lett.* **53**, 771 (1984).
10. M. A. Lennon, D. S. Elliot, and A. Crowe, Critical Survey of Electron Impact Ionization Data for Selected Molecules, The Queen's University of Belfast, Department of Computer Science Report (unpublished), 1988.
11. M. Hayashi, in *Non-Equilibrium Processes in Partially Ionized Gases*, NATO ASI Series, Ser. B, No. 320 (M. Capitelli and J. N. Bardsley, eds.), Plenum Press, New York (1990), p. 333; M. Hayashi, in *Swarm Studies and Inelastic Electron Molecule Collisions* (L. C. Pitchford, B. V. McKoy, A. Chutjian, and S. Trajmar, eds.), Springer-Verlag, Berlin (1987), p. 167.
12. R. Browning and H. B. Gilbody, *J. Phys. B* **1**, 1149 (1968).
13. H. Tawara, Y. Itikawa, H. Nishimura, and M. Yoshino, *J. Phys. Chem. Ref. Data* **19**, 617 (1990).
14. V. Grill, G. Walder, D. Margreiter, T. Rauth, H. U. Poll, P. Scheier, and T. D. Märk, *Z. Phys. D* **25**, 217 (1993).
15. H. Winter, *J. Chem. Phys.* **63**, 3462 (1975); *Chem. Phys.* **36**, 353 (1979).
16. J. Perrin, J. P. Schmitt, G. de Rosny, B. Drevillon, J. Huc, and A. Lloret, *Chem. Phys.* **73**, 383 (1982).
17. C. E. Melton and P. S. Rudolph, *J. Chem. Phys.* **47**, 1771 (1967).
18. B. Adamczyk, A. J. H. Boerboom, and J. Kistemaker, *J. Chem. Phys.* **44**, 4640 (1966).
19. T. Nakano, H. Toyoda, and H. Sugai, *Jpn. J. Appl. Phys.* **30**, 2912 (1991).
20. C. Winstead, Q. Sun, V. McKoy, J. L. da Silva Lino, and M. A. P. Lima, *Z. Phys. D* **24**, 141 (1992).
21. D. P. Wang, L. C. Lee, and S. K. Srivastava, *Chem. Phys. Lett.* **152**, 513 (1988).
22. V. Grill, G. Walder, P. Scheier, M. Kurdel, and T. D. Märk, *Int. J. Mass Spectrom. Ion Processes* **129**, 31 (1993).
23. H. Chatham, D. Hils, R. Robertson, and A. Gallagher, *J. Chem. Phys.* **81**, 1770 (1984).
24. C. Melton, *J. Chem. Phys.* **37**, 562 (1962).
25. D. Rapp and P. Englander-Golden, *J. Chem. Phys.* **43**, 1464 (1965).
26. H. Nishimura and H. Tawara, *J. Phys. B*, **27**, 2063 (1994).
27. A. Gaudin and R. Hagmann, *J. Chim. Phys.* **64**, 917 (1967); *J. Chim. Phys.* **64**, 1209 (1967).
28. J. T. Tate and P. T. Smith, *Phys. Rev.* **39**, 270 (1932).
29. H. Nishimura and H. Tawara, *J. Phys. B* **24**, L363 (1991).
30. C. Winstead, Q. Sun, and V. McKoy, *Chem. Phys.* **96**, 4246 (1992).
31. R. Locht and J. Momigny, *Chem. Phys.* **49**, 173 (1980).

32. K. Ito, N. Oda, Y. Hatano, and T. Tsuboi, *Chem. Phys.* **21**, 203 (1977).
33. T. Ogawa, J. Kurawaki, and M. Higo, *Chem. Phys.* **61**, 181 (1981).
34. N. Yonekura, K. Nakashima, and T. Ogawa, *J. Chem. Phys.* **97**, 6276 (1992); N. Yonekura, T. Tsuboi, H. Tomura, K. Nakashima, and T. Ogawa, *Jpn. J. Appl. Phys.* **32**, 3296 (1993).
35. T. G. Finn, B. L. Carbahan, W. C. Wells, and E. C. Zipf, *J. Chem. Phys.* **63**, 1596 (1975).
36. J. A. Schiavone, D. E. Donahue, and R. S. Freund, *J. Chem. Phys.* **67**, 759 (1977).
37. J. B. A. Mitchell, *Phys. Rep.* **186**, 215 (1990).
38. H. Takagi, N. Kosugi, and M. Le Dourneuf, *J. Phys. B* **24**, 711 (1991); H. Takagi, in *The Physics of Electronic and Atomic Collisions* (T. Andersen, B. Fastrup, F. Folkmann, H. Knudsen, and N. Andersen, eds.), AIP Conf. Proc., No. 245, American Institute of Physics, New York (1993), p. 442.
39. P. Forck, C. Broude, M. Grieser, D. Habs, J. Kenntner, J. Liebmann, R. Repnow, A. Amitay, and D. Zaifman, *Phys. Rev. Lett.* **72**, 2002 (1994).
40. P. M. Mul, J. B. A. Mitchell, V. S. D'Angelo, P. Defrance, J. W. McGowan, and H. R. Froelich, *J. Phys. B* **14**, 1353 (1981).
41. D. Gregory and H. Tawara, Abstract Book of XVIth International Conference on Physics of Electronic and Atomic Collisions, New York, 1989, p. 352.
42. J. F. M. Aarts, C. I. M. Beenakker, and F. J. de Heer, *Physica* **53**, 32 (1971).
43. C. I. M. Beenakker and F. J. de Heer, *Chem. Phys.* **6**, 291 (1974).
44. K. D. Pang, J. M. Ajello, B. Franklin, and D. E. Shemannsky, *J. Chem. Phys.* **86**, 2750 (1987).
45. I. V. Sushanin and S. M. Mishko, *Sov. Astron.* **18**, 265 (1974).
46. C. I. M. Beenakker, P. J. Verbeek, G. R. Möhlmann, and F. J. de Heer, *J. Quant. Spectrosc. Radiat. Transfer* **15**, 333 (1975).
47. M. Tokeshi, K. Nakajima, and T. Ogawa, *Chem. Lett. (Jpn.)* **1993**, 995.
48. A. Pospieszczyk, Y. Ra, Y. Hirooka, R. W. Conn, D. M. Goeble, B. LaBombard, and R. E. Nygren, Spectroscopic Studies of Carbon Containing Molecules and Their Breakup in PISCES-A, University of California Los Angeles, Report UCLA-PPG-1251, 1989; A. Pospieszczyk, J. Hogan, Y. Ra, Y. Hirooka, R. W. Conn, D. Goebel, B. LaBombard, and R. E. Nygren, in *6th European Conference on Controlled Fusion and Plasma Physics*, Vol. 13B (S. Segre, H. Knoepfel, and E. Sindoni, eds.), European Physical Society, Geneva (1989), Part III, p. 987.
49. T. Ogawa, Y. Ueda, and M. Higo, *Bull. Chem. Soc. Jpn.* **56**, 3033 (1983).
50. *Nucl. Fusion, Supplement* **3** (1992).
51. H. Tawara, Atomic and Molecular Data for H_2O, CO and CO_2 Relevant to Edge Plasma Impurities, National Institute for Fusion Science, NIFS-DATA-19, 1992.

Index

Animated crossed beams method, 331
Associative ionization, 331
    in hydrogen/helium ion–atom (molecule) collisions, 332, 333
Atom transfer reactions, 413
Atomic-orbital expansion method, 355
Atomic structure data, 17
Atomic transition probabilities, 18, 19
Auger transition probability, 96
Autoionization, 93, 160

Belt limiter, 3
Bethe approximation, 68
Bethe–Born formula, 66
Born approximation, 42

Charge exchange: *See* Electron capture
Chemical sputtering, 464
Classical phase space method, 438
Classical-trajectory Monte Carlo method, 356
Collision strength, 120
    effective, 134
Condenser plate method, 63
Coulomb–Born approximation, 124, 156
Crossed beams method, 167, 401

Dielectronic recombination, 92; *see also* Electron–ion recombination
Dissociative electron attachment, 199

Dissociative electron capture, 408, 409
Dissociative excitation
    in electron–molecule collisions, 211
        cross section data for, 212–215
    in electron–molecular ion collisions, 238, 251, 258
        cross section data for, 239, 240, 250, 257, 259
Dissociative ionization
    in electron–molecule collisions, 216, 240
        cross section data for, 218, 221, 477–481
    in electron–molecular ion collisions, 240, 484
        cross section data for, 241
Dissociative recombination
    in electron–molecular ion collisions, 230, 245, 252, 256, 484
        cross section data for, 231–238, 246–249, 253, 257
Distorted-wave approximation, 95, 120, 124, 156
Divertor, 4
    plates, 4

Edge plasma, 2
    composition, 6, 15
    parameters, 7
Elastic scattering
    amplitude, 284

Elastic scattering (cont.)
  atom–atom, 299
    cross section data for, 299–303
  Coulomb phase shift for, 284
  cross section definition, 33, 282
  electron–atom, 31
    cross section data for, 39–41, 50
  ion–atom, 281
    cross section data for, 296–298
  ion–molecule, 303
    cross section data for, 304
Electron attachment to H_2/D_2
  dissociative, 199
Electron beam ion source, 169, 345
Electron beam ion trap, 170
Electron capture collisions
  in hydrogen/helium ion–atom systems, 309
    total cross section data for, 315–330
  of impurity ions with H, 341
    total cross sections for, 357–367
    state-selective cross sections for, 384, 385
  of impurity ions with He, 367
    total cross sections for, 368–372
    state-selective cross sections for, 386–388
  of impurity ions with H_2, 367, 377
    total cross sections for, 378–380
    state-selective cross sections, 389
Electron emission spectroscopy, 349
Electron energy loss spectroscopy, 469
Electron impact excitation
  of atoms, 44, 50
  of ions, 119
    collision strength, 120
    cross section data for, 130–133, 143–146
  of molecules, 200, 467
    cross section data for, 203–210, 470–474
Electron impact ionization
  of atoms, 59, 60
    partial cross section data for, 78–85
    total cross section data for, 69–73
  of ions, 153
    cross section data for, 174–180, 183–190
    cross section formulae for, 158–160
    Lotz formula for, 159
  of molecules, 216
    cross section data for, 217, 219, 470–474
Electron–ion recombination; see also Dissociative recombination
  general theory of, 94–96
  dielectronic, 92
    rate coefficient data for, 104–110

Electron–ion recombination (cont.)
  radiative, 92, 97
    cross section data for, 98, 99
    rate coefficient data for, 100
    plasma field effects on, 110
Electron transfer: See Electron capture
Energy level tables, 23
Excitation-autoionization, 154, 160

Fano plot, 68
Fast neutral detection method, 400
Franck–Condon factor, 201

Green's function, 96
Guided beams method, 400, 435

Helium exhaust, 12
Hydrocarbon molecule collisions
  with electrons, 461
    cross section data, 470–474
Hydrocarbon molecule generation, 462–463

Ion-beam-gas-cell method, 399
Ion–molecule collisions, 397
  dissociative, 398, 434
  electron transfer in, 398, 404; see also Electron capture
    cross section data for, 405–407
  Langevin mechanism in, 409
  particle interchange in, 398, 409, 433
    state-selective cross sections for, 424–428
    total cross sections for, 412–422, 440–453
Ion-pair formation
  in atom–atom (molecule) collisions, 315, 316
  in electron–molecular ion collisions, 241, 242, 252
Ion sources, 436
Isolated resonance approximation, 95

Kramers formula, 99

Langevin cross section, 289, 409
Lorentzian line profile, 101

Massey–Mohr approximation, 286–288
Merged beams method, 346, 400
Molecular-orbital-expansion method, 353
Momentum transfer
  in atom–atom collisions, 299
    cross section data for, 299–303
  in electron–atom collisions, 31
    cross section data for, 39–41, 50

Index

Momentum transfer (*cont.*)
  in ion–atom collisions, 285, 288
    cross section data for, 296–298, 305, 306
Multiple ionization
  in electron–atom collisions, 60, 81–86
  in electron–ion collisions, 164

Optical line emission, 16
Oscillator strength, 17
  differential dipole continuum, 68, 157

Particle rearrangement processes, 309, 403, 409
Particle recycling, 5, 11
Penning ionization, 331, 333
Photon emission spectroscopy, 349, 469
Physical sputtering, 462
Plasma edge
  configurations, 3
  impurities, 8
  radiative cooling, 7, 11
Plasma power exhaust, 11
Plasma spectroscopy, 16
Potential energy curves, 227, 245
Proton transfer reactions, 413

Radiative lifetime, 20
Reactive collisions, 397
Resonance-enhanced multiphoton ionization, 403
Resonant excitation
  of ions by electron impact, 154, 182
  of molecules by electron impact, 197, 198
R-matrix method, 177

Secondary electrons
  from electron impact ionization, 263
    angular distribution of, 264, 266, 269, 273, 275
    energy distribution of, 264, 267, 272, 274
Semiclassical approximation, 286
Slow ion detection method, 400
Spectral transitions, 16
Spectroscopic databases, 22
State-selective electron capture: *See* Electron capture

Tandem mass spectrometry method, 400
Transfer-excitation, 342, 376
Transfer-ionization, 342, 374
Translational energy spectroscopy, 349
Two-electron capture, 341, 372, 374, 380–382

Vibrational excitation
  of H_2/D_2 by electron impact, 197, 198
Vibrational level populations
  of molecular hydrogen ions, 228, 229, 243, 244
  of molecular oxygen ion, 253, 254
Viscosity cross section, 285
  data for atom-atom systems, 299, 302–304
  data for ion-atom systems, 296–298, 305, 306

Wannier threshold law, 158
Wavelength tables, 23